INTRODUCTION TO ORGANIC CHEMISTRY

SECOND EDITION

William H. Brown
Beloit College

Willard Grant Press
Boston, Massachusetts

Willard Grant Press is a division of Wadsworth, Inc.

Third printing: August 1979

Library of Congress Cataloging in Publication Data

Brown, William Henry
    Introduction to organic chemistry.

    Includes index
    1. Chemistry, Organic.   I. Title.
QD251.B76   1978   547   78-2208
ISBN 0-87150-725-0

This book was composed in Times Roman and Univers by
J.W. Arrowsmith Ltd., Bristol, England. Text and cover design by
John Servideo in collaboration with the production staff of
Willard Grant Press. The cover illustration is of Acadia National Park;
Owen Franklin photographer, Stock, Boston. Printing and binding by
Halliday Lithograph Corporation; cover printing by New England Book
Components, Inc.

Preface

This second edition provides an introduction to organic chemistry and assumes as background a course in general chemistry. Although students take a short organic course for different reasons, most have one thing in common. They are aiming toward careers in science, but few if any intend to become professional chemists. Rather, they are preparing for careers in areas that require a grounding in the fundamentals of organic chemistry. Here is the place to examine the structure, properties, and reactions of rather simple organic molecules. The student can then build on this knowledge in later course work and in professional life.

This text can be divided roughly into two parts. The first, Chapters 1 to 10, lays the foundation in organic chemistry by discussing the structures and typical reactions of the important functional groups with which the student must be familiar. Although the basic approach and content of these chapters is retained from the first edition, many suggestions and comments from users have been incorporated into the second. One of the most obvious changes is the redistribution of material from the former Chapter 2, "Chemical and Physical Properties," to appropriate places in subsequent chapters. For example, polarity of bonds and molecules is now treated in Chapter 1 along with the discussions of covalent bonding. Intermolecular forces, hydrogen bonding, and the solubility of organic molecules in polar solvents now appear in Chapter 5, along with the discussion of physical properties of alcohols and ethers. Acidity and basicity of organic substances has been integrated into the sections dealing with chemical and physical properties.

Regrouping material in this way has two distinct advantages. First, students will move quickly from the discussion of structure and bonding in Chapter 1 to what is for them the substance of organic chemistry—alkanes, alkenes, alcohols, etc. Second, students can master better the relationships between structure and physical and chemical properties now that these important concepts are placed within the context of appropriate chapters.

Added to Chapter 3 on the chemistry of alkenes is a discussion of the free radical polymerization of ethylene and substituted ethylenes. In Chapter 5 greater emphasis is placed on the fact that conversions of alcohols to aldehydes, ketones, and carboxylic acids are oxidations, rather than dwelling on how such oxidations are carried out in the laboratory. The discussion of the Williamson ether synthesis in the same chapter has been expanded to show how substitution can be maximized and elimination minimized by the proper choice of starting materials.

The discussion of fatty acids, soaps, and detergents appears now in Chapter 8, "Carboxylic Acids." Students find this material very interesting, and it is easily presented at this point. The structure and important reactions of acyl halides are presented in Chapter 9, "Functional Derivatives of Carboxylic Acids."

Chapters 11 to 14, comprising the second part of the book, provide a comprehensive coverage of the structure and function of four key classes of biomolecules: carbohydrates, lipids, amino acids and proteins, and nucleic acids. A number of changes have been made in these chapters in the second edition.

A section on the structure of ascorbic acid and its role in the cross-linking of collagen fibers has been added to the carbohydrates chapter. Chapter 12 now contains: expanded discussions of fats, oils, waxes, and phospholipids; a new section treating the organization of phospholipids in cell membranes; a fuller discussion of the structure and function of the fat-soluble vitamins, steroid hormones, and bile acids.

Chapter 13 offers a more balanced introduction to the structure and physical properties of amino acids and proteins. Discussion has been expanded on the acid-base properties of amino acids, the zwitterion structure, isoelectric points and the physical basis for electrophoresis. Reference to the Sanger method for end group determination has been deleted in favor of the newer and more versatile Edman degradation. Finally, the discussion of quaternary structure of proteins and the physical basis for protein denaturation has been enlarged.

Throughout the book problems have been revised, and their number has been increased substantially. One new type of exercise emphasizes transformations—how to convert one substance to another. Most require no more than two or three steps. The intent of these problems is to provide one more way for students to practice using the important reactions discussed within the chapters. In my experience the investment of time and effort in doing them leads to a better understanding of organic chemistry.

All of the mini-essays retained from the first edition have been revised and updated. Two new ones, entitled "Ethylene" and "Clinical Chemistry—The Search for Specificity," have been added. The purpose of these short, optional articles is to reveal to students the role of the organic chemistry just learned in areas of concern and interest to them. Offering a glimpse of the human involvement in research and discovery, the essays demonstrate that organic chemistry is an exciting and creative field.

The question of whether or not to include material on spectroscopy in a short introductory book is a difficult one to answer. While more and more instructors are introducing this topic, others, for very good reasons, choose not to do so. Spectroscopy is clearly a part of the workday life of almost everyone in the health and biological sciences. For this reason I have chosen to include spectroscopy, and I have tried to do this in a way that will be useful to those who want to use it but easily omitted by those who do not want to use it.

All discussions are qualitative in nature and require no more background than what students have from general chemistry. Any section or problem containing spectral information is marked with the symbol ∿ so that it can be easily identified.

A chapter at the end of the book introduces infrared, ultraviolet-visible, and nuclear magnetic resonance spectroscopy. This chapter may be taken up at any time once Chapter 2 has been completed. Free-standing sections on the major spectral characteristics of each functional group have been added to appropriate chapters.

Problems requiring the use of spectral information are included at the ends of problem sets. Most ask the student to list the characteristics that will allow him or her to identify the functional group or distinguish between pairs of substances. Frequently these problems are presented with ones asking the student for a chemical means for making the same determination.

A second edition of the *Study Guide* is also available. Its purpose is to guide students in their approach to solving organic chemistry problems and to provide complete and detailed solutions to all problems in the text.

Acknowledgments

The second edition of this text has been shaped by the suggestions and opinions expressed by users of the first, particularly: Kenneth Andersen, University of New Hampshire; Warren Biggerstaff, California State University, Fresno; Robert Caret, Towson State University; George Clemans, Bowling Green State University; William Dolbier, University of Florida; John Gilje, University of Hawaii; Robert Ingham, Ohio University; Ronald Johns, Spokane Falls Community College; George Levy, Florida State University; Robert Nagler, Western Michigan University; Roy Pointer, Bloomsburg State University; James Rudesill, North Dakota State University; Robert Smalley, Emporia Kansas State College; L.G. Wade, Colorado State University; George Wahl, North Carolina State University.

To Jack Leonard, Texas A & M University; Layton McCoy, University of Missouri, Kansas City; J.L. Pettus, California State University, Long Beach; and Melvyn Usselman, University of Western Ontario, for their valuable reviews of the manuscript; and to Paul Jones, University of New Hampshire, and Scott Mohr, Boston University, for their careful checking of the mini-essays, I want to express my special thanks.

Finally, my appreciation to John Servideo of Willard Grant Press for his patient guidance and careful attention to editorial detail throughout the many phases of this revision.

William H. Brown

Contents

2 ⬡ Alkanes and Cycloalkanes

3 ⬡ Alkenes and Alkynes

Mini-Essay I

**Stereoisomerism
and Optical Activity**

 Alcohols, Ethers, and Thiols

 **Benzene
and the Concept of Aromaticity**

7 Aldehydes and Ketones

Mini-Essays II

8 Carboxylic Acids

Mini-Essay III

9 Functional Derivatives of Carboxylic Acids

Mini-Essay IV

10 Amines

Mini-Essays V

11 Carbohydrates

Mini-Essay VI

12 Lipids

13 Amino Acids and Proteins

Mini-Essays VII

14 Nucleic Acids

⟨15⟩ Spectroscopy

The Covalent Bond and the Geometry of Molecules

1.1 Introduction

An introduction to organic chemistry must begin with a review of atomic structure and bonding. Much of organic chemistry is the chemistry of carbon and only a few other elements: hydrogen, nitrogen, oxygen, and the halogens. Therefore, in this chapter we will describe how atoms of these elements combine to form molecules by sharing electron pairs, and then we will examine the three-dimensional shapes of some simple organic molecules. Finally we will develop the concept of constitutional isomerism and functional groups. In subsequent chapters we will turn to the reactions organic molecules undergo, the conditions under which certain bonds can be broken and new ones formed, and the ways to convert one molecule into another.

Although discussions of this type can become quite sophisticated, we must not lose sight of the fact that they are based on the application of a few fundamental and logical principles. By the time you finish this course, you should have a good working knowledge of these principles.

1.2 Electronic Structure of Atoms

You should already know certain fundamental principles about the electronic structure of atoms and ionic and covalent bonding from a previous course in chemistry. Let us review some of these briefly.

TABLE 1.1   The electronic configuration of the first 18 elements.

Element	Atomic Number	1s	2s	$2p_x$	$2p_y$	$2p_z$	3s	$3p_x$	$3p_y$	$3p_z$
H	1	1								
He	2	2								
Li	3	2	1							
Be	4	2	2							
B	5	2	2	1						
C	6	2	2	1	1					
N	7	2	2	1	1	1				
O	8	2	2	2	1	1				
F	9	2	2	2	2	1				
Ne	10	2	2	2	2	2				
Na	11	2	2	2	2	2	1			
Mg	12	2	2	2	2	2	2			
Al	13	2	2	2	2	2	2	1		
Si	14	2	2	2	2	2	2	1	1	
P	15	2	2	2	2	2	2	1	1	1
S	16	2	2	2	2	2	2	2	1	1
Cl	17	2	2	2	2	2	2	2	2	1
Ar	18	2	2	2	2	2	2	2	2	2

An <u>atom</u> of an element consists of a dense <u>nucleus</u> surrounded by <u>electrons</u>. The nucleus bears a positive charge that is numerically equal to the number of electrons that surround it. The mass of an atom is concentrated in the nucleus and is equal to the sum of the masses of the protons and neutrons in the nucleus.

The electrons of an atom are concentrated in certain three-dimensional regions called orbitals. These electron orbitals are grouped in <u>shells</u> or <u>principal energy levels</u> identified by the principal quantum numbers 1, 2, 3, and so on. These are also sometimes referred to by the letters K, L, M, etc. The principal quantum number 1 shell consists of a single orbital, the $1s$ orbital; the 2 shell consists of four orbitals, the $2s$, $2p_x$, $2p_y$, and $2p_z$; the 3 shell consists of nine orbitals, one $3s$ orbital, three $3p$ orbitals, and five $3d$ orbitals.

In addition to the number and kind of orbitals, we need to remember that an orbital can accommodate a maximum of two electrons. Thus the maximum number of electrons that can occupy the first shell is two; the second shell, eight; and the third shell, eighteen. If we "build" an atom by surrounding the nucleus with just enough electrons to neutralize its positive charge, the first orbital to fill will be the $1s$, that is, the orbital of lowest energy (the one closest to the nucleus). Next to fill will be the $2s$ orbital, then the $2p$, etc. Table 1.1 shows the electronic configuration of the first eighteen elements of the periodic table.

We generally focus our attention on the electrons in the outermost or <u>valence shell</u> for it is these electrons that participate in chemical bonding and reaction. The <u>valence electrons</u> are represented by one or more dots surrounding the usual symbol for the atom. Each dot represents one valence electron. In the second-period elements, lithium through neon, we consider the nucleus and the two $1s$ electrons as a unit called the <u>kernel</u>. In the third-period elements, sodium through argon, we consider the two $1s$

H·							He:
Li·	Be:	Ḃ:	·Ċ:	·N̈:	:Ö:	:F̈:	:N̈e:
Na·	Mg:	Al̈:	·Si̇:	·P̈:	:S̈:	:C̈l·	:Är:

TABLE 1.2 Valence electrons for the first 18 elements.

electrons, the two 2s, and the six 2p electrons as the kernel. Thus we would write the valence electrons for the first eighteen elements as shown in Table 1.2.

You should compare these valence electron representations with the electron configurations given in Table 1.1. For example, notice that in Table 1.2 beryllium is shown with two paired valence electrons; these are the two paired 2s electrons listed in Table 1.1. Carbon is shown with four valence electrons, two of which are paired and two of which are unpaired; these represent the two paired 2s electrons and the single $2p_x$ and $2p_y$ electrons listed in Table 1.1. Notice also that carbon and silicon each have four valence electrons, nitrogen and phosphorus each have five valence electrons, oxygen and sulfur each have six valence electrons, and fluorine and chlorine each have seven valence electrons.

Although the number of valence electrons for each of these pairs of atoms is the same, the shells in which these valence electrons are found are different. For C, N, O, and F, the valence electrons belong to the principal quantum number 2 shell. With eight electrons this shell is completely filled. For Si, P, S, and Cl, the valence electrons belong to the principal quantum number 3 shell. This shell is only partially filled with eight electrons; the 3s and 3p orbitals are fully occupied but the five 3d orbitals can accommodate an additional ten valence electrons. Because of this difference between the number and kind of orbitals in shells 2 and 3, we should expect differences in the covalent bonding of oxygen and sulfur, and of nitrogen and phosphorus. Such differences do exist, as we shall see.

1.3 The Lewis Model of the Covalent Bond

In 1916, Gilbert N. Lewis, Professor of Chemistry at the University of California, came up with a beautifully simple hypothesis that unified many of the apparently disparate facts about reactions of the chemical elements. Lewis pointed out that the chemical inertness of the noble gases indicates a high degree of stability of the electronic complements of these elements: helium with a shell of two electrons, neon with shells of two and eight electrons, and argon with shells of two, eight, and eight electrons.

According to the Lewis model of bonding, elements other than the noble gases can achieve a greater degree of stability by gaining, losing, or sharing electrons, and atoms will tend to undergo reactions to acquire a noble gas electronic configuration in their outer (valence) shell. For example, sodium (which has one too many valence electrons for the electronic configuration of neon) and chlorine (which has one too few valence electrons for the electronic configuration of argon) can achieve a mutually advantageous state by the transfer of one electron from the sodium atom to the chlorine atom.

$$Na\cdot \ +\ \cdot \ddot{C}l\colon \ \longrightarrow \ Na^+ \ +\ \colon\!\ddot{C}l\colon^-$$

As a result of this transfer, sodium acquires a unit positive charge and eight electrons in its outermost filled shell; chlorine acquires a unit negative charge and eight valence electrons in its outermost filled shell. Because oppositely charged particles attract, in the crystal of sodium chloride there is a force of attraction between the sodium ion and the chloride ion. This electrostatic attractive force is called an ionic bond.

For elements near the center of the periodic table, the complete transfer of electrons is not energetically favorable; such a transfer would result in too great a concentration of either positive or negative charge on the atom. For example, if carbon were to form purely ionic bonds, it would either have to gain four electrons to become C^{4-} (and achieve an electronic configuration resembling that of neon) or lose four electrons to become C^{4+} (and achieve an electronic configuration resembling that of helium). Rather, atoms near the center of the periodic table tend to acquire filled outer shells by a process of sharing electrons.

The chemical bond arising from the sharing of electrons is called a covalent bond. The simplest example of the formation of a covalent bond is found in the case of the hydrogen molecule. When two hydrogen atoms combine, the single electrons from each combine to form an electron pair.

$$H\cdot \ +\ \cdot H \ \longrightarrow \ H\colon\!H \ +\ 104\ kcal/mole$$

In terms of our model of the covalent bond, we consider this pair of electrons to function in two ways simultaneously; it is shared equally by the two hydrogens and at the same time fills the outer shell of each hydrogen. In other words, for the purposes of acquiring a noble gas electron configuration, we consider each atom to "own" completely all electrons it shares in covalent bonds with other atoms. The gain in stability of the combination is evidenced by the large quantity of energy liberated (as heat) when the hydrogen molecule is formed from two separated atoms. The covalent bond is a strong bond, for it would require 104 kilocalories to dissociate one mole of H_2 into hydrogen atoms.

What is the reason for this great stability of the covalent bond? We can still imagine each electron as belonging to one nucleus. In the Lewis picture, the electron pair occupies the region between the two nuclei and interacts with both of them. The electron pair forming the covalent bond serves to shield one positively charged nucleus from the repulsive force of the other nucleus and at the same time, the electron pair attracts both nuclei. In other words, putting an electron pair in the space between two nuclei bonds them together and fixes the distance between atoms to within very narrow limits. We call this distance the bond length.

By way of additional examples of chemical bond formation through electron pair sharing, the molecules HCl, H_2O, NH_3, CH_4, and C_2H_6 can be formulated as shown in Figure 1.1. In these compounds, each nucleus has an electron atmosphere resembling that of a noble gas. For example, each hydrogen atom of the water molecule shares one electron pair with the oxygen atom and therefore has a filled outer shell resembling that of helium. The oxygen atom of water is surrounded by eight electrons, and therefore has a filled outer shell resembling that of neon. Note that of the

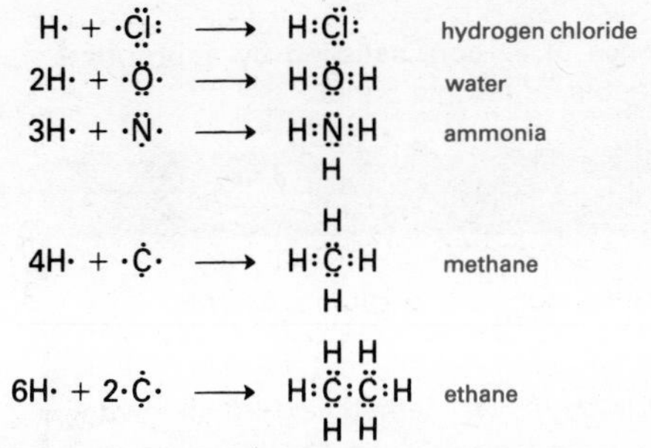

H· + ·C̈l: ⟶ H:C̈l:　　hydrogen chloride

2H· + ·Ö· ⟶ H:Ö:H　　water

3H· + ·N̈· ⟶ H:N̈:H　　ammonia
　　　　　　　　H

4H· + ·Ċ· ⟶ H:C̈:H　　methane
　　　　　　　　H

6H· + 2·Ċ· ⟶ H:C̈:C̈:H　　ethane
　　　　　　　　H H

FIGURE 1.1 Electronic formulas for some simple molecules.

eight electrons surrounding oxygen, two pairs are involved in covalent bonding and are called bonding electrons. The other two pairs of electrons are not involved in bonding and are called nonbonding electrons, or unshared electron pairs.

Indicating each individual electron pair in a covalently bonded compound, as we have done in Figure 1.1, is sometimes tedious. Therefore it is common practice to represent a shared electron pair (a covalent bond) by a dash. In addition it is common to show only shared electron pairs. Thus we customarily represent the molecules H_2O and NH_3 as

$$H-O-H \qquad H-\underset{\underset{H}{|}}{N}-H$$

water　　　　　ammonia

Two atoms may share more than a single pair of electrons. If two pairs of electrons are shared, we speak of a double bond. If three pairs of electrons are shared between two atoms, we speak of a triple bond.

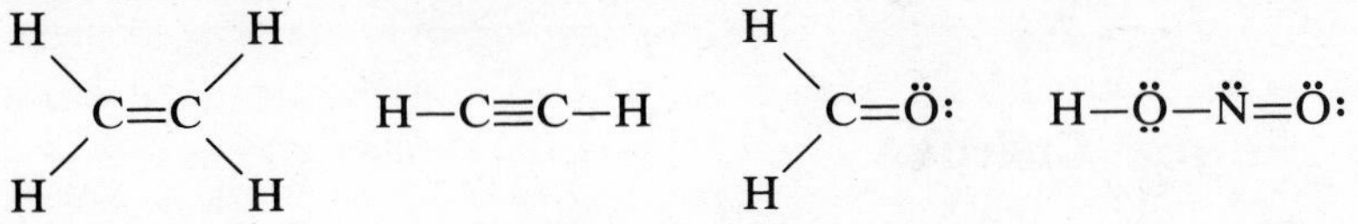

ethylene　　　　acetylene　　　formaldehyde　　nitrous acid

Even in the case of compounds containing double and triple bonds, we still find carbon with its characteristic four bonds, oxygen with two bonds, and nitrogen with three bonds.

1.4 Atomic Valences

As a guide to writing structural formulas for organic compounds it is useful to define valence for the various elements. Simply stated, the valence of an element is the number of bonds that it can form. That the concept of valence applies to organic as well as inorganic compounds was first recognized in 1859 by the brilliant German chemist, August Kekulé. In 1916, G. N. Lewis stated that the characteristic valence of an element can be correlated with its electronic structure if it is postulated that in the process of bonding each atom acquires a complete shell of electrons. Hydrogen has one valence electron and in the process of bonding acquires one more electron; hydrogen has a valence of one. Oxygen has six valence electrons and in the process of bonding acquires two electrons; oxygen has a valence of two. By similar reasoning, nitrogen has a valence of three, carbon a

5

TABLE 1.3 The tetravalence of carbon, satisfied by appropriate combinations of single, double, and triple bonds.

Type	Example	Name
$-\overset{\displaystyle\mid}{\underset{\displaystyle\mid}{C}}-$	CH_4, CH_3-CH_3	methane, ethane
$\overset{\diagdown}{\underset{\diagup}{C}}=$	$CH_2{=}CH_2$, $H_2C{=}O$	ethylene, formaldehyde
$=C=$	$O{=}C{=}O$	carbon dioxide
$-C\equiv$	$H-C{\equiv}C-H$	acetylene

valence of four, and the halogens a valence of one. As we have already seen, each of the valences must be used in bonding, and they are used by some appropriate combination of single, double, and triple bonds. In the case of carbon, there are four different ways the tetravalence may be satisfied (Table 1.3).

Knowing these simple rules will help you in writing proper structural formulas. For example, you would expect methane, CH_4, to be a stable molecule because the normal tetravalence of carbon is satisfied. However, you would not expect compounds of formula CH_2 or CH_3 to be stable molecules because they do not satisfy the normal tetravalence of carbon. Your expectations are fully justified in fact. Compounds of these formulas have been shown to exist, but they are so highly unstable that except under extraordinary circumstances they exist for only fractions of seconds.

1.5 Formal Charges

Sometimes a covalent bond is formed in which one atom provides both electrons. In this case the atom providing the pair of electrons acquires a formal positive charge. Examples are the reaction of a water molecule with a proton to form the hydronium ion, H_3O^+, and the reaction of an ammonia molecule with a proton to form the ammonium ion, NH_4^+

$$H-\overset{\displaystyle\mid}{\underset{\displaystyle H}{\ddot{O}}}{:} + H^+ \longrightarrow H-\overset{\displaystyle\mid}{\underset{\displaystyle H}{\ddot{O}}}{\overset{\pm}{}}-H$$

water   hydronium ion

$$H-\overset{\displaystyle H}{\underset{\displaystyle H}{\overset{\mid}{\underset{\mid}{N}}}}{:} + H^+ \longrightarrow H-\overset{\displaystyle H}{\underset{\displaystyle H}{\overset{\mid}{\underset{\mid}{N}}}}{\overset{\pm}{}}-H$$

ammonia   ammonium ion

While it is obvious that the hydronium and ammonium ions are positively charged, we must carry our thinking a bit farther and ask, "Which *atom* in each of these molecules bears the positive charge?" Formal positive or formal negative charges are derived by assigning all unshared (nonbonding) electrons and half of the shared (bonding) electrons to a particular atom. Comparison of this number with the normal complement of valence electrons in the neutral, unbonded atom gives the formal charge. In the case of the hydronium ion, the formal charge on oxygen is derived by assigning five electrons (two nonbonding and half of three bonding pairs) to oxygen. This is one fewer than oxygen's normal complement of six valence electrons and accordingly the oxygen atom is said to have a formal positive charge. By applying the same electronic bookkeeping you should be able to show that the nitrogen atom in the NH_4^+ ion bears a formal positive charge.

1.6 The Need for Another Model of Covalent Bonding

As much as the Lewis model of the covalent bond has helped us to formulate a clearer picture of chemical bonding, nonetheless it leaves many important questions unanswered. One of these is that of molecular geometry. The fact is that the Lewis simple electronic structures tell us nothing about the geometry of bonding. A second and much more important question, at least for our purposes, is that of the relationship between molecular structure and chemical reactivity. For example, the carbon-carbon double bond is quite different in reactivity from a carbon-carbon single bond. Most carbon-carbon single bonds are quite unreactive, but carbon-carbon double bonds react with a wide variety of reagents under a variety of experimental conditions. The Lewis model gives us no way to account for these differences. Yet to discuss modern organic chemistry at even the most elementary level, we must have a clear understanding of how the chemist accounts for them. Therefore, let us now approach the question of covalent bonding on a different and more sophisticated level—in terms of atomic orbitals and covalent bond formation by the overlap of atomic orbitals.

1.7 Extension of Atomic Orbitals in Space

Atomic orbitals are regions in space centered around the nucleus of an atom. The orbital of lowest energy (greatest stability) is the 1s orbital. An electron in a 1s orbital is the closest to the positively charged nucleus, experiences the greatest attractive force, and therefore is the most difficult to remove from the atom. The 1s orbital has no definite boundary since there is a probability, although very small, of finding the electron at large distances from the nucleus, and in fact even separated from the atom. However, the probability of finding the electron decreases very rapidly beyond a certain distance from the nucleus. It is common practice to represent the orbital as a boundary surface, that is, to enclose by a surface that region of space where there is a 95% probability of finding the electron. Figure 1.2 shows a boundary surface for a 1s orbital. Note that this orbital is actually a sphere. Also shown in Figure 1.2 is a cross section of a 1s orbital.

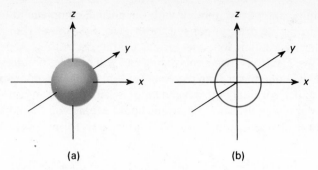

FIGURE 1.2 Atomic orbitals. (a) 1s orbital. Nucleus at the center. (b) Spherical cross section of 1s atomic orbital.

Next in energy is the 2s orbital. This too is a sphere with its center at the nucleus. The boundary surface for the 2s orbital is several times as large as the boundary surface for the 1s orbital.

Next there are three orbitals of equal energy, the 2p orbitals. Each of these is dumbbell shaped with the center of the dumbbell at the nucleus. The axis of each 2p orbital is perpendicular to that of the other two. They are designated as $2p_x$, $2p_y$, and $2p_z$, where x, y, and z refer to the three coordinate axes. The three 2p orbitals are shown schematically in Figure 1.3.

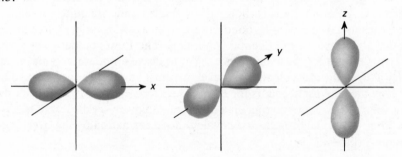

FIGURE 1.3 Atomic orbitals: p orbitals. Axes mutually perpendicular.

We will stop at this point in our consideration of atomic orbitals, but this is by no means the whole range of orbitals. As mentioned, elements in the third row of the periodic table like sulfur and phosphorus have 3d orbitals available for covalent bonding. Fourth-row elements have 4f orbitals available for bonding. However, there will be no need for us, in our introduction to organic chemistry, to go beyond the s and p orbitals, for these are the only ones used in covalent bonding in the compounds of hydrogen, carbon, oxygen, and nitrogen.

1.8 Covalent Bond Formation by the Overlap of Atomic Orbitals

In Section 1.3 we described the covalent bond of the hydrogen molecule according to the Lewis formulation. Each hydrogen atom donates one valence electron to form a two-electron covalent bond.

$$H\cdot + \cdot H \longrightarrow H{:}H \quad \text{or} \quad H{-}H$$

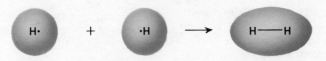

FIGURE 1.4   Bond formation. The overlap of two $1s$ atomic orbitals to form a molecular orbital.

Now let us re-examine this covalent bond and see how it can be interpreted in terms of our more sophisticated model.

The formation of a covalent bond between two atoms amounts to bringing the atoms up to each other in such a way that an atomic orbital of one atom overlaps with an atomic orbital of the other atom. For example, in forming the covalent bond in the molecule H_2, the two hydrogen atoms approach each other so that their $1s$ atomic orbitals overlap.

When this is the case the two atomic orbitals combine to form two new orbitals, called molecular orbitals, which encompass both nuclei. In one of these, the bonding molecular orbital, the electrons are concentrated in the region between the two nuclei and hold the nuclei together. In the other, the antibonding molecular orbital, the electrons are not concentrated between the two nuclei and no bonding results. In this text we shall be concerned only with bonding molecular orbitals.

Like the atomic orbital, a molecular orbital can accommodate at most two electrons. This is why the single covalent bond involves the sharing of only two electrons. The molecular orbital resulting from the overlap of two $1s$ atomic orbitals is cylindrically symmetrical about the axis joining the two nuclei. Molecular orbitals which have this electron distribution are called sigma (σ) orbitals, and the bond is called a sigma bond.

Note that the sigma bond in the hydrogen molecule is formed by the overlap of atomic $1s$ orbitals. Such bonds can also be formed by the overlap of other combinations of s and p orbitals as we will see very soon. The essential feature of the sigma bond is the overlap or fusion of two orbitals that lie on the bond axis, that is, on the line joining the two nuclei which are bonded together (Figure 1.4).

1.9   The Tetrahedral Carbon Atom

The formation of four bonds to carbon, that is, the tetravalency of carbon, is one of the central facts of organic chemistry. Now we must ask what is the geometry of bonding? More specifically what is the geometry of bonding in CH_4 and other molecules in which carbon is bonded by single bonds to four other groups? It has been found experimentally that CH_4 is symmetrical and that the disposition of all four bonds in CH_4 and in other compounds of the type CX_4 is tetrahedral; all bond angles are $109.5°$ (Figure 1.5).

Can we account for this tetrahedral geometry using the orbital theory of bonding? As a first attempt, we might expect carbon to form the four bonds in methane, CH_4, by overlap of four hydrogen $1s$ orbitals with one $2s$ orbital and three $2p$ orbitals of carbon. Since the three $2p$ orbitals are at right angles to each other, we would predict three of the H—C—H angles

FIGURE 1.5 Bond formation in methane, CH_4. The four bonds of carbon are directed toward the corners of a regular tetrahedron. Hydrogen atoms are located at these corners for maximum overlap of atomic orbitals.

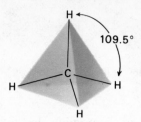

to be 90°. Since the $2s$ orbital is spherically symmetrical, we cannot predict the geometry of attachment of the fourth hydrogen.

Clearly bonding by one spherically symmetrical $2s$ orbital and three highly directional $2p$ orbitals at angles of 90° does not correspond to the actual geometry of the molecule. How can we reconcile the observed tetrahedral angle with our orbital theory of bonding? We do it by devising a new set of four equivalent, tetrahedrally oriented atomic orbitals. If we assume that atomic orbitals on the same atom can combine with each other, then the mathematical combination of one $2s$ orbital and three $2p$ orbitals results in the formation of four new and equivalent atomic orbitals (Figure 1.6).

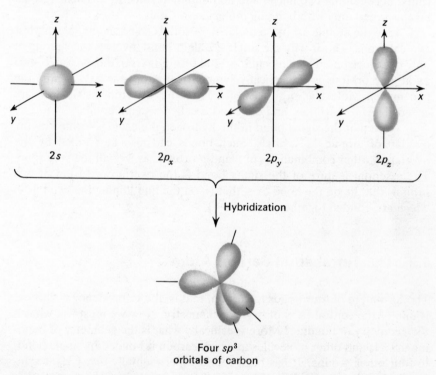

FIGURE 1.6 Hybridization of one atomic $2s$ orbital and three atomic $2p$ orbitals to form four sp^3 hybrid orbitals.

This process of combining atomic orbitals is called hybridization. The new orbitals in this case are called sp^3 orbitals. These sp^3 hybrid orbitals are directed from the carbon nucleus toward the corners of a regular tetrahedron, and the resulting molecular orbitals bonding each carbon and hydrogen are also tetrahedrally oriented. Accordingly we would predict all

FIGURE 1.7 Bond formation in water and ammonia showing the unshared electron pairs in filled sp^3 hybrid orbitals.

bond angles in methane to be 109.5°, a prediction in complete agreement with the experimentally measured angle.

Actually, we can account for the shape of the methane molecule using a more direct, almost intuitive approach. Note from the Lewis structure of CH_4 that there are four separate pairs of electrons around the carbon atom. How can these four pairs be arranged so that the repulsive interactions between them are minimized? A little experimentation with molecular models or a paper-and-pencil calculation involving some solid geometry will convince you that these pairs will be distributed around carbon to form a regular tetrahedron with hydrogen nuclei at the corners of the tetrahedron and the carbon nucleus embedded in the center. The bond angles for this distribution are 109.5°, a value identical with the experimentally determined angle and also identical with that predicted by the orbital theory of bonding.

Next, let us consider the bonding in H_2O in terms of the overlap of sp^3 hybridized atomic orbitals. We know that oxygen has six valence electrons. Filling two sp^3 orbitals accounts for four of the six valence electrons, and placing one electron in each of the other two sp^3 orbitals accounts for the remaining two. Each partially filled sp^3 orbital forms a molecular orbital with a $1s$ orbital of hydrogen, and the hydrogen atoms will occupy two corners of a regular tetrahedron. The other two corners of the tetrahedron will be occupied by the unshared pairs of electrons (Figure 1.7). Thus, we predict an H—O—H bond angle of 109.5°. Although the experimentally measured angle of 104.5° is somewhat less than we had predicted, we can rationalize this difference by assuming that the unshared (nonbonding) pairs of electrons occupy a larger region in space than the shared or bonding pairs. This seems very reasonable because the shared pairs are highly localized between two nuclei (oxygen and hydrogen). The greater repulsive interaction of the unshared electron pairs causes contraction of the H—O—H angle.

Finally let us look at the covalent bonding in ammonia, NH_3, in terms of hybridized atomic orbitals. The nitrogen atom has five valence electrons. In the bonding state, one of the sp^3 orbitals is filled with a pair of electrons while each of the other three sp^3 hybrid orbitals has but one electron. Overlapping each of these three sp^3 hybrid orbitals with a $1s$ orbital of hydrogen gives the molecule NH_3. The fourth sp^3 hybrid orbital contains the unshared pair of electrons. Figure 1.7 shows a representation of the NH_3 molecule.

Because we have used sp^3 hybrid orbitals of nitrogen for bonding, we would predict the H—N—H angle to be 109.5°. From experimental measurements the H—N—H bond angle in ammonia is known to be 107°. This small difference between the predicted and the measured bond angle can be accounted for by assuming, as we did in our discussion of the bonding in the water molecule, that the single pair of nonbonding electrons fills a larger region in space than the bonding pairs. Note that this distortion from 109.5° is larger in H_2O which contains two unshared pairs of electrons, and is smaller in NH_3 which contains only one unshared pair of electrons. There is no distortion in CH_4.

1.10   The Trigonal Carbon Atom

In Section 1.3 we saw that ethylene, C_2H_4, contains a carbon-carbon double bond; that is, the carbon atoms share two pairs of electrons. If we had been asked at that point to predict the H—C—H angles by arranging the two single pairs and one double pair of electrons about one carbon so as to minimize repulsive interactions, we would have arrived, sooner or later, at a prediction of 120° for all bond angles in the molecule. These are quite close to the observed angles, and so we would judge our prediction a good one. But while this approach to the geometry of bonding is useful, it gives us little or no understanding of the nature of the double bond itself. For that we must turn to atomic orbitals and to the hybridization of atomic orbitals.

   To form bonds with three other atoms, carbon uses three equivalent sp^2 hybrid orbitals formed by mixing the $2s$ and two $2p$ orbitals (arbitrarily designated $2p_x$ and $2p_y$ orbitals). After hybridization we have one $2p_z$ orbital and three equivalent sp^2 orbitals which lie in a plane and are directed toward the corners of an equilateral triangle; the angle between the sp^2 hybrid orbitals is 120°. This trigonal arrangement maximizes the separation of the hybrid orbitals and accordingly minimizes their electrostatic interaction. The remaining $2p_z$ orbital, which is not involved in the hybridization, consists of two lobes lying perpendicular to the plane of the sp^2 hybrid orbitals. Note that this arrangement puts the unhybridized $2p_z$ orbital the maximum distance from the three sp^2 hybrid orbitals. Figure 1.8 shows the three equivalent sp^2 hybrid orbitals and the $2p_z$ orbital of carbon.

   To form a carbon-carbon double bond in the C_2H_4 molecule we arrange the two carbons and the four hydrogens so that the carbons are

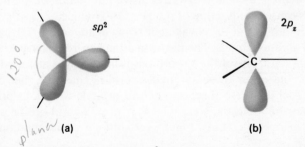

FIGURE 1.8 Atomic orbitals: hybrid sp^2 orbitals. (a) The three equivalent sp^2 hybrid orbitals lying at angles of 120°. (b) Unhybridized $2p_z$ orbital lying perpendicular to the sp^2 hybrid orbitals.

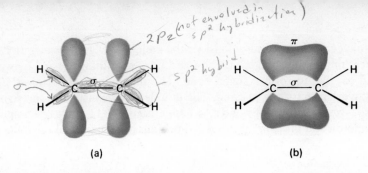

(a)   (b)

FIGURE 1.9 The ethylene molecule and the carbon-carbon double bond. (a) Sigma bonds shown; $2p_z$ orbitals not overlapping. (b) $2p_z$ orbitals overlap to form the pi bond above and below the plane of the atoms.

bonded to each other by overlapping sp^2 orbitals and so that each carbon is bonded to two hydrogens by the overlap of an sp^2 orbital of carbon and a $1s$ orbital of hydrogen. Each of these bonds is cylindrically symmetrical about the line joining the nuclei and is called a sigma bond.

The remaining $2p_z$ atomic orbitals overlap and the two electrons of this new molecular orbital form a second bond, called a pi (π) bond. The pi bond consists of two sausage-shaped regions of electron density, one on either side of the plane formed by the carbon and hydrogen atoms of ethylene. Because of the lesser degree of orbital overlap, the pi bond joining the two carbons is weaker than the sigma bond. As we can see from Figure 1.9, this overlap of $2p_z$ orbitals to give the pi bond can occur only if the $2p_z$ orbitals are parallel. Hence all six atoms of the ethylene molecule must lie in a plane. There is abundant experimental evidence that such is the case.

To form a carbon-oxygen double bond as in the formaldehyde molecule, CH_2O, the carbon atom is joined to three other atoms by overlapping the three equivalent sp^2 hybrid orbitals with two hydrogen $1s$ orbitals and one sp^2 orbital of oxygen (Figure 1.10a). The remaining $2p_z$ orbital of carbon overlaps with a $2p_z$ orbital of oxygen to form a pi bond (Figure 1.10b). Here, as in ethylene, the double bond consists of two types of molecular orbitals: sp^2 orbitals overlapping to form a sigma bond and $2p_z$ orbitals overlapping to form a pi bond.

We can describe the bonding in carbon-nitrogen double bonds in much the same manner as we have already done for the carbon-carbon and

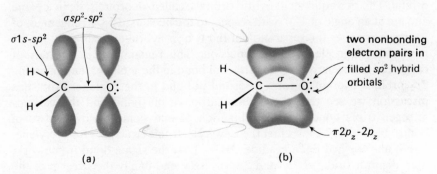

(a)   (b)

FIGURE 1.10 The formaldehyde molecule. (a) Sigma bonds shown; $2p_z$ orbitals not overlapping. (b) $2p_z$ orbitals overlap to form a pi bond.

13

carbon-oxygen double bonds. Consider the molecule $CH_3CH{=}NH$. In this substance, the double-bonded carbon and nitrogen atoms are sp^2 hybridized and overlap to form a sigma bond. The remaining unhybridized $2p_z$ orbitals overlap to form a pi bond. The double bond consists of one sigma bond $(sp^2\text{-}sp^2)$ and one pi bond $(2p_z\text{-}2p_z)$. Figure 1.11 shows the sigma bond skeleton and the unhybridized $2p_z$ orbitals overlapping to form the pi bond.

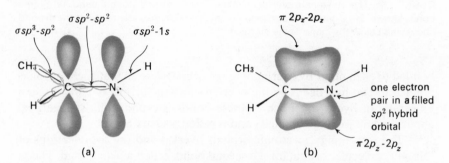

(a)   (b)

FIGURE 1.11   Covalent bonding in $CH_3CH{=}NH$. (a) Sigma bonds shown; $2p_z$ orbitals not overlapping. (b) $2p_z$ orbitals overlap to form a pi bond.

Before going on, we should note that the hybridized atomic orbital model provides us with a much clearer picture of the nature of the carbon-carbon, carbon-oxygen, and carbon-nitrogen double bonds than does the Lewis model. The double bond is not just a combination of two identical bonds. Rather, the double bond consists of one sigma bond and one pi bond. It is this approach to bonding that we shall use in subsequent chapters to help us understand the chemistry of compounds containing double bonds.

1.11   The Linear Carbon Atom

To form triple bonds with other atoms, carbon and nitrogen use two equivalent hybrid orbitals formed by mixing the $2s$ orbital and one $2p$ orbital. The two equivalent hybrid orbitals, called sp orbitals, lie in a plane and are at an angle of $180°$ with respect to the nucleus. This arrangement of course maximizes the separation of the two sp hybrid orbitals and thereby minimizes their electrostatic repulsion. The remaining $2p_y$ orbitals of carbon or nitrogen overlap to form a pi bond in the y plane; the remaining $2p_z$ orbitals overlap to form a second pi bond in the z plane. From this discussion we see that the carbon-carbon triple bond and the carbon-nitrogen triple bond are a combination of one sigma bond and two pi bonds. Figure 1.12 shows first the Lewis electronic structures for acetylene, C_2H_2, and for hydrogen cyanide, HCN; next the sigma bond framework; then the unhybridized $2p_y$ and $2p_z$ orbitals; and finally the overlap of the unhybridized $2p$ orbitals to form the two pi bonds.

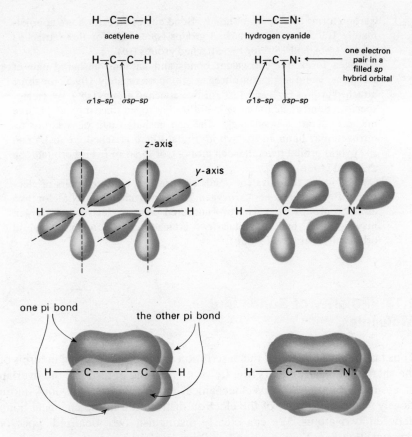

FIGURE 1.12 Covalent bonding of acetylene and hydrogen cyanide.

1.12   Covalent Bonding—An Interim Summary

Following are the Lewis electronic formulas for several of the molecules we have discussed.

By examining together these examples of covalent bonding involving hydrogen, carbon, nitrogen, and oxygen we can make certain generalizations. In these stable, uncharged molecules:

15

1. Carbon forms four covalent bonds. Bond angles on carbon are approximately 109.5° for four attached groups (sp^3), 120° for three attached groups (sp^2), and 180° for two attached groups (sp).

2. Nitrogen forms three covalent bonds and has one unshared pair of electrons. Bond angles on nitrogen are approximately 109.5° for three attached groups (sp^3) and 120° for two attached groups (sp^2). We cannot specify a bond angle for an sp hybridized nitrogen since it requires three nuclei to create a bond angle. The one unshared pair of electrons on nitrogen may lie in an sp^3 hybrid orbital (three attached groups), in an sp^2 hybrid orbital (two attached groups), or in an sp hybrid orbital (one attached group).

3. Oxygen forms two covalent bonds and has two unshared pairs of electrons. Bond angles about oxygen are approximately 109.5° for two attached groups (sp^3). The two unshared pairs of electrons on oxygen may lie in sp^3 hybrid orbitals (two attached groups) or in sp^2 hybrid orbitals (one attached group).

1.13 Polarity of Bonds and Molecules

Thus far we have examined the description of the covalent bond in terms of the sharing of electron pairs (the Lewis model) and in terms of the overlap of atomic orbitals (the wave mechanical model). Now let us look more closely at the uniformity of the electron distribution in the chemical bond between two atoms. We can clearly distinguish two idealized types of chemical bonding: the ionic bond and the covalent bond.

Ionic bonds are usually formed between elements of widely different electronegativities and involve a complete transfer of one or more electrons from one atom to another. The bond strength results from the electrostatic interaction of the positive and the negative ions. Sodium chloride is a typical example of an ionic compound formed by this process (Section 1.3). Sodium and chloride ions in the crystal are held together by electrostatic forces.

Of more interest to the organic chemist is the covalent bond. In the case of the covalent bond in the hydrogen molecule, each of the atoms involved is identical in electronegativity, and the pair of bonding electrons is shared equally between the two atoms. Between the extremes of the ionic bond and the pure covalent bond there are many compounds in which the bonds involve unequal sharing of the electrons between atoms of intermediate differences in electronegativity. These are called polar bonds, and we sometimes speak of the polarity of a covalent bond, or the partial ionic character of a covalent bond. In the water molecule the hydrogen and oxygen atoms are joined together by covalent bonds. However, because oxygen is more electronegative than hydrogen, the sharing of electrons in the bond is not equal; the center of electron density is not midway between the two atoms. The more electronegative oxygen atom acquires slightly more than one-half of the shared pair of electrons. As a result oxygen has a slightly negative charge and hydrogen a slightly positive charge. We represent this bond polarity diagrammatically by an arrow as shown in Figure 1.13a. The head of the arrow points to the atom or region of greater

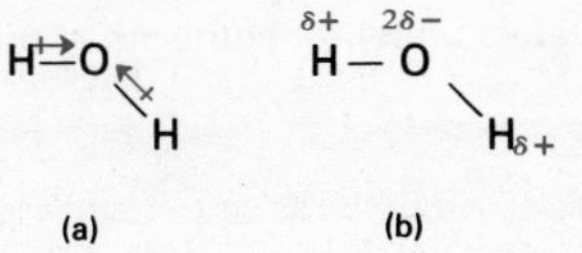

H⇌O_↘H   (a) $\overset{\delta+\ \ 2\delta-}{H-O}$_↘H_{δ+}   (b)

FIGURE 1.13 The partial
ionic character of the
covalent bond in water.

concentration of negative charge, and the tail of the arrow (here drawn in the shape of a plus sign) indicates the atom or region of greater concentration of positive charge. Alternatively, we may indicate a partial charge, as shown in Figure 1.13b, by the signs $\delta+$ and $\delta-$ (read delta plus and delta minus).

To determine which end of an unsymmetrical bond is negative and which is positive, we rely on our knowledge of the periodic table and the electronegativities of atoms (Table 1.4). As we proceed from left to right across a period, electronegativity increases because the increasing nuclear charge exerts a greater attractive force for the valence electrons. Thus fluorine is the most electronegative element in the first period. Oxygen, nitrogen, and carbon show decreasing electronegativities.

H 2.1						
Li 1.0	Be 1.5	B 2.0	C 2.5	N 3.0	O 3.5	F 4.0
Na 0.9	Mg 1.2	Al 1.5	Si 1.8	P 2.1	S 2.5	Cl 3.0

TABLE 1.4
Electronegativities of
atoms. (Values relative to
an assigned value of 4.0 for
fluorine, the most
electronegative element.)

As we proceed from top to bottom in a given family of the periodic table, the electronegativity decreases. Although the nuclear charge increases in a given family, the increasing number of electron shells provides a more effective screening of the valence electrons from the nuclear charge. Thus the electronegativity of fluorine is greater than that of chlorine.

Electronegativity   F > O > Cl,N > C,S,H

Since the electronegativity of carbon is about the same as that of hydrogen, the C—H bond is essentially nonpolar.

As we shall see, a knowledge of the polarity of bonds will serve as a valuable guide to understanding a great deal of the descriptive chemistry of organic compounds.

A molecule is said to be dipolar if the centers of negative and positive charge do not coincide, and such a molecule possesses a dipole moment which can be measured by appropriate experimental techniques. The dipole moment is the product of the magnitude of the charge times the distance between the centers of the charge and is reported in Debye units (D).

Notice from Table 1.5 that carbon tetrachloride, CCl_4, has a dipole moment of zero even though each of the C—Cl bonds is itself polar. How can it be that a molecule has polar bonds and yet no dipole moment? If we examine a three-dimensional structure of CCl_4 we see that the centers of positive and negative charge in the molecule coincide in the center of the tetrahedron where the carbon nucleus is located. Since the separation

17

TABLE 1.5 Dipole moments (in Debye units) for some small molecules.

Molecule	Dipole Moment	Molecule	Dipole Moment
O_2	0	CH_3CH_2OH	1.69
N_2	0	CH_3NH_2	1.31
H_2	0	CH_3Cl	1.86
Cl_2	0	CH_2Cl_2	1.60
CO_2	0	$CHCl_3$	1.01
H_2O	1.84	CCl_4	0
NH_3	1.46	CH_3OCH_3	1.30
CH_4	0	CH_3CO_2H	1.74
CH_3OH	1.70	CH_3COCH_3	2.88

between the centers of positive and negative charge is zero, the dipole moment is zero.

1.14 Common Organic Functional Groups

In the preceding sections we have examined the different types of covalent bonds formed by carbon with hydrogen, nitrogen, and oxygen. These bonds combine in various ways to form certain unique structural features known as functional groups. The concept of the functional group is important to organic chemistry for several reasons. First, functional groups serve as a basis for nomenclature (naming) of organic compounds. Second, they serve to classify organic compounds into families based on the presence of one or more functional groups. Finally and perhaps most important, they are the sites of chemical reaction or function. A particular functional group will have very similar chemical and physical properties whenever it is found in an organic molecule. Collected in Table 1.6 are the functional groups that will be of most importance to us in this introduction to organic chemistry. It is important to realize that this list does not include all of the functional groups in organic chemistry. However, by mastering the chemistry of these groups, you will acquire a good grounding in organic chemistry, and you will be in a position to apply this understanding to the study of more advanced work in the physical and life sciences.

1.15 Constitutional Isomerism

A molecular formula is simply a listing of the number and kind of atoms in a substance. Notice from the examples given in Table 1.6 that the molecular formula of ethanol is C_2H_6O. The molecular formula of dimethyl ether is also C_2H_6O. Dimethyl ether and ethanol have the same molecular formula but different structural formulas; that is, the order of attachment of the atoms in each is different. Notice also that propanal and acetone have the same molecular formula, C_3H_6O, but different structural formulas. Likewise propanoic acid and methyl acetate each have the molecular

TABLE 1.6 Important functional groups.

SECTION 1.16
Resonance

Functional Group Name	Functional Group (attached to carbon)	Example	
acid chloride	—C̈—C̈l: (with :O: double bonded to C)	CH_3CH_2—C—C̈l: (with :O: double bonded to C)	propanoyl chloride
acid anhydride	—C—Ö—C— (with :O: double bonded to each C)	CH_3—C—Ö—C—CH_3 (with :O: double bonded to each C)	acetic anhydride
alcohol and phenol	—ÖH	CH_3CH_2—ÖH	ethanol
aldehyde	—C—H (with :O: double bonded to C)	CH_3CH_2—C—H (with :O: double bonded to C)	propanal
alkene or olefin	\C=C/	H\C=C/H with H below each C	ethylene
alkyne	—C≡C—	H—C≡C—H	acetylene
amide	—C—N̈— (with :O: double bonded to C)	CH_3—C—N̈H_2 (with :O: double bonded to C)	acetamide
amine, primary	—N̈H_2	CH_3—N̈H_2	methylamine
amine, secondary	—N̈H—	CH_3—N̈H—CH_3	dimethylamine
amine, tertiary	—N̈—	CH_3—N̈—CH_3 with CH_3 below N	trimethylamine
carboxylic acid	—C—Ö—H (with :O: double bonded to C)	CH_3CH_2—C—Ö—H (with :O: double bonded to C)	propanoic acid
ester	—C—Ö— (with :O: double bonded to C)	CH_3—C—Ö—CH_3 (with :O: double bonded to C)	methyl acetate
ether	—Ö—	CH_3—Ö—CH_3	dimethyl ether
halide	—Ẍ:	CH_3CH_2—F̈:	ethyl fluoride
ketone	—C— (with :O: double bonded to C)	CH_3—C—CH_3 (with :O: double bonded to C)	acetone
thiol or mercaptan	—S̈H	CH_3CH_2—S̈H	ethanethiol
sulfide	—S̈—	CH_3CH_2—S̈—CH_2CH_3	diethyl sulfide

formula $C_3H_6O_2$ but different structural formulas. Substances which have the same molecular formula but different order of attachment of atoms are called constitutional isomers.

1.16 Resonance

Structural theory in organic chemistry has been developed to aid us in understanding the ways atoms combine to form molecules, the geometry of covalent bonding, and the correlation between structure and the physical and chemical properties of molecules. Where structural theory has been

19

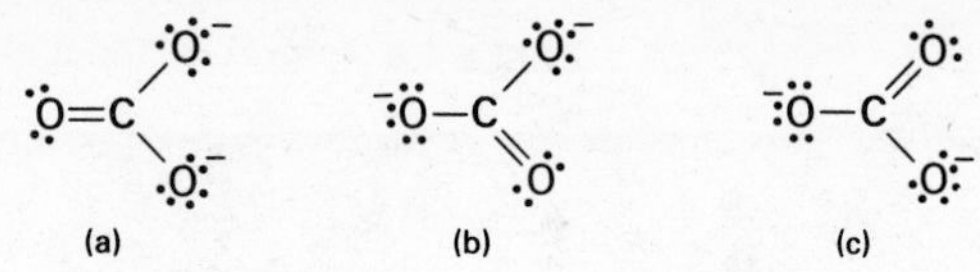

FIGURE 1.14 Three Lewis electronic structural formulas for the carbonate ion.

(a)   (b)   (c)

inadequate, chemists either have attempted to modify and extend current theory or, as we have already seen, have discarded one model of bonding and created another. For example, we began our discussion of the covalent bond and the geometry of molecules by first drawing Lewis electronic structures. Then, of necessity, we moved to the more sophisticated discussion of covalent bond formation by the overlap of atomic orbitals. This led us quite naturally to hybridization of atomic orbitals and a better understanding of single as well as double and triple bonds.

We have also developed the concept of the functional group and used it so far to classify molecules into families or groups according to the presence of a particular combination of atoms. As noted in Section 1.14, a particular functional group will have very similar chemical and physical properties whenever it is found in an organic molecule. In other words, if we can write a structural formula for a molecule and recognize the particular functional group or groups, this immediately implies a certain set of chemical and physical properties. Slowly, as the study of organic chemistry unfolded, it became clear that for a great many molecules no single Lewis structural formula provided a truly accurate representation. For many molecules, certain properties such as bond length, bond strength, acidity, basicity, dipole moment, or stability, seemed to be quite different from the predictions that would be made by considering a single classical (Lewis) structure. All of the molecules for which structural theory was inadequate seemed to have two or more functional groups close together.

For example, suppose you are asked to draw a Lewis electronic formula for the carbonate ion, CO_3^{2-}. You would probably begin by writing the carbon and three oxygen atoms attached in the correct order, then determining the proper number of valence electrons, and finally building the molecule by adding valence electrons to arrive at an acceptable Lewis structure. As you do this surely you would realize that any one of the three carbon-oxygen bonds could be written as a double bond with the other two carbon-oxygen bonds written as single bonds. These three possible Lewis structures for the carbonate ion are shown in Figure 1.14. Each of these structures implies that one carbon-oxygen bond is different from the other two. However, this is not so since it has been shown by X-ray analysis that all three bonds are identical in length. In the early days of structural theory, the actual molecule was considered as something intermediate between these three classical structures (Figure 1.14).

To describe molecules such as this more adequately it became necessary to refine still further the models for covalent bonding. Chemists use two such approaches. One is the molecular orbital approach which we have already discussed; the second is the resonance or valence-bond approach. While most chemists will admit that the molecular orbital approach is more sophisticated and more useful for calculating molecular properties, the valence-bond approach is pictorially more useful for the organic chemist interested in understanding reactions of organic compounds. Therefore it is this approach that we will use throughout most of the text.

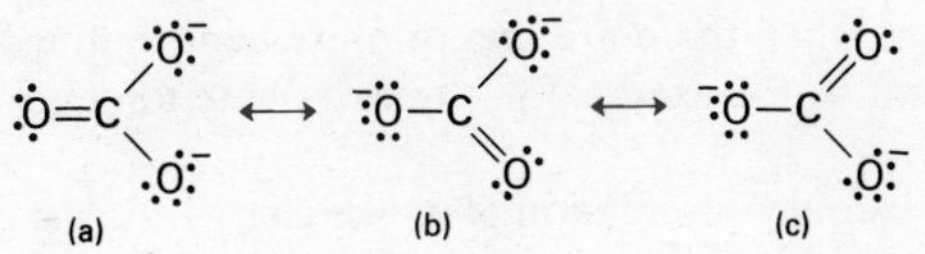

(a)   (b)   (c)

FIGURE 1.15 The carbonate ion represented as a resonance hybrid of three contributing structures.

The <u>resonance</u> or <u>valence-bond method</u> was developed in this country by <u>Linus Pauling</u>. According to this method, the molecule is best described by writing two or more valence-bond structures and considering the real molecule as a <u>hybrid</u> of these classical structures. These valence-bond structures are known as <u>contributing</u> or <u>resonance structures</u>. We consider the real molecule as a hybrid of the written structures by interconnecting them with <u>double-headed arrows</u>. Using this description, we would represent the resonance hybrid for the carbonate ion as shown in Figure 1.15.

It is important to remember that the molecule has one and only one real structure. The problem is in how to best represent and describe this real structure. The resonance method is an attempt to describe the real molecule and at the same time retain the classical electron structures with electron pair bonds. Thus, while we fully realize that the carbonate ion is not accurately represented by contributing structure (a) or (b) or (c), we will continue to represent it as one of these contributing structures for convenience. Of course we will understand that what is intended is the resonance hybrid.

Following are drawn resonance hybrids of acetone (a ketone) and acetate ion (a carboxylate anion).

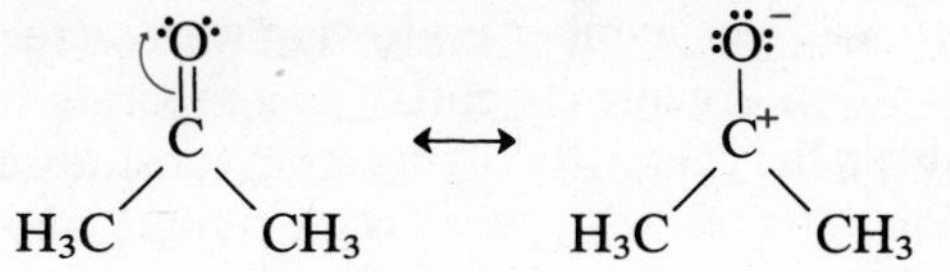

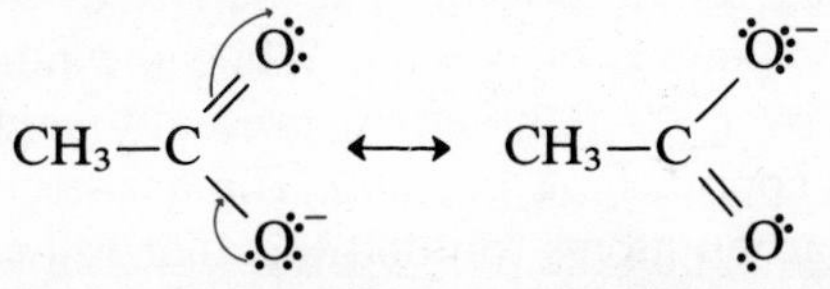

In drawing these structures, we introduce a <u>curved arrow</u> to show how one contributing structure may be converted into another. The curved arrow indicates the movement of an electron pair, with the pair moving from the position indicated by the tail of the arrow to the position indicated by the head of the arrow. Note that in the case of the second contributing structure for acetone, the molecule bears a formal positive charge on carbon and a formal negative charge on oxygen.

In this section we have seen that certain molecules are best represented as resonance hybrids and have practiced writing contributing structures. The next question we must ask is how can we predict when resonance is and is not important. A few guidelines will help in making such predictions.

1. In resonance contributing structures the position of all nuclei must remain fixed in space. In other words, contributing structures differ only in the position of valence electrons.

21

2. Resonance is more important when there are two or more contributing structures that are equivalent, as for example in the carbonate ion and the acetate ion.

3. Contributing structures that involve the creation of unlike charge are less important than those that do not involve the separation of unlike charge. For example, of the two contributing structures drawn for acetone on page 21 the second (separation of unlike charge) will be less important in the hybrid than the first (no separation of unlike charge).

Ultimately we must look at collections of experimental observations in order to gain insight into the importance of resonance. We will do this repeatedly in the following chapters as we attempt to understand the physical and chemical properties of organic molecules in terms of modern structural theory.

1.17 Organic Chemistry—The Uniqueness of Carbon

According to the most simple definition, organic chemistry is the chemistry of compounds of carbon. While the term *organic* itself does remind us that a great many of the compounds of carbon are either of animal or plant origin, by no means is that the limit. With ever increasing skill and ease, man has synthesized thousands upon thousands of new "organic" compounds in the laboratory. It is estimated that there are over 3 million known organic compounds, either isolated from nature or synthesized in the laboratory, and this number is growing with increasing speed.

Why establish organic chemistry as a separate branch of chemistry? Or, why establish the chemistry of one atom for study as a special branch of chemistry? Put another way, what is so unique about the chemistry of carbon? Of course the answer is in part due to the exceptionally large number of compounds of carbon. While the number of known organic compounds is over 3,000,000 at the present time, the number of inorganic compounds is approaching 100,000. The answer is also due in part to the tendency of carbon atoms to bond together in long chains and the possibilities for structural isomerism. But at a more fundamental level, the answer rests on the special stability of compounds containing carbon-carbon and carbon-hydrogen single bonds. To see this in more concrete terms, let us look at Table 1.7 and some representative bond dissociation energies (BDE).

For diatomic molecules, the bond dissociation energy is defined as the energy necessary to split one mole of gaseous molecules into separate atoms at 25°C and at one atmosphere pressure. For more complex poly-atomic molecules only average bond dissociation energies are listed. While the BDE for the C—C bond is given as 83 kcal/mole, the BDE for any particular C—C bond may be larger or smaller depending on the location of the C—C bond in the molecule and on the other atoms attached to the carbons. But while there is a certain amount of variation in particular bonds, we can nonetheless make two generalizations: (1) With the exception of C—C and H—H, single bonds between identical atoms are relatively weak. Compare C—C and H—H with N—N, Cl—Cl, etc. (2) C—H

TABLE 1.7 Representative bond dissociation energies (BDE).

Molecule	BDE (kcal/mole)	Bond	Average BDE (kcal/mole)
H_2	104	C—C	83
Cl_2	58	N—N	50
Br_2	46	O—O	34
I_2	36	C—H	94
		C—N	73
		C—O	85

bonds are on the average stronger than C—N and C—O bonds. Another way of stating generalization (1) is that single bonds between carbon (C—C) are notably stronger than single bonds between atoms with unshared pairs of electrons (N—N, O—O).

We might wonder why this is so. One possible explanation lies in the fact that the unshared pairs of electrons of nitrogen and oxygen repel each other and thereby substantially weaken the respective single bonds. The fact that the O—O bond is considerably weaker than the N—N bond is consistent with this explanation.

Another fact of organic chemistry is that C—C and C—H single bonds show little tendency to participate in chemical reactions. The C—O and C—N bonds, on the other hand, readily undergo a variety of chemical reactions. Why? Again the explanation lies in the presence of unshared pairs of electrons on the nitrogen and oxygen atoms. It is the presence of these unshared pairs of electrons that makes oxygen and nitrogen atoms susceptible to attack by electron-deficient atoms. Carbon, on the other hand, when it is tetrahedrally bonded by four single bonds to four other carbons or hydrogens, has no unshared pairs of electrons and is not susceptible to attack by electron-deficient atoms. Furthermore, since tetravalent carbon has a complete outer shell of electrons, it is not susceptible to attack by electron-rich reagents.

Herein lie the reasons for the uniqueness of carbon compounds and of organic chemistry: the particular strength of the carbon-carbon single bond, and the resistance of carbon-carbon and carbon-hydrogen bonds to attack by electron-deficient or electron-rich reagents.

PROBLEMS

1.1 Following is the electron configuration of O^{2-}.

$$O^{2-} \quad 1s^2 \quad 2s^2 \quad 2p^6$$

Using the same notation, also write the electron configuration for F^-, Ne, Na^+, and Mg^{2+}. Compare these five electron configurations.

1.2 Write a Lewis structure for the following molecules. Be certain to show all valence electrons.

(a) H_2O_2

(b) N_2H_4

(c) CH_3OH

(d) CH_3SH

(e) CH_3NH_2

(f) CH_3Cl

(g) CH_3OCH_3 **(h)** C_2H_6 **(i)** C_2H_4

(j) C_2H_2 **(k)** CO_2 **(l)** H_2CO_3

(m) CH_2O **(n)** CH_3CO_2H **(o)** CH_3COCH_3

(p) CH_3NNCH_3 **(q)** HCN **(r)** HNO_2

1.3 Write a Lewis structure for the following ions. Be certain to show all valence electrons. Assign formal charges as appropriate.

(a) OH^- **(b)** H_3O^+ **(c)** NH_4^+

(d) NH_2^- **(e)** HCO_3^- **(f)** CO_3^{2-}

(g) Cl^- **(h)** Cl^+ **(i)** NO_2^-

1.4 Following the rule that each atom of carbon, oxygen, nitrogen, and the halogens reacts to achieve a complete outer shell of eight valence electrons, add unshared pairs of electrons as necessary to complete valence shells in the following molecules or ions. Assign formal positive or negative charges to each as appropriate.

(a)
$$H-O-\overset{\overset{\displaystyle O}{\|}}{C}-O$$

(b) CH_3CH_2-O

(c)
$$CH_3-\overset{\overset{\displaystyle CH_3}{|}}{\underset{\underset{\displaystyle CH_3}{|}}{N}}-CH_3$$

(d)
$$H-\overset{\overset{\displaystyle H}{|}}{\underset{\underset{\displaystyle H}{|}}{C}}-\overset{\overset{\displaystyle H}{|}}{\underset{\underset{\displaystyle H}{|}}{C}}$$

(e)
$$CH_3-O-N\overset{\nearrow O}{\searrow_O}$$

(f)
$$H-\overset{\overset{\displaystyle H}{|}}{\underset{\underset{\displaystyle H}{|}}{N}}-\overset{\overset{\displaystyle H}{|}}{\underset{\underset{\displaystyle H}{|}}{C}}-\overset{\overset{\displaystyle O}{\|}}{C}-O$$

1.5 The covalence of carbon is 4. Draw the structural formula for an organic compound of carbon in which this covalence is satisfied by:

(a) four single bonds   **(b)** two single bonds and one double bond

(c) two double bonds   **(d)** one single and one triple bond

1.6 In discussing the reactions of organic compounds, we will encounter reactive intermediates in which the electron configuration around a carbon atom looks like one of the following:

$$
\underset{1}{CH_3-\overset{\overset{\displaystyle CH_3}{|}}{\underset{\underset{\displaystyle CH_3}{|}}{C}}}
\qquad
\underset{2}{CH_3-\overset{\overset{\displaystyle CH_3}{|}}{\underset{\underset{\displaystyle CH_3}{|}}{C}}\cdot}
\qquad
\underset{3}{CH_3-\overset{\overset{\displaystyle CH_3}{|}}{\underset{\underset{\displaystyle CH_3}{|}}{C}}:}
$$

(a) For each example, determine the formal charge on the central carbon.

(b) Which structures are positively charged; negatively charged; uncharged; electron deficient?

1.7 As early as the beginning of the 19th century it was recognized that chemical substances isolated from living materials contained carbon and hydrogen, and in fact, organic chemistry was defined as a branch of chemistry dealing with compounds obtained from plant and animal sources. Since organic substances were derived from living organisms, they were thought to be different from inorganic substances in that they contained some kind of essential "vital force."

In 1828, the German chemist, Friedrich Wöhler, discovered that urea could be made by heating ammonium cyanate. Urea had previously been obtained only from urine, whereas ammonium cyanate was a typical inorganic or "nonorganic" salt. This experiment was significant in the history of organic chemistry for it demonstrated the interconversion of inorganic and organic substances.

Draw Lewis formulas for urea (a substance containing only covalent bonds), and ammonium cyanate (a substance containing both ionic and covalent bonds).

$$\underset{\text{urea}}{H_2N-\overset{\overset{\displaystyle O}{\|}}{C}-NH_2} \qquad \underset{\text{ammonium cyanate}}{NH_4^+OCN^-}$$

1.8 The following substances contain both ionic and covalent bonds. Draw a Lewis structure for each.

(a) NaOH   (b) CH₃ONa   (c) NaHCO₃
(d) Na₂CO₃   (e) NH₄Cl   (f) CH₃NH₃Cl

1.9 According to the atomic orbital theory of the covalent bond, sigma and pi bonds are similar in that each is formed by the overlap of atomic orbitals of adjacent atoms. In what way(s) do sigma and pi bonds differ?

1.10 What shape would you predict for the H_2O molecule if covalent bond formation involved the overlap of unhybridized $2p_y$ and $2p_z$ orbitals of oxygen with $1s$ orbitals of hydrogen; for the NH_3 molecule if covalent bond formation involved overlap of unhybridized $2p_x$, $2p_y$, and $2p_z$ orbitals of nitrogen with $1s$ orbitals of hydrogen? Compare your predictions with the experimentally determined bond angles in these two molecules.

1.11 Structural formulas for the following molecules are shown in Table 1.6. For each describe the bonding in terms of the orbitals involved and predict all bond angles.

(a) acetic anhydride   (b) ethanol   (c) propanal
(d) ethylene   (e) acetylene   (f) acetamide
(g) methylamine   (h) dimethylamine (i) trimethylamine
(j) propanoic acid   (k) methyl acetate (l) dimethyl ether
(m) ethyl fluoride   (n) acetone

1.12 Following are five structural formulas:

(a) $CH_3-CH_2-\overset{\overset{O}{\|}}{C}-OH$

(b) $CH_3-O-CH_2-\overset{\overset{O}{\|}}{C}-H$

(c) $CH_3-\overset{\overset{O}{\|}}{C}-O-CH_3$

(d) $HO-CH_2-CH_2-\overset{\overset{O}{\|}}{C}-H$

(e) $CH_3-\overset{\overset{O}{\|}}{C}-CH_2-OH$

By counting the number of atoms and noting the order of attachment, verify for yourself that each has the same molecular formula, $C_3H_6O_2$, but a different structural formula. Name the functional groups present in each of these structural isomers.

1.13 Write a structural formula for a compound of molecular formula C_3H_7NO that contains the following functional groups:

(a) an amide
(b) a ketone and a primary amine
(c) an aldehyde and a primary amine
(d) an aldehyde and a secondary amine

1.14 Write a structural formula for a compound of molecular formula C_4H_8O that contains the following functional groups:

(a) a carbon-carbon double bond and an alcohol
(b) a ketone
(c) an aldehyde
(d) a carbon-carbon double bond and an ether

1.15 Write structural formulas for all:

(a) amines of molecular formula $C_4H_{11}N$
(b) alcohols of molecular formula $C_5H_{12}O$
(c) ethers of molecular formula $C_5H_{12}O$
(d) aldehydes of molecular formula $C_6H_{12}O$

25

(e) ketones of molecular formula $C_6H_{12}O$

(f) carboxylic acids of molecular formula $C_6H_{12}O_2$

(g) esters of molecular formula $C_6H_{12}O_2$

1.16 Draw structural formulas for each of the molecules in Table 1.5. For each indicate the direction of bond polarity (if any) and of molecular polarity (if any). How would you account for the fact that the dipole moment of $CHCl_3$ is smaller than that of CH_2Cl_2?

1.17 Suppose that water molecules were found to be linear, that is, the H—O—H angle is 180°. Would you expect water to be a polar molecule?

1.18 Draw the contributing structure indicated by the curved arrows. Assign formal charges as appropriate.

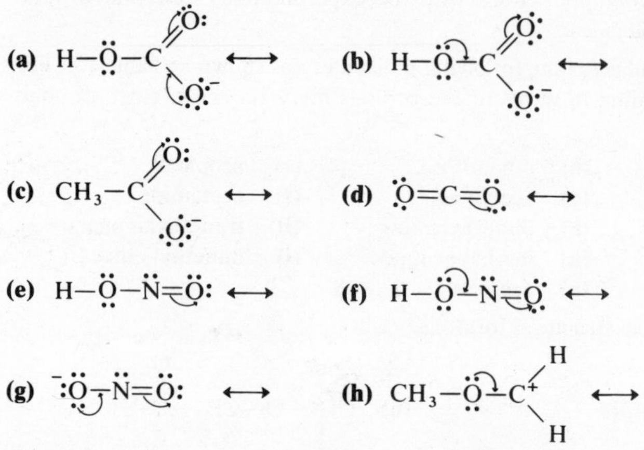

1.19 Following are drawn structural formulas for pairs of molecules. In which molecule of each pair will resonance be more important? Explain your reasoning. (Note that none of the structures shows valence electrons. You should begin by first showing all valence electrons.)

(a) $CH_3CH_2-O^-$ or $CH_3-\overset{\overset{\textstyle O}{\|}}{C}-O^-$

(b) $CH_3-CH_2-CH_2^+$ or $CH_3-O-CH_2^+$

(c) $CH_3-\overset{\overset{\textstyle O}{\|}}{C}-OH$ or $CH_3-\overset{\overset{\textstyle O}{\|}}{C}-O^-$

1.20 How might you account for the fact that C—C single bonds are notably stronger than N—N or O—O single bonds?

1.21 How might you account for the fact that compounds containing C—O and C—N single bonds react readily with H^+, whereas those containing only C—C and C—H single bonds do not?

2

Alkanes
and Cycloalkanes

2.1 Introduction

The compounds consisting solely of carbon and hydrogen are called
hydrocarbons. If the carbon atoms in a hydrocarbon are joined together
only by single covalent bonds, the compounds are called saturated hydrocarbons or alkanes. If any of the carbon atoms of the hydrocarbon are
bonded together by one or more double or triple bonds, the compounds
are called unsaturated hydrocarbons. In this chapter we shall discuss the
saturated hydrocarbons. They are the simplest organic substances from the
structural point of view and therefore a good place to begin the study of
organic chemistry. We shall discuss unsaturated hydrocarbons in later
chapters, alkenes and alkynes in Chapter 3, and aromatic hydrocarbons in
Chapter 6.

The terms *saturated* and *unsaturated* classify hydrocarbons according
to the presence or absence of carbon-carbon double and triple bonds.
Another set of terms classifies hydrocarbons according to the presence or
absence of carbon atoms joined together in such a way as to form rings:

1. Aliphatic hydrocarbons are composed of chains of carbon atoms and do
   not contain carbon rings. Aliphatic hydrocarbons are sometimes called
   open-chain or acyclic hydrocarbons.
2. Alicyclic hydrocarbons contain one or more carbon rings. They are also
   called cyclic hydrocarbons.
3. Aromatic hydrocarbons are derived from benzene and will be considered
   separately in Chapter 6.

FIGURE 2.1 Methane and
ethane. All bond angles
predicted to be 109.5°.

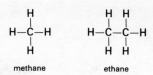

methane   ethane

We have already encountered the first two members of the alkane family, methane and ethane. Remember that the representation of ethane as in Figure 2.1 makes no attempt to depict bond angles. All bond angles in ethane are 109.5° rather than either 90° or 180° as the diagram might suggest. In even more abbreviated form, we can write a condensed structural formula for ethane as $CH_3—CH_3$ or CH_3CH_3.

By increasing the number of carbon atoms in the chain, we can form the next members of the series: propane, C_3H_8; butane, C_4H_{10}; and pentane, C_5H_{12}

$$CH_3CH_2CH_3 \qquad CH_3CH_2CH_2CH_3 \qquad CH_3CH_2CH_2CH_2CH_3$$
propane   butane   pentane

If we examine the structures of these alkanes, we see that ethane contains one C and two H more than methane; propane contains one C and two H more than ethane, etc. A series of compounds wherein each member differs from the previous member by a constant increment is called a homologous series. In the case of the alkanes, the constant increment is the unit $—CH_2—$, the methylene group. We refer to the members of such a series as homologs.

TABLE 2.1 Names, molecular formulas, and condensed structural formulas of the first ten alkanes.

Name	Molecular Formula	Condensed Structural Formula	Number of Possible Isomers
methane	CH_4	CH_4	1
ethane	C_2H_6	CH_3CH_3	1
propane	C_3H_8	$CH_3CH_2CH_3$	1
butane	C_4H_{10}	$CH_3(CH_2)_2CH_3$	2
pentane	C_5H_{12}	$CH_3(CH_2)_3CH_3$	3
hexane	C_6H_{14}	$CH_3(CH_2)_4CH_3$	5
heptane	C_7H_{16}	$CH_3(CH_2)_5CH_3$	9
octane	C_8H_{18}	$CH_3(CH_2)_6CH_3$	18
nonane	C_9H_{20}	$CH_3(CH_2)_7CH_3$	35
decane	$C_{10}H_{22}$	$CH_3(CH_2)_8CH_3$	75

Inspection of the molecular formulas of the alkanes reveals that each member of the series has the general formula C_nH_{2n+2}. As we shall see in subsequent chapters, each homologous series has its own characteristic general formula. The names of the first ten alkanes in Table 2.1 should be committed to memory for they are the basis of the systematic nomenclature of alkane derivatives.

2.2 Constitutional Isomerism in Alkanes

We have already encountered examples of <u>constitutional isomerism</u> (Section 1.15). Recall that two or more compounds having the same molecular formula but differing in the order of attachment of atoms are called constitutional isomers; the phenomenon is known as constitutional isomerism.

For methane, ethane, and propane there is only one structural formula that can be drawn once the molecular formula has been established. However, as the number of carbon atoms increases, so too does the possibility for constitutional isomerism. Butane has the molecular formula C_4H_{10}, and we can draw two different structural formulas corresponding to this molecular formula. In one of the structures of composition C_4H_{10}, the four carbons are attached in a chain; in the other they are attached three in a chain with the fourth carbon as a branch on the chain.

$$CH_3CH_2CH_2CH_3 \qquad \overset{\textstyle CH_3CHCH_3}{\underset{\textstyle CH_3}{|}}$$

butane
bp −0.5°C

isobutane
bp −10.2°C

We can distinguish between these two constitutional isomers by the names <u>butane</u> and <u>isobutane</u>. Notice that the boiling points of butane and isobutane differ by over 9 degrees.

There are three constitutional isomers of C_5H_{12}, 18 of C_8H_{18}, and a total of 75 for $C_{10}H_{22}$. It should be obvious that for even a rather modest number of carbon and hydrogen atoms, a very large number of constitutional isomers is possible. In fact, the potential for structural and functional individuality open to nature from just the basic building blocks of carbon, hydrogen, oxygen, and nitrogen is practically limitless. We shall continue to develop this theme of constitutional isomerism in this and later chapters.

2.3 Nomenclature of Organic Compounds

As the number and complexity of known organic compounds increased, it became abundantly clear that the problem of nomenclature is of singular importance. Ideally every organic substance should have a name that is both systematic and unique, and, at least for the simpler substances, a name from which a structural formula can be deduced. Various committees and commissions have met over the last century with the intent of devising such a system. A thorough and comprehensive set of rules for systematic nomenclature was recommended by the <u>International Union of Pure and Applied Chemistry</u> (IUPAC), and the system has been generally accepted by chemists throughout the world. Using the IUPAC system of nomenclature, it has been possible to develop indices for the voluminous catalogs of factual information about organic compounds and thereby facilitate chemical communication.

But in spite of the precision of the IUPAC system, routine communication in organic chemistry still relies on a hodgepodge of trivial, semisystematic, and systematic names. The reasons for this situation are rooted in both convenience and historical development.

For example, for the relatively simple compound $CH_3CHOHCH_3$ there are three acceptable names: <u>isopropyl alcohol</u>, <u>2-propanol</u>, and <u>dimethylcarbinol</u>.

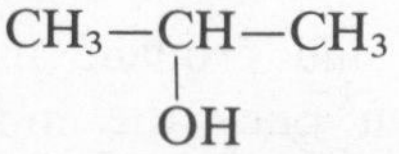

$$CH_3-CH-CH_3$$
$$|$$
$$OH$$

isopropyl alcohol
2-propanol
dimethylcarbinol

The first of these, and the name most commonly used, is derived from a semisystematic system which is easily applicable to only low-molecular-weight hydrocarbons and their derivatives. The second is the IUPAC name. The third is a derived name and is rarely if ever used today.

Also, many of the organic compounds isolated from the biological world have been given names derived from the natural source of the material (for example, penicillin from the mold *Penicillium notatum*, palmitic acid from palm oil).

We shall strive as far as possible to use systematic names in this text, but quite unavoidably there will be some crossing between the systems.

In the older, semisystematic nomenclature, the total number of carbon atoms in the alkane, regardless of their arrangement, determines the name. The first three alkanes are methane, ethane, and propane. All alkanes of formula C_5H_{12} are called pentanes, all alkanes of formula C_6H_{14} are hexanes, etc. For those alkanes beyond propane, "normal" or the prefix "n-" is sometimes used to indicate that all carbons are joined in a continuous chain. The prefix "iso-" indicates that one end of an otherwise continuous chain terminates in the $(CH_3)_2CH-$ group. The prefix "neo-" indicates that one end of an otherwise continuous chain of carbon atoms terminates in the $(CH_3)_3C-$ group. Examples using this system of nomenclature are shown below.

$$CH_3CH_2CH_2CH_2CH_3$$

pentane
or *n*-pentane

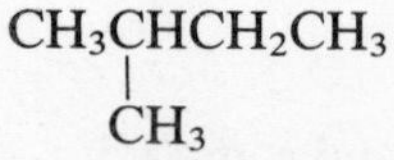

$$CH_3CHCH_2CH_3$$
$$|$$
$$CH_3$$

isopentane

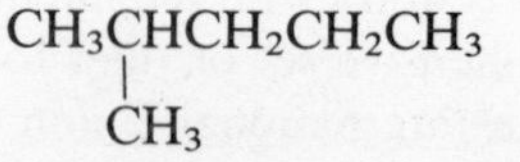

$$CH_3CHCH_2CH_2CH_3$$
$$|$$
$$CH_3$$

isohexane

$$CH_3$$
$$|$$
$$CH_3CCH_2CH_2CH_3$$
$$|$$
$$CH_3$$

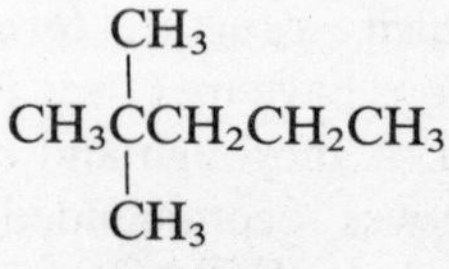

neoheptane

This semisystematic system has no good way of handling other branching patterns, and for more complex alkanes it is necessary to use the more flexible IUPAC system of nomenclature.

2.4 The IUPAC System of Nomenclature

The IUPAC rules for naming saturated hydrocarbons are essentially as follows:

1. The general name of the saturated hydrocarbon is alkane.
2. For branched-chain hydrocarbons, the hydrocarbon derived from the longest continuous chain of carbon atoms is taken as the parent compound and the root name is that of the parent compound.
3. Each substituent group is given a name and a number. The number refers to the carbon of the parent compound to which the substituent is attached. The name of the hydrocarbon substituent is derived from the alkane of the same number of carbons by changing the -ane ending to -yl. Saturated hydrocarbon groups are commonly called alkyl groups and are commonly represented by the symbol R—.
4. If two or more substituents are present, the numbering of the parent alkane starts from the end that gives the lower number to the substituent that is encountered first.
5. If the same substituent occurs more than once, the number of the carbon atom to which each is attached is given. The number of identical groups is indicated by the Greek prefixes di-, tri-, tetra-, etc.
6. If there are several different alkyl substituents, they may be named either (a) in the order of increasing complexity or (b) in alphabetical order. Whichever of the two systems is used, the names are placed in order before the name of the parent compound.
7. The last alkyl group named is prefixed to the name of the parent alkane, forming one word.

2.5 Alkyl Groups

As we have indicated in rule 3, alkyl groups are named simply by dropping the -ane of the parent alkane and adding -yl. The first two alkyl groups are methyl and ethyl (Figure 2.2). There are two isomeric alkyl groups of molecular formula C_3H_7—, namely propyl and isopropyl (Figure 2.3). There are four isomeric butyl groups, two derived from the straight-chain butane and two derived from branched chains (Figure 2.4).

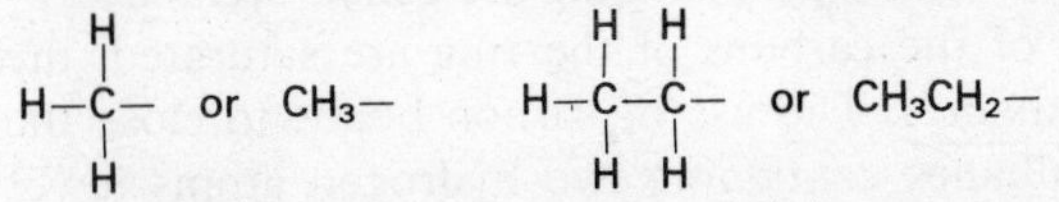

methyl ethyl

FIGURE 2.2 Alkyl groups: methyl and ethyl.

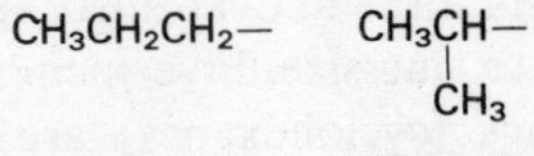

propyl isopropyl

FIGURE 2.3 Alkyl groups: propyl and isopropyl.

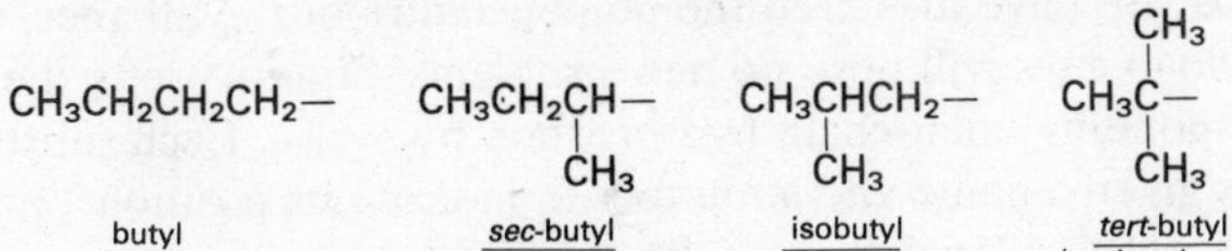

butyl · *sec*-butyl (secondary butyl) · isobutyl · *tert*-butyl (tertiary butyl)

FIGURE 2.4 Alkyl groups: the four butyl groups.

Beyond butyl the number of isomeric alkyl groups derived from each alkane becomes so large that it is impractical to designate them by prefixes such as iso-, sec-, tert- and neo-. For more complicated alkyl group substituents, IUPAC names are used.

Below are some examples of nomenclature of alkanes.

(a) $\overset{1}{C}H_3-\overset{2}{C}H-\overset{3}{C}H_2-\overset{4}{C}H_3$
         $|$
         CH_3

    2-methylbutane
    not 3-methylbutane

(b) $\overset{1}{C}H_3-\overset{2}{C}H-\overset{3}{C}H_2-\overset{4}{C}H-\overset{5}{C}H_2-\overset{6}{C}H_3$
         $|$ $|$
         CH_3 CH_2CH_3

    2-methyl-4-ethylhexane
    not 3-ethyl-5-methylhexane

(c) CH_3
               $|$
         $\overset{1}{C}H_3-\overset{2}{C}-\overset{3}{C}H_2-\overset{4}{C}H-\overset{5}{C}H_3$
               $|$ $|$
               CH_3 CH_3

    2,2,4-trimethylpentane
    not 2,4-trimethylpentane
    not 2,2,4-methylpentane

(d) CH_3
                                   $|$
         $\overset{1}{C}H_3-\overset{2}{C}H-\overset{3}{C}H_2-\overset{4}{C}H_2-\overset{5}{C}H-\overset{6}{C}H-CH_3$
               $|$ $|$
               CH_3 $_6CH_2$
                                 $|$ $_8$
                              $_7CH_2-\overset{8}{C}H_3$

    2-methyl-5-isopropyloctane
    not 2,6-dimethyl-3-propylheptane

(e) $CH_2-CH_2-CH_3$
                           $|$
         $\overset{1}{C}H_3-\overset{2}{C}H_2-\overset{3}{C}H_2-\overset{4}{C}-\overset{5}{C}H_2-\overset{6}{C}H_2-\overset{7}{C}H_3$
                           $|$
                           $CH-CH_3$
                           $|$
                           CH_3

    4-propyl-4-isopropyl-heptane
    not 2-methyl-3,3-di-propylhexane

2.6 Cycloalkanes

So far we have considered only linear arrangements of carbons atoms, often having one or more branches. The ends of these chains can be folded and joined together to form a ring or cycle of carbon atoms. Molecules of this type containing only carbon atoms in the ring are called cyclic hydrocarbons. Further, when all of the carbons of the ring are saturated, the molecules are called <u>cycloalkanes</u>. The use of carbon bonds to close the ring means that the cycloalkanes containing two hydrogen atoms fewer than the corresponding open-chain alkanes.

Cyclic hydrocarbons of ring size from three to over thirty are found in nature and in principle there is no limit to ring size. Five-membered rings (<u>cyclopentanes</u>) and six-membered rings (<u>cyclohexanes</u>) are especially abundant in nature and therefore have received special attention.

Once you have mastered the nomenclature of the alkanes, the naming of cycloalkanes will pose no new problems. Simply prefix the name of the corresponding open-chain hydrocarbon by cyclo-. Each substituent on the ring is given a name and a number to indicate its position.

As a matter of convenience the organic chemist does not usually write out the structural formulas of the cycloalkanes in the manner shown on the

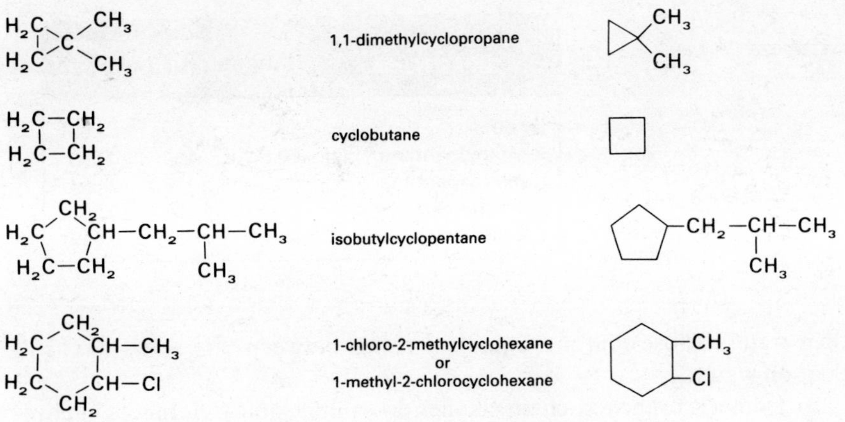

H₂C CH₃
  C
H₂C CH₃

1,1-dimethylcyclopropane

H₂C—CH₂
| |
H₂C—CH₂

cyclobutane

CH₂
H₂C CH—CH₂—CH—CH₃
H₂C—CH₂   CH₃

isobutylcyclopentane

CH₂
H₂C CH—CH₃
H₂C CH—Cl
   CH₂

1-chloro-2-methylcyclohexane
or
1-methyl-2-chlorocyclohexane

FIGURE 2.5 Examples of cycloalkanes.

left side of Figure 2.5. Rather the rings are represented by the appropriate polygon. Using this convention the same compounds are drawn on the right side of Figure 2.5.

The chemical and physical properties of the cycloalkanes resemble very closely the properties of the corresponding linear hydrocarbons. However, some significant differences do arise because of the reduced flexibility when the carbons are constrained in a cyclic structure. In Sections 2.8 and 2.9 we shall discuss some of the consequences of this constraint.

2.7 Physical Properties of Alkanes

Table 2.2 lists the boiling points, melting points, and densities for a number of alkanes. As can be seen, the melting points and boiling points increase as the number of carbons increases for the linear-chain alkanes. Up to a point, densities increase with the size of the alkanes and then tend to level off at about 0.8 grams per milliliter. Hence all of the alkanes are less dense

TABLE 2.2 Physical properties of alkanes.

Name	Formula	mp (°C)	bp (°C)	Density at 20°C
methane	CH_4	−182	−164	
ethane	CH_3CH_3	−183	−88	
propane	$CH_3CH_2CH_3$	−190	−42	
butane	$CH_3(CH_2)_2CH_3$	−138	0	
pentane	$CH_3(CH_2)_3CH_3$	−130	36	0.626
hexane	$CH_3(CH_2)_4CH_3$	−95	69	0.659
heptane	$CH_3(CH_2)_5CH_3$	−90	98	0.684
octane	$CH_3(CH_2)_6CH_3$	−57	126	0.703
nonane	$CH_3(CH_2)_7CH_3$	−51	151	0.718
decane	$CH_3(CH_2)_8CH_3$	−30	174	0.730

33

IUPAC Name	bp (°C)	mp (°C)	Density at 20°C
hexane	68.7	−95	0.659
2-methylpentane	60.3	−154	0.653
3-methylpentane	63.3	−118	0.664
2,3-dimethylbutane	58.0	−129	0.661
2,2-dimethylbutane	49.7	−98	0.649

TABLE 2.3 Physical properties of the isomeric hexanes.

than water. Those that are liquids or solids between 0°C and 100°C will float on water.

Isomeric branched-chain alkanes do exhibit some differences in physical properties. Table 2.3 lists the physical properties of the five isomeric hexanes.

Notice that each of the branched-chain isomers has a boiling point lower than hexane itself and that the more branching there is, the lower the boiling point. Particularly striking is the difference of 19° between the boiling points of hexane and its isomer 2,2-dimethylbutane. Why should branching have the effect of lowering the boiling point? Perhaps the most reasonable explanation is that with branching the shape of the molecule tends to become more compact and spherical. As this happens, surface area decreases and the degree of intermolecular interaction between molecules decreases. Thus it will require less energy (lower temperature) to separate a molecule from its neighbors in the liquid state. This effect of branching on boiling point is observed in all families of organic compounds.

2.8 Conformations of Alkanes and Cycloalkanes

In recent years, the actual shapes of molecules defined by distances between atoms and the angles between bonds have assumed greater and greater importance in the thinking of chemists and biochemists. Structural formulas such as we have been using so far are at best inadequate and frequently downright misleading with respect to the three-dimensional molecular geometry. At this point, therefore, as we take our first look at conformational questions, we must simultaneously introduce new graphical representations.

In a simple molecule such as ethane, there is an infinite number of possible arrangements of the atoms depending on the angle of rotation of one carbon with respect to the other. Such arrangements are called conformations. The Newman projection is especially convenient for showing molecular conformations of alkanes. In this type of diagram, shown for ethane (Figure 2.6), you are looking at the C—C bond head-on. The three hydrogens nearer your eye are shown as lines from the center of the circle at angles of 120°. The three hydrogens of the carbon farther from your eye are shown by lines extending from the circumference of the circle. Remember, of course, that the bond angles about each of these carbon

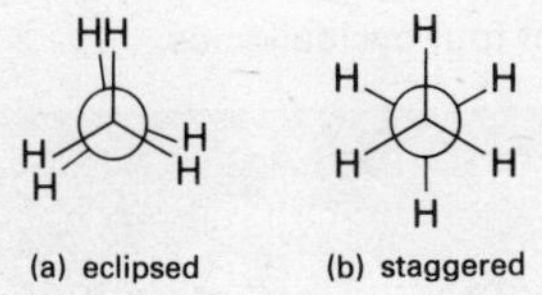

(a) eclipsed   (b) staggered

FIGURE 2.6 Newman projection formulas for two conformations of ethane: (a) eclipsed and (b) staggered.

atoms are 109.5° and not 120° as the Newman projection formula might at first suggest.

The conformations of ethane are interconvertible by rotation about the carbon-carbon bond. At room temperature, ethane molecules undergo collision with sufficient energy that rotation about the carbon-carbon single bond from one conformation to another is easily possible. In fact for a long time chemists believed there was completely free rotation about the carbon-carbon single bond. However, more recent studies of ethane and other molecules have revealed that the rotation is not completely free, but rather is hindered by the size and the electrical character of the atoms joined to one or both of the connected carbon atoms. In the case of ethane, there is a preference for the staggered conformation over the eclipsed because the interaction between the hydrogen atoms is minimized.

Of the three conformations shown for butane (Figure 2.7) the staggered-anti form (a) is preferred because it minimizes the repulsive forces between the larger methyl groups. Conformation (c), the eclipsed form, is least preferred since it maximizes the interactions of the methyl groups and of the eclipsed hydrogens.

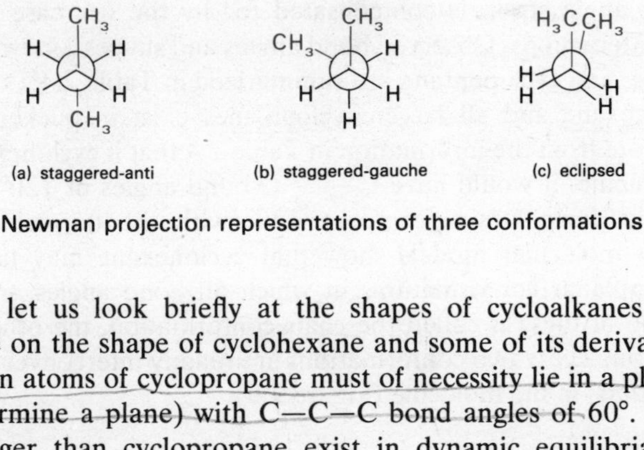

(a) staggered-anti   (b) staggered-gauche   (c) eclipsed

FIGURE 2.7 Newman projection representations of three conformations of butane.

Next, let us look briefly at the shapes of cycloalkanes and then concentrate on the shape of cyclohexane and some of its derivatives. The three carbon atoms of cyclopropane must of necessity lie in a plane (three points determine a plane) with C—C—C bond angles of 60°. All cycloalkanes larger than cyclopropane exist in dynamic equilibria between puckered conformations. This can be illustrated using cyclopentane as an example. Cyclopentane can be drawn as a planar molecule with all C—C—C bond angles of 108° (Figure 2.8a). However, it is known that

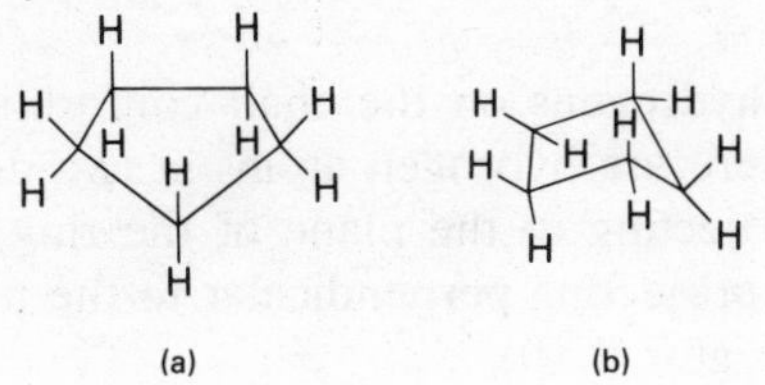

FIGURE 2.8 (a) Planar conformation of cyclopentane; little or no angle strain, but 10 fully eclipsed C—H bond interactions. (b) Puckered or envelope conformation of cyclopentane; only 6 fully eclipsed C—H bond interactions.

(a)   (b)

TABLE 2.4 Shapes of four cycloalkanes.

Name	Planar Representation	C—C—C Bond Angle in Planar Molecule	Actual C—C—C Bond Angle	Ring Shape
cyclopropane	△	60°	60°	planar
cyclobutane	□	90°	88°	slightly puckered
cyclopentane	⬠	108°	105°	slightly puckered
cyclohexane	⬡	120°	109.5°	puckered. Chair conformation most stable

cyclopentane is not a planar molecule but rather is slightly puckered or folded (Figure 2.8b).

Inspection of a molecular model of a planar cyclopentane ring will show that each hydrogen is fully eclipsed with two adjacent hydrogens. This strain due to the interaction of eclipsed hydrogens can be at least partially relieved by a slight puckering of the ring. In the puckered conformation of cyclopentane shown in Figure 2.8b, four carbon atoms are planar and one is bent slightly out of the plane, rather like an envelope with its flap bent downward. In this conformation, C—C—C bond angles are compressed to about 105°, thus increasing internal angle strain. This increase in angle strain is compensated for by the decrease in eclipsed hydrogen interactions. Observed bond angles and shapes for cyclopropane, cyclobutane, and cyclopentane are summarized in Table 2.4.

Cyclohexane and all larger cycloalkanes exist in puckered conformations. Note from the information in Table 2.4 that if cyclohexane were a planar molecule, it would have C—C—C bond angles of 120° and hence considerable angle strain (compare 120° with the normal tetrahedral angle). Yet molecular models show that cyclohexane may take up two distinct nonplanar conformations in which all bond angles are precisely 109.5°. One of these is called the chair conformation, the other the boat conformation. These two conformations are readily interconverted without breaking bonds in the molecule (Figure 2.9).

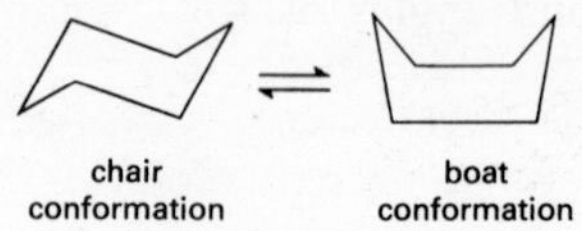

FIGURE 2.9 Two conformations of cyclohexane. (Hydrogen atoms not shown.) In each nonplanar conformation, the C—C—C bond angles are 109.5°.

chair
conformation

boat
conformation

When the arrangement of hydrogens on the chair conformation is considered it can be seen that there are hydrogen atoms in two different geometrical situations—those projecting in the plane of the ring, called equatorial hydrogens, and those projecting perpendicular to the plane of the ring, called axial hydrogens (Figure 2.10).

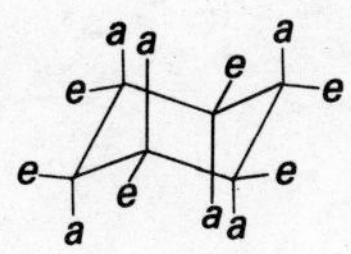

FIGURE 2.10 Chair conformation of cyclohexane. Equatorial hydrogens indicated by *e* and axial hydrogens indicated by *a*.

Note that in the chair conformation adjacent pairs of hydrogens are completely staggered with respect to each other just as they are in the ethane molecule (Figure 2.6b).

When one chair is converted to the other chair, the equatorial hydrogens become axial and vice versa. Thus in cyclohexane itself, where the two chair forms are readily interconverted, each hydrogen atom is axial half of the time and equatorial the other half of the time (Figure 2.11).

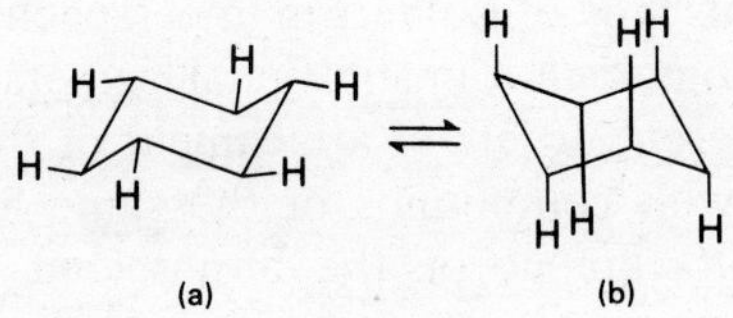

(a)   (b)

FIGURE 2.11 Two chair conformations of cyclohexane (the intermediate boat conformation is not shown). Hydrogen atoms equatorial in conformation (a) are axial in (b).

In the boat conformation (Figure 2.12), the hydrogens of carbons 2–3 and 5–6 are eclipsed (as in ethane) and the hydrogens of carbons 1 and 4 jut forward toward each other. Because of these interactions the boat form of cyclohexane is of higher energy than the chair form.

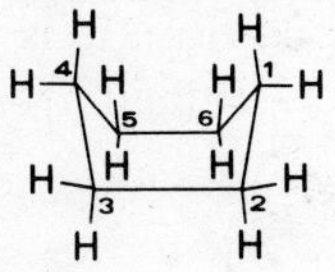

FIGURE 2.12 Interactions in the boat form of cyclohexane.

If one of the hydrogen atoms is replaced by a methyl group or other substituent, the group will occupy an axial position in one chair and an equatorial position in the other chair. This means that the two chair conformations are no longer equivalent and are no longer of equal stability. In general, bulky substituents do not occupy axial positions because of the strain arising from the interaction with the two other axial groups (hydrogens in the case of methylcyclohexane itself) on the same side of the ring. On the other hand, a substituent group in an equatorial position is as far away as possible from the other atoms on the ring, causing the chair form with the substituent equatorial to be favored at equilibrium. When the substituent is methyl, most of the molecules exist in the chair conformation with the —CH$_3$ in the less crowded equatorial position, and only a small percentage (about 5%) of the molecules of methylcyclohexane have the methyl group in the axial position. The two chair conformations of methylcyclohexane are shown in Figure 2.13.

As the size of the substituent is increased, the preference for the conformation with the group equatorial is increased. When the group is as large as *tert*-butyl, the equatorial conformation is 10,000 times more

37

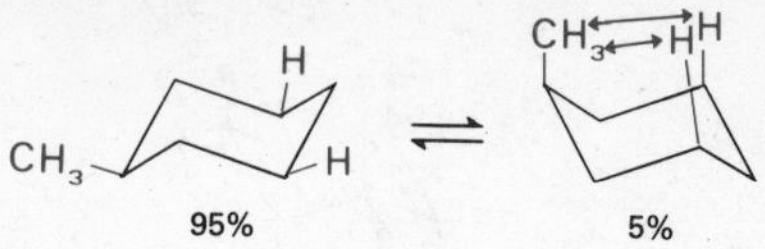

FIGURE 2.13 The two chair conformations of methylcyclohexane.

abundant at room temperature than the axial and, in effect, the ring is "locked" into this chair conformation.

2.9 Stereoisomerism in Cycloalkanes

Because of the restricted rotation about the carbon-carbon single bonds imposed by the ring structure, cycloalkanes show a type of isomerism called stereoisomerism. To put this type of isomerism in perspective, recall our discussion of constitutional isomerism. Constitutional isomers have the same molecular formula but different orders of attachment of the atoms. With stereoisomerism we deal with compounds of the same molecular formula, the same order of attachment of the atoms, but different arrangements of the atoms in space.

The term *cis-trans* isomerism is applied to the type of stereoisomerism that depends on the arrangement of substituent groups, either on a cyclic structure as we shall discuss here, or on a double bond as will be shown in the following chapter.

The principles of *cis-trans* isomerism in cyclic structures can be illustrated by looking at models of 1,2-dimethylcyclopentane. (For our purposes in describing this type of isomerism it is quite sufficient to consider the cyclopentane ring as a planar pentagon.)

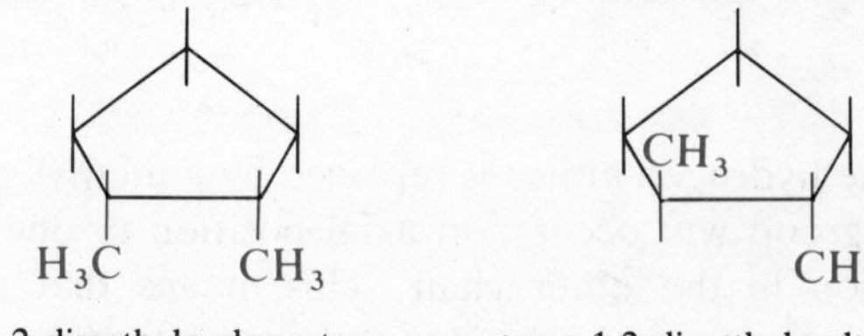

cis-1,2-dimethylcyclopentane   trans-1,2-dimethylcyclopentane

There are two *cis-trans* isomers of 1,2-dimethylcyclopentane. In one of these isomers, both of the —CH_3 groups are on the same side of the ring, whereas in the other isomer they are on opposite sides of the ring. The prefix *cis-* indicates that the substituents are on the same side of the ring; *trans-* indicates that they are on opposite sides.

The *cis* and *trans* forms cannot be superimposed on each other, and no amount of rotation about the carbon-carbon bonds will convert one into the other.

All cycloalkanes show this type of isomerism. In the case of cyclohexane and the larger rings, the situation is somewhat more complicated by the existence of various ring conformations. We will consider only the case of cyclohexane. There are two *cis-trans* isomers of 1,2-dimethylcyclohexane. In the *cis* isomer both —CH_3 groups are on the same side of the ring, and in the *trans* isomer they are on opposite sides. If the cyclohexane ring is represented as a planar hexagon, these isomers are shown below.

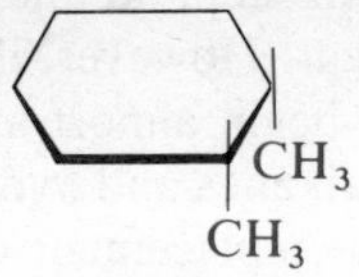

cis-1,2-dimethylcyclohexane 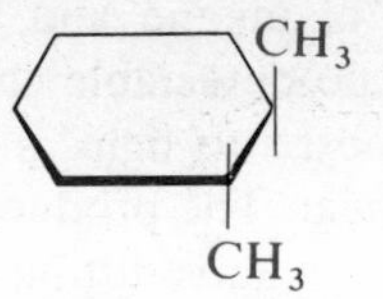   *trans*-1,2-dimethylcyclohexane

Yet as we saw in Section 2.8, a planar hexagon does not accurately represent the cyclohexane ring. Rather we should use the chair representation.

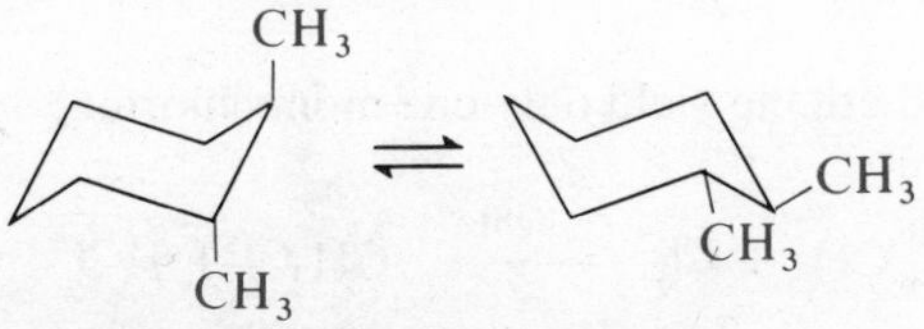

trans-1,2-dimethylcyclohexane

Clearly, when the two —CH$_3$ groups are in axial positions, they are *trans* to each other. Since one chair is readily interconvertible to the other chair by rotation about carbon-carbon bonds, the two —CH$_3$ groups are also *trans* to each other when they are attached to 1,2-diequatorial positions. Of these two conformations, the *trans* diequatorial is by far the more stable.

In the *cis* configuration, one of the —CH$_3$ groups will occupy an equatorial position on the ring, and the other will occupy an axial position.

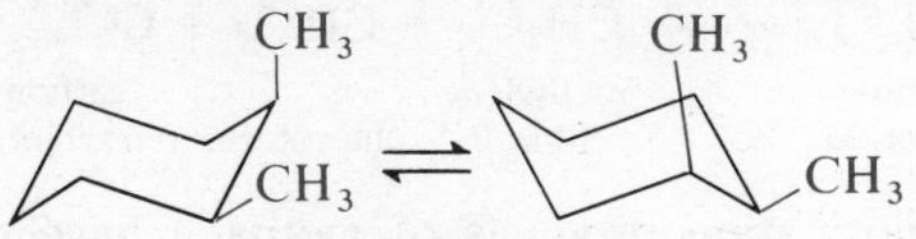

cis-1,2-dimethylcyclohexane

These two chair conformations are of equal stability for in each, one methyl group is equatorial and one is axial.

2.10 Reactions of Alkanes

As we have already seen in Section 1.17, the saturated hydrocarbons are quite inert because (1) carbon-carbon and carbon-hydrogen bond dissociation energies are high and the bonds themselves are nonpolar, and (2) since carbon has no unshared pairs of electrons, it is not susceptible to attack by electron-deficient or electron-rich reagents. Saturated hydrocarbons are quite resistant to attack by most strong acids and bases, and to powerful oxidizing and reducing agents. However, the saturated hydrocarbons do react with halogens and with oxygen. We will consider both of these reactions, the first because it is useful in the preparation of substituted alkanes and the second because it is the basis for the use of saturated hydrocarbons (and unsaturated hydrocarbons as well) as fuel.

2.11 Halogenation of Alkanes

If a mixture of alkane and chlorine gas is kept in the dark at room temperature, no detectable change will occur. However, if the mixture is heated or exposed to light, a reaction will begin almost at once with the evolution of heat. The products are chloroalkanes and hydrogen chloride. What occurs is a <u>substitution reaction</u>—the replacement of one or more hydrogen atoms on the alkane by chlorine and the production of an equivalent amount of hydrogen chloride. Recall from Section 2.4 (rule 3) that the symbol R— represents an alkyl group.

$$R-H + Cl-Cl \xrightarrow{\text{heat or light}} R-Cl + HCl$$

Both methane and ethane yield only one monochlorination product.

$$CH_4 + Cl_2 \xrightarrow{\text{light}} \underset{\substack{\text{methyl} \\ \text{chloride}}}{CH_3Cl} + HCl$$

$$CH_3CH_3 + Cl_2 \xrightarrow{\text{light}} \underset{\text{ethyl chloride}}{CH_3CH_2Cl} + HCl$$

If methyl chloride is allowed to react with more chlorine, further chlorination produces a mixture of methylene chloride, chloroform, and carbon tetrachloride, each of which is widely used as a solvent.

$$\underset{\substack{\text{methyl} \\ \text{chloride}}}{CH_3Cl} \xrightarrow{Cl_2} \underset{\substack{\text{methylene} \\ \text{chloride}}}{CH_2Cl_2} + \underset{\text{chloroform}}{CHCl_3} + \underset{\substack{\text{carbon} \\ \text{tetrachloride}}}{CCl_4}$$

These various chlorination products of methane have different boiling points and may be readily separated from one another by distillation.

Both propane and butane can yield two isomeric monochlorination products.

$$CH_3CH_2CH_3 + Cl_2 \xrightarrow{\text{light}} CH_3CH_2CH_2Cl + CH_3\underset{\underset{Cl}{|}}{C}HCH_3 + HCl$$

1-chloropropane   2-chloropropane
n-propyl chloride   isopropyl chloride
48%   52%

$$CH_3CH_2CH_2CH_3 + Cl_2 \xrightarrow{\text{light}} CH_3CH_2CH_2CH_2Cl + CH_3CH_2\underset{\underset{Cl}{|}}{C}HCH_3 + HCl$$

1-chlorobutane   2-chlorobutane
n-butyl chloride   *sec*-butyl chloride

As you might expect, higher alkanes and those with more branching yield more complex mixtures of products, and for this reason halogenation of higher alkanes is not of great synthetic use in the laboratory.

2.12 Commercially Important Halogenated Hydrocarbons

Because of their physical and chemical properties, several of the halogenated hydrocarbons have found wide commercial use. These include applications as commercial solvents, refrigerants, dry cleaning agents, local and general anesthetics, and insecticides.

Carbon tetrachloride, CCl_4, is a dense, nonflammable liquid, bp 77°C, which is remarkably inert to most common reagents and laboratory conditions. It is immiscible with water but is a good solvent for oils and greases, and at one time found wide use in the dry cleaning industry. It is somewhat toxic, readily absorbed through the skin, and like all organic solvents should be used only with adequate ventilation. Prolonged exposure to carbon tetrachloride vapors results in liver and renal damage.

Chloroform, $CHCl_3$, is a colorless, dense, rather sweet-smelling liquid, bp 61°C. It too is a widely used solvent for organic substances. In the past chloroform was used extensively as a general anesthetic for surgery but it is rarely used for this purpose now because it is known to cause extensive liver damage.

Ethyl chloride, CH_3CH_2Cl, is used as a fast-acting, topically applied local anesthetic. It owes its anesthetic property more to its physical than chemical characteristics. It boils at 13°C, and unless under pressure, is a gas at room temperature. When sprayed on the skin, it evaporates rapidly and because evaporation is a cooling process, it cools the skin surface and nerve endings.

Halothane, $C_2HBrClF_3$, is a recently discovered and now widely used inhalation anesthetic. It has distinct advantages over other general inhalation anesthetics (as for example diethyl ether and cyclopropane) in that it is nonflammable, nonexplosive, and causes minimum discomfort to the patient. Although there have been a few cases of liver damage caused by its use, its record as a safe anesthetic is impressive.

Some of the more complex chlorinated hydrocarbons are poisons and, as you well know, have found wide use as insecticides. Undoubtedly, the best known, the cheapest, and the most astonishingly effective of these is DDT.

DDT (dichlorodiphenyltrichloroethane)

Other widely known and used polychlorinated hydrocarbons are Dieldrin, Aldrin, Chlordane, and Lindane (Problem 2.14).

Dieldrin   Aldrin   Chlordane

41

Of all the fluorinated hydrocarbons, those manufactured under the trade name Freon (du Pont) have had the most dramatic impact. The first of the Freons was developed in a search for new refrigerants. Of all the refrigerants in use prior to 1930, none was without serious disadvantage. Some, like ethylene, were flammable. Others like sulfur dioxide were corrosive and quite toxic. Ammonia combined all three of these hazards. Carbon dioxide would have been ideal except that it had to be used at relatively high pressure and the equipment required for its use was almost prohibitively bulky. It was against this background that Thomas Midgley, Albert Henne, and others at General Motors set out to find an ideal refrigerant—a compound that would be nontoxic, nonflammable, odorless, and noncorrosive. In 1930 they announced the discovery of just such a compound, dichlorodifluoromethane, which was marketed under the trade name Freon-12.

The Freons are manufactured by reacting a chlorinated hydrocarbon with hydrofluoric acid in the presence of an antimony pentafluoride or antimony chlorofluoride catalyst. Both Freon-11 and Freon-12 can be prepared from carbon tetrachloride. Freon-22, monochloro-difluoromethane, is made from chloroform.

$$CCl_4 \xrightarrow[SbF_5]{HF} \underset{\text{Freon-11}}{CCl_3F} \xrightarrow[SbF_5]{HF} \underset{\text{Freon-12}}{CCl_2F_2}$$

The new refrigerant, Freon-12, went into commercial production in 1931 and very shortly thereafter the refrigeration industry began to modify its equipment to use the new product. By 1935 the product line included five Freons, three derivatives of methane and two of ethane. A major new use of the Freons came during World War II with the development of aerosol insecticides for which they served as propellants. By 1974, U.S. production of Freons had grown to more than 1.1 billion pounds annually, almost one-half the world production.

Concern about the environmental impact of fluorocarbons like Freon-11 and Freon-12 arose in 1974 when Drs. Sherwood Rowland and Mario Molina of the University of California, Irvine, announced their theory of ozone destruction by these substances. When used as aerosol propellants and refrigerants, these fluorocarbons escape to the lower atmosphere, but because of their general inertness do not decompose there. Slowly, they find their way to the stratosphere where they absorb ultraviolet radiation from the sun and then decompose. As they decompose, they set up a chemical reaction that may also lead to the destruction of the stratospheric ozone layer. What makes this a serious problem is that the stratospheric ozone layer acts as a shield for the earth against excess ultraviolet radiation. The depletion of this shield and the increased levels of ultraviolet radiation striking the earth could result in damage to certain crops and agricultural species, climate modification, and even increased incidence of skin cancer in sensitive individuals. Controversy continues and for this reason it is critical that the scientific community, along with government and consumer agencies, take steps to determine the real potential of fluorocarbons for ozone depletion and its impact on the environment.

2.13 Saturated Hydrocarbons for Heat and Power

The reaction of alkanes with oxygen to form carbon dioxide and water—and most important, heat—is the basis for the use of hydrocarbons as a source of heat (natural gas) and power (fuel for the internal combustion engine).

$$CH_4 + 2O_2 \longrightarrow CO_2 + 2H_2O \qquad \Delta H = -212 \text{ kcal/mole}$$

$$\underset{\substack{| \\ CH_3}}{CH_3}\overset{\substack{CH_3 \quad CH_3 \\ | \qquad |}}{C}CH_2CHCH_3 + \tfrac{25}{2}O_2 \longrightarrow 8CO_2 + 9H_2O \qquad \Delta H = -1304 \text{ kcal/mole}$$

2,2,4-trimethylpentane
"isooctane"

These hydrocarbons are obtained commercially from natural gas, coal, petroleum, and from shale rock.

Natural gas consists of methane mixed with varying amounts of ethane, propane, butane, and isobutane. These last three components can be liquefied under pressure at room temperature. Liquid propane (LPG) can be stored easily and shipped in metal tanks, and it is a convenient source of gaseous fuel.

2.14 Petroleum and Petroleum Refining

Petroleum is a liquid mixture of literally thousands of substances, most of them hydrocarbons, formed from the decomposition of marine plants and animals. Petroleum and petroleum-derived products fuel automobiles, aircraft, and trains. They provide heat for buildings and fuel for electric generating plants. They provide most of the greases and lubricants required for the machinery of our highly industrialized society. Furthermore, petroleum, along with natural gas, provides close to 90% of the organic raw materials for the synthesis and manufacture of synthetic fibers, plastics, detergents, and a multitude of other products.

It is the task of the petroleum refinery to produce usable products, with a minimum of waste, from the thousands of different hydrocarbons in this liquid mixture. The various physical and chemical processes for this purpose fall into two broad categories: separation processes which simply separate the complex mixture into various fractions and conversion processes which alter the molecular structure of the hydrocarbon components themselves. An example of how these two processes have been used over the years to respond to changing economic and market demand can be seen in the yield of gasoline per barrel of crude oil. In 1920 the average yield of gasoline that could be obtained by separation processes was 11 gallons per 42-gallon barrel of crude, or 26%. With the introduction of new and more sophisticated conversion processes, the yield increased to approximately 50%.

43

The fundamental process in refining and the one most widely used is distillation. Practically all crude oil that enters a refinery today goes to distillation units where it is heated to temperatures as high as 370°C to 425°C. The vapors rise through a fractionating column, cooling as they rise and condensing when they have cooled to approximately their boiling points. The resulting liquids are drawn off at predetermined heights on the columns. Each fraction or "cut" contains a mixture of hydrocarbons that boil within a corresponding range. Following are the common names associated with several of these fractions along with the major uses of each.

1. Gases boiling below 20°C are taken off highest on the column. This fraction is a mixture of low-molecular-weight hydrocarbons, predominantly methane, ethane, propane, and butanes.
2. Naphthas, bp 20 to 200°C, are a mixture of C_4 to C_{10} alkanes and cycloalkanes. The naphthas may also contain some aromatic hydrocarbons such as benzene, toluene, and xylene. The light naphtha fraction, bp 20 to 150°C, is the source of what is known as straight run gasoline and averages approximately 25% of crude petroleum. In a sense, the naphthas are the most valuable distillation fractions for they are useful not only as fuel but also as a source of raw materials for the organic chemical industry.
3. Kerosene, bp 175 to 275°C, is a mixture of C_9 to C_{15} hydrocarbons and is used for fuel and heat.
4. Gas oil, bp 200 to 400°C, is a mixture of C_{15} to C_{25} hydrocarbons. It is from this fraction that diesel fuel is obtained.
5. Lubricating oil and heavy fuel oil distill from the column at temperatures over 350°C.
6. Asphalt is the name given to the black, tarry residue remaining after the removal of the other volatile fractions.

The second major separation process is solvent extraction. In this process a solvent is used to selectively dissolve and extract a desired hydrocarbon component. The solvent is then separated from the desired hydrocarbon by distillation. In one of the first solvent extraction processes, liquid sulfur dioxide was used to refine kerosene. In 1970, the petroleum industry used various extraction processes to separate some 3 billion gallons of lubricating oils. Solvent extraction is also used to separate the aromatic or benzene-toluene-xylene (BTX) fraction.

Early in the 20th century it became obvious that conventional separation techniques could not meet the needs of the radically changing market for petroleum products. The automobile showed every sign of creating an enormous demand for gasoline. Yet distillation and other separation techniques could only produce as much gasoline as crude oil contained naturally. Refiners could meet this growing demand either by processing uneconomically large volumes of crude oil or by devising some means to convert other hydrocarbon fractions into hydrocarbons in the gasoline range. This need to adapt processing of crude oil stocks was made more imperative by the new demands for aviation fuels in the 1930s and 1940s and the demand for jet fuel beginning in the 1950s.

Gasoline, as we have already indicated, is a complex mixture of C_4 to C_{10} hydrocarbons. The quality of gasoline, as measured by octane rating, varies greatly with composition and the particular amount of branched-chain and aromatic hydrocarbons in the blend. A high-octane fuel is one that burns smoothly and delivers power smoothly to the piston. A low-octane fuel tends to "explode" in the cylinder, that is, it undergoes too rapid combustion and leads to engine "knocking." When the scale of octane ratings was set up, 2,2,4-trimethylpentane ("isooctane"), which has excellent knock properties, was arbitrarily assigned an octane rating of 100. Heptane, which causes a great deal of knocking, was assigned an octane rating of 0. The octane rating of a particular gasoline is that percent isooctane in a mixture of isooctane and heptane that has equivalent knock properties. For example, the knock properties of 2-methylhexane are the same as those of a mixture of 42% isooctane and 58% heptane; therefore, the octane rating of 2-methylhexane is 42. Octane itself has an octane rating of −20.

What was needed by the petroleum refining industry was a way to sharply increase the yield of gasoline from crude oil and at the same time significantly improve octane ratings.

Both of these goals were achieved by a <u>catalytic cracking</u> process developed by the French engineer, Eugene Houdry. The first commercial plant to use the Houdry process was built in 1937 by Sun Oil. By 1976 catalytic cracking capacity in the United States reached 150 million gallons of feedstock a day. In the original Houdry process, the hydrocarbons were cracked as they flowed over a stationary bed of catalyst at roughly 500°C and at pressures up to 2 atmospheres. The catalysts were natural and synthetic silica-aluminas. The catalysts used today are basically the same materials but with much improved chemical and physical performance characteristics. Catalytic cracking processes also produce substantial amounts of low-molecular-weight hydrocarbons such as ethylene, propene and butenes. As we will see presently, each of these in turn is used as a raw material for other commercial products. Ethylene, for example, is used in the production of polyethylene and polystyrene plastics, ethanol, ethylene glycol, polyester fibers, and many more items.

As demand for gasoline grew steadily, petroleum chemists began to seek ways to convert these low-molecular-weight by-product hydrocarbons into larger hydrocarbons in the gasoline range. Of the processes developed, <u>alkylation</u> is the most significant in the United States. The alkylation process can be illustrated by the reaction of 2-methylpropene and 2-methylpropane to form a branched-chain C_8 hydrocarbon.

$$
\underset{\substack{\text{2-methyl-}\\\text{propene}}}{CH_3-\overset{\overset{\displaystyle CH_3}{|}}{C}=CH_2} + \underset{\substack{\text{2-methyl-}\\\text{propane}}}{CH_3-\overset{\overset{\displaystyle CH_3}{|}}{CH}-CH_3} \xrightarrow[\text{or HF}]{H_2SO_4} \underset{\text{2,2,4-trimethylpentane (isooctane)}}{CH_3-\overset{\overset{\displaystyle CH_3}{|}}{\underset{\underset{\displaystyle CH_3}{|}}{C}}-CH_2-\overset{\overset{\displaystyle CH_3}{|}}{CH}-CH_3}
$$

The first commercial alkylation plant was built in 1938 by Humble Oil. By the end of World War II, refiners were producing more than 3 million gallons of alkylate daily and current production of alkylate exceeds 30 million gallons per day. Alkylate along with the BTX aromatics is a prime ingredient in no-lead high-octane motor gasoline.

<u>Catalytic reforming</u> supplements catalytic cracking as a means of converting low-octane hydrocarbons into high-octane components.

Catalytic reforming, for example, converts hexane into cyclohexane and then into benzene, a stable high-octane aromatic.

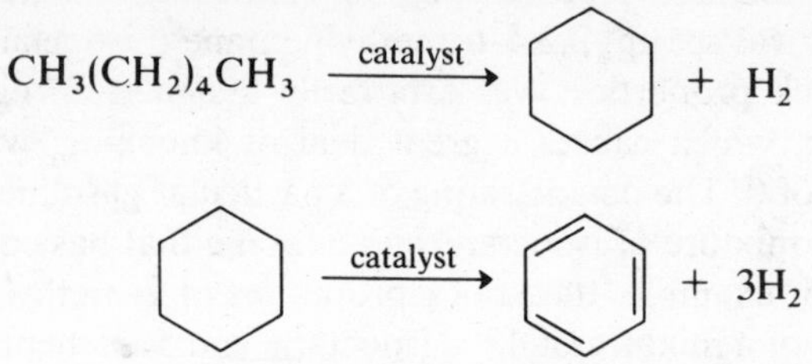

$$CH_3(CH_2)_4CH_3 \xrightarrow{\text{catalyst}} \bighexagon + H_2$$

$$\bighexagon \xrightarrow{\text{catalyst}} \bighexagon + 3H_2$$

benzene

The first catalytic reforming process came into use in 1940 and used a silica-molybdena catalyst in the presence of hydrogen gas. In 1949 Universal Oil Products introduced a platinum catalyst. This process, called Platforming, is extremely effective and remains the dominant catalytic reforming process in use today. The petroleum industry now uses catalytic reforming to treat more than 120 million gallons per day of feedstock, or close to 25% of the crude oil that enters refineries.

PROBLEMS

2.1 Write names for the following structural formulas.

(a) CH$_3$CHCH$_2$CH$_2$CH$_3$
       |
       CH$_3$

(b) CH$_3$CHCH$_2$CH$_2$CHCH$_3$
       | |
       CH$_3$ CH$_3$

(c) CH$_3$CH$_2$CHCH$_2$CHCH$_3$
            | |
            CH$_3$ CH$_2$CH$_3$

(d) (CH$_3$)$_3$CH

(e) CH$_3$CH$_2$CHCH$_2$CH$_2$CH$_2$CH$_3$
            |
            CH$_3$CHCH$_3$

(f) CH$_3$CH$_2$CH$_2$CHCH$_3$
            |
            CH$_2$CH$_2$CH$_3$

(g) CH$_3$(CH$_2$)$_8$CH$_3$

(h) (CH$_3$)$_2$CHCH$_2$CH$_2$C(CH$_3$)$_3$

(i) 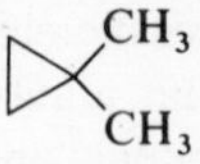 ▷⟨ CH$_3$ / CH$_3$

(j) 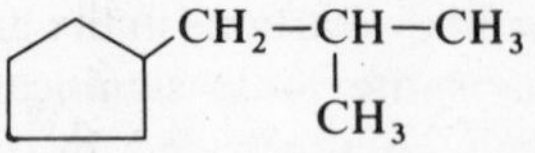 ⬠ CH$_2$—CH—CH$_3$ / CH$_3$

(k) 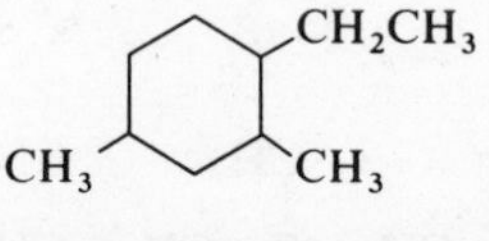 ⬡ CH$_2$CH$_3$ / CH$_3$ ... CH$_3$

(l) 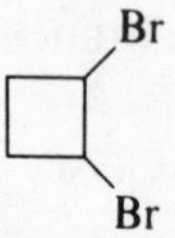 ☐ Br / Br

2.2 Write structural formulas for the following compounds.

(a) 2,2,4-trimethylhexane
(b) 1,1,2-trichlorobutane
(c) 2,2-dimethylpropane
(d) 2,4,5-trimethyl-3-ethyloctane
(e) 2-bromo-2,4,6-trimethyloctane
(f) 2,4-dimethyl-5-butylnonane
(g) 4-isopropyloctane
(h) 3,3-dimethylpentane
(i) 1,1,1-trichloroethane
(j) *trans*-1,3-dimethylcyclopentane
(k) *cis*-1,2-diethylcyclobutane
(l) 1,1-dichlorocycloheptane

2.3 Explain why each of the following names is incorrect. Write a correct name.

(a) 1,3-dimethylbutane   **(b)** 4-methylpentane
(c) 2,2-diethylbutane   **(d)** 2-ethyl-3-methylpentane
(e) 4,4-dimethylhexane   **(f)** 2-propylpentane
(g) 2,2-diethylheptane   **(h)** 5-butyloctane
(i) 2-dimethylpropane   **(j)** 2-sec-butyloctane
(k) 4-isopentylheptane   **(l)** 1,3-dimethyl-6-ethylcyclohexane

2.4 Name and draw structural formulas for all isomeric alkanes of formula C_6H_{14}.

2.5 Name and draw structural formulas for all cycloalkanes of molecular formula C_5H_{10}. Be certain to include *cis-trans* as well as constitutional isomers.

2.6 The general formula for alkanes is C_nH_{2n+2}. What is the general formula for cycloalkanes?

2.7 There are 35 constitutional isomers of formula C_9H_{20}. Name and draw the eight isomers that have five carbons in the longest chain.

2.8 Name and draw structural formulas for:

(a) the two monochloro derivatives of 2-methylpropane
(b) the two dichloro derivatives of ethane
(c) the four dichloro derivatives of propane
(d) the one monochloro derivative of cyclopentane
(e) all monochloro derivatives of hexane
(f) all monochloro derivatives of 2-methylpentane
(g) all monochloro derivatives of 2,3-dimethylbutane
(h) all monochloro derivatives of 2,2,5-trimethylhexane

2.9 There are three isomeric alkanes of molecular formula C_5H_{12}. Isomer *A* gives a mixture of four monochlorination products when reacted with chlorine gas at 300°C. Under the same conditions, isomer *B* gives a mixture of three monochlorination products while isomer *C* gives only one monochlorination product. From this information assign structural formulas to isomers *A*, *B*, and *C*.

2.10 If the chlorination of propane were completely random, that is, equally probable at any one of the eight hydrogens, predict the relative percentage yields of 2-chloropropane and 1-chloropropane. How does this compare with experimentally observed percentages?

2.11 Draw and name the *cis* and *trans* isomers of dimethylcyclopropane.

2.12 Draw the alternative chair conformations for the *cis* and *trans* isomers of 1,2-dimethylcyclohexane; of 1,3-dimethylcyclohexane; and of 1,4-dimethylcyclohexane. Label axial and equatorial positions.

(a) For which isomers are the two chair conformations of equal stability?
(b) For which isomers is one chair conformation more stable than the other chair?

2.13 **(a)** How many *cis-trans* isomers are there of 2-isopropyl-5-methyl-1-cyclohexanol? One of these isomers is the fragrant oil menthol.

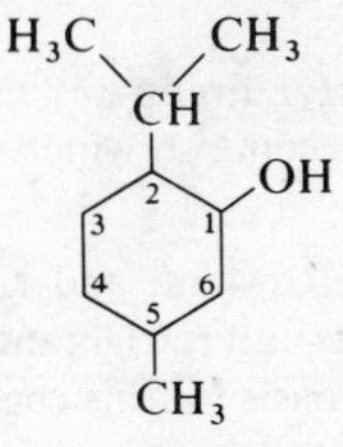

2-isopropyl-5-methyl-1-cyclohexanol

(b) Draw a chair conformation of the isomer you predict to be the most stable, that is, the isomer with minimal interactions between the atoms of the molecule. (If

47

you have drawn the correct answer, you have drawn the structural formula of menthol.)

(c) Draw the less stable chair conformation of the isomer shown in part **a**.

2.14 "Benzene hexachloride," more properly named 1,2,3,4,5,6-hexachloro-cyclohexane, is a mixture of various *cis-trans* isomers. The crude mixture is sold as the insecticide benzene hexachloride (BHC).

   The insecticidal properties of the mixture arise from one isomer known as the γ-isomer (gamma isomer) which is marketed under the name Lindane or Gammexane. Below is a representation of the gamma isomer with the cyclohexane ring shown as a planar hexagon.

benzene hexachloride (BHC)   the γ-isomer of BHC

(a) Draw a chair conformation of the γ-isomer and label the chlorine substituents either axial or equatorial.

(b) Draw the other chair conformation of the γ-isomer and again label chlorine substituents axial or equatorial.

(c) Which of these two chair conformations of the γ-isomer would you predict to be the more stable? Why?

2.15 What is the major component of natural gas? of bottled or LP gas?

2.16 What generalization can you make about the densities of alkanes relative to that of water?

2.17 In a handbook of chemistry or other suitable reference, look up the densities of methylene chloride, chloroform, and carbon tetrachloride. Which of these substances are more dense than water; which are less dense?

2.18 What straight-chain alkane has about the same boiling point as water? (Refer to Table 2.2 for data on the physical properties of alkanes.) Calculate the molecular weight of this alkane and compare it with water.

2.19 Account for the fact that saturated hydrocarbons are quite inert, that is, they are resistant to attack by most strong acids and bases as well as most oxidizing and reducing agents.

2.20 Complete and balance the following combustion reactions. Assume that each hydrocarbon is converted completely to carbon dioxide and water.

(a) propane + O_2 $\longrightarrow$
(b) octane + O_2 $\longrightarrow$
(c) 2,2,4-trimethylpentane + O_2 $\longrightarrow$
(d) benzene (C_6H_6) + O_2 $\longrightarrow$

2.21 Draw structural formulas for Freon-11, Freon-12, and Freon-22. Explain why Freons such as these have become so widely used as refrigerants and aerosol propellants.

2.22 In 1974, Drs. Sherwood Rowland and Mario Molina proposed that the Freons used as aerosol propellants and refrigerants may have a very harmful effect on the environment. Explain the basis for this concern.

3

Alkenes
and Alkynes

3.1 Introduction

Alkenes and alkynes are hydrocarbons that contain one or more multiple bonds. The alkenes, ethylene and its homologs, contain one or more carbon-carbon double bonds. The alkynes, acetylene and its homologs, contain one or more carbon-carbon triple bonds. Reaction of these hydrocarbons with hydrogen gives alkanes or saturated hydrocarbons.

$$CH_2{=}CH_2 + H_2 \xrightarrow{\text{catalyst}} CH_3{-}CH_3$$
ethylene   ethane

$$CH_3{-}C{\equiv}CH + 2H_2 \xrightarrow{\text{catalyst}} CH_3{-}CH_2{-}CH_3$$
propyne   propane

Because alkenes and alkynes contain fewer hydrogen atoms than their alkane counterparts, these two classes of organic compounds are commonly referred to as unsaturated hydrocarbons. Strictly speaking, aromatic hydrocarbons are also unsaturated hydrocarbons. However, because of their distinctive physical and chemical properties, they will be discussed separately in Chapter 6.

In this chapter we will look at the structure and bonding in alkenes and alkynes, typical reactions of alkenes, one group of alkenes widely distributed in the plant world (the terpene hydrocarbons), and another group of alkenes important in the commercial world (the alkene raw materials used in the large-scale industrial synthesis of addition polymers such as Teflon, Orlon, Lucite, and Saran).

3.2 Structure of Alkenes

Alkenes form a homologous series of compounds with the general formula C_nH_{2n}. Ethylene or ethene, C_2H_4 is the first member of the alkene family. The second member is propylene or propene, C_3H_6

$$CH_2{=}CH_2 \qquad CH_3{-}CH{=}CH_2$$

ethene   propene
ethylene   propylene

For the next member of the alkene family, there are four alkenes of formula C_4H_8. The structural formulas for three of these should be obvious; two isomers with four carbons in a continuous chain and differing only in the position of the double bond, and one isomer with a branched chain. According to the IUPAC rules, these are named 1-butene, 2-butene, and 2-methylpropene, respectively.

$$CH_3{-}CH_2{-}CH{=}CH_2 \qquad CH_3{-}CH{=}CH{-}CH_3 \qquad CH_3{-}\underset{\underset{CH_3}{|}}{C}{=}CH_2$$

1-butene   2-butene   2-methylpropene
isobutylene

The existence of the fourth isomer, or more accurately the third and fourth isomers, depends on the spatial arrangement of the atoms about the carbon-carbon double bond. There are two isomers for the structure represented as 2-butene. These are designated as *cis*-2-butene and *trans*-2-butene.

cis-2-butene   *trans*-2-butene

Recall our discussion in Section 1.10, where we noted that, for the pi bond to have a maximum stability, the two $2p_z$ orbitals must have maximum overlap. This necessity for orbital overlap places a constraint on the flexibility of the molecule and the result is hindered rotation of the two carbons of the double bond. The compounds *cis*-2-butene and *trans*-2-butene are not readily interconvertible, and they are not superimposable. They are *cis-trans* isomers. Table 3.1 lists the boiling and melting points of the four isomeric alkenes of formula C_4H_8.

We have now seen *cis-trans* isomerism in two classes of organic compounds—the cycloalkanes and the alkenes. Remember that isomers of this kind have the same molecular formula and the same order of attachment of the atoms, but differ in the orientation of the atoms in space. The key structural feature responsible for the existence of *cis-trans* isomerism is restricted rotation about a bond. This type of restraint may be introduced into a molecule in either of two ways: by the presence of a carbon-carbon

Name	bp (°C)	mp (°C)
2-methylpropene	−7	−141
1-butene	−6	−185
trans-2-butene	1	−106
cis-2-butene	4	−139

TABLE 3.1 Physical properties of the isomeric alkenes of formula C_4H_8.

double bond as we have just seen, or by the presence of a cyclic structure as we saw in the previous chapter.

There are six isomeric alkenes of formula C_5H_{10}, three with straight chains and three with branched chains.

straight chain

$CH_3CH_2CH_2CH{=}CH_2$

1-pentene

$$\underset{H}{\overset{CH_3CH_2}{\diagdown}} C{=}C \underset{CH_3}{\overset{H}{\diagup}}$$

trans-2-pentene

$$\underset{H}{\overset{CH_3CH_2}{\diagdown}} C{=}C \underset{H}{\overset{CH_3}{\diagup}}$$

cis-2-pentene

branched chain

$CH_3CH_2\underset{\underset{CH_3}{|}}{C}{=}CH_2$

2-methyl-1-butene

$CH_3\underset{\underset{CH_3}{|}}{C}HCH{=}CH_2$

3-methyl-1-butene

$CH_3\underset{\underset{CH_3}{|}}{C}{=}CHCH_3$

2-methyl-2-butene

The number of isomeric alkenes increases rapidly as the number of carbon atoms increases, for in addition to variations in chain length and branching, there are variations in both the position of the double bond and the cis-trans arrangement (i.e., the stereochemistry) of the groups attached to the double bond.

3.3 Nomenclature of Alkenes

The two commonly used methods of naming alkenes are illustrated in Table 3.2. Many alkenes, particularly the smaller ones, are known almost exclusively by their common names, for example, ethylene, propylene, isobutylene. The use of common names is generally avoided for alkenes of more than four carbons because of the large number of isomers possible. The IUPAC system is the most versatile of the two nomenclature systems and is readily extended to substances having more than one double bond.

TABLE 3.2 Naming of alkenes. The name preferred in common usage is indicated by the asterisk.

Formula	Common Name	IUPAC Name
$CH_2{=}CH_2$	ethylene*	ethene
$CH_3CH{=}CH_2$	propylene*	propene
$CH_3CH_2CH{=}CH_2$	α-butylene	1-butene*
$CH_3CH{=}CHCH_3$	β-butylene	2-butene*
$(CH_3)_2C{=}CH_2$	isobutylene*	2-methylpropene

The IUPAC names of alkenes are formed by changing the -ane of the parent alkane to -ene. Hence C_2H_4 becomes ethene. There is no chance for ambiguity in naming the first two members of this homologous series because ethene and propene can contain a double bond in only one position. Since in butene and all higher alkenes there are isomers that differ in the location of the double bond, a coding system must be used to indicate the location of the double bond. According to the IUPAC system, the longest continuous carbon chain that contains the double bond is numbered in such a manner as to give the doubly bonded carbons the lowest possible numbers. The position of the double bond is then indicated by the number of the first carbon of the double bond. Branched or substituted alkenes are named in a manner similar to alkanes. The carbon atoms are numbered, substituent groups are located and named, the double bonds are located, and the main chain is named.

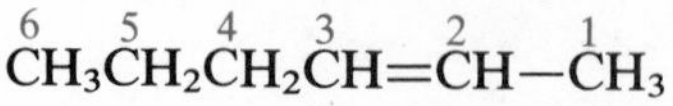

$$\overset{6}{C}H_3\overset{5}{C}H_2\overset{4}{C}H_2\overset{3}{C}H=\overset{2}{C}H-\overset{1}{C}H_3 \qquad \overset{6}{C}H_3\overset{5}{C}H_2\overset{4}{C}H\overset{3}{C}H=\overset{2}{C}H\overset{1}{C}H_3$$
$$\underset{CH_3}{|}$$

2-hexene   4-methyl-2-hexene

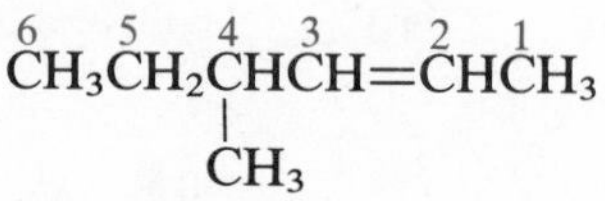

(*cis* and *trans*)   (*cis* and *trans*)

Note that alkenes are named by selecting the longest continuous carbon chain that contains the double bond. In the following compound there is a continuous chain of five carbon atoms, but the longest chain that contains the double bond is a four-carbon chain. The proper name for this compound then is 2-ethyl-3-methyl-1-butene.

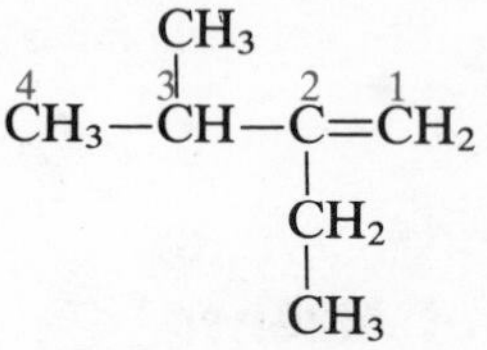

$$\begin{array}{c} CH_3 \\ | \\ \overset{4}{C}H_3-\overset{3}{C}H-\overset{2}{C}=\overset{1}{C}H_2 \\ | \\ CH_2 \\ | \\ CH_3 \end{array}$$

2-ethyl-3-methyl-1-butene

In naming cyclic alkenes, the carbons of the ring are numbered in such a direction that the carbons of the double bond are numbered 1 and 2, and the substituents receive the smallest numbers possible.

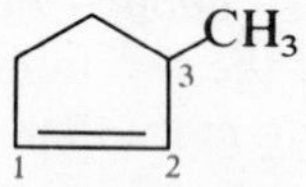

3-methylcyclopentene 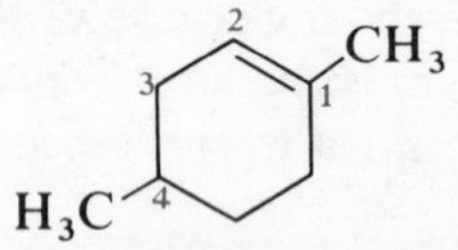   1,4-dimethylcyclohexene

(not 2,5-dimethylcyclohexene)

Alkenes that contain more than one double bond are called alkadienes, alkatrienes, or more simply dienes, trienes, etc. If two or more double bonds are separated by only one single bond, as in the case of isoprene shown below, then the double bonds are said to be <u>conjugated</u>. If the two double bonds are separated by more than one single bond, as in the case of 1,4-pentadiene shown on page 53, the double bonds are said to be <u>non-conjugated</u>.

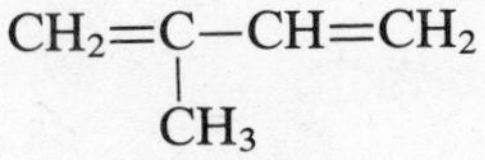

$$CH_2{=}C{-}CH{=}CH_2 \qquad CH_2{=}CH{-}CH_2{-}CH{=}CH_2$$
$$\underset{\displaystyle CH_3}{|}$$

2-methyl-1,3-butadiene
(isoprene)
conjugated

1,4-pentadiene
non-conjugated

In complex molecules containing one or more double bonds, we generally pick the longest continuous chain containing the double bond and designate the stereochemistry of the molecule based simply on whether the groups making up the main chain are on the same side (*cis*) or the opposite side (*trans*) of the double bond.

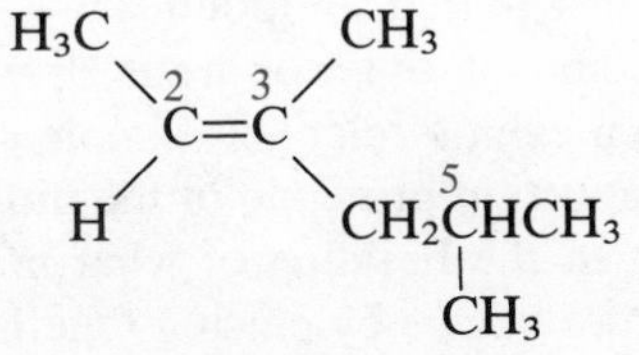

3,5-dimethyl-*trans*-2-hexene

Double bonds can be introduced into five- and six-membered rings and we have already seen several examples of cyclohexenes and cyclo-pentenes. In these rings, the double bond must have the *cis* configuration. It is only in the larger cycloalkenes that it is possible for a double bond to have the *trans* configuration. This is best understood by working with molecular models.

3.4 Physical Properties of Alkenes

The physical properties of alkenes are much like those of the correspond-ing alkanes. The first four members of the alkene family are gases at room temperature. The pentenes and higher homologs are colorless liquids, all less dense than water.

TABLE 3.3   Physical properties of some alkenes.

Name	Structural Formula	mp (°C)	bp (°C)	Density at 20°C
ethene	$CH_2{=}CH_2$	−169	−102	
propene	$CH_3CH{=}CH_2$	−185	−48	
1-butene	$CH_3CH_2CH{=}CH_2$	−185	−6	
2-methylpropene	$CH_3C{=}CH_2$ $\quad\ \ |$ $\quad\ \ CH_3$	−140	−7	
1-pentene	$CH_3(CH_2)_2CH{=}CH_2$	−165	30	0.641
1-hexene	$CH_3(CH_2)_3CH{=}CH_2$	−141	64	0.673
cyclohexene	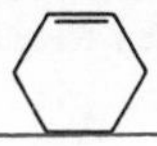 ⬡	−104	83	0.811

3.5 Naturally Occurring Alkenes—The Terpene Hydrocarbons

A wide variety of substances in the plant and animal world contain one or more carbon-carbon double bonds. In many of these compounds, the double bond is but one of several functional groups, and we will see a variety of such compounds in later chapters. At this point, however, we will focus our attention on one group of compounds—the terpene hydrocarbons—in which many individual members of the group contain one or more double bonds as the only functional group.

Aside from the fact that terpene hydrocarbons are examples of naturally occurring alkenes, there are other perhaps more important reasons for looking at this group of organic compounds. First, they are among the most widely distributed compounds in nature. Second, they will provide a glimpse at some of the wondrous diversity of structure that nature can generate from even a relatively simple carbon skeleton. Third, terpenes illustrate an important principle of the molecular logic of organic chemistry, namely, that in the building of what might seem to be rather complex molecules, nature begins by piecing together small, readily available subunits to produce an intriguingly complex but logically designed skeletal framework. At this point we will be concerned only with identifying this framework. Later we shall study reactions by which the framework can be modified to include other functional groups.

Before we examine structural formulas of particular terpene hydrocarbons, we should point out that virtually all of these substances have one structural feature in common: their carbon skeletons can be divided into two or more units of isoprene.

$$\underset{\text{isoprene}}{CH_2=\overset{\overset{\textstyle CH_3}{|}}{C}-CH=CH_2}$$

This does not mean that in nature terpenes are built from isoprene (actually they are built from units of acetate). It does mean, however, that the carbon skeleton of essentially every terpene can be divided into two or more subunits that are identical to the carbon skeleton of isoprene. This generalization is known as the isoprene rule.

With this background let us now turn to look at some particular terpene hydrocarbons of plant origin. Probably the terpenes most familiar to you, at least by odor, are components of the so-called "essential oils" which can be obtained by either steam distillation or ether extraction of various parts of plants. These essential oils contain relatively low-molecular-weight, volatile substances which are in large part responsible for the characteristic odor of the plant. Among typical essential oils are those from mint and eucalyptus leaves, pine needles, rose petals, sandalwood, cedar, and oil of cloves. Myrcene, $C_{10}H_{16}$, can be obtained from bayberry wax and from the oils of bay and verbena.

As you can see from structural formula (a), myrcene contains 10 carbon atoms and 3 carbon-carbon double bonds. For convenience in drawing and in order to show structural features more clearly, the organic chemist sometimes represents structures such as this by using a shorthand notation.

(a) (b) (c)

Carbon-carbon bonds are represented as lines and carbon atoms are not shown but are understood to be at junctions of lines and where lines end. Structures (a) and (b) are equivalent representations of the structural formula for myrcene. As you can easily see from the position of the dashed lines in (a) or (b) or from the skeletal framework shown in (c), myrcene is divisible into two isoprene units joined from carbon 1 (head) of one unit to carbon 4 (tail) of a second unit. This head-to-tail linkage of isoprene units is vastly more common in nature than the alternative head-to-head (1,1) or tail-to-tail (4,4) linkages. Figure 3.1 shows further examples of acyclic, monocyclic, and bicyclic terpene hydrocarbons. These structural formulas are drawn to show clearly the isoprene-derived skeleton and, except for the one *trans* ring double bond in caryophyllene, do not necessarily show the correct *cis-trans* geometry about double bonds.

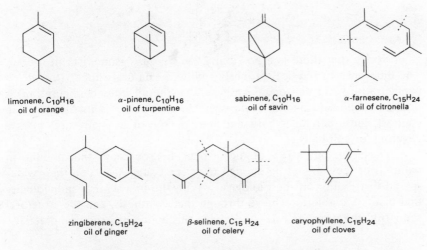

limonene, $C_{10}H_{16}$
oil of orange

α-pinene, $C_{10}H_{16}$
oil of turpentine

sabinene, $C_{10}H_{16}$
oil of savin

α-farnesene, $C_{15}H_{24}$
oil of citronella

zingiberene, $C_{15}H_{24}$
oil of ginger

β-selinene, $C_{15}H_{24}$
oil of celery

caryophyllene, $C_{15}H_{24}$
oil of cloves

FIGURE 3.1 Representative terpene hydrocarbons.

Actually terpenes themselves are the simplest members of a group of compounds more properly called terpenoids. Depending on the number of isoprene units, the various subgroups are known as:

terpenes	C_{10}	two isoprene units
sesquiterpenes	C_{15}	three isoprene units
diterpenes	C_{20}	four isoprene units
triterpenes	C_{30}	six isoprene units
polyterpenes		many isoprene units

55

One diterpene, familiar to all of us, is <u>vitamin A</u>, $C_{20}H_{30}O$. The function of this fat-soluble vitamin is discussed in Chapter 12.

vitamin A

The tetraterpene plant pigment β-carotene, $C_{40}H_{56}$, has an orange-red color and is commonly used as food coloring. It also has some vitamin A activity because it is apparently cleaved in the body at the central carbon-carbon double bond to give a vitamin A-derived molecule. Notice that β-carotene can be divided into two regular diterpenes with each of the units then joined together in a tail-to-tail manner.

β-carotene
(all *trans* double bonds)

<u>Natural rubber</u> is an acyclic polyterpene, that is, it is composed of a continuous chain formed from head-to-tail linkages of n (where n is very large) isoprene units.

$$-(CH_2-\overset{\overset{\displaystyle CH_3}{|}}{C}=CH-CH_2)_n$$

Notice that there is a possibility for *cis-trans* isomerism in rubber. The difference in physical properties between natural rubber (the all *cis* configuration) and its isomer <u>gutta-percha</u> (the all *trans* configuration) is dramatic. The outstanding feature of natural rubber is its elasticity. In contrast, gutta-percha is hard and horny. One of its many uses is as a covering for golf balls.

These are but a few of the terpene hydrocarbons that abound in nature, but they should be enough to at least suggest to you their wide-spread distribution in the biological world, the biochemical individuality that plants are able to achieve through their synthesis, and the structural similarity (the isoprene rule) underlying this apparent structural diversity.

3.6 Preparation of Alkenes

Alkenes are most often synthesized by a process called an <u>elimination reaction</u>. In this process, some small molecule (most commonly HOH, HCl, or HBr) is split out or eliminated from adjacent carbons of a larger molecule.

$$R-\underset{\underset{\displaystyle H}{|}}{C}H-\underset{\underset{\displaystyle OH}{|}}{C}H-R \xrightarrow{\overset{\text{dehydrating}}{\text{agent}}} R-CH=CH-R + H_2O$$

$$R-\underset{\underset{H}{|}}{C}H-\underset{\underset{Br}{|}}{C}H-R \xrightarrow{\text{strong base}} R-CH=CH-R + HBr$$

The first reaction shown here involves removal of a molecule of water from adjacent carbons of an alcohol and is called dehydration. The second involves removal of a molecule of HBr and is called dehydrobromination or dehydrohalogenation. These elimination reactions are frequently the reverse of the addition reactions we will encounter later in this chapter.

In the laboratory, dehydration of an alcohol is usually brought about by heating the alcohol with either 85% phosphoric acid or concentrated sulfuric acid.

$$CH_3\underset{\underset{OH}{|}}{C}HCH_2CH_3 \xrightarrow[\text{heat}]{85\%\ H_3PO_4} CH_3CH=CHCH_3 + CH_2=CHCH_2CH_3 + H_2O$$

2-butanol
(sec-butyl alcohol)

2-butene   1-butene
80%   20%

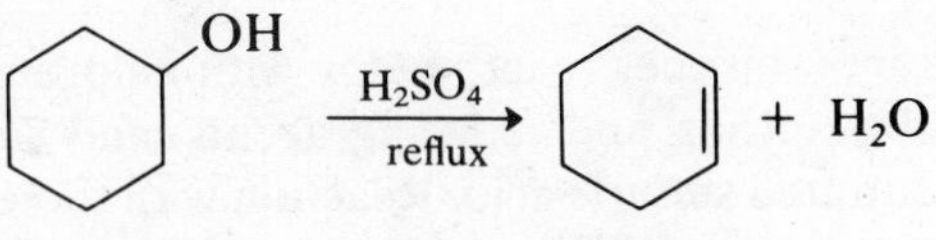

cyclohexanol   cyclohexene
                        85%

Acid-catalyzed dehydration of alcohols can usually lead to formation of isomeric alkenes; for example, dehydration of 2-butanol produces 80% 2-butene and 20% 1-butene. From a study of these and many other reactions we can make the following generalization: where it is possible to obtain isomeric alkenes from the acid-catalyzed dehydration of an alcohol, the alkene having more substituents on the double bond will generally predominate. Following this rule, you would predict that acid-catalyzed dehydration of 2-methylcyclohexanol would give mostly 1-methyl-cyclohexene with only minor amounts of the isomeric 3-methyl-cyclohexene or methylenecyclohexane.

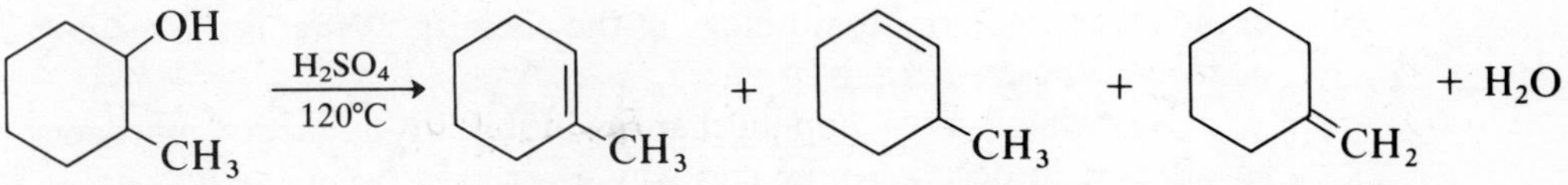

2-methylcyclohexanol   1-methylcyclohexene   3-methylcyclohexene   methylenecyclohexane
                        (major product)   (minor product)   (minor product)

This is precisely what is observed experimentally. The mechanism for acid-catalyzed dehydration will be discussed in Section 5.9.

Dehydration can also be brought about by passing an alcohol vapor over aluminum oxide (Al_2O_3), thorium oxide (ThO_2), or other metal oxide. High temperature is required to bring about this type of dehydration.

$$CH_3(CH_2)_5CH_2CH_2OH \xrightarrow[350°C]{Al_2O_3} CH_3(CH_2)_5CH=CH_2 + H_2O$$

1-octanol   1-octene

Dehydrohalogenation is the removal of an acid compound (such as HBr and HCl) and requires the action of strong base.

57

$$CH_3CH_2CHCH_3 + NaOH \longrightarrow CH_3CH=CHCH_3 + NaBr + H_2O$$
$$\underset{Br}{|}$$

2-bromobutane   2-butene
(sec-butyl bromide)

$$\underset{\underset{Br}{|}}{\overset{\overset{CH_3}{|}}{CH_3CCH_2CH_3}} + CH_3CH_2ONa \longrightarrow \overset{\overset{CH_3}{|}}{CH_3C}=CHCH_3 + CH_3CH_2OH + NaBr$$

2-bromo-2-methyl-   2-methyl-2-butene
butane

Where dehydrohalogenation can lead to two or more isomeric alkenes, it is the more heavily substituted alkene that is formed in greatest amount.

3.7 Reactions of Alkenes

In contrast to alkanes, alkenes react readily with halogens, certain strong acids, a variety of oxidizing and reducing agents, and even water in the presence of concentrated sulfuric acid. Reaction with these and most other reagents is characterized by <u>addition</u> to the double bond.

$$\overset{\diagdown}{\diagup}C=C\overset{\diagup}{\diagdown} + A-B \longrightarrow \overset{\diagdown}{\diagup}\underset{A}{\overset{|}{C}}-\underset{B}{\overset{|}{C}}\overset{\diagup}{\diagdown}$$

In the general reaction shown, there is rupture of one sigma bond (A—B) and one pi bond, and the formation of two sigma bonds. Consequently, addition reactions to the double bond are almost always energetically favorable because there is net conversion of one pi bond into one sigma bond.

In the following sections, we shall look at typical alkene reactions. As we examine these reactions and those of other functional groups in later chapters, we will be asking constantly, "How do these reactions occur?" or in the more common terminology of the chemist, "What is the mechanism of each of these reactions?"

Just what is a <u>reaction mechanism</u>? Literally, a reaction mechanism is an attempt to describe how and why a reaction occurs as it does. It is a description of how covalent bonds are broken and reformed, the role of catalysts and solvents, and the rates at which the various changes take place during the course of the reaction. Further, it is a description of the role of <u>reactive intermediates</u>, species formed during the reaction by breaking a carbon bond and having a carbon atom in an abnormal valence state. In this chapter we will encounter two different types of reactive intermediates, <u>carbocations</u> and <u>free radicals</u>. In Chapter 7 we will encounter a third type of reactive intermediate, the <u>carbanion</u>.

A complete specification of all of these factors is more than we know about any reaction at the present time. Therefore, we will be content in our description of a reaction mechanism if we can specify (1) the steps in the reaction including the formation of any reactive intermediates, (2) the relative rates of the various steps, and (3) the structure of transition states.

How do we begin to describe a reaction mechanism? First, we assemble all of the available experimental observations and facts about the particular chemical reaction under consideration. Next, through a combination of chemical sophistication, creative insight, and guesswork, we propose several sets of steps, or mechanisms, each of which will account for the overall chemical transformation. Then, we test each of these mechanisms against the experimental observations and exclude those mechanisms that are not consistent with the facts. A mechanism becomes generally established by excluding reasonable alternatives and by showing that the mechanism is consistent with every test that the scientist can devise. This of course does not mean that a generally accepted mechanism is in fact a completely true and accurate description of the chemical events, but only that it is consistent with the mass of experimental evidence.

Before we go on to consider reactions and reaction mechanisms, we might also ask, why is it worth the trouble of chemists to establish them and students to learn about them? Certainly one answer lies in the understanding and intellectual satisfaction derived from constructing models that accurately reflect the behavior of chemical systems. Problems of this kind present a particular fascination and exciting challenge to the creative scientist. But there are other, more practical reasons as well. Mechanisms provide us with a theoretical framework within which to organize a great deal of descriptive chemistry. For example, with some insight into how certain reagents add to particular alkenes, we can then generalize about how these same reagents might react with other alkenes and make predictions about how different reagents might also react with the same alkenes. Finally, to the practicing chemist, a knowledge of mechanisms is often helpful in selecting conditions of temperature, pressure, concentration, reaction time, etc. so as to better achieve desired results in a specific reaction.

3.8 Addition of Hydrogen—Hydrogenation

The addition of hydrogen to a carbon-carbon double bond converts an alkene to an alkane. This "saturation" or hydrogenation of the double bond is a very important reaction in the laboratory as well as in nature.

$$CH_2{=}CH_2 + H_2 \xrightarrow[\text{catalyst}]{\text{metal}} CH_3{-}CH_3$$

1-methylcyclohexene   methylcyclohexane

Although the addition of hydrogen to an alkene is strongly exothermic, these reactions are immeasurably slow at moderate temperatures in the gas phase, but they occur readily in the presence of finely divided platinum, palladium, or nickel. Separate experiments have shown that these and other metals near the center of the periodic table are able to absorb large quantities of hydrogen gas by incorporating it into the metallic lattice as hydrogen atoms. These hydrogen atoms are electron-deficient and therefore highly reactive.

Another type of experimental observation that must be incorporated into any acceptable mechanism is the known stereospecificity of the catalytic hydrogenation. For example, addition of hydrogen to 1,2-dimethylcyclopentene yields mostly cis-1,2-dimethylcyclopentane. From this and other observations of the stereochemistry of the addition of hydrogen it was concluded that both hydrogen atoms are added simultaneously and from the same side of the alkene molecule.

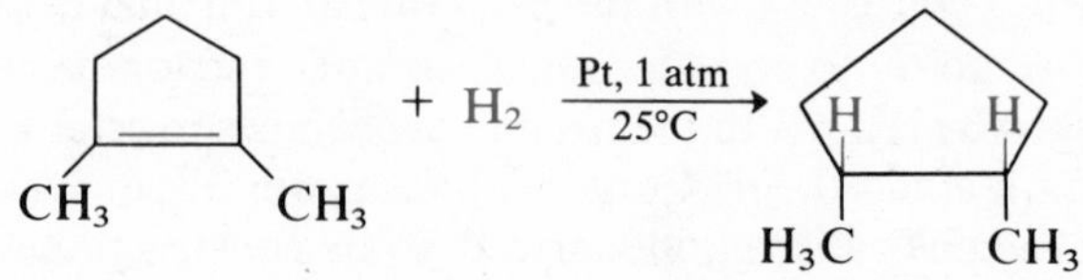

| 1,2-dimethylcyclopentene | cis-1,2-dimethylcyclopentane |

The generally accepted mechanism for a catalytic hydrogenation reaction postulates that molecular hydrogen is first adsorbed on the surface of the metal and the bond between the hydrogen atoms is weakened by interaction with the metal. When the alkene is also adsorbed on the surface of the metal, addition occurs—both hydrogen atoms are delivered simultaneously and the resulting alkane is desorbed. A schematic representation of these steps is shown in Figure 3.2.

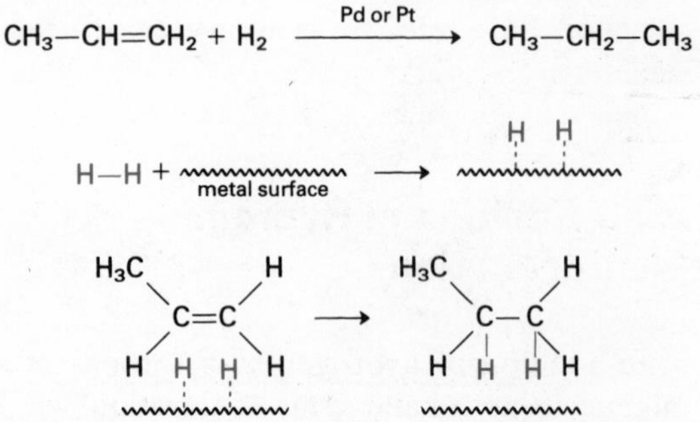

$$CH_3{-}CH{=}CH_2 + H_2 \xrightarrow{\text{Pd or Pt}} CH_3{-}CH_2{-}CH_3$$

FIGURE 3.2 Hydrogenation of propene involving a metal catalyst.

This concept of the interaction of one or more molecules on a catalytic surface is one which underlies a great deal of our thinking on the catalytic action of enzymes.

3.9 Addition of Hydrogen Halides

Dry HCl, HBr, or HI add readily to alkenes. These reactions may be carried out either with the pure reagents or in the presence of a polar solvent such as acetic acid.

$$CH_2{=}CH_2 + HCl \longrightarrow CH_3CH_2Cl$$

chloroethane

$$CH_3CH{=}CH_2 + HBr \longrightarrow CH_3\underset{\underset{Br}{|}}{C}HCH_3 + CH_3CH_2CH_2Br$$

| propene | 2-bromopropane
isopropyl bromide
(*major product*) | 1-bromopropane
n-propyl bromide
(*minor product*) |

$$CH_3\underset{\underset{CH_3}{|}}{C}{=}CH_2 + HI \longrightarrow CH_3\underset{\underset{CH_3}{|}}{\overset{\overset{I}{|}}{C}}CH_3 + CH_3\underset{\underset{CH_3}{|}}{C}HCH_2I$$

| 2-methyl-2-
propene | 2-iodo-2-methylpropane
tert-butyl iodide
(*major product*) | 1-iodo-2-methylpropane
isobutyl iodide
(*minor product*) |

The reaction of ethylene involves the addition of hydrogen to one of the carbon atoms and chlorine to the other and only one product is possible. In addition of HBr to propene we encounter a different situation. Without either experimental evidence or a theoretical model to guide us, we would have to expect the formation of two isomeric products, isopropyl bromide and *n*-propyl bromide, depending on the orientation of the addition. True to our unsophisticated expectation, each of these isomers is formed, but the fact that isopropyl bromide predominates in the product mixture clearly indicates that the addition is by no means random.

After studying a large number of reactions involving the addition of acids to alkenes, in 1871 the Russian chemist, Vladimir Markovnikov, made the following empirical generalization: in the addition of an unsymmetrical reagent to an unsymmetrical alkene, the positive part of the reagent adds to the carbon of the double bond that contains the greater number of hydrogen atoms. We should remember that while this rule systematizes many of the observed reactions of addition to alkenes, it is not an explanation for the observations.

The challenge for the modern organic chemist is to try to understand how these acids add to alkenes, what if any reactive intermediates are formed during the course of the reaction, and why one of the possible products predominates over the others.

3.10 Electrophilic Attack on the Carbon-Carbon Double Bond

It became apparent to chemists studying these and other alkene addition reactions that the initial attack on the carbon-carbon double bond is by H^+, or, in more general terms, the initial attack on the double bond is by an electrophilic (electron-loving or -seeking) reagent. This susceptibility of the double bond to attack by an electrophile is certainly consistent with our electronic formulation, which pictures the double bond as a center of high electron (negative) density.

To account for these reactions of alkenes, organic chemists have proposed a two-step ionic mechanism which correlates much that is known about these electrophilic additions. Let us consider first the addition of HBr to ethylene.

$$CH_2{=}CH_2 + HBr \longrightarrow CH_3{-}CH_2Br$$

The first step in the generally accepted mechanism involves the reaction of a pair of electrons of the double bond with H^+ and the formation of a new covalent bond between carbon and hydrogen.

Step 1 $H^+ + CH_2{=}CH_2 \longrightarrow CH_3{-}CH_2^+$

Step 2 $CH_3{-}CH_2^+ + :\ddot{B}r:^- \longrightarrow CH_3{-}CH_2\ddot{B}r:$

Step 1 results in the formation of a reactive intermediate in which one of the carbon atoms has only six electrons in its valence shell and, therefore, bears a positive charge. In step 2, this reactive intermediate completes its valence octet by forming a new covalent bond with an unshared pair of electrons on the bromide ion. Combination of steps 1 and 2 gives the observed stoichiometry.

The terminology for naming such carbon-containing reactive intermediates is undergoing reexamination at the present time. Until very recently, they have been called carb<u>onium</u> <u>ions</u> by analogy with amm<u>onium</u> <u>ions</u> (e.g., NH_4^+) and ox<u>onium</u> <u>ions</u> (e.g., H_3O^+). However the appropriateness of the term carbonium ion is now being questioned for, in ammonium and oxonium ions, both nitrogen and oxygen have complete octets of valence electrons. Yet in the so-called carbonium ion, the carbon bearing the positive charge is electron deficient for it has only six valence electrons. Therefore, it is reasoned, we should use another suffix for naming these electron-deficient carbon-containing cations. The term <u>carbocation</u> (<u>carbon</u> <u>cation</u>) has been proposed. We shall use the term carbocation throughout the text. Nevertheless, you should be aware of the term carbonium ion for you will undoubtedly encounter it in the literature of chemistry and in your discussions with other chemists.

But how does the carbocation mechanism account for the observations that the addition of HBr to propene gives mainly isopropyl bromide and very little *n*-propyl bromide? How does the theory account for the type of observations generalized in <u>Markovnikov's rule</u>? The answer is that the mechanism as elaborated thus far does not account for it. At this point, we have two alternatives; either we discard the idea of a carbocation intermediate and start again, or we can refine the mechanism by introducing a new concept. In this case it is more convenient and fruitful to refine the mechanism and introduce a new concept, namely, the <u>relative ease of formation of carbocations</u>.

The carbocation mechanism proposes the addition of a proton to an unsymmetrical alkene such as propene can give either of two isomeric carbocations.

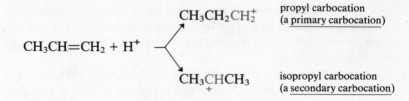

$CH_3CH_2CH_2^+$ propyl carbocation (a <u>primary</u> carbocation)

$CH_3CH{=}CH_2 + H^+$

$CH_3\underset{+}{C}HCH_3$ isopropyl carbocation (a <u>secondary</u> carbocation)

In the same manner the reaction of 2-methylpropene with a proton can give two isomeric carbocations:

$$CH_3\overset{\underset{\textstyle |}{CH_3}}{C}=CH_2 + H^+ \longrightarrow$$

$$CH_3\overset{\underset{\textstyle |}{CH_3}}{CH}-CH_2^+ \qquad \text{isobutyl carbocation (a primary carbocation)}$$

$$CH_3\overset{\underset{\textstyle |}{CH_3}}{C^+}-CH_3 \qquad \textit{tert}\text{-butyl carbocation (a tertiary carbocation)}$$

The propyl carbocation will react with bromide ion to give *n*-propyl bromide, and the isopropyl carbocation will react with bromide ion to form isopropyl bromide. Hence the alkyl halide formed depends on the carbocation formed. It is clear from a detailed study of the addition of HBr to propene and other alkenes that the slow step in the addition reaction is the formation of the carbocation. Once the carbocation is formed, it reacts with bromide ion to give the observed product.

Since isopropyl bromide is the major product from the addition of HBr to propene, the isopropyl carbocation must be formed more easily than the propyl carbocation. Since *tert*-butyl iodide predominates over isobutyl iodide the *tert*-butyl carbocation must be formed more easily than the isobutyl carbocation. From the study of these and many other reactions which proceed via carbocations, we infer that the ease of formation of carbocations is of the following order:

$$CH_3-\overset{\underset{\textstyle |}{CH_3}}{\overset{\textstyle |}{C^+}} > CH_3-\overset{\underset{\textstyle |}{CH_3}}{\overset{\textstyle |}{C^+}} > CH_3-\overset{\underset{\textstyle |}{H}}{\overset{\textstyle |}{C^+}} > H-\overset{\underset{\textstyle |}{H}}{\overset{\textstyle |}{C^+}}$$

The ease of formation of a carbocation increases as the ion becomes more highly substituted by alkyl groups at the positive carbon. To generalize, we can write the following order for the ease of formation of carbocations.

$$3° > 2° > 1° > methyl$$

Carbocations are involved as reactive intermediates in many different types of organic reactions. We will see other examples of their formation and reaction in this and the next several chapters.

3.11   Addition of Water—Hydration

In the presence of an acid catalyst, often 60% aqueous sulfuric acid, water adds to the more reactive alkenes in accordance with Markovnikov's rule to produce alcohols. The addition of the elements of water to an alkene is called hydration.

$$CH_3CH=CH_2 + H_2O \xrightarrow{H^+} CH_3\overset{\underset{\textstyle |}{\textstyle OH}}{CH}CH_3 + CH_3CH_2CH_2OH$$

propene

2-propanol
isopropyl alcohol
(*major product*)

1-propanol
n-propyl alcohol
(*minor product*)

$$CH_3\overset{\underset{|}{CH_3}}{C}=CH_2 + H_2O \xrightarrow{H^+} CH_3\overset{\underset{|}{CH_3}}{\underset{OH}{C}}CH_3 + CH_3\overset{\underset{|}{CH_3}}{C}HCH_2OH$$

2-methylpropene   2-methyl-2-propanol   2-methyl-1-propanol
                              tert-butyl alcohol   isobutyl alcohol
                              (*major product*)   (*minor product*)

The mechanism for the acid-catalyzed hydration of alkenes follows much the same scheme as the addition of hydrogen halides. As we have noted, these reactions require the presence of a strong acid. The first step of the accepted mechanism involves the reaction of a pair of electrons of the alkene with a proton, forming a carbocation. In step 2, this reactive intermediate completes its valence octet by forming a new covalent bond with an unshared electron pair on the oxygen of a water molecule. Finally, loss of a proton results in the formation of ethanol and regeneration of a proton.

Step 1 $H^+ + CH_2{=}CH_2 \longrightarrow CH_3{-}CH_2^+$

Step 2 $CH_3{-}CH_2^+ + :\overset{\underset{|}{}}{\underset{H}{O}}{-}H \longrightarrow CH_3{-}CH_2{-}\overset{+}{\underset{H}{O}}{-}H$

Step 3 $CH_3{-}CH_2{-}\overset{+}{\underset{H}{O}}{-}H \longrightarrow CH_3{-}CH_2{-}\overset{..}{O}{-}H + H^+$

Of course, we must ask whether our proposed mechanism is in agreement with the experimental observations? First, the mechanism does account for the observation that, although H^+ is not a part of the stoichiometry of the reaction, it is necessary as a catalyst; acid is involved in the initiation of the reaction and is regenerated in the final step. Second, given the concept of the relative ease of formation of primary, secondary, and tertiary carbocations, this mechanism is consistent with the observations that hydration of propene gives mostly 2-propanol and only very little 1-propanol.

There is a second and in many ways more useful method for converting an alkene to an alcohol. In this two-step process, an alkene is first reacted with diborane, B_2H_6, to yield a trialkylborane, R_3B. Subsequent oxidation of the trialkylborane by hydrogen peroxide yields alcohols of the type $R{-}OH$. Let us examine each of these steps more closely.

H—B bonds add readily to alkenes. Note that while the molecular formula of the reactive boron-hydrogen compound is B_2H_6, it behaves in this particular reaction as if it were borane, BH_3, and is therefore shown as such in the following sequence. BH_3 adds first to one molecule of alkene, then to a second, and finally to a third to form a trialkylborane.

$$3CH_3CH{=}CH_2 + BH_3 \longrightarrow CH_3CH_2CH_2{-}B\overset{\displaystyle CH_2CH_2CH_3}{\underset{\displaystyle CH_2CH_2CH_3}{}}$$

propene   borane   tripropylboron

The reaction of BH_3 and an alkene to form a trialkylborane is called hydroboration. Hydroboration does not appear to involve a carbocation intermediate. <u>Rather, there is simultaneous addition of hydrogen and boron to the alkene in such a way that the boron atom becomes attached to the least substituted (least hindered) carbon of the double bond.</u> Trialkylboranes are generally not isolated but are oxidized directly with hydrogen peroxide in aqueous sodium hydroxide to yield the alcohol and sodium borate.

$$(CH_3CH_2CH_2)_3B + H_2O_2 + 3NaOH \longrightarrow 3CH_3CH_2CH_2OH + 3Na_3BO_3$$

| tripropylborane | hydrogen peroxide | 1-propanol n-propyl alcohol | sodium borate |

The net reaction from hydroboration and subsequent oxidation is <u>anti-Markovnikov hydration</u> of the double bond. To appreciate the value of this sequence, we might compare the product of acid-catalyzed hydration of 2-methylpropene with that obtained from hydroboration and hydrogen peroxide oxidation.

Thus, hydroboration-oxidation is so useful because it gives alcohols not obtainable by any other methods from alkenes.

3.12 Halogenation of Alkenes

Bromine and chlorine add readily to alkenes to form single covalent bonds on adjacent carbons. While iodine is generally too unreactive to add, the more reactive iodine monochloride (ICl) and iodine monobromide (IBr) do add readily. (See problem 3.16 for a quantitative application of iodine monobromide addition.) <u>Halogenation</u> with bromine or chlorine is carried out either with the pure reagents (neat) or by mixing the reagents in CCl$_4$ or some other inert solvent.

As illustrated by the second example, the addition of halogen to a simple alkene is quite specific in producing only one dihalide. Where *cis-trans* isomerism is possible, it is the *trans* dihalide which is formed as the major product.

Addition of bromine is a particularly useful qualitative test for the detection of alkenes. A solution of bromine in carbon tetrachloride is red. Alkenes and dibromoalkanes are usually colorless. The rapid discharge of red color of a solution of bromine in carbon tetrachloride is a characteristic property of alkenes.

We might pause a moment to compare and contrast the halogenation of alkenes with that of alkanes (Section 2.11). Recall that chlorine and bromine do not react with alkanes unless the halogen-alkane mixture is exposed to strong light or heated to temperatures of 250–400°C. The reaction which then occurs is one of substitution of halogen for hydrogen and the formation of an equivalent amount of HCl or HBr. The halogenation of most higher alkanes inevitably gives a complex mixture of monosubstitution products. In contrast, chlorine and bromine react readily with alkenes at room temperature in the dark. Further, the reaction is the addition of halogen to the two carbon atoms of the double bond with the formation of two new carbon-halogen bonds. If we know the position of the double bond, we can predict with certainty the location of the halogen atoms in the product.

The addition of bromine to an alkene can be formulated in terms of a two-step mechanism and a carbocation intermediate. The nonpolar halogen molecule X_2 becomes partially polarized as it approaches the double bond, a region of high electron density. The polarized bromine or chlorine molecule is then capable of acting as an electrophile and initiating attack on the olefinic double bond. In step 1, the bromine atom which first bonds with carbon leaves behind the pair of electrons which formerly bonded it to the other bromine. It is for this reason that the latter acquires a negative charge and becomes a bromide ion. The carbocation formed in the initial attack reacts with the halide ion also formed in the first step, and the final product is the simple adduct of the halogen and olefin. This mechanism is illustrated in the reaction of bromine and ethylene.

Step 1 $:\ddot{B}r-\ddot{B}r: + CH_2=CH_2 \longrightarrow :\ddot{B}r-CH_2-CH_2^+ + :\ddot{B}r:^-$

Step 2 $:\ddot{B}r-CH_2-CH_2^+ + :\ddot{B}r:^- \longrightarrow :\ddot{B}r-CH_2-CH_2-\ddot{B}r:$

3.13 Oxidation of Alkenes

The electrons of the carbon-carbon double bond are readily attacked by oxidizing agents. In this section we shall consider two of these agents: potassium permanganate ($KMnO_4$) and ozone (O_3).

Treatment of an alkene with a dilute basic solution of potassium permanganate in water or an aqueous organic solvent oxidizes an alkene to a 1,2-dialcohol (a glycol).

$$3CH_3CH=CH_2 + 2KMnO_4 + 4H_2O \longrightarrow 3CH_3\underset{\underset{\text{OH}}{|}}{CH}-\underset{\underset{\text{OH}}{|}}{CH_2} + 2MnO_2 + 2KOH$$

propylene glycol   manganese dioxide

cis-1,2-cyclopentanediol

The *cis* geometry of the product is accounted for by the formation of a cyclic intermediate.

This reaction is the basis of a qualitative test for the presence of an alkene because in the course of the reaction, the purple color of the permanganate ion is discharged and a brown precipitate of MnO_2 forms. Unfortunately, this test is not completely specific for alkenes since certain other easily oxidized functional groups will also discharge the permanganate color.

Even at low temperatures most alkenes react with ozone, O_3, to cleave the carbon-carbon double bond. This reaction is useful in the preparation of aldehydes and ketones (Chapter 7) and as a means of locating the position of the double bond within an alkene.

In practice, the alkene is dissolved in an inert solvent such as CCl_4 and a stream of ozone gas is bubbled through the solution. Since the ozonide intermediates are usually unstable and explosive in pure form, they are generally not isolated. The reaction mixture is poured into water in the presence of a reducing agent, usually powdered metallic zinc. The resulting aldehydes or ketones are isolated and purified. Through determination of the structures of the products of ozonolysis, it is possible to work backwards and deduce the structure of an unknown alkene.

3.14 Polymerization of Substituted Ethylenes

From the perspective of the chemical industry, the single most important reaction of alkenes is that of polymerization, the building together of many small units known as <u>monomers</u> (Greek: *mono + meros*, single part) into very large, high-molecular-weight <u>polymers</u> (Greek: *poly + meros*, many parts). In this section we will be concerned with <u>addition polymerization</u> in which monomer units are joined together without loss of atoms in the process. The alternative process, known as <u>condensation polymerization</u>, involves the loss of one or more atoms during the polymerization process. Condensation polymerization is illustrated in the mini-essay *Nylon and Dacron*.

Polymers are very large molecules that consist of many repeating subunits. The subunits themselves are called monomers. Two monomers attached to one another form a <u>dimer</u>, three monomers attached together form a <u>trimer</u>, and many monomers attached together form a polymer. Polymerization of ethylene yields <u>polyethylene</u>.

$$
\begin{array}{ccc}
\underset{\substack{\text{ethylene}\\\text{(monomer)}}}{\overset{\displaystyle H}{\underset{\displaystyle H}{}}C=C\overset{\displaystyle H}{\underset{\displaystyle H}{}}} & \longrightarrow & \underset{\substack{\text{polyethylene}\\\text{(polymer)}}}{\left(\!\!\begin{array}{c} H\ H\\ -C-C-\\ H\ H \end{array}\!\!\right)_{n}}
\end{array}
$$

The subscript, n, indicates that the monomer-derived unit repeats n times in the polymer chain. Molecular weights of polyethylenes range from 50,000 to over 1,500,000.

Much of the early interest in polymers arose from the desire to make synthetic rubber and relieve the near-total dependence on natural rubber. In the mid-1920s, several companies, most notably du Pont in the United States and I.G. Farbenindustrie in Germany, began research programs in this area. By the mid-1930s, du Pont was producing the first synthetic rubber, Neoprene, on a commercial basis. Neoprene synthetic rubber is polychloroprene.

$$
n\text{CH}_2=\overset{\displaystyle \text{Cl}}{\overset{\displaystyle |}{\text{C}}}-\text{CH}=\text{CH}_2 \xrightarrow{\ \text{polymerization}\ } -(\text{CH}_2-\overset{\displaystyle \text{Cl}}{\overset{\displaystyle |}{\text{C}}}=\text{CH}-\text{CH}_2)_n-
$$

chloropropene
2-chlorobutadiene   Neoprene
polychloroprene

The years since have seen extensive research and development in polymer chemistry and physics, and an almost explosive growth in plastics, coatings, and rubber technology has created a world-wide multibillion dollar industry. A few basic characteristics account for this phenomenal growth. First, the raw materials for plastics are derived mainly from petroleum. With the development of efficient cracking processes (Section 2.15) the raw materials became generally cheap and plentiful. Second, within broad limits, scientists have learned how to tailor-make polymers to the requirements of the end use. Third, many plastics can be fabricated more cheaply than the competing materials. For example, plastic technology created the water-based (latex) paints that have revolutionized the coatings industry; plastic films and foams have done the same for the packaging industry. The list could go on and on as we think of the manufactured items that surround us in our daily lives.

TABLE 3.4 Polymers derived from substituted ethylene monomers.

Monomer	Monomer Name	Polymer Name or Trade Name
$CH_2{=}CH_2$	ethylene	polyethylene, Polythene, for unbreakable containers and tubing
$CH_2{=}CHCH_3$	propylene	polypropylene, Herculon, fibers for carpeting and clothes
$CH_2{=}CHCl$	vinyl chloride	polyvinyl chloride, PVC, Koroseal
$CH_2{=}CCl_2$	1,1-dichloroethylene	Saran, food wrappings
$CH_2{=}CHCN$	acrylonitrile	polyacrylonitrile, Orlon, Acrylics
$CF_2{=}CF_2$	tetrafluoroethylene	polytetrafluoroethylene, Teflon
$CH_2{=}CHC_6H_5$	styrene	polystyrene, Styrofoam, for insulation
$CH_2{=}\underset{\underset{CH_3}{\vert}}{C}{-}CO_2CH_3$	methyl methacrylate	polymethyl methacrylate, Lucite, Plexiglas, for glass substitutes
$CH_2{=}CHCO_2CH_3$	methyl acrylate	polymethyl acrylate, Acrylics, for latex paints

Table 3.4 lists several important polymers of substituted ethylenes along with their common names and uses.

The alkene polymers, mainly polyethylene and polypropylene are the largest tonnage plastics in the world.

The tetrafluoroethylene polymers were discovered accidentally in 1938 by du Pont chemists during the search for new refrigerants (the Freons described in Section 2.12). One morning a cylinder of tetrafluoroethylene appeared to be empty (no gas escaped when the valve was opened) and yet the weight of the cylinder indicated it was full. The cylinder was opened and inside was found a waxy solid, the forerunner of Teflon. The solid proved to have very unusual properties: extraordinary chemical inertness, outstanding heat resistance, very high melting point and unusual frictional properties. In 1941 du Pont began limited production of Teflon. The small amount of polymer was preempted at once by the Manhattan Project where it was used in equipment to contain the highly corrosive UF_6 during the separation of the isotopes of uranium. In 1948 du Pont built the first commercial Teflon plant and the product was used to make gaskets, bearings for automobiles, nonstick equipment for candy manufacturers and commercial bakers, seals for rotating equipment, and a number of other items. Teflon became a household word in 1961 with the introduction of nonstick frying pans in the U.S. market.

Polymerization of ethylene and substituted ethylenes can be initiated by trace amounts of cations, anions, or free radicals. It is on the free radical polymerization of ethylene that we shall concentrate. A <u>free radical</u> is any atom or molecule that contains one or more unpaired or odd electrons. Following are structural formulas for the chlorine radical (actually, a chlorine atom), the ethyl radical, and an alkoxide radical.

$$:\overset{..}{\underset{..}{Cl}}\cdot \qquad CH_3-\overset{\displaystyle H}{\underset{\displaystyle H}{C}}\cdot \qquad R-\overset{..}{\underset{..}{O}}\cdot$$

| chlorine | ethyl | alkoxide |
| radical | radical | radical |

Thus far we have seen structural formulas for two classes of reactive intermediates, carbocations and free radicals. Let us compare the electronic structures of each. Consider for example the ethyl carbocation and the ethyl radical.

$$CH_3-\overset{\displaystyle H}{\underset{\displaystyle H}{C^+}} \qquad CH_3-\overset{\displaystyle H}{\underset{\displaystyle H}{C}}\cdot$$

| ethyl | ethyl |
| carbocation | radical |

These reactive intermediates are similar in that each has a carbon atom in an abnormal valence state. In the carbocation, the carbon has only six valence electrons in its outer shell. In the free radical it has only seven. However, they differ in charge; the carbocation bears a positive charge while the free radical has no charge, that is, it is neutral.

Free radicals from any number of sources will <u>initiate</u> the polymerization of ethylene. Organic peroxides are exceptionally good sources of radicals for this process because the weak —O—O— bond undergoes rupture at relatively low temperatures.

$$R-\overset{..}{\underset{..}{O}}-\overset{..}{\underset{..}{O}}-R \quad \xrightarrow{\text{heat}} \quad 2R-\overset{..}{\underset{..}{O}}\cdot \qquad \text{initiation}$$

The chain-lengthening reactions are shown:

$$R-\overset{..}{\underset{..}{O}}\cdot + \ \overset{H}{\underset{H}{C}}=\overset{H}{\underset{H}{C}} \ \longrightarrow \ R-O-\overset{H}{\underset{H}{C}}-\overset{H}{\underset{H}{C}}\cdot$$

$$R-O-\overset{H}{\underset{H}{C}}-\overset{H}{\underset{H}{C}}\cdot + \ \overset{H}{\underset{H}{C}}=\overset{H}{\underset{H}{C}} \ \longrightarrow \ R-O-\overset{H}{\underset{H}{C}}-\overset{H}{\underset{H}{C}}-\overset{H}{\underset{H}{C}}-\overset{H}{\underset{H}{C}}\cdot$$

chain lengthening
or
chain propagation

The <u>chain-lengthening</u> sequence occurs at a very rapid rate, often as high as thousands of additions per second, depending on the experimental conditions. The characteristic feature of this type of polymerization is a series of reactions each of which consumes a reactive particle (in this case, a free radical) and produces another reactive particle (another free radical). Such processes are called <u>chain reaction polymerizations</u>.

In principle, a polymer chain might continue to grow until all monomer units are used up. In practice, however, the chain lengths generally do not exceed 10^4 or 10^5 monomer units (often they are far less). Chain growth or propagation is halted by what are called <u>chain termination</u> steps. One such chain termination process is <u>radical combination</u>.

$$\text{\tiny www}CH_2\cdot \ \cdot CH_2\text{\tiny www} \ \longrightarrow \ \text{\tiny www}CH_2-CH_2\text{\tiny www}$$

| growing chains | a radical |
| | combination product |

The other major chain termination process is <u>disproportionation</u>, in which a radical of one chain transfers a hydrogen atom (H·) to another chain to generate two non-radical containing molecules.

$$\text{mmCH}_2{-}\text{CH}_2\cdot + \cdot\text{CH}_2{-}\text{CH}_2\text{mm} \longrightarrow \text{mmCH}_2\text{CH}_3 + \text{CH}_2{=}\text{CHmm}$$

growing chains   chain disproportionation
                                               products

3.15 Alkynes

The <u>alkynes</u> contain one or more <u>carbon-carbon triple</u> bonds. <u>Acetylene</u> is the first member of this class, which is often referred to by the common name of acetylenes.

Often, simple alkynes are named as derivatives of acetylene. In the IUPAC nomenclature the ending, -yne, indicates the presence of a triple bond. The following examples illustrate both systems of nomenclature.

$$\text{H}{-}\text{C}{\equiv}\text{C}{-}\text{H} \qquad \text{CH}_3{-}\text{C}{\equiv}\text{C}{-}\text{H} \qquad \text{CH}_3{-}\text{C}{\equiv}\text{C}{-}\text{CH}_3$$

ethyne   propyne   2-butyne
or   or   or
acetylene   methylacetylene   dimethylacetylene

The acetylenes are not widely distributed in nature. Their chief value lies in the ready availability of acetylene itself and its great reactivity under a wide variety of experimental conditions. It is of value as an inexpensive starting material for a number of synthetic processes of commercial importance.

<u>In many ways the reactions of alkynes are like those of alkenes.</u> Reagents like Br_2 and HBr add to alkynes, though somewhat more slowly than to alkenes.

$$\text{HC}{\equiv}\text{CH} + \text{HCl} \longrightarrow$$

H   H
 \\   /
  C=C
 /   \\
H   Cl

vinyl chloride

$$\text{CH}_3\text{C}{\equiv}\text{CH} + \text{HBr} \longrightarrow$$

H$_3$C   H
  \\   /
   C=C
  /   \\
Br   H

2-bromopropene

$$\text{CH}_3\text{C}{\equiv}\text{CH} + \text{Br}_2 \longrightarrow$$

H$_3$C   Br
  \\   /
   C=C
  /   \\
Br   H

trans-1,2-dibromopropene

$$\text{CH}_3\text{C}{\equiv}\text{CH} + 2\text{Br}_2 \longrightarrow \text{CH}_3{-}\underset{\underset{\text{Br}}{|}}{\overset{\overset{\text{Br}}{|}}{\text{C}}}{-}\underset{\underset{\text{Br}}{|}}{\overset{\overset{\text{Br}}{|}}{\text{C}}}{-}\text{H}$$

1,1,2,2,-tetrabromopropane

71

$$CH_3C\equiv CH + 2HCl \longrightarrow CH_3-\overset{\overset{\displaystyle Cl}{|}}{\underset{\underset{\displaystyle Cl}{|}}{C}}-CH_3$$

2,2-dichloropropane

Acetylene also reacts with water and acetic acid.

$$HC\equiv CH + H_2O \xrightarrow[H_2SO_4]{HgSO_4} \left[\begin{array}{c} H \\ H \end{array} \diagdown C=C \diagup \begin{array}{c} OH \\ H \end{array} \right] \longrightarrow CH_3-\overset{\overset{\displaystyle O}{||}}{C}-H$$

acetaldehyde

$$HC\equiv CH + HO-\overset{\overset{\displaystyle O}{||}}{C}-CH_3 \xrightarrow{Hg^{2+}} \begin{array}{c} H \\ H \end{array} \diagdown C=C \diagup \begin{array}{c} O-\overset{\overset{\displaystyle O}{||}}{C}-CH_3 \\ H \end{array}$$

vinyl acetate

The hydration of acetylene, catalyzed by sulfuric acid and mercuric sulfate, produces an intermediate which rearranges to acetaldehyde. This type of rearrangement is known as keto-enol tautomerism and we shall study it in Chapter 7. Addition of acetic acid produces vinyl acetate. Addition reactions of acetylenes are important ways of synthesizing many of the substituted ethylene monomers shown in Table 3.4.

3.16 Spectroscopic Properties of Alkenes

Both carbon-carbon double bonds and hydrogen atoms attached to carbon-carbon double bonds, often referred to as vinyl hydrogens, show absorption in the infrared region of the spectrum. The characteristic absorption frequencies for the stretching of these bonds are:

Type of bond	Frequency (cm^{-1})	Wavelength (μm)
=C—H (stretching)	3000 to 3200	3.13 to 3.33
C=C (stretching)	1620 to 1680	5.95 to 6.17

Each of these stretching frequencies can be seen in the infrared spectrum of 1-octene (Figure 3.3).

Notice the peak of medium intensity at 3080 cm^{-1} corresponding to stretching of the vinyl hydrogens and the peak of medium intensity at 1670 cm^{-1} corresponding to stretching of the carbon-carbon double bond. Also note the three closely spaced peaks around 2900 cm^{-1} corresponding to stretching of the alkane carbon-hydrogen bonds. You might compare the infrared spectrum of 1-octene, an alkene (Figure 3.3), with that of octane, an alkane (Figure 15.2), and note the similarities and differences.

Isolated carbon-carbon double bonds do not absorb near ultraviolet (200 to 400 nm) or visible (400 to 700 nm) radiation. However, conjugated

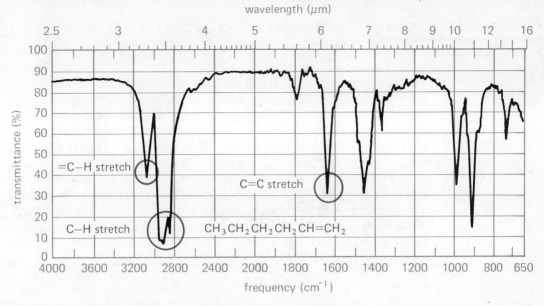

FIGURE 3.3 An infrared spectrum of 1-octene.

double bond systems do absorb strongly in this region of the spectrum. Butadiene, the simplest conjugated diene, shows an absorption maximum in the ultraviolet spectrum at 217 nm. In general, as the number of double bonds in the conjugated system increases, the more the absorption maximum is shifted toward the visible region of the spectrum. 1,3,5-hexatriene with three conjugated carbon-carbon double bonds absorbs at 258 nm.

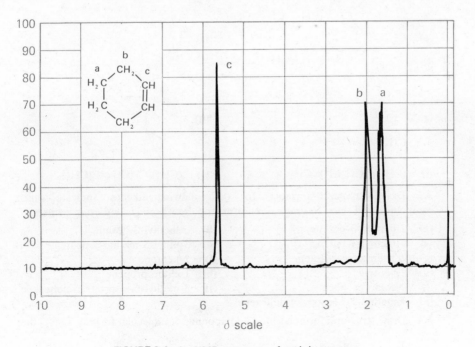

FIGURE 3.4 An NMR spectrum of cyclohexene.

$$CH_2=CH-CH=CH_2 \qquad CH_2=CH-CH=CH-CH=CH_2$$

butadiene

217 nm

1,3,5-hexatriene

258 nm

The absorption maxima for molecules with five or more conjugated carbon-carbon double bonds generally occur in the visible region of the spectrum, and accordingly we say that these substances are colored. For example, the diterpene vitamin A alcohol (p. 56) which contains five conjugated double bonds, is yellow. The tetraterpene β-carotene (p. 56) which contains eleven conjugated double bonds is red.

Vinyl hydrogens give rise to signals in the nuclear magnetic resonance spectrum in the range $\delta = 4.5$–6.5. In the case of cyclohexene (Figure 3.4), the signal due to the two vinyl hydrogens appears at $\delta = 5.7$. Notice that in this molecule there are three sets of chemically equivalent hydrogens and therefore you would predict three signals. The NMR spectrum indeed shows three signals in the ratio $4:4:2$.

PROBLEMS

3.1 Give an acceptable name for the following alkenes.

(a)

$$\overset{\overset{\displaystyle CH_3}{|}}{CH_2=C}-CH_2-CH_3$$

(b)

$$CH_3-\overset{\overset{\displaystyle CH_3}{|}}{C}=CH-CH_2-\overset{\underset{\displaystyle CH_3}{|}}{CH}-CH_3$$

(c)

$$\begin{array}{ccc} Cl & & H \\ & C=C & \\ H & & Cl \end{array}$$

(d)

$$CH_2=\overset{\overset{\displaystyle CH_3}{|}}{C}-CH=CH_2$$

(e)

$$CH_2=C\overset{\displaystyle CH_2-CH_2-CH_2-CH_3}{\underset{\displaystyle CH_2-\overset{\underset{\displaystyle CH_3}{|}}{CH}-CH_3}{}}$$

(f)

(g)

$$Cl \quad CH_3 \atop CH_3$$

(h)

$CH=CH_2$

(i) $(CH_3)_3CCH_2CH=CH_2$

(j) $(CH_3)_2CHCH=C(CH_3)_2$

3.2 Draw structural formulas for the following alkenes. Indicate which compounds will show *cis-trans* isomerism, and draw both *cis* and *trans* isomers.

(a)   2-methyl-3-hexene
(b)   2-methyl-2-hexene
(c)   2-methyl-1-butene
(d)   3-ethyl-3-methyl-1-pentene
(e)   2,3-dimethyl-2-butene
(f)   1-pentene
(g)   2-pentene
(h)   1-chloropropene
(i)   2-chloropropene
(j)   3-methyl-6-isopropylcyclohexene
(k)   4-methyl-1-isopropylcyclohexene

3.3 Draw structural formulas for all compounds of molecular formula C_5H_{10} that are:

(a)   alkenes that will not show *cis-trans* isomerism

(b) alkenes that will show *cis-trans* isomerism
(c) cycloalkanes that will not show *cis-trans* isomerism
(d) cycloalkanes that will show *cis-trans* isomerism

3.4 The general formula of an alkane is C_nH_{2n+2}; that of a cycloalkane is C_nH_{2n}. What is the general formula of an alkene? of a cycloalkene? of an alkadiene?

3.5 Draw structural formulas for the four isomeric chloropropenes (C_3H_5Cl).

3.6 There are four *cis-trans* isomers of 2,4-heptadiene. Name and draw structural formulas for all four.

3.7 (a) Make a model of *cis*-3-hexene. By twisting about the carbon-carbon single bonds, show that carbon atoms 1 and 6 can be brought close enough to form a ring without introducing any appreciable angle strain.
(b) Make a model of *trans*-3-hexene and show that carbon atoms 1 and 6 cannot be brought to within bonding distance without introducing a great deal of angle strain.
(c) Attempt to construct molecular models for *trans*-cycloheptene, *trans*-cyclooctene, and *trans*-cyclononene. Which of these *trans*-cycloalkenes would you predict could be prepared in the laboratory? Is your prediction consistent with the fact that *trans*-cyclooctene is the smallest cycloalkene that has yet been made?

3.8 The three isoprene units of α-farnesene and β-selinene are shown by dashed lines in Figure 3.1. In the same manner show the isoprene units of the other terpenes shown in Section 3.5 and verify for yourself that, with the exception of one tail-to-tail linkage in β-carotene, they are all head-to-tail linkages.

3.9 Note that α-farnesene, zingiberene, β-selinene, and caryophyllene all have the same molecular formula, $C_{15}H_{24}$, and all are divisible into three isoprene units hooked head-to-tail. Show how the carbon chain of α-farnesene can be coiled and then cross linked by one or more carbon-carbon bonds to give each of these structures. In doing this do not be concerned with the location of the double bonds.

3.10 Santonin, $C_{15}H_{18}O_3$, can be isolated from the flower heads of certain species of *Artemisia*. Work on the structure of santonin was carried out over a fifty-year period by groups of Italian, German, Swiss, and English chemists, and the correct structural formula was first proposed in 1930. The first successful laboratory synthesis of santonin was achieved in 1954. Santonin is an anthelmintic, i.e., a drug used to rid the body of worms (helminths). It has been estimated that over one-third of the world's population is infested with these parasites. Santonin is used as an anthelmintic for roundworm (*Ascaris lumbricoides*) in oral doses of 60 to 200 mg.

santonin

Santonin is a sesquiterpene. Locate the three isoprene units and show how the carbon skeleton of α-farnesene might be coiled and then cross-linked to give santonin. (There are actually two different coiling patterns that could lead to santonin. See if you can find them both.)

3.11 A structural formula for a section of polypropylene derived from three units of propylene monomer follows:

polypropylene

75

Draw structural formulas for comparable sections of

(a) polyvinyl chloride   **(b)** Saran
(c) Teflon   **(d)** Orlon
(e) Styrofoam   **(f)** Plexiglas

3.12 Predict the product(s) of the acid-catalyzed hydration of the following alkenes.

(a) 2-butene   **(b)** 2-methyl-2-butene
(c) 2-methyl-1-butene   **(d)** cyclohexene
(e) 1-methylcyclohexene   **(f)** *cis*-2-pentene
(g) *trans*-2-pentene   **(h)** 2,3-dimethyl-2-butene

3.13 Predict the product(s) of the hydroboration-oxidation of the alkenes listed in the previous problem.

3.14 Terpin hydrate is prepared commercially by the addition of two moles of water to limonene in the presence of dilute sulfuric acid. Limonene is one of the main components of lemon, orange, caraway, dill, bergamot, and some other oils. Terpin hydrate is used medicinally as an expectorant for coughs. It may be given as terpin hydrate and codeine. Propose a structure for terpin hydrate and a reasonable mechanism to account for the formation of the product you have predicted.

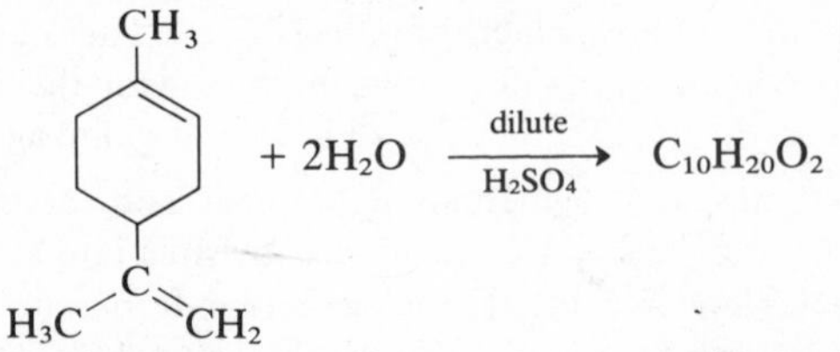

$$\text{limonene} + 2H_2O \xrightarrow[\text{H}_2\text{SO}_4]{\text{dilute}} C_{10}H_{20}O_2$$

limonene   terpin hydrate

3.15 The hypochlorous acids HOCl, HOBr, and HOI add to alkenes. The addition of hypochlorous acid to cyclohexene produces 2-chlorocyclohexanol.

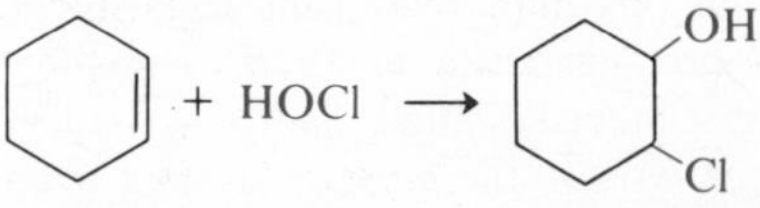

2-chlorocyclohexanol

Propose a mechanism for this reaction starting with the assumption that the addition is initiated by a chloronium ion, Cl^+.

3.16 Before the recent development of sensitive instrumental techniques, a number of methods were developed to measure the degree of unsaturation of fats and oils. One such method was to experimentally determine an "iodine number." In this procedure, equivalent amounts of I_2 and Br_2 are mixed in acetic acid to produce the highly reactive iodine monobromide, IBr. This reagent adds to alkenes as shown below:

$$R-CH{=}CH-R + IBr \longrightarrow R-\underset{\underset{I}{|}}{C}H-\underset{\underset{Br}{|}}{C}H-R$$

(a) Propose a mechanism to account for the addition of IBr to an alkene. Based on your mechanism would you expect the addition of IBr to propene to give

$$CH_3-\underset{\underset{I}{|}}{C}H-\underset{\underset{Br}{|}}{C}H_2 \quad \text{or} \quad CH_3-\underset{\underset{Br}{|}}{C}H-\underset{\underset{I}{|}}{C}H_2?$$

(b) The iodine number is defined as the number of grams of iodine that adds to 100 grams of a fat or oil. (For definition of the terms fat and oil see Section 12.2.)

76

Given below are average molecular weights of three oils and one fat and their corresponding iodine numbers.

Fat or Oil	Average Molecular Weight (g/mole)	Iodine Number
corn oil	870–900	115–130
soybean oil	860–890	125–140
linseed oil	855–895	175–205
butter fat	700–720	25–40

Which of these substances is the most highly unsaturated; the least highly unsaturated? Explain your reasoning.

3.17 Which is the best means for the preparation of 1,2-dichloropentane? Explain your reasoning.

(a) $CH_3-CH_2-CH_2-CH_2-CH_3 + 2Cl_2 \xrightarrow{300°} CH_3-CH_2-CH_2-\underset{\underset{Cl}{|}}{CH}-\underset{\underset{Cl}{|}}{CH_2} + 2HCl$

(b) $CH_3-CH_2-CH_2-CH=CH_2 + Cl_2 \xrightarrow[CCl_4]{} CH_3-CH_2-CH_2-\underset{\underset{Cl}{|}}{CH}-\underset{\underset{Cl}{|}}{CH_2}$

3.18 Write an equation to illustrate the reaction of 2-methyl-2-butene with each of the following reagents:

(a) H_2/Pt **(b)** Cl_2
(c) Br_2 **(d)** IBr
(e) H_2O/H_2SO_4 **(f)** $KMnO_4$ (dilute, basic solution)
(g) O_3, then $Zn + H_2O$ **(h)** B_2H_6 followed by H_2O_2

3.19 Repeat problem 3.18 using cyclohexene.

3.20 Reaction of 2-methylpropene with methanol in the presence of H_2SO_4 yields a compound of formula $C_5H_{12}O$.

$$\underset{\underset{CH_3}{|}}{CH_3-C}=CH_2 + CH_3OH \xrightarrow{H_2SO_4} C_5H_{12}O$$

Propose a structural formula for this compound. Also propose a mechanism to account for its formation.

3.21 Draw the structural formula of the alkene which will give the indicated products on ozonolysis.

(a) $C_6H_{12} \longrightarrow CH_3CH_2CHO$ as the only product
(b) $C_6H_{12} \longrightarrow CH_3CHO + CH_3COCH_2CH_3$ in equal amounts
(c) $C_6H_{12} \longrightarrow CH_3COCH_3$ as the only product
(d) $C_7H_{12} \longrightarrow CH_3\overset{O}{\overset{||}{C}}CH_2CH_2CH_2\overset{O}{\overset{||}{C}}CH_3$

(e) $C_{10}H_{16} \longrightarrow$

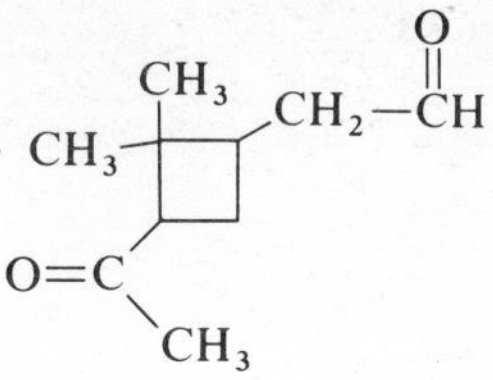

77

3.22 What product would you expect to isolate after ozonolysis of natural rubber? (Hint: draw a section of the carbon chain including several double bonds. You should then be able to see what small molecule will be produced.)

3.23 Write an equation to show the reaction of 1-butyne with each of the following reagents:

(a) H_2/Pt (one mole)
(b) H_2/Pt (two moles)
(c) HBr (one mole)
(d) HBr (two moles)
(e) Br_2 (one mole)
(f) Br_2 (two moles)
(g) H_2O/H_2SO_4, $HgSO_4$

3.24 Draw structural formulas for the starting materials for each reaction. Note that in several instances, there is more than one possible starting material that will give the indicated product.

(a) $C_5H_{10}O$ (an alcohol) $\xrightarrow[\text{heat}]{85\% \ H_3PO_4}$ $CH_3-\overset{\overset{\displaystyle CH_3}{|}}{C}=CH-CH_3 + H_2O$

(b) $C_6H_{11}Cl$ (an alkyl halide) $\xrightarrow{KOH}$ [cyclopentane ring with =CH₂] + KCl

(c) $C_6H_{11}Cl$ (an alkyl halide) $\xrightarrow{KOH}$ [cyclopentene ring with CH₃]

(d) C_5H_{10} (an alkene) + H_2 $\xrightarrow{Pt}$ $CH_3-\underset{\underset{\displaystyle CH_3}{|}}{CH}-CH_2-CH_3$

(e) C_5H_{10} (an alkene) + H_2O $\xrightarrow{H_2SO_4}$ $CH_3-\overset{\overset{\displaystyle CH_3}{|}}{\underset{\underset{\displaystyle OH}{|}}{C}}-CH_2-CH_3$

(f) C_5H_{10} (an alkene) $\xrightarrow[\text{(2) } H_2O_2 \text{ in aqueous base}]{\text{(1) } B_2H_6}$ $CH_2-\overset{\overset{\displaystyle CH_3}{|}}{\underset{\underset{\displaystyle OH}{|}}{CH}}-CH_2-CH_3$

3.25 Show the reagents and reaction condition you might use to transform the given starting material into the desired product. Note that some transformations require only one step, while others will require two steps.

(a) $CH_3-\overset{\overset{\displaystyle CH_3}{|}}{CH}-CH_2-OH$ $\xrightarrow{\text{(1 step)}}$ $CH_3-\overset{\overset{\displaystyle CH_3}{|}}{C}=CH_2$

(b) $CH_3-\overset{\overset{\displaystyle CH_3}{|}}{CH}-CH_2-OH$ $\xrightarrow{\text{(2 steps)}}$ $CH_3-\overset{\overset{\displaystyle CH_3}{|}}{\underset{\underset{\displaystyle OH}{|}}{C}}-CH_3$

(c) $CH_3-\overset{\overset{\displaystyle CH_3}{|}}{CH}-CH_2-CH_2-CH_2-Cl$ $\xrightarrow{\text{(2 steps)}}$ $CH_3-\overset{\overset{\displaystyle CH_3}{|}}{CH}-CH_2-\overset{\overset{\displaystyle }{}}{\underset{\underset{\displaystyle Cl}{|}}{CH}}-CH_3$

(d) [cyclohexane ring with CH_3 and OH] $\xrightarrow{\text{(2 steps)}}$ [cyclohexane ring with CH_3 and OH]

(e)

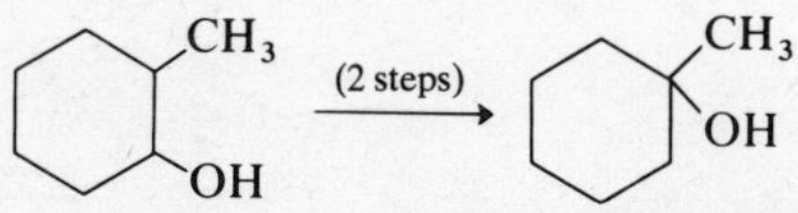

$\xrightarrow{\text{(2 steps)}}$

(f)

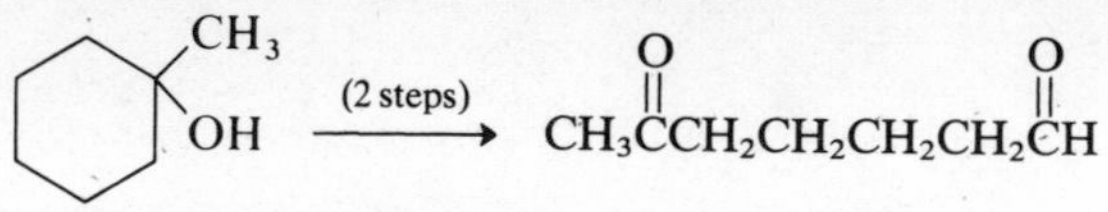

$\xrightarrow{\text{(2 steps)}}$ $CH_3\overset{O}{\overset{\|}{C}}CH_2CH_2CH_2CH_2\overset{O}{\overset{\|}{C}}H$

(g) $CH_3CH_2\underset{\underset{OH}{|}}{CH_2}$ $\xrightarrow{\text{(2 steps)}}$ $CH_3-\underset{\underset{HO}{|}}{CH}-\underset{\underset{OH}{|}}{CH_2}$

(h) $CH_3-\underset{\underset{}{|}}{\overset{\overset{CH_3}{|}}{CH}}-CH=CH_2$ $\xrightarrow{\text{(2 steps)}}$ $CH_3-\overset{\overset{CH_3}{|}}{C}=CH-CH_3$

3.26 Show how you might distinguish between the following pairs of compounds by a simple, chemical test. In each case, tell what test you would perform, what you would expect to observe, and write an equation for each positive test.

(a) cyclohexane and 1-hexene
(b) 1-hexene and 2-chlorohexane
(c) 1,1-dimethylcyclopentane and 2,3-dimethyl-2-butene

3.27 List one major spectral characteristic that will enable you to distinguish between the following pairs of compounds.

(a) cyclohexane and 1-hexene (IR and NMR)
(b) 1-hexene and 2-chlorohexane (IR and NMR)
(c) 1,1-dimethylcyclopentane and 2,3-dimethyl-2-butene (NMR)
(d) 2-methyl-1,3-butadiene and 2-methyl-2-butene (UV)
(e) 2,3-dimethyl-1-butene and 2,3-dimethyl-2-butene (IR and NMR)

3.28 An alkyl bromide of molecular formula C_3H_7Br shows to signals in the NMR spectrum, a doublet at $\delta = 1.7$ and a septet at $\delta = 4.3$. The relative areas of these two signals are in the ratio $6:1$. Draw a structural formula for this alkyl bromide.

3.29 There are four constitutional isomers of molecular formula $C_3H_6Cl_2$.

(a) Draw structural formulas for each isomer.
(b) Show how you could distinguish between these four by the presence or absence of an NMR signal for $-CH_3$ hydrogens and the splitting pattern of any $-CH_3$ signals. (Hint: for one isomer the $-CH_3$ signal will be a triplet, for another it will be a doublet, and for a third it will be a singlet. For the fourth, there will be no signal due to $-CH_3$ hydrogens.)

I

Mini-Essay

Ethylene

Ethylene

On a pound per pound basis, the U.S. chemical industry produces more ethylene than any other organic chemical—some 24 billion pounds of it in 1977. That is more than 120 pounds per year for every man, woman, and child in this country. Next in total production among the organics are propylene and benzene, each at about 13 billion pounds in 1977. These three substances, ethylene, propylene, and benzene are among the six or seven most important starting materials from which the U.S. chemical industry manufactures 250 billion pounds of organic chemicals per year. Of this vast output, approximately 40% are derived from ethylene alone and another 35% are derived from propylene and benzene. Clearly, the impact of these three primary organics on our economy is enormous. In this mini-essay, we will focus our attention on ethylene.

First, how do we obtain it? The answer is we don't. There are only insignificant amounts of ethylene in either natural gas or refinery gas, the lowest boiling fraction from the distillation of crude petroleum. If we do not find ethylene in nature, then how do we make it from the raw materials available to us? The answer to this question is not an easy one and in fact the answer differs depending on geography, availability of raw materials, and economic demands. And, as we will see presently, how we make it today may be quite different from how we will be forced to make it in the future.

In the United States where there are vast reserves of natural gas, the major process for ethylene production has been the thermal cracking of the small quantities of ethane, propane, and butane present along with methane in natural gas. It is for this reason that the ethylene-generating plants in this country are concentrated in Louisiana and along the Gulf coast—areas rich in natural gas reserves.

$$CH_3-CH_3 \xrightarrow{\text{thermal cracking}} CH_2{=}CH_2 + H_2$$

Written as a balanced equation, the thermal cracking of ethane seems simple enough. Actually it is very complicated and there are several other substances produced along with ethylene. For example, for every 1000 pounds of ethylene produced from ethane, there is also obtained 36 pounds of coproduct propylene and 35 pounds of coproduct butadiene. As the percentage of propane and butane in the feedstock increases, so does the quantity of coproduct propylene and butadiene. For every 1000 pounds of ethylene from the thermal cracking of propane, there is about 400 pounds of coproduct propylene and 100 pounds of coproduct butadiene. Although several low molecular weight alkanes can be cracked to yield ethylene, the cracking of ethane yields the highest percentage and highest purity ethylene. We will not discuss these coproduct gases further except to say that it has been a challenge to inventiveness and technology to turn them into starting materials for the manufacture of useful and marketable chemicals. It has been done, but that is another story.

Let us continue with the story of ethylene. As we have said, in the United States natural gas is the major raw material for the manufacture of ethylene. Approximately 10% of the natural gas consumed each year is used for this purpose. In Europe and Japan, however, supplies of natural gas are much more limited and, as a result, these countries have had to depend almost entirely on the catalytic cracking of light naphtha fractions for their ethylene. Thus, these countries produce not only aromatic hydrocarbons such as benzene, toluene, and xylene from naphtha (just as we do), but they also produce ethylene and propylene from it as well. The problem is that naphtha fractions from petroleum distillation are in heavy demand as sources of straight run gasoline, as feedstocks for catalytic cracking and reforming to produce higher octane blend stocks (including the benzene, toluene, xylene), and as a source of starting materials (again including benzene, toluene, xylene) for the chemical industry. In other words, the naphthas are in high demand for two very important reasons—for fuel and for chemicals. The worldwide demand for naphtha has increased steadily over the years and so has its price, in step with demand. There has been incentive in Europe and Japan, and more recently in the United States as well, to develop an ethylene-generating technology that will use higher boiling petroleum fractions, particularly gas oil. Thus, while we have depended in the past on natural gas and naphtha as raw materials for ethylene manufacture, it seems almost certain that in the future we will depend more and more on gas oil to meet this need. And we may also come to depend on coal as coal gasification and liquification technology advances.

Now that we have seen how we make ethylene, let us turn to the second question, namely, how do we use it. In answering this question, let us look at several of the major factors that spurred the demand for ethylene. One of the first technological breakthroughs in the use of ethylene came in the mid-1920s with the development of industrial processes to convert ethylene into synthetic ethyl alcohol and ethylene glycol. Ethylene glycol was and still is in demand as a solvent and as an antifreeze for automobile cooling systems. Even today, approximately 10% of all ethylene produced in this country is converted to ethylene glycol and used as permanent antifreeze. The decades following 1940 saw an ever increasing demand for ethylene in the manufacture of tetraethyl lead, a gasoline additive used to improve octane ratings. At one time, the demand for ethylene for the production of tetraethyl lead seemed on the verge of outstripping capacity. World War II created two new and enormous demands for ethylene. One of these was for polyethylene, important because of its electrical insulation properties. The other was for styrene (a monomer made from ethylene and benzene) for the manufacture of styrene-butadiene synthetic rubbers.

Today, ethylene is the starting material for the synthesis each year of almost 100 billion pounds of chemicals and polymers. Following are structural formulas for seven of the largest volume chemicals synthesized from ethylene. After the formula of each is given its name and yearly production from ethylene in billions of pounds.

Chemical	Production (Billions of Pounds)	Name
Cl—CH$_2$—CH$_2$—Cl	7.9	1,2-dichloroethane or ethylene dichloride
⬡—CH=CH$_2$	6.1	styrene
CH$_2$=CH—Cl	5.7	vinyl chloride
CH$_2$—CH$_2$ (O)	4.2	ethylene oxide
HO—CH$_2$—CH$_2$—OH	3.4	ethylene glycol
CH$_3$—CO$_2$H	2.4	acetic acid
CH$_3$—CH$_2$—OH	1.2	ethanol or ethyl alcohol

(CH$_2$=CH$_2$ →)

1,2-Dichloroethane is an intermediate used for the synthesis of a variety of industrial solvents and lubricants. Ethylene glycol is of course used as a permanent antifreeze in automotive cooling systems. It is also one of two essential building blocks for the synthesis of Dacron polyester textile fibers and for Mylar films for the recording industry. However, by far the largest fraction of ethylene produced in the U.S. is used for the synthesis of polymers. Forty-five percent, or 10.8 billion pounds in 1977, went into the production of polyethylene. Another 35% went into the synthesis of polyvinyl chloride, polystyrene, and related polymers. These plastics were in turn fabricated into materials for use in packaging, appliances, construction, synthetic rubbers, toys, radio and TV, furniture, housewares, luggage, water-based paints, etc.

Clearly, our dependence on ethylene is enormous, and given the nature of our industrialized society, it will continue to be enormous.

REFERENCES

American Chemical Society. "Chemistry in the Economy." American Chemical Society, Washington, D.C., 1973.

Billmeyer, F. W., Jr. *Textbook of Polymer Science*, 2nd ed. John Wiley and Sons, New York, 1971.

Chemical and Engineering News, "Ethylene." 7 November 1977. C&EN is a valuable reference source for current and future trends in the chemical industry.

Stereoisomerism and Optical Activity

4.1 Introduction

All compounds that have the same structural formula but different orientations of the atoms in space are grouped collectively under the classification <u>stereoisomers</u>. *Cis-trans* isomerism is one type of stereoisomerism; conformational isomerism is a second type; and optical isomerism is still a third type of stereoisomerism. In this chapter we will deal with optical isomers known as enantiomers, diastereomers, and meso compounds.

We shall first examine how optical isomerism can be detected in the laboratory, the structural features giving rise to this type of isomerism, the intimate relationship between optical and *cis-trans* isomerism, and finally the significance of stereoisomerism in the biological world.

4.2 The Polarimeter

The study of optical isomerism began in the early 1800s with the observation that most compounds isolated from natural sources are able to rotate the plane of polarized light, or to use the more common terminology, they are <u>optically active</u>. <u>Ordinary light consists of waves vibrating in all possible planes perpendicular to its path</u> (Figure 4.1). Certain substances such as <u>Polaroid</u> sheet (a plastic film containing properly oriented crystals of an organic substance embedded in it) or a <u>Nicol prism</u> (a device constructed from two crystals of Icelandic spar) will selectively transmit light waves vibrating only in a specific plane. Light in which vibrations occur in only a specific plane is said to be plane polarized. No doubt you are familiar with the effect of Polaroid on light from experience with sunglasses made of this material. If two Polaroid sheets are placed in a light path in such a way that their polarizing axes are parallel to each other, then a maximum intensity of light will pass through the second disc to you. If, however, their polarizing axes are oriented perpendicular to each other (that is, at an angle of 90°), then no light will pass through the second polarizing disc.

A polarimeter is an instrument used to measure quantitatively the effect of an optically active substance on the plane of polarized light. This device consists essentially of two polarizing prisms, one designated as the polarizer, the other as the analyzer (Figure 4.1).

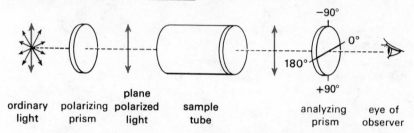

plane
ordinary polarizing polarized sample   analyzing   eye of
light   prism   light   tube   prism   observer

FIGURE 4.1 Schematic diagram of polarimeter, sample tube empty.

If the sample tube is empty, the intensity of the light reaching you will be a maximum when the polarizing axes of the two prisms are parallel. If the analyzer is turned either clockwise or counterclockwise, less light will be transmitted to the observer. When the analyzer is at right angles to the polarizer, the field of view in the instrument will be dark. We take this as the zero point, or 0° on the optical scale. If an optically active substance is placed in the sample tube, the plane of polarized light will be rotated by the substance, and a certain amount of light will pass through the analyzer to the observer. Turning the analyzer clockwise or counterclockwise a few degrees will restore the dark field of view. The number of degrees, α, that the analyzer is turned will be equal to the number of degrees that the optically active substance has rotated the plane of the polarized light. If the analyzer must be turned to the right (clockwise) to restore the dark field, we say the substance is dextrorotatory. If the analyzer must be turned to the left (counterclockwise) to restore the dark field, we say the substance is levorotatory. In either case the substance is optically active. (See Figure 4.2).

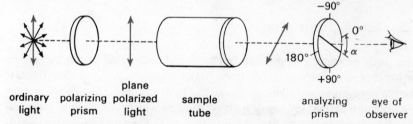

plane
ordinary polarizing polarized   sample   analyzing   eye of
light   prism   light   tube   prism   observer

FIGURE 4.2 Schematic diagram of polarimeter with sample tube containing an optically active substance. To restore the dark field of view the analyzer has been rotated clockwise by α degrees.

The term α is known as the <u>observed rotation</u> and is dependent on the structure and concentration of the compound, the length of the light path through the sample tube, the temperature, the wavelength of the light source, and the nature of the solvent. To facilitate tabulation and communication of data on optically active compounds, chemists have established a set of arbitrarily chosen standard conditions. This commonly reported rotation value, the <u>specific rotation</u>, is defined as the rotation caused by a substance at a concentration of one gram per cubic centimeter and in a sample tube of one decimeter length. Of course, the specific rotation still depends on the temperature and the wavelength of the light, and these values must be reported.

$$\frac{\text{specific}}{\text{rotation}} = [\alpha]_{\text{wavelength}}^{\text{temperature}} = \frac{\text{observed rotation (degrees)}}{\text{length (decimeters)} \times \text{concentration (g/ml)}}$$

The most commonly used wavelength is the D line of sodium, which is responsible for the yellow color of excited sodium vapor. In reporting either the observed or the specific rotation, it is common practice to indicate dextrorotatory by a positive sign (+) and levorotatory by a negative sign (−). Using these conventions, the specific rotation of <u>sucrose</u> (a common sugar) at 25°C using the D line of sodium as the light source is reported as follows:

$$[\alpha]_D^{25°C} = +66.5 \ (H_2O)$$

4.3 Structure and Optical Activity

Now that we are able to detect and measure optical activity, the question arises as to why certain compounds are optically active while others are not. Stated more specifically, what structural features will give rise to optical activity?

By the middle of the 19th century a number of optically active compounds had been isolated from natural sources, and their structures had been determined. Among the compounds known at that time were:

$$\begin{array}{ccc}
\text{OH} & \text{CH}_3 & \text{CO}_2\text{H} \\
| & | & | \\
\text{CH}_3\text{CHCO}_2\text{H} & \text{CH}_3\text{CH}_2\text{CHCH}_2\text{OH} & \text{CHOH} \\
& & | \\
& & \text{CH}_2 \\
& & | \\
& & \text{CO}_2\text{H} \\
\end{array}$$

lactic acid   2-methyl-1-butanol   malic acid
"active" amyl alcohol

<u>Lactic acid</u> was one of the compounds most intensively investigated. Scheele (1780) discovered lactic acid in sour milk. Subsequently the same lactic acid was found to arise from the bacterial fermentation of milk sugar (lactose) and other naturally occurring sugars. The lactic acid made by fermentation, at least as originally obtained, is optically inactive. Berzelius (1807) discovered a similar acid substance could be extracted from muscle. It was soon established that this too was lactic acid, except for the very important difference that the muscle lactic acid is dextrorotatory, for which reason it is designated (+)-lactic acid. Both acids have the same structure

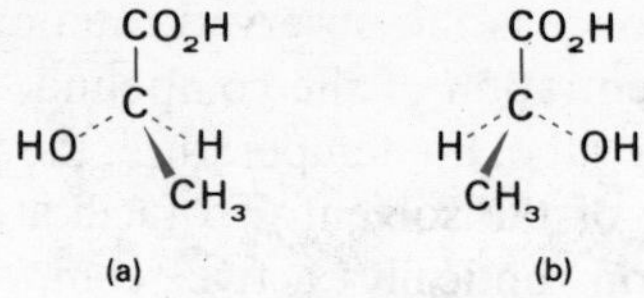

$$CO_2H \qquad\qquad CO_2H$$

(a)   (b)

FIGURE 4.3 Perspective formulas of lactic acid.

➤ represents a bond projecting in front of the plane of the paper.
— represents a bond projecting in the plane of the paper.
··· represents a bond projecting behind the plane of the paper.

and yet are different. By the third quarter of the 19th century a great many optically active compounds had been isolated and their structures determined, and yet up to that time no satisfactory explanation for this phenomenon existed.

We must not forget that even as late as the 1870s, theories of structure and bonding in organic molecules were only in their infancy. Yet, the foundations for a solution to the problem of optical activity and optical isomerism were being developed in the period following 1860 as chemists attempted to understand something of the three-dimensional shapes of molecules and possible relationships between molecular shapes and chemical and physical properties. A solution came in 1874 with a bold hypothesis advanced about the geometry of organic molecules, namely, that if the four valences of a carbon atom were directed toward the corners of a regular tetrahedron, then the four different groups can assume two and only two different spatial arrangements. These two possible arrangements are related as a person's left and right hands are related, by reflection. One is the mirror image of the other, but the two are not superimposable. Molecules that are not identical with their mirror images are said to be chiral (pronounced ki'-ral, to rhyme with spiral; from the Greek *cheir*, "hand"). In such a molecule, the carbon that has four different groups attached to it is called an asymmetric carbon, or better still, a chiral carbon. It is important to realize that chirality is a property of an object, the property of being different from its mirror image.

Consider for example the arrangement of the four different groups around the carbon 2 of lactic acid (Figure 4.3). Examination of the perspective formulas of the two lactic acids shows that they are different; (a) is the mirror image of (b) and is not superimposable on (b). Compounds that are related in this way are termed enantiomers (Greek *enantio*, "opposite" + *meros*, "part"). In general terms, enantiomers are mirror images that are not superimposable.

Many of the properties of enantiomers are identical: they have the same melting points, the same boiling points, the same solubilities in various solvents. Yet they are isomers and we can expect them to show some differences in their properties. One important difference is optical activity. One member of a pair of enantiomers will be dextrorotatory and the other member will be levorotatory. For each, the absolute magnitude of the optical activity will be the same, but the rotations will differ in sign. We shall see other differences in enantiomeric pairs later. For example, one enantiomer may taste different from the other, or smell different. One enantiomer may be physiologically active whereas the other is physiologically inactive. We will suggest why this might be, later in the chapter.

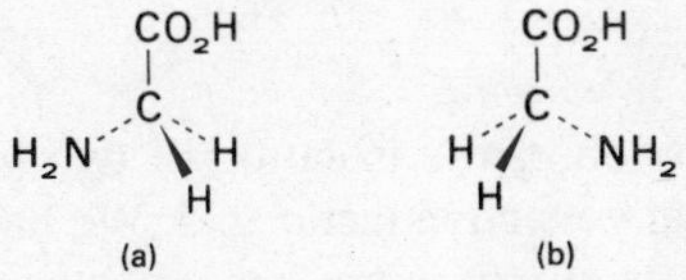

CO$_2$H   CO$_2$H
C   C
H$_2$N   H   H   NH$_2$
H   H
(a)   (b)

FIGURE 4.4 Glycine. (a) and (b) are mirror images.

It is a necessary and sufficient condition for enantiomerism that a compound and its mirror image be nonsuperimposable. Conversely, if a compound and its mirror image are superimposable, then the two are identical and the compound will not show enantiomerism. Consider for example the amino acid, glycine (Figure 4.4).

In the perspective representations of glycine, (a) and (b) are mirror images. Are they superimposable? The answer is yes. To see this, simply rotate (b) by 120° around the C—CO$_2$H bond axis. It is now possible to place (b) directly on (a) and have all groups correspond. Therefore the mirror images are identical and (a) and (b) do not constitute a pair of enantiomers. Glycine is achiral and cannot show optical activity.

Let us emphasize again that every compound has a mirror image. The key question is whether the mirror images are or are not superimposable.

4.4 Molecular Symmetry

Any molecule that is superimposable on its mirror image is symmetrical in some way. There are three recognized types of molecular symmetry, but for our purposes we shall consider only one, the existence of a plane of symmetry. A plane of symmetry is defined as a plane (often visualized as a mirror) cleaving a molecule in such a way that one side of the molecule is the mirror image of the other.

Inspection of the projection formula of glycine (Figure 4.5a) shows that it possesses a plane of symmetry running through the molecule on the axis of the C—C—N bonds. Alternatively, we might rotate the molecule in space into a different projection formula (b), and see the plane of symmetry again running through the axis of the C—C—N bonds, this time oriented in a plane perpendicular to the plane of the paper.

If you are interested in deciding whether a given structure is symmetrical (and therefore identical with its mirror image and optically inactive), it is often easier to look for a plane of symmetry than to construct a mirror image and compare it with the original.

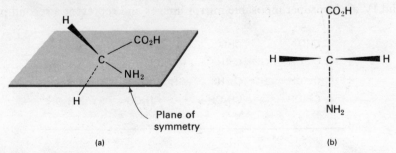

Plane of symmetry

(a)   (b)

FIGURE 4.5 Glycine. A plane of symmetry running through the axis of the C—C—N bond.

4.5 Racemic Mixtures

Let us return to the earlier discussion of the isolation of two lactic acids, designated as (+)-lactic and optically inactive lactic acid. We have already demonstrated that lactic acid and its mirror image are not superimposable (that is, it is a chiral molecule) and that lactic acid will show optical activity. How is it that the lactic acid from fermentation is optically inactive while that from muscle shows optical activity? The answer is straightforward and can be demonstrated easily in the laboratory (Section 4.10). The inactive form of lactic acid is an equal mixture of (+)-lactic acid and (−)-lactic acid. Because the mixture contains equal numbers of molecules that rotate the plane of polarized light to the right and to the left, the mixture does not rotate the plane of polarized light and is optically inactive. Such a mixture containing equal amounts of a pair of enantiomers is called a racemate, a racemic mixture, or simply a ±-mixture.

4.6 Multiple Chiral Centers

Compounds that contain two or more chiral centers can exist in more than two stereoisomeric modifications. Each additional chiral carbon atom doubles the number of possible enantiomers. In fact, it can easily be shown that the total number of enantiomers is 2^n, where n is the number of different chiral carbon atoms. Actually this number is a maximum number of enantiomers for any molecule of n chiral centers. Many molecules with n chiral carbons have fewer than 2^n enantiomers if two or more of the chiral carbons are similar, or if ring formation reduces the number of configurations capable of existence.

As an example of a molecule with two different chiral centers, consider 2,3,4-trihydroxybutanal.

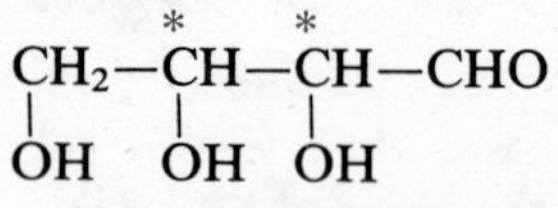

$$\overset{\displaystyle *}{} \qquad \overset{\displaystyle *}{}$$
$$CH_2-CH-CH-CHO$$
$$\;|\quad\;\;|\quad\;\;|$$
$$OH\quad OH\quad OH$$

2,3,4-trihydroxybutanal

The two chiral carbons are marked by asterisks. Thus four stereoisomers can be written as given in Figure 4.6, each with a unique configuration. Inspection of the formulas shows that structures I and II are nonsuperimposable mirror images and therefore one pair of enantiomers. Likewise III and IV are nonsuperimposable mirror images and represent a second pair

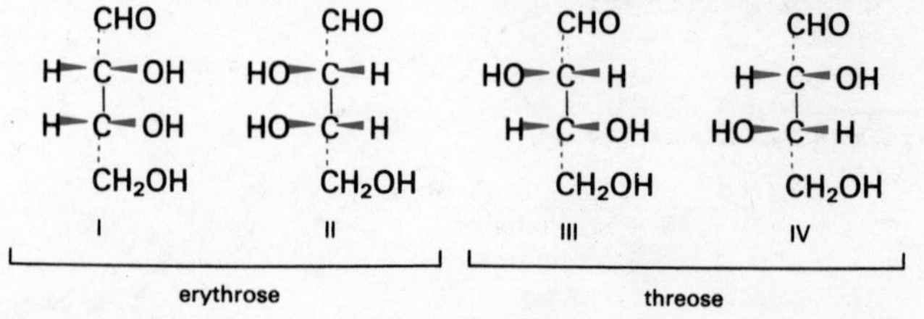

FIGURE 4.6 Stereoisomers of a substance having two dissimilar chiral carbons. Trihydroxybutanal.

of enantiomers. Taken in equal amounts, I and II form one racemate, and III and IV form another racemate.

Neither I nor II is a mirror image of III or IV, nor for that matter are either I or II superimposable on III or IV. Stereoisomer I is a diastereomer of III and IV. Diastereomers are stereoisomers that are not mirror images of each other. Diastereomers have different chemical and physical properties. The diastereomers of trihydroxybutanal are designated by the common names erythrose and threose.

As we indicated earlier, the 2^n isomer number represents the maximum number of stereoisomers. Often molecules have special symmetry properties and the isomer number is reduced. Tartaric acid is a classic example of a molecule possessing two similar chiral carbon atoms. The 2^n rule would predict four stereoisomers, while in fact only three are known (Figure 4.7).

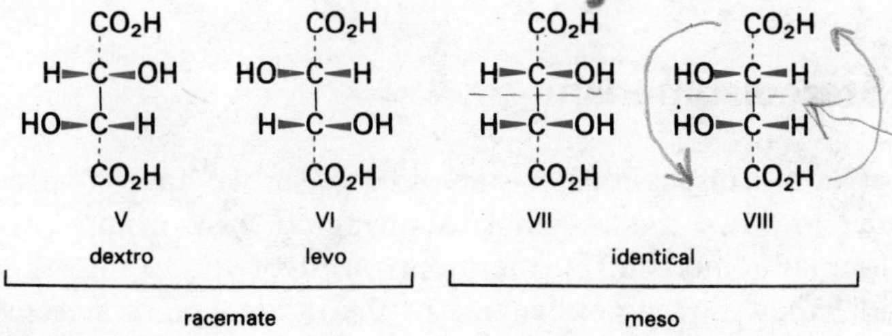

after flipping one or the other VII = VIII

FIGURE 4.7 Stereoisomers of a substance having two similar chiral carbons—tartaric acid.

Structures V and VI are nonsuperimposable mirror images and constitute a pair of enantiomers. VII and VIII are also mirror images but they are superimposable; they represent the same compound. Note that by rotating VII by 180° in the plane of the paper, it can be superimposed on VIII. In accord with our earlier statement that any molecule superimposable on its mirror image must be symmetrical in some way, we see that there is a plane of symmetry perpendicularly bisecting the central carbon-carbon bond. Symmetrical substances that contain chiral carbon atoms are called meso compounds. Meso tartaric acid is a diastereomer of (−)-tartaric acid and of (+)-tartaric acid. Physical properties of the three stereoisomers of tartaric acid are given in Table 4.1.

Acid	mp (°C)	$[\alpha]_D^{25°C}$
dextro	170	+12°
levo	170	−12°
racemic	206	inactive
meso	146	inactive

TABLE 4.1 Physical properties of the tartaric acids.

4.7 Predicting Enantiomerism

There are three methods that you can use to determine whether a molecule will show enantiomerism. Because it is a necessary and sufficient condition for enantiomerism that a molecule and its mirror image be nonsuperimposable, the first and most direct test is to build a model of the molecule and one of its mirror image. If the two are superimposable, then the substance is symmetric (or achiral) and will not show enantiomerism. If they are not superimposable, then the molecule will show enantiomerism.

A second method is to look for a plane of symmetry. If the molecule has a plane of symmetry, then the mirror images are identical, the molecule is achiral and will not show enantiomerism.

Third, you can look for the presence or absence of a chiral center (an asymmetric carbon atom). For each chiral center there will be a maximum of 2^n enantiomers. While looking for chiral centers is perhaps the easiest way to predict enantiomerism, you must remember that a molecule may have two or more chiral centers and still have a plane of symmetry and not show enantiomerism. Such molecules are said to be meso compounds.

4.8 Stereoisomerism

Thus far in our discussions of stereoisomerism we have clearly separated molecules into two classes—those showing *cis-trans* isomerism and those showing enantiomerism. Furthermore, most of the examples have dealt with molecules showing either one or the other type of stereoisomerism. This separation has been arbitrary, for, in fact, vast numbers of molecules in the natural world have one or more chiral centers and contain one or more sites for *cis-trans* isomerism. The question we should be asking about these molecules is, "How many stereoisomers are possible?"

Following is the structural formula for 4-hydroxy-2-pentene.

$$CH_3-\overset{*}{C}H-CH=CH-CH_3$$
$$|$$
$$OH$$

4-hydroxy-2-pentene

How many stereoisomers are there of this molecule? You first notice that there is one chiral center and also one double bond about which *cis-trans* isomerism is possible. Without too much trouble you should arrive at the conclusion that there are four possible stereoisomers. These are drawn in Figure 4.8 as two pairs of enantiomers.

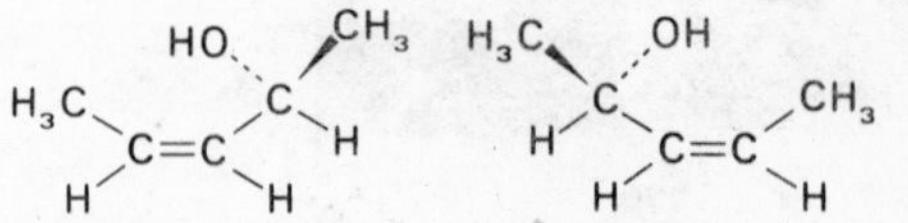

(a)   (b)

(+) and (−) enantiomers of *cis*-4-hydroxy-2-pentene

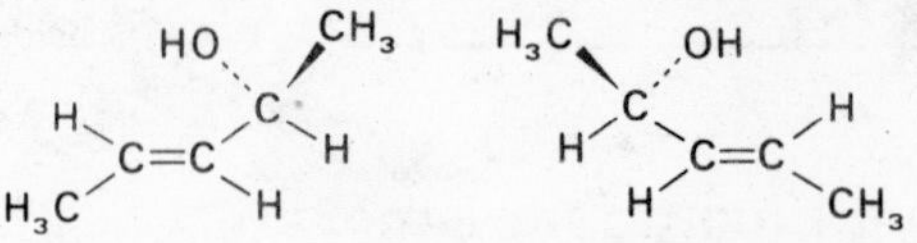

(c)   (d)

(+) and (−) enantiomers of *trans*-4-hydroxy-2-pentene

FIGURE 4.8 Stereoisomers of 4-hydroxy-2-pentene.

From inspection of these drawings it should be obvious that stereoisomers (a) and (b) are one pair of enantiomers; (c) and (d) are a second pair of enantiomers; stereoisomer (a) is the diastereomer of (c) and (d); stereoisomer (c) is the diastereomer of (a) and (b), etc.

As a second example consider 2-methylcyclohexanol.

2-methylcyclohexanol

How many stereoisomers are there of this molecule? The answer is four:

(e)	(+)-*cis*-2-methylcyclohexanol
(f)	(−)-*cis*-2-methylcyclohexanol
(g)	(+)-*trans*-2-methylcyclohexanol
(h)	(−)-*trans*-2-methylcyclohexanol

Here (e) and (f) are one pair of enantiomers; (g) and (h) are a second pair of enantiomers. Stereoisomer (e) is the diastereomer of (g) and (h), etc.

Earlier in this chapter we stated that the total number of enantiomers is given by the formula 2^n where n is the number of chiral centers. This is of course a maximum number and the actual number may be smaller due to the existence of meso compounds. Does this same rule apply to predicting the total number of stereoisomers? The answer is yes, if we redefine the rule slightly. For the molecule 2-methylcyclohexanol, there are two chiral centers and therefore the rule predicts four stereoisomers; in fact there are four. For 4-hydroxy-2-pentene, which contains only one chiral center, the 2^n rule would seem to predict only two stereoisomers. Yet there are four. The 2^n rule can be used in cases such as this if we redefine n to be equal to the number of chiral centers plus the number of double bonds that are capable of *cis-trans* isomerism. The 2^n rule now predicts four stereoisomers, and in fact there are four.

4.9 Conventions for Representing Absolute Configuration

We have seen that the most readily observed difference between enantiomers is the interaction with plane polarized light. For over a century after the discovery of optical activity and this type of stereoisomerism, the only method for distinguishing between enantiomers was by the sign of the rotation, hence the designations (+) and (−) in the naming of particular enantiomers. Furthermore, although it was possible to draw absolute configurations for a pair of enantiomers, there was no way to determine which enantiomer had which absolute configuration. As an example, the following are structural formulas for the lactic acid enantiomers.

These are said to represent absolute configurations because each specifies fully and precisely how the four groups attached to the chiral center are oriented in space. Which of these is the absolute configuration of (+)-lactic acid and which is the absolute configuration of (−)-lactic acid? Until recently there seemed no way to answer this question. The problem was further compounded when it was realized that there is not even a simple and necessary relationship between the sign of rotation and the absolute configuration of a molecule. For example, (+)-lactic acid can be converted into an ester, methyl lactate, by chemical procedures which in no way change the absolute configuration of the chiral center.

$$
\begin{array}{ccc}
\text{OH} & & \text{OH} \\
| & \text{conversion} & | \\
\text{CH}_3-\text{C}-\text{CO}_2\text{H} & \xrightarrow{\text{to an ester}} & \text{CH}_3-\text{C}-\text{CO}_2\text{CH}_3 \\
| & & | \\
\text{H} & & \text{H}
\end{array}
$$

(+)-lactic acid   (−)-methyl lactate
$[\alpha] = +3.82°$ $[\alpha] = -8.25°$

This methyl ester is found to be levorotatory and is written as (−)-methyl lactate. Whatever the absolute arrangement of the four groups on the chiral center is in (+)-lactic acid, it is the same in (−)-methyl lactate. In other words, these two substances have the same absolute configuration and yet both the direction and the magnitude of the specific rotation are different.

So, while we do not know the absolute configuration of either (+)-lactic acid or (−)-methyl lactate, we do at least know that they both have the same configuration at the chiral carbon atom.

The question of the absolute configuration of lactic acid and a great many other organic substances has been solved, but only recently. Just how this has been done will be discussed in Chapter 11 along with the stereochemistry of carbohydrates. For now we shall simply say that the absolute configuration of the enantiomers of lactic acid and methyl lactate are as follows:

(+)-lactic acid   (−)-lactic acid   (−)-methyl lactate   (+)-methyl lactate

In these and in all other instances where we are discussing either one or the other absolute configuration, we would like to be able to specify which configuration we mean. Of course, one way to do this is to draw a stereorepresentation of the particular absolute configuration. While such a practice is completely clear and unambiguous on paper, it is rather cumbersome and is certainly impractical for verbal discussion. There are two currently used conventions for specifying absolute configuration. One, involving the use of the letters D and L, was proposed by Emil Fischer near the turn of the century to deal with the absolute configurations of carbohydrates. We will discuss the use of this convention in Chapter 11. More recently another set of rules has been proposed called the Cahn-Ingold-Prelog convention after the three chemists who proposed it. This system focuses on the chiral center and assigns an arbitrary order of priority to each of the atoms or groups bonded to it. Priority is assigned on the basis of the atomic number of the atom bonded directly to the chiral center; the higher the atomic number, the higher the priority. Thus typical substituents in order of decreasing priority are:

$$I > Br > Cl > SH > F > OH > NH_2 > CH_3 > H$$

Obviously hydrogen has the lowest priority of any substituent. If two of the atoms attached directly to the chiral center are identical, then the priority is determined by referring to the next closest atoms.

CH₃—CH—   higher priority than $CH_3CH_2CH_2$—
     |
    CH₃

CH_3CH_2—   higher priority than CH_3—
CH_3O—   higher priority than   HO—

Isopropyl takes precedence over *n*-propyl; ethyl takes precedence over methyl; and methoxyl takes precedence over hydroxyl. For the purposes of determining priority, a double-bonded atom is considered to have two single bonds to the same atom. Thus

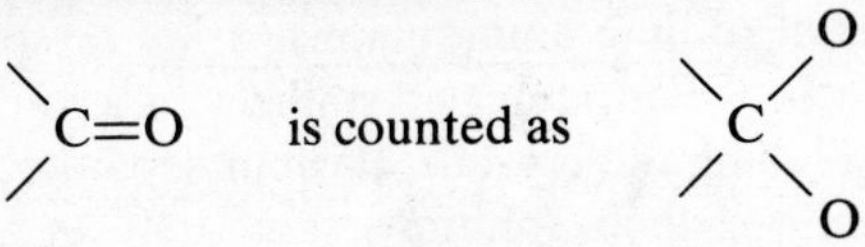

$$\ce{>C=O}\quad\text{is counted as}\quad$$

According to this last rule, an aldehyde group has a higher priority than a primary alcohol.

$$\overset{\displaystyle O}{\underset{}{\overset{\|}{-C-H}}}\qquad -CH_2OH$$

an aldehyde   a primary alcohol

Using these rules, the order of priority of the four groups attached to the chiral carbon atom of lactic acid, ranked from highest (1) to lowest (4), are

—OH $-CO_2H$ $-CH_3$ —H
(1)   (2)   (3)   (4)

Once the priorities have been assigned, then the molecule is turned so that it can be viewed from the side directly opposite the group of lowest priority. For (+)-lactic acid, we would orient the molecule as shown below.

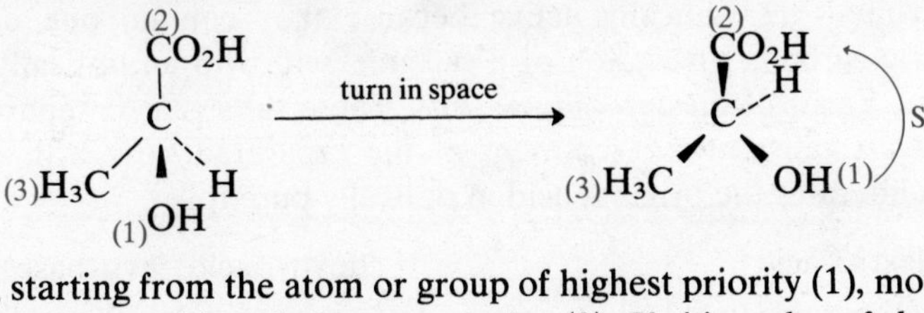

turn in space

Finally, starting from the atom or group of highest priority (1), move to the group of priority (2) and then to priority (3). If this order of descending priority travels clockwise, then the absolute configuration of the chiral center is said to be R (from the Latin *rectus* meaning right). If the order of descending priority travels counterclockwise, then the absolute configuration of the chiral center is said to be S (from the Latin *sinister* meaning left).

Returning to (+)-lactic acid, we see that the molecule has the S configuration, and it is properly designated as S-lactic acid.

Before leaving this discussion of absolute configuration and optical activity, let us repeat: to write "R-lactic acid" specifies precisely the absolute configuration. To write "(+)-lactic acid" specifies precisely which enantiomer is intended. However, remember that there is no simple or necessary relationship between the absolute configuration (R or S) and the sign of the rotation (dextrorotatory or levorotatory). Only by isolating S-lactic acid and determining its interaction with polarized light can we determine that it is indeed dextrorotatory and therefore properly written S-(+)-lactic acid. It follows also that the enantiomer of S-(+)-lactic is R-(−)-lactic acid.

4.10 Resolution of Racemic Mixtures

The process of separating a pair of enantiomers into (+)- and (−)-isomers is called resolution. The first demonstration of this process was the historic resolution of racemic tartaric acid by Louis Pasteur in 1848. Below 25°C the (+)-enantiomer of sodium ammonium tartrate crystallizes in one type of crystal while the (−)-enantiomer crystallizes in a mirror-image crystal. By carefully hand-picking the crystals, Pasteur separated the mixture into two piles and examined their solutions separately in a polarimeter. He made the exciting observation that a solution of one form rotated the plane of polarized light to the right and the other to the left. When equal weights of the two kinds of crystals were dissolved in water, the solution of the mixture, like the starting material, had no effect on the plane of polarized light. This initial resolution by Pasteur is particularly remarkable, for since that time very few additional examples have been encountered in which crystallization produces enantiomorphic crystals large enough that manual separation is possible. Fortunately other methods of resolution are now known.

A second and more generally useful method of resolution is chemical, and based on the fact that diastereomers have different physical properties. Pasteur observed that the salts of (+)-tartaric and of (−)-tartaric acid with metals or ammonia were identical in their solubilities. However, salts of (+)- and (−)-tartaric acid derived from certain naturally occurring amines such as strychnine or quinine no longer have the same solubilities. (The amines named are optically active because they contain one or more asymmetric carbons.) Reaction of a racemic acid with an optically active base forms a pair of diastereomeric salts. These salts can be separated by fractional crystallization. Treatment of the separated salts with mineral acid then liberates the original acid in optically pure form.

$$\left.\begin{matrix} \text{dextro-acid} \\ \text{levo-acid} \end{matrix}\right\} + \text{levo-base} \longrightarrow \left\{\begin{matrix} \text{dextro-acid: levo-base} \\ \text{levo-acid: levo-base} \end{matrix}\right.$$

$$\qquad\text{enantiomers}\qquad\qquad\qquad\qquad\qquad\text{diastereomers}$$

If the starting racemate is an amine, then an optically active acid can be used for the process. This method of resolution is both practical and has wide application.

4.11 The Significance of Asymmetry in the Biological World

Thus far we have looked at ways of detecting and measuring optical activity in the laboratory, predicting chirality from molecular structure, specifying absolute configuration, and resolving a mixture of enantiomers in the laboratory.

Why is it so important that we be able to describe stereoisomerism and recognize it in molecules? The reason stems from the fact that we are interested in the reactions between molecules, and particularly between organic substances in the biological world. Living organisms, plant or animal, consist largely of asymmetric and therefore optically active substances. Except for molecules like water, inorganic salts, and a relatively few low-molecular-weight organic molecules, most of the compounds of nature are chiral. Although, in principle, these molecules can exist as a mixture of stereoisomers, almost invariably only one stereoisomer is found in nature. There are of course instances where both enantiomers can be found in nature but they do not seem to exist together in the same biological system. For example, both dextrorotatory and levorotatory lactic acids are found in nature; the (+)-lactic acid is found in living muscle, while the (−)-lactic acid is found in sour milk. We can make the further generalization that not only is just one enantiomer found in nature but also only one enantiomer can be used or assimilated. In fact this latter observation is a basis for a sometimes-used biological resolution technique.

Pasteur himself discovered in 1858–1860 that when the microorganism *Penicillium glaucum*, the green mold found in aging cheese and rotting fruit, is grown in a medium containing racemic tartaric acid, the solution slowly becomes levorotatory. The microorganism preferentially consumes or metabolizes the (+)-tartaric acid. If the process is interrupted at the right time, the (−)-tartaric acid can be crystallized from solution in pure form. If the process is allowed to continue, the microorganism will eventually consume the (−)-tartaric acid as well. Thus while both enantiomers of tartaric acid are metabolized by the microorganism, the (+)-form is metabolized at a much greater rate. As another example, when racemic mevalonic acid is fed to rats, one enantiomer, R-mevalonic acid is totally absorbed while almost all of the other enantiomer, S-mevalonic acid, is excreted in the urine.

$$H_3C \quad OH \qquad\qquad H_3C \quad OH$$
$$C \qquad\qquad\qquad C$$
$$H_2C \quad CH_2 \qquad\qquad H_2C \quad CH_2$$
$$H_2COH \quad CO_2H \qquad\qquad H_2COH \quad CO_2H$$

R-mevalonic acid
(metabolized)

S-mevalonic acid
(excreted)

Many other examples are known in which a mold or other microorganism will use one enantiomer of a racemic mixture. Resolution by this technique is generally referred to as <u>microbiological resolution</u> or separation. This method of resolution, just as with the hand-picking of crystals, is seldom of preparative value. One of the enantiomers, usually the more interesting natural or biologically active form, is sacrificed. The nonmetabolized form may be toxic. Furthermore, since the process must be conducted in dilute solution and in a nutrient medium conducive to the growth of the organism, recovery of the nonmetabolized enantiomer is often difficult. But we have not discussed this method only to conclude that it is seldom of preparative value; we have discussed it for the insight it gives into the operation of biological systems on enantiomeric compounds.

The observations that only one enantiomer is found in a given biological system and that only one enantiomer can be metabolized should be enough to convince us that it is an asymmetric world in which we live. At least it is asymmetric at the molecular level. Essentially all chemical reactions in the biological world take place in an asymmetric environment. Let us develop this last point a bit further. Perhaps the most conspicuous examples of asymmetry among biological molecules are the enzymes, all of which have a very high degree of chirality. Just to illustrate this point, consider the enzyme chymotrypsin which functions so efficiently in the intestine at pH 7–8 to catalyze the digestion of proteins. This enzyme is made up of 241 amino acids. Of these, 218 have one chiral center and therefore can exist as R or S centers of chirality. The number of potential stereoisomers is then 2^{218}, a number that surely seems beyond comprehension. Fortunately, nature does not squander its precious resources and energies unnecessarily; only one of these is made in any given organism.

Enzymes catalyze biological reactions by first adsorbing on their surface the small molecule or molecules about to undergo reaction. They may be bound by an accumulation of hydrogen bonds or by actual covalent bond formation. Thus, whether these smaller molecules are asymmetric or not, they are now held for reaction in an asymmetric environment. Let us look in more detail at just two examples, one to illustrate how an enzyme might distinguish between a pair of enantiomers, the second to illustrate how an enzyme might catalyze the conversion of a symmetrical molecule into one pure enantiomer uncontaminated by its mirror image.

Consider first glyceraldehyde, a key intermediate in the metabolism of carbohydrates. How might an enzyme discriminate between one enantiomer of glyceraldehyde and the other. It is generally agreed that an enzyme with specific receptor or binding sites for three of the four substituents on the chiral center can distinguish readily between two enantiomers. Assume for example that the enzyme involved in the catalysis of a glyceraldehyde reaction has three binding sites, one specific for —H, another for —OH, and a third for —CH_2OH, and that three sites are arranged on the enzyme surface as shown in Figure 4.9.

The enzyme can "recognize" R-(+)-glyceraldehyde (the natural or biologically active form) in the presence of S-(−)-glyceraldehyde since the correct enantiomer can be adsorbed with three groups attached to the appropriate receptor or binding sites while the other enantiomer can, at best, bind to only two of these sites.

The stereospecific course of an enzyme-catalyzed reaction which converts a symmetrical starting material into one pure enantiomer can also

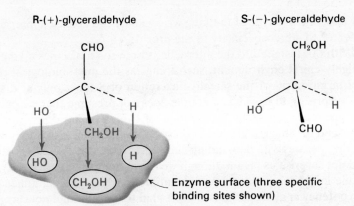

S-(−)-glyceraldehyde

Enzyme surface (three specific
binding sites shown)

FIGURE 4.9 Enzyme stereospecificity. A schematic diagram of an enzyme surface capable of interacting with R-(+)-glyceraldehyde at three binding sites but with S-(−)-glyceraldehyde at only two of the three potential binding sites.

be understood with the same type of model. Consider the enzyme-catalyzed reduction of pyruvic acid to S-(+)-lactic acid.

$$CH_3-\overset{\overset{\textstyle O}{\|}}{C}-CO_2H \xrightarrow{\text{enzyme}} CH_3-\overset{\overset{\textstyle OH}{|}}{CH}-CO_2H$$

pyruvic acid   S-(+)-lactic acid

Pyruvic acid is superimposable on its mirror image and therefore is not a chiral molecule. However, because of the requirements of very specific and precise interactions between it and the enzyme surface, it may be possible that the chemical reducing agent can approach the enzyme-bound pyruvic acid molecule from only one direction. Figure 4.10 is a schematic diagram that illustrates how a pyruvic acid molecule might be held on an enzyme surface and the approach of a reducing agent from the "top" of the molecule to form S-lactic acid. Of course if the reducing agent were to approach pyruvic acid from the other side, it would lead to the formation of R-lactic acid.

Quite understandably we do not know the intimate details of how these three molecules (pyruvic acid, enzyme, and reducing agent) interact,

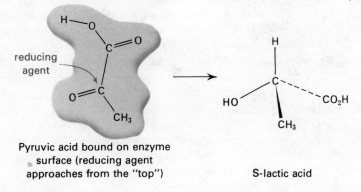

reducing
agent

Pyruvic acid bound on enzyme
surface (reducing agent
approaches from the "top")

S-lactic acid

FIGURE 4.10 Enzyme stereospecificity. A schematic diagram for the enzyme-catalyzed reduction of pyruvic acid to S-lactic acid.

but by applying this type of thinking we certainly can at least appreciate the high degree of stereospecificity of the process.

With this insight into the interactions between molecules taking place in a highly chiral environment, such things as the microbiological resolution of tartaric acid or the selective excretion of S-mevalonic acid should not be surprising to us. This of course does not presume that we have any means to make accurate predictions of which enantiomer might be metabolized or which might be synthesized, but the fact is that the biological world operates with demanding and precise stereospecificity. Further, it should not surprise us (again, though we might not have predicted it) that a molecule might function as a potent antibiotic while its enantiomer either has no potency at all or perhaps is somewhat toxic. It should not surprise us to learn that the physiological and psychological effects of LSD are not shared by its enantiomer. Neither should it surprise us that a molecule and its enantiomer may have markedly different tastes. For example, (+)-leucine tastes sweet while its enantiomer, (−)-leucine, is bitter.

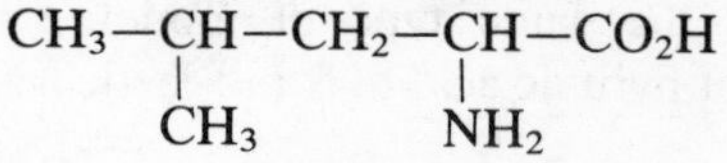

$$CH_3-CH-CH_2-CH-CO_2H$$
$$\quad\;\; | \qquad\qquad |$$
$$\quad\;\; CH_3 \qquad\quad NH_2$$

leucine

The fact that the interactions of molecules in the biological world are so very specific in geometry and chirality is not surprising, but just how these interactions are accomplished with such high precision and efficiency is one of the great challenges that modern science has only recently begun to unravel.

PROBLEMS

4.1 Examine the following structures and mark each chiral center (asymmetric carbon atom) with an asterisk. Build a molecular model for each molecule you have marked with an asterisk; also build one of its mirror image. Verify for yourself that the molecule and its mirror image are not superimposable.

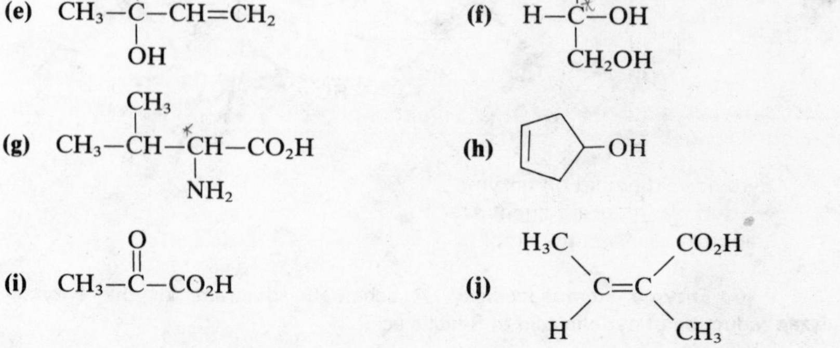

(a) $CH_2=CH-\overset{*}{C}H-CH_3$
$\qquad\qquad\quad |$
$\qquad\qquad\quad OH$

(b) $CH_2=CH-\overset{*}{C}H-CH_2-CH_3$
$\qquad\qquad\qquad |$
$\qquad\qquad\qquad OH$

(c) $CH_3-CH_2-CH-CH_2-CH_3$
$\qquad\qquad\qquad |$
$\qquad\qquad\qquad OH$

(d) $CH_2-CH_2-\overset{*}{C}H-CH_2-CH_3$
$\quad\;\; |\qquad\qquad\;\; |$
$\quad\;\; OH\qquad\qquad OH$

(e) $CH_3-\overset{\overset{\textstyle CH_3}{|}}{C}-CH=CH_2$
$\qquad\quad |$
$\qquad\quad OH$

(f) $H-\overset{\overset{\textstyle CO_2H}{|}}{\underset{\underset{\textstyle CH_2OH}{|}}{C}}-OH$

(g) $CH_3-\overset{\overset{\textstyle CH_3}{|}}{C}H-\overset{*}{C}H-CO_2H$
$\qquad\qquad\qquad |$
$\qquad\qquad\qquad NH_2$

(h) [cyclopentane ring]—OH

(i) $CH_3-\overset{\overset{\textstyle O}{||}}{C}-CO_2H$

(j) $\underset{H}{\overset{H_3C}{\diagdown}}C=C\underset{CH_3}{\overset{CO_2H}{\diagup}}$

(k)

$$CH_2OH$$
$$H\!-\!\overset{|}{\underset{|}{C}}\!-\!OH$$
$$CH_2OH$$

(l)

$$\overset{CH_3}{\underset{}{}}$$
$$CH_3\!-\!CH\!-\!\overset{*}{CH}\!-\!CH_3$$
$$\underset{OH}{|}$$

(m)

$$CH_2\!-\!CO_2H$$
$$HO\!-\!\overset{|}{\underset{|}{C}}\!-\!CO_2H$$
$$CH_2\!-\!CO_2H$$

(n)

$$\overset{*}{}$$
$$HO\!-\!CH_2\!-\!CH\!-\!CO_2H$$
$$\underset{NH_2}{|}$$

(o)

Cl on cyclohexene ring (*)

(p)

Cl on cyclohexene ring

4.2 The following are absolute configurations for a series of small molecules each containing only one chiral center. Under each is written the name of the molecule. Determine whether the enantiomer drawn is R or S.

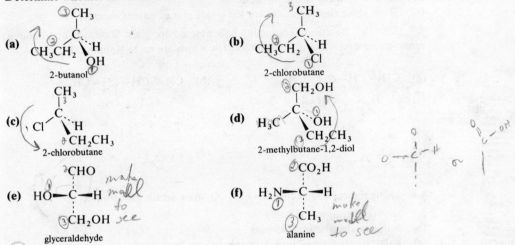

(a) 2-butanol

(b) 2-chlorobutane

(c) 2-chlorobutane

(d) 2-methylbutane-1,2-diol

(e) glyceraldehyde

(f) alanine

4.3 Drawn here are several stereorepresentations of lactic acid. Taking (a) as a reference structure, note that the other structures are stereorepresentations of lactic acid viewed from other perspectives. Which of the alternative representations are identical to (a) and which are mirror images of (a)?

(a)   (b)   (c)   (d)   (e)

4.4 Explain the difference in molecular structure between meso tartaric acid and racemic tartaric acid.

4.5 Below the structural formula of each of the following molecules is given the number of stereoisomers, and in parentheses the number of these stereoisomers that are meso compounds. For each meso compound, draw an appropriate projection formula and indicate clearly the plane of symmetry.

(a)

HO, OH on cyclohexane ring; CH with H₃C, CH₃

4 (2 meso)

(b) $CH_3CH_2CH\!-\!CHCH_2CH_3$ with HO and OH

3 (1 meso)

(c) HO_2C — cyclopentane with H_3C CH_3 — CO_2H

3 (1 meso)

4.6 Inositol is widely distributed in plants and animals and is a growth factor for animals and microorganisms. It is used medically for treatment of cirrhosis, hepatitis, and fatty infiltration of the liver. Inositol has nine possible stereoisomers, seven of which are meso and two of which are enantiomers. The most prevalent natural form is *cis*-1,2,3,5-*trans*-4,6-cyclohexanehexol.

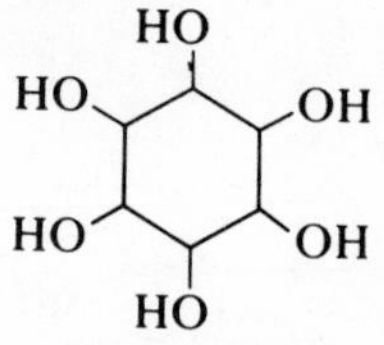

inositol

(a) Draw a chair conformation of the prevalent natural isomer and determine whether it is optically active or meso.
(b) Draw chair conformations for the single pair of enantiomers.

4.7 Draw all stereoisomers for the following compounds. State which structures are enantiomers; which are diastereomers; which are meso compounds.

(a) CH$_3$—CH—CH—CH$_3$
       | |
      H$_2$N   OH

(b) CH$_3$—CH—CH—CH$_3$
       | |
      HO   OH

(c)

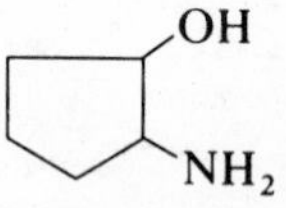

(d)

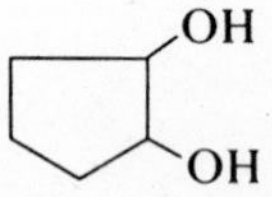

4.8 Using the molecular formula C$_6$H$_{12}$O, draw structural formulas for:
(a) four constitutional isomers
(b) four functional group isomers
(c) a pair of acyclic *cis-trans* isomers
(d) a pair of cyclic *cis-trans* isomers
(e) a pair of conformational isomers
(f) a pair of enantiomers
(g) a pair of diastereomers

4.9 Draw the structural formula of at least one alkene of molecular formula C$_5$H$_9$Br that will show:
(a) neither *cis-trans* isomerism nor enantiomerism
(b) *cis-trans* isomerism but not enantiomerism
(c) enantiomerism but not *cis-trans* isomerism
(d) both *cis-trans* isomerism and enantiomerism

4.10 Draw the stereoisomers of grandisol, a sex hormone, secreted by the hind gut of the male boll weevil (*Anthonomus grandis*).

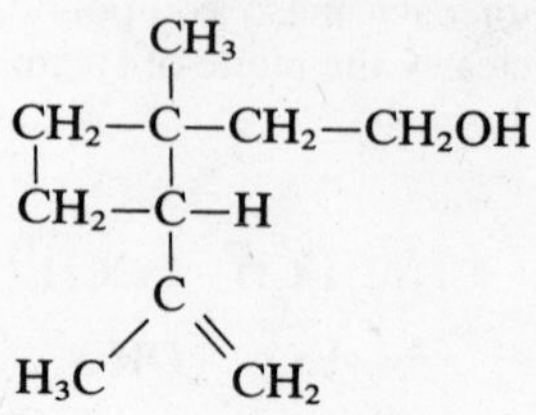

grandisol

4.11 State the total number of stereoisomers for the following molecules.

(a) $CH_3-CH-CH=CH-CH-CO_2H$
            | |
          CH_3 CH_3

2,5-dimethyl-3-hexenoic acid

(b)
$$\begin{array}{c} CHO \\ | \\ H-C-OH \\ | \\ CH_2OH \end{array}$$

glyceraldehyde

(c)
$$\begin{array}{c} CH_2-CO_2H \\ | \\ H-C-CO_2H \\ | \\ HO-CH-CO_2H \end{array}$$

isocitric acid

(d)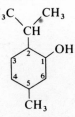

2-isopropyl-5-methyl-1-cyclohexanol
(the all-equatorial isomer is menthol)

(e)

terpineol
(camphor oil)

(f)

juvabione
(a juvenile hormone analog
of balsam fir, page 198)

(g) $CH_3CH=CHCO_2H$

2-butenoic acid

(h) vitamin A (see page 56)

(i) santonin (page 75)

(j)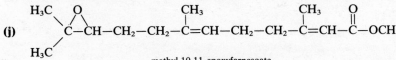

methyl 10,11-epoxyfarnesoate
(a synthetic insect juvenile hormone, page 196)

4.12 Calculate the observed rotation of a solution prepared by dissolving 5.0 g of (+)-tartaric acid in 100 ml of water and placing a 1-decimeter sample tube filled with this solution in a polarimeter.

4.13 A sample of an optically active sugar is prepared by dissolving 0.120 g in 100 ml of water. When this solution is placed in a 1-decimeter tube, the observed rotation is −0.49°. Calculate the specific rotation of this sugar. If the concentration of the sugar were doubled, what would be the observed rotation? the specific rotation?

4.14 List three experimental methods for the resolution of racemic mixtures and explain each briefly. Which of these methods is the most widely used?

4.15 How might you explain the following?

(a) An enzyme is able to distinguish between a pair of enantiomers and catalyze a biochemical reaction of one enantiomer but not of its mirror image.

(b) An enzyme is able to catalyze the conversion of a symmetrical, optically inactive molecule into one pure enantiomer uncontaminated by its mirror image.

(c) The microorganism *Penicillium glaucum* preferentially metabolizes (+)-tartaric acid rather than (−)-tartaric acid.

5

Alcohols, Ethers, and Thiols

5.1 Introduction

In this chapter we will discuss the organic chemistry of alcohols, ethers, and thiols. Alcohols and ethers are derivatives of water in which one or both hydrogens are replaced by alkyl groups. Thiols are derivatives of hydrogen sulfide in which one hydrogen is replaced by an alkyl group.

$$R-\overset{..}{\underset{..}{O}}-H \qquad R-\overset{..}{\underset{..}{O}}-R \qquad R-\overset{..}{\underset{..}{S}}-H$$
an alcohol   an ether   a thiol

We will discuss some of the physical properties of each of these classes of compounds, how each can be synthesized in the laboratory, and the important reactions of each. In addition, we will use one method of alcohol synthesis to introduce a fundamental type of organic reaction, namely nucleophilic substitution at a saturated carbon atom.

5.2 Nomenclature of Alcohols and Ethers

Several systems are used for naming alcohols. The most versatile of these, the IUPAC system, first selects the longest continuous chain that contains the —OH group. The alcohol is then named by replacing the -e of the parent alkane by -ol and adding a number to show the position of the —OH group. Common names for alcohols are derived by naming the alkyl

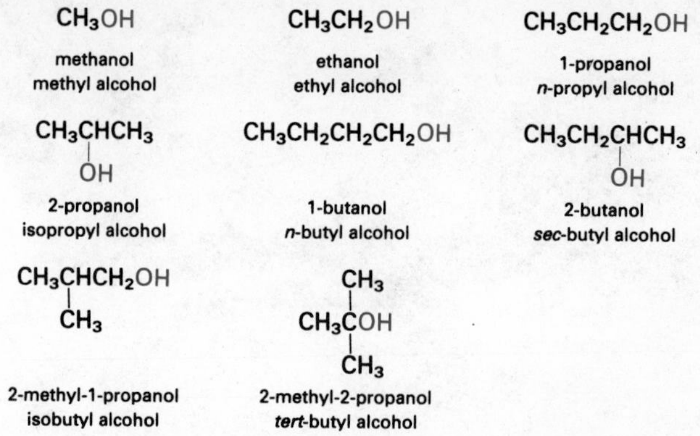

CH_3OH

methanol
methyl alcohol

CH_3CH_2OH

ethanol
ethyl alcohol

$CH_3CH_2CH_2OH$

1-propanol
n-propyl alcohol

CH_3CHCH_3
|
OH

2-propanol
isopropyl alcohol

$CH_3CH_2CH_2CH_2OH$

1-butanol
n-butyl alcohol

$CH_3CH_2CHCH_3$
|
OH

2-butanol
sec-butyl alcohol

CH_3CHCH_2OH
|
CH_3

2-methyl-1-propanol
isobutyl alcohol

CH_3
|
CH_3COH
|
CH_3

2-methyl-2-propanol
tert-butyl alcohol

FIGURE 5.1   Structural formulas of alcohols.

group attached to —OH and then adding the word alcohol. Figure 5.1 lists both IUPAC and common names for a variety of simple alcohols.

Compounds containing two hydroxyl groups are called <u>diols</u>, those containing three hydroxyl groups are called <u>triols</u>, and so on. Several examples follow. Under each is given the IUPAC name and the common name.

$CH_2—CH_2$
| |
OH OH

1,2-ethanediol
ethylene glycol

$CH_2—CH—CH_2$
| | |
OH OH OH

1,2,3-propanetriol
glycerol, glycerine

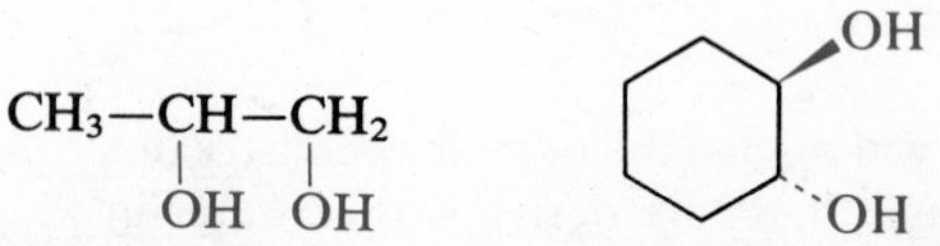

$CH_3—CH—CH_2$
| |
OH OH

1,2-propanediol
propylene glycol

trans-1,2-cyclohexanediol

Of course, all of these diols and triols can be named as derivatives of the parent alkane. Yet as with so many organic compounds, common names have persisted. Both ethylene glycol and propylene glycol can be prepared by controlled oxidation of ethylene and propylene—hence their common names. <u>Glycerol is a major by-product of the manufacture of soaps and is</u> <u>related in structure to glyceraldehyde and glyceric acid, both key inter-</u> <u>mediates in the metabolism of carbohydrates.</u>

Molecules that contain two hydroxyl groups on the same carbon, 1,1-glycols, are almost never isolated.

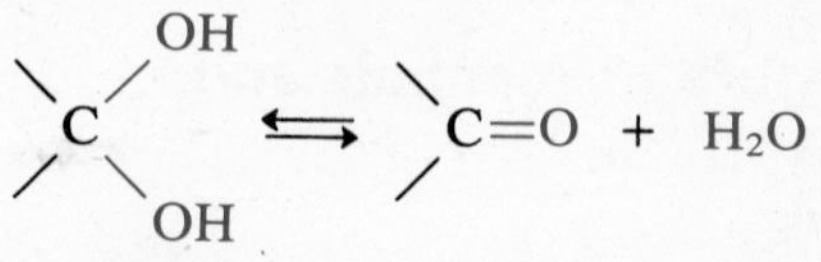

$$\diagdown \!\!\!\!\!\!\!\diagup OH \quad\rightleftharpoons\quad \diagdown C{=}O \;+\; H_2O$$

a 1,1-glycol

As we shall see in Section 7.7, compounds of this type are in equilibrium with the corresponding aldehyde or ketone, and equilibrium generally lies very far to the right.

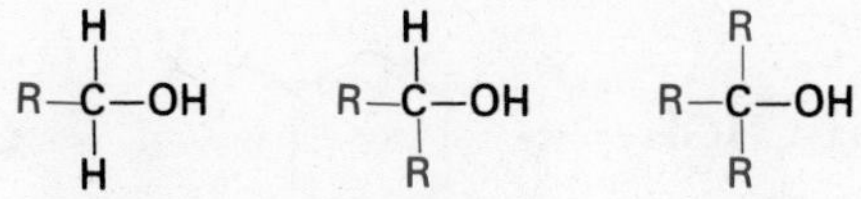

FIGURE 5.2 Classification of alcohols: primary, secondary and tertiary.

primary 1° secondary 2° tertiary 3°

We often refer to alcohols as being primary (1°), secondary (2°), or tertiary (3°). This classification depends on the number of alkyl groups on the carbon bearing the —OH function. (Figure 5.2.) (Compare this classification with that of carbocations, page 63.)

One last comment on the nomenclature of alcohols. The suffix -ol is generic to alcohols and although names such as glycerol, menthol, and cholesterol contain no clues to their carbon skeletons, the names do indicate that each compound contains one or more hydroxyl groups.

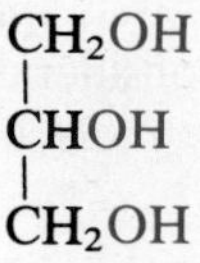

glycerol 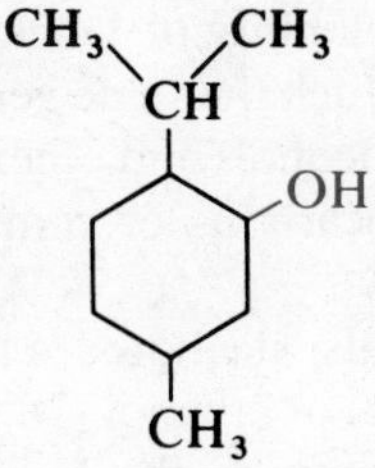   menthol 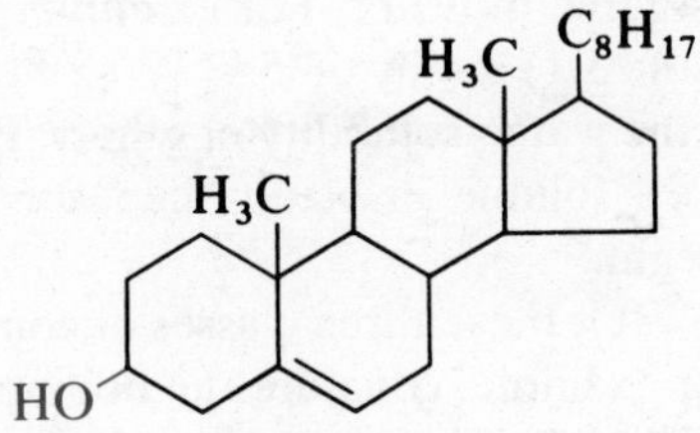   cholesterol

Ethers are generally named by attaching the names of the alkyl or aryl groups to the generic name ether. In naming simple ethers, the prefix di- is sometimes not used.

CH_3-O-CH_3 $CH_3CH_2-O-CH_2CH_3$

—$O-CH_3$

dimethyl ether   diethyl ether
methyl ether   ethyl ether   cyclohexyl methyl ether

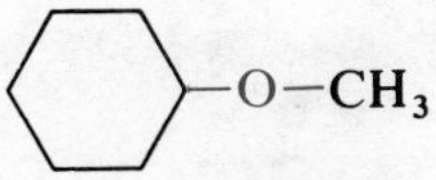

Heterocyclic ethers, that is, cyclic compounds in which the ether oxygen is one of the atoms in a ring, are given special names.

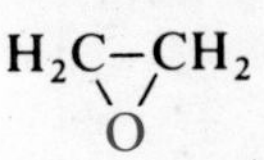

ethylene   tetrahydrofuran 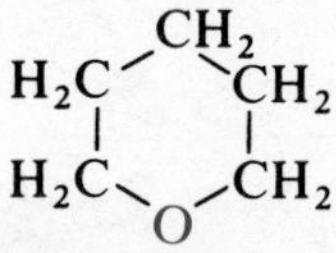   tetrahydropyran 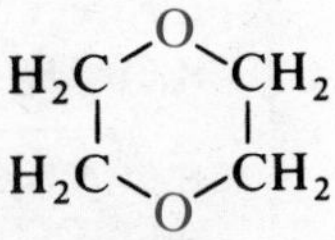   1,4-dioxane
oxide

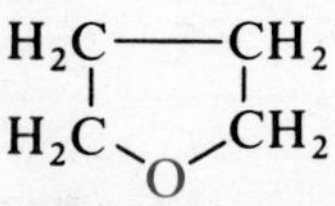

The chemical reactivity of ethylene oxide is quite different from that of other cyclic and acyclic ethers and will be discussed separately in Section 5.13.

In ethers of more complex structure where there is no simple name for one of the attached alkyl groups or where the ether is only one of several functional groups, it may be necessary to indicate the —OR group as an alkoxy group; methoxy- for —OCH_3, ethoxy- for —OCH_2CH_3, and so on.

109

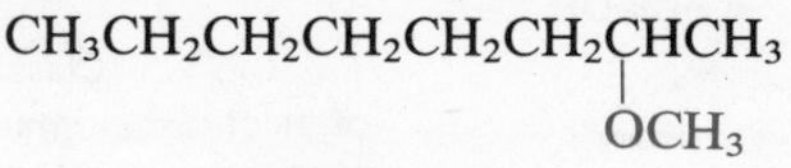

$$CH_3CH_2CH_2CH_2CH_2CH_2CHCH_3$$
$$\underset{OCH_3}{|}$$

2-methoxyoctane

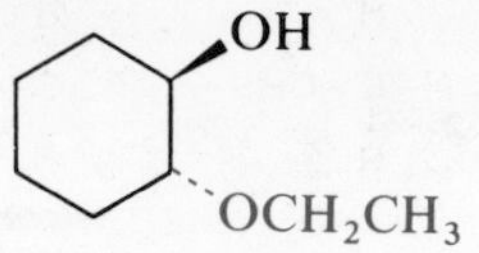

trans-2-ethoxycyclohexanol

5.3 Physical Properties of Alcohols and Ethers

Listed in Table 5.1 are boiling points and solubilities in water of several groups of alcohols, ethers, and hydrocarbons of similar molecular weight.

Methanol, ethanol, and 1-propanol are soluble in water in all proportions. Alcohols of higher molecular weight are either slightly soluble or insoluble in water. For example, the solubility of 1-octanol (not shown in Table 5.1) is only 0.05 g/100 g water. Much the same generalizations apply to the water-solubility of ethers. Both alcohols and ethers are considerably more soluble in water than are hydrocarbons of comparable molecular weight.

Of these three classes of compounds, alcohols are the highest boiling. For example, compare the boiling points of ethanol (78°C), dimethyl ether (−24°C), and propane (−42°C). Or compare the boiling points of 1,4-butanediol (230°C), ethylene glycol dimethyl ether (85°C), and hexane

TABLE 5.1 Boiling points and solubilities in water of several groups of alcohols, ethers, and hydrocarbons of similar molecular weight.

Compound	Molecular weight	bp (°C)	Solubility in water (g/100 g H_2O)
CH_3OH	32	65	soluble (∞)
CH_3CH_3	30	−89	insoluble
CH_3CH_2OH	46	78	soluble (∞)
CH_3OCH_3	46	−24	soluble (7 g/100 g)
$CH_3CH_2CH_3$	44	−42	insoluble
$CH_3CH_2CH_2OH$	60	97	soluble (∞)
$CH_3CH_2OCH_3$	60	11	soluble
$CH_3CH_2CH_2CH_3$	58	0	insoluble
$CH_3CH_2CH_2CH_2OH$	74	117	soluble (8 g/100 g)
$CH_3CH_2OCH_2CH_3$	74	35	soluble (8 g/100 g)
$CH_3CH_2CH_2CH_2CH_3$	72	36	insoluble
$CH_3CH_2CH_2CH_2CH_2OH$	88	138	slightly soluble (2.3 g/100 g)
$CH_3CH_2CH_2CH_2OCH_3$	88	71	slightly soluble
$HOCH_2CH_2CH_2CH_2OH$	90	230	soluble (∞)
$CH_3OCH_2CH_2OCH_3$	90	85	soluble (∞)
$CH_3CH_2CH_2CH_2CH_2CH_3$	86	69	insoluble

(69°C). Notice that while the boiling points of alcohols are considerably higher than those of hydrocarbons of comparable molecular weight, the boiling points of ethers are only slightly higher.

To understand these relationships between molecular structure and physical properties, we must first review the nature of intermolecular forces, that is, the forces between molecules. The most important of these are dipole-dipole, ion-dipole, and hydrophobic interactions. While each of these forces is weak in comparison to the energy of covalent bonds, they are nevertheless very important in determining both physical and chemical properties of molecules. Furthermore, they are important in determining the three-dimensional shapes of biomolecules. In this chapter we shall discuss dipole-dipole and ion-dipole interactions as they affect the solubility of polar organic molecules in water and other polar solvents. In later chapters dealing with lipids, proteins, and nucleic acids, we shall discuss hydrophobic interactions and how they affect the properties of nonpolar organic molecules.

Dipole-dipole interaction is the attraction of a positive part of one polar molecule by the negative part of another. You may have already encountered this phenomenon in general chemistry in the study of the physical properties of water and hydrogen sulfide. The formula weight of hydrogen sulfide is 34, nearly twice that of water. Yet its boiling point is 160° below that of water. This is due to the high degree of association of water molecules in the liquid state, by dipole-dipole interaction.

Since the proton is very small, an adjacent water molecule can approach to within a short distance of it. The electrostatic force of attraction between the oppositely charged hydrogen and oxygen is appreciable at this short distance, and the molecules tend to form aggregates in solution. Each water molecule interacts directly with several other water molecules (Figure 5.3).

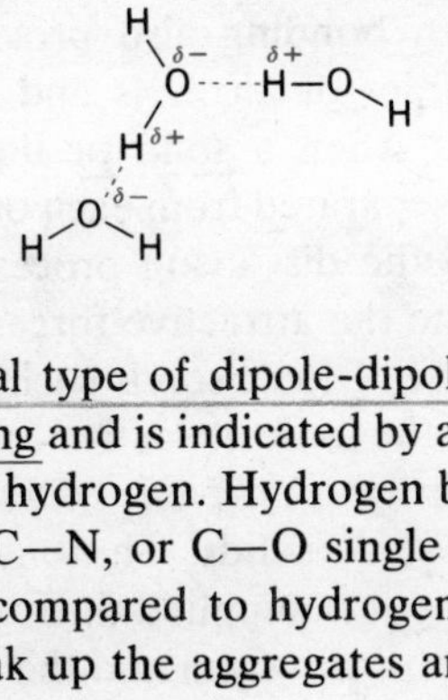

FIGURE 5.3 Hydrogen bonding in water. The formation of molecular aggregates.

This special type of dipole-dipole interaction is given the name of hydrogen bonding and is indicated by a dashed line connecting the interacting oxygen and hydrogen. Hydrogen bonds are about 5% as strong as the average C—C, C—N, or C—O single covalent bonds. The higher boiling point of water compared to hydrogen sulfide is due to the extra energy required to break up the aggregates and allow the individual water molecules to enter the vapor state.

This type of association is also possible with alcohols, as illustrated by the association of methanol, CH_3OH, in Figure 5.4.

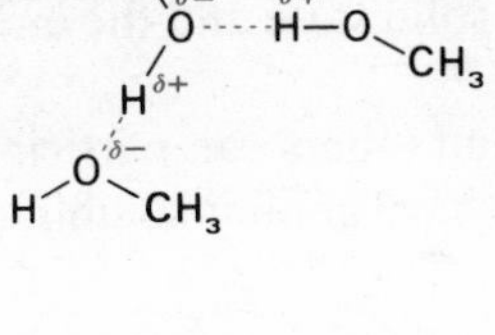

FIGURE 5.4 The association of methanol in the liquid state.

111

The effect of hydrogen bonding on boiling points is dramatically illustrated by comparing ethanol, CH_3CH_2OH, bp 78°C, and its isomer dimethyl ether, CH_3OCH_3, bp −24°C. Each of these compounds has the same molecular formula, C_2H_6O, and hence the same molecular weight. Therefore differences in boiling point must be due to differences in the degree of association between molecules in the pure liquid. Note from Table 1.5 that both ethanol and dimethyl ether are polar molecules, but there is no large difference in dipole moment between ethanol (1.69 D) and dimethyl ether (1.30 D). The remarkably high association of CH_3CH_2OH compared to CH_3OCH_3 is due to the fact that both the shape and the structure of the molecule allow close approach of the centers of positive and negative charge (compare Figure 5.4). While there is certainly bond and molecular polarity in dimethyl ether, the centers of positive and negative charge are buried more deeply within the molecule than in ethanol, with the result that the positively charged carbon atom and the negatively charged oxygen atom cannot approach one another closely, and therefore the resulting electrostatic interaction is very small (Figure 5.5).

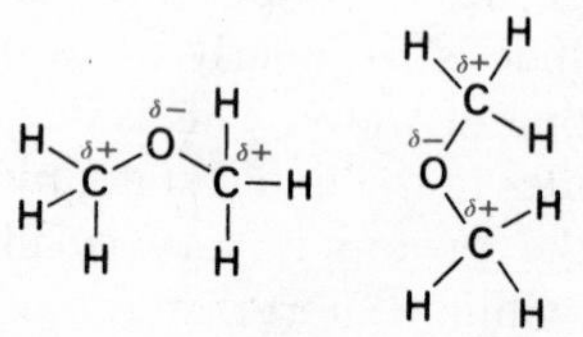

FIGURE 5.5 Hydrogen bonding is not possible in dimethyl ether for there are no highly polarized hydrogens. Dipole-dipole interaction is not appreciable because the centers of partial negative and positive charge cannot come close enough together.

The presence of one or more additional —OH groups in a molecule further increases this affect of hydrogen bonding.

The phenomenon of hydrogen bonding also provides us with an explanation for the relative solubilities of alcohols and ethers in water. Recall from general chemistry that when a solid or liquid dissolves in water, the solute molecules become separated from each other and become surrounded by water molecules. In the dissolving process, as in boiling, energy must be supplied to overcome the attractive forces between molecules of the solid or liquid. In the case of ionic solids such as NaCl, a great deal of energy is required to disrupt the ionic attractive forces. Only water and a few other solvents of high polarity are capable of dissolving appreciable quantities of NaCl and other ionic solids. The energy required to break the attractive forces between ions is supplied through the formation of new attractive forces; between the positive ion and the negative ends of water molecules, and between the negative ion and the positive ends of water molecules. Such forces are called ion-dipole interactions. In solution, each ion is surrounded by a cluster of water molecules and is said to be hydrated or solvated (Figure 5.6).

The solubility of an ionic solid in water depends on the relative sizes of the energy holding the ions in the solid state and the energy of solvation (hydration).

The oxygen atom of alcohols and ethers can participate in hydrogen bonding and hence these compounds have greater solubilities in water than

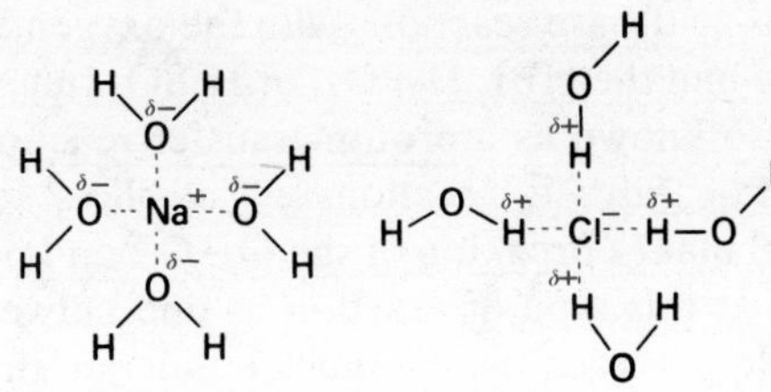

FIGURE 5.6 Hydration of sodium and chloride ions in water.

do nonpolar hydrocarbon molecules. Alcohols can participate in hydrogen bonding with water molecules at two sites, that is, at both atoms of the —OH group (Figure 5.7). Ethers can form hydrogen bonds with water at only the oxygen atom.

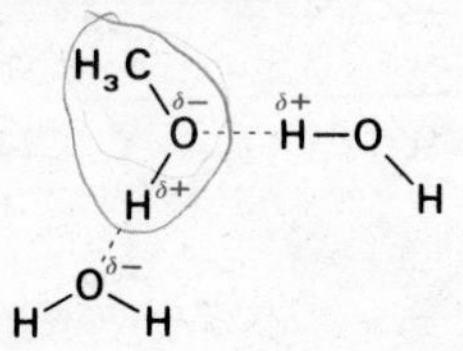

FIGURE 5.7 Hydrogen bonding between water and methanol.

5.4 Acidity and Basicity of Alcohols and Ethers

We can begin a discussion of the chemical properties of the R—OH group by comparing it with H—OH. Like water, alcohols are sufficiently acidic to react with active metals to liberate hydrogen and form metal salts.

$$2H_2O + 2Na \longrightarrow 2Na^+OH^- + H_2$$

$$2CH_3OH + 2Na \longrightarrow 2CH_3O^-Na^+ + H_2$$

$$2CH_3CH_2OH + Mg \longrightarrow (CH_3CH_2O^-)_2Mg^{2+} + H_2$$

These salts are named as metal alkoxides and are bases comparable in strength to sodium hydroxide. Specific alkoxides are named using as a root the alkyl group attached to oxygen. The three alkoxides shown below are commonly used in organic reactions requiring a strong base in a nonaqueous solvent.

$$CH_3—\overset{..}{\underset{..}{O}}:^-Na^+ \qquad CH_3CH_2—\overset{..}{\underset{..}{O}}:^-Na^+ \qquad \overset{\displaystyle CH_3}{\underset{\displaystyle CH_3}{CH_3\overset{|}{\underset{|}{C}}—\overset{..}{\underset{..}{O}}:^-K^+}}$$

sodium methoxide   sodium ethoxide   potassium *tert*-butoxide

Alcohols are also bases, although less so than water. In the presence of a strong acid, the oxygen atom of water will accept a proton to form an oxonium ion. Similarly alcohols also form alkoxonium ions.

$$H—\overset{..}{\underset{..}{O}}—H + H—\overset{..}{\underset{..}{B}}r: \longrightarrow H—\overset{+}{\underset{\underset{\displaystyle H}{|}}{\overset{..}{O}}}—H + :\overset{..}{\underset{..}{B}}r:^-$$

oxonium ion

$$CH_3—\overset{..}{\underset{..}{O}}—H + H_2SO_4 \longrightarrow CH_3—\overset{+}{\underset{\underset{\displaystyle H}{|}}{\overset{..}{O}}}—H + HSO_4^-$$

methoxonium
ion

We can look at these reactions as <u>acid-base reactions</u> with the oxygen atom of ROH group acting as the base and the HBr, H_3PO_4, or H_2SO_4 functioning as the acid. This process, also known as a <u>proton-transfer reaction</u>, is fundamental to all of the acid-catalyzed reactions of alcohols for it "activates" the —OH group and makes breaking of the C—O bond possible. Alternatively, we can look at this type of reaction as one between a <u>nucleophile and an electrophile.</u> Oxygen is the nucleophile in that it possesses a pair of electrons that can be shared with another atom to form a new covalent bond (in this case a new O—H bond). The proton is said to be the electrophile in that it is electron-deficient and therefore accepts a pair of electrons in forming the new covalent bond.

<u>Ethers are also weak bases and react with strong acids to form oxonium ions.</u>

$$CH_3CH_2-\overset{..}{\underset{..}{O}}-CH_2CH_3 + H_2SO_4 \longrightarrow CH_3CH_2-\overset{\overset{+}{|}}{\underset{H}{\overset{..}{O}}}-CH_2CH_3 + HSO_4^-$$

diethyl ether   diethyloxonium ion

5.5 Preparation of Alcohols

<u>Methanol</u> (<u>methyl alcohol</u>), commonly called <u>wood alcohol</u>, was prepared by the destructive distillation of wood, at least until 1923. When wood is heated to temperatures above 250°C without access to air, it decomposes to charcoal and a volatile fraction which partially condenses on cooling. This condensate contains methyl alcohol, acetic acid, and traces of acetone. At the present time methanol is made on a large scale synthetically by the high-pressure hydrogenation of carbon monoxide over a chromic oxide-zinc oxide catalyst.

$$CO + 2H_2 \xrightarrow[\text{350°C, 200 atm}]{\text{ZnO-Cr}_2\text{O}_3} CH_3OH$$

<u>Ethanol</u> (<u>ethyl alcohol</u>), or simply "alcohol" in nonscientific language, has been prepared since antiquity by the fermentation of sugars and starches by yeast, and is the basis for the preparation of alcoholic "spirits." Extensive biochemical investigations have established that the fermentation of sugars (for example, glucose) proceeds through an elaborate series of steps, each catalyzed by a particular and specific enzyme.

$$\underset{\text{glucose}}{C_6H_{12}O_6} \xrightarrow{\text{enzymes}} 2CH_3CH_2OH + 2CO_2$$

The sugars for the fermentation process may come from a variety of sources: blackstrap molasses, a residue from the refining of cane sugar; various grains, hence the name "<u>grain alcohol</u>"; grape juice; various vegetables. The immediate product of the fermentation process is an aqueous solution containing up to about 15% alcohol. This alcohol may be concentrated by distillation. Beverage alcohol may contain traces of flavor derived from the source (brandy from grapes, whiskeys from grains) or may be essentially flavor free (vodka). The most important synthetic method for the preparation of ethanol is the hydration of ethylene (Section 3.11).

Whatever the method of preparation of ethanol, be it the hydration of ethylene or fermentation of grains, it is first obtained mixed with water. This mixture must be concentrated and the alcohol separated from the other materials in the reaction mixture by fractional distillation. If we examine the boiling points of pure ethanol (78.5°C) and of water (100°C), we would predict that the first material to distill would be ethanol, the component with the lower boiling point. However, in a mixture of ethanol and water, the material of lowest boiling point is neither alcohol nor water but rather a mixture composed of 95% alcohol and 5% water. Although the boiling point of 95% alcohol is 78.2°C, only 0.3°C below the boiling point of pure alcohol, this small difference makes it impossible to produce absolute ethanol by any kind of fractional distillation of mixtures of ethanol containing more than 5% water. As a result, commercial ethanol as used for extracts, medicines, beverages, and so on, never contains over 95% of the essential component and usually slightly less. This 95% alcohol is said to be a constant-boiling mixture, an azeotrope, or an azeotropic mixture. An azeotrope is a mixture of liquids of a certain definite composition that distills at a constant temperature without change in composition. The boiling point of an azeotropic mixture is usually lower than that of the lowest boiling component, but in some cases it is higher than the boiling point of the highest boiling component. The 95% alcohol azeotrope contains two components and is known as a binary azeotrope.

Since 95% alcohol behaves as if it were a pure compound on distillation, pure or "absolute" alcohol cannot be obtained from it by this technique and other methods must be employed to remove the water from the mixture. One widely used method takes advantage of the fact that a mixture of benzene-water-ethanol forms a ternary azeotrope that boils at 64.6°C, below the boiling point of any of the three pure components of the azeotrope. The mixture contains 74.1% benzene, 18.5% ethanol, and 7.4% water. Absolute ethanol is made by adding benzene to the 95% alcohol and removing the water as the volatile azeotrope.

Ethylene glycol is prepared commercially from air oxidation of ethylene at high temperature over a silver catalyst to give ethylene oxide (Section 5.13), followed by acid-catalyzed hydrolysis.

$$CH_2{=}CH_2 + \tfrac{1}{2}O_2 \xrightarrow[300°C]{Ag} CH_2{-}CH_2 \xrightarrow[H^+]{H_2O} CH_2{-}CH_2$$

It is also manufactured by the reaction of ethylene with hypochlorous acid followed by hydrolysis of ethylene chlorohydrin in aqueous sodium carbonate.

$$CH_2{=}CH_2 + HOCl \longrightarrow CH_2{-}CH_2 \xrightarrow[Na_2CO_3]{OH^-} CH_2{-}CH_2$$

Ethylene glycol is used as a permanent antifreeze in automotive cooling systems because it is completely miscible with water in all proportions, and a 50% solution with water freezes at −34°C (−29°F). Among its many other important commercial applications is its use as a monomer in the copolymer Dacron (see the mini-essay "Nylon and Dacron").

Glycerol is a major by-product of the manufacture of soaps (Section 8.8) which produces roughly 60 pounds of it per ton of soap. An alternative means for the production of glycerol became available in the 1930s when Shell Development Company discovered that the intermediate allyl chloride could be made from propylene by direct chlorination of propylene (readily available from the cracking of petroleum and natural gas) at high temperatures. Hydrolysis in aqueous sodium hydroxide to form allyl alcohol, addition of hypochlorous acid, and further hydrolysis yields glycerol.

$$CH_3-CH=CH_2 \xrightarrow[400-500°C]{Cl_2} \underset{\underset{Cl}{|}}{CH_2}-CH=CH_2 \xrightarrow[NaOH,H_2O]{OH^-} \underset{\underset{OH}{|}}{CH_2}-CH=CH_2$$

3-chloro-1-propene   2-propene-1-ol

allyl chloride   allyl alcohol

$$\underset{\underset{OH}{|}}{CH_2}-CH=CH_2 \xrightarrow{HOCl} \underset{\underset{OH}{|}}{CH_2}-\underset{\underset{OH}{|}}{CH}-\underset{\underset{Cl}{|}}{CH_2} \xrightarrow[NaOH,H_2O]{OH^-} \underset{\underset{OH}{|}}{CH_2}-\underset{\underset{OH}{|}}{CH}-\underset{\underset{OH}{|}}{CH_2}$$

glycerol

Notice that the first step in this sequence is a substitution reaction rather than the more usual addition of chlorine to the double bond. Reaction of a substituted alkene in the vapor phase and at high temperature favors this type of substitution. Glycerol is a sweet-tasting, syrupy liquid. Because of the presence of the three hydroxyl groups on such a small carbon skeleton, glycerol is very high boiling (290°C) and is miscible with water and alcohol in all proportions. It is widely used in the manufacture of resins and nitroglycerine, in the pharmaceutical and cosmetic industries, and as a moistening agent.

Recall from Section 3.11 that alcohols can also be prepared from alkenes by acid-catalyzed hydration or hydroboration followed by peroxide oxidation.

Alkyl halides may be converted into the corresponding alcohols by heating in water or in aqueous alkali as illustrated by the conversion of allyl chloride into allyl alcohol (above) and methyl bromide into methyl alcohol.

$$CH_3-Br + OH^- \longrightarrow CH_3-OH + Br^-$$

In general, the substitution of hydroxyl for halogen in an alkyl halide is not a good preparative method for the synthesis of alcohols. Consequently there would seem to be little need for us to examine this reaction in any detail. Yet in another sense there is very good reason to study this reaction carefully, for it is a specific example of a general reaction class known as nucleophilic substitution at a saturated carbon atom. This is one of the simplest and most important classes of organic reactions, and a variety of industrial and laboratory transformations can be performed through the use of specific reactions belonging to this class.

5.6 Nucleophilic Substitution at a Saturated Carbon

Chemists have uncovered two idealized types of reaction mechanisms for nucleophilic substitution at a saturated carbon atom. In this search, two general tools have been of great usefulness: kinetic studies and stereo-chemical studies. Kinetic studies, that is, the measurement of reaction rates, can often tell us something about the timing of the steps in the reaction sequence. Stereochemical studies can often tell us something about the relationship between the configuration of the starting material and the products of a reaction. Let us examine some experimental evidence and then formulate these two general reaction mechanisms.

Methyl, ethyl, isopropyl, and *tert*-butyl bromides can be converted to the corresponding alcohols by reaction with sodium hydroxide in aqueous ethanol. Furthermore, the rates of these reactions can be measured by analyzing the reaction mixture at known time intervals. When such studies were first carried out in the early 1930s, the results were remarkable in that certain of the findings were totally unexpected. Let us consider first the reaction of 2-bromo-2-methylpropane (*tert*-butyl bromide) with hydroxide ion.

$$(CH_3)_3CBr \quad + OH^- \xrightarrow{\text{aqueous}\atop\text{ethanol}} (CH_3)_3COH \quad + Br^-$$

2-bromo-2-methylpropane   2-methyl-2-propanol
tert-butyl bromide   *tert*-butyl alcohol

It was found that the rate of conversion of *tert*-butyl bromide to *tert*-butyl alcohol depended only on the concentration of the tertiary butyl halide; it was completely independent of the concentration of the hydroxide ion. Doubling or halving the concentration of sodium hydroxide had no effect on the rate of conversion to the alcohol. The rate-determining step there-fore cannot involve hydroxide ion; it must involve only one reactant, the *tert*-butyl bromide.

This reaction is clearly a nucleophilic substitution in that hydroxide seeks a nucleus to which it can donate a pair of electrons and form a new covalent bond. In the process of donating an electron pair, hydroxide ion displaces bromide ion. Finally, this reaction is unimolecular (or first order) in that only one molecular species is involved in the rate-determining step. We summarize all of these statements by referring to such a reaction as S_N1 (Substitution; Nucleophilic; Unimolecular). This designation of course does not tell us how the reaction takes place, only that it is a unimolecular nucleophilic substitution at a saturated carbon atom.

Next let us look at the reactions of methyl and ethyl bromides with hydroxide ion.

$$CH_3-Br + OH^- \longrightarrow CH_3-OH + Br^-$$

methyl
bromide

methanol

$$CH_3CH_2-Br + OH^- \longrightarrow CH_3CH_2-OH + Br^-$$

ethyl bromide

ethanol

In contrast to the reaction of *tert*-butyl bromide, the rates of reaction of these two compounds were found to be directly proportional to the concentration of both alkyl halide and hydroxide ion. Doubling the concentration of either the alkyl halide or the hydroxide ion doubles the rate of reaction; decreasing the concentration of either by one-half results in a decrease in the rate of reaction by one-half. This reaction is a nucleophilic substitution at a saturated carbon atom and the rate-determining step is bimolecular (or second order) in that it involves two reacting species. We summarize these statements by referring to such reactions as S_N2 (*S*ubstitution; *N*ucleophilic; *Bi*molecular). Again, the designation S_N2 does not tell us how the reaction takes place, only that it is a bimolecular nucleophilic substitution at saturated carbon.

The rate of reaction for isopropyl bromide was found to be directly proportional to the concentrations of isopropyl bromide and hydroxide ion when hydroxide ion was about 1 molar, but independent of hydroxide when hydroxide ion concentration was low. We would then describe this reaction as either S_N1 or S_N2 depending on the experimental conditions.

Studies of the stereochemical course of nucleophilic displacement were first reported around 1900 by Paul Walden, who observed that nucleophilic displacement sometimes, but not always, is accompanied by inversion of configuration. This phenomenon is known as the Walden inversion. In an S_N2 reaction of an optically active alkyl halide, if the carbon undergoing substitution is the center of chirality, the product will be optically active and opposite in configuration from the starting material. The stereochemical results of first-order substitution reactions are completely different. In an S_N1 reaction of an optically active alkyl halide, if the carbon undergoing the substitution is the center of chirality, the product is most often optically inactive; that is, it is a racemic mixture.

With the experimental observations from kinetic and stereochemical studies, we can now examine the two different mechanisms that have been proposed for nucleophilic substitution reactions at saturated carbon. To account for the S_N2 reaction, chemists propose a one-step mechanism in which the new covalent bond is formed at the same time the old one is broken. In the displacement the nucleophile attacks the saturated carbon from the back side, that is, from the side opposite the group being displaced. Consider the reaction of hydroxide with optically active 2-bromobutane (Figure 5.8).

In this reaction, the attack of the hydroxide ion is facilitated by the polarity of the carbon-bromine bond and the reaction begins by the attraction of unlike charge. Hydroxide ion approaches and begins to supply electrons to the carbon. As the new carbon-oxygen bond is formed, the

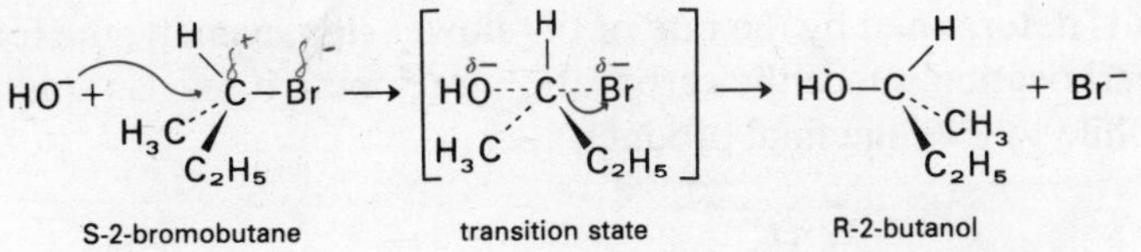

HO$^-$ + ... → [...]$^{\delta-}$ → ... + Br$^-$

S-2-bromobutane   transition state   R-2-butanol

FIGURE 5.8 Inversion of configuration during the S_N2 reaction.

carbon-bromine bond is broken. We speak of this as a concerted or simultaneous process. The final result is the formation of a new carbon-oxygen bond and the loss of a bromine as bromide ion.

In this formulation of the S_N2 reaction at a saturated carbon atom, both hydroxide ion and alkyl halide must collide. However, not all collisions between these reactants will be effective; only those which meet two stringent requirements will lead to reaction. First, the molecules must collide with sufficient energy to attain a state in which a chemical bond may be made or broken. Second, the molecules colliding must be suitably oriented with respect to each other. Since, in this displacement reaction, hydroxide becomes bonded to the carbon bearing the bromine atom, it is not likely that any collision of the nucleophile with the bromine atom itself or with either of the alkyl substituents will lead to substitution. Furthermore, even collision with proper orientation will not lead to substitution if the energy of collision is insufficient. A useful model for discussing reaction mechanisms is that of the transition state. The transition state is defined as the state of a reacting system that meets both the energy and orientation requirements for effective collision. The transition state for the reaction of hydroxide ion and 2-bromobutane is shown in Figure 5.8.

Note that in the transition state three groups are fully bonded to carbon by single covalent bonds and are planar. This is exactly what we saw earlier for an sp^2 hybridized carbon. This leaves the $2p_z$ orbital to form partial bonds with both the leaving group and the entering group. Because of the geometry of the $2p_z$ orbital, the entering group must approach the carbon atom from the side opposite the leaving group and the result is inversion of configuration. This is represented schematically in Figure 5.9.

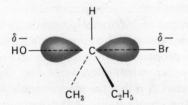

H
$\delta-$ | $\delta-$
HO — C — Br

CH$_3$ C$_2$H$_5$

FIGURE 5.9 Transition state for an S_N2 reaction showing a $2p_z$ orbital interacting with the leaving group and the entering group.

To account for the S_N1 reaction, chemists have proposed a different type of mechanism. Consider the conversion of *tert*-butyl bromide to *tert*-butyl alcohol and remember that although the reaction does involve substitution of hydroxide for bromide, the rate of the reaction does not depend at all on the concentration of the hydroxide ion.

For this type of nucleophilic substitution the first step is proposed to be slow ionization of the C—Br bond to give a carbocation, followed by a second and rapid step, the reaction of the carbocation with the nucleophile to give the alcohol. In this two-step process, the rate of formation of product is determined by the rate of the slower step, namely, the formation of the carbocation. Once the carbocation is formed, it reacts rapidly with a nucleophile to give the final product.

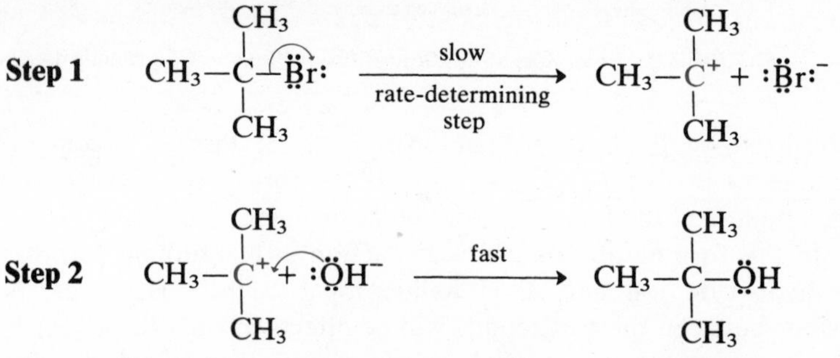

Step 1

$$CH_3-\underset{\underset{CH_3}{|}}{\overset{\overset{CH_3}{|}}{C}}-\ddot{B}r: \xrightarrow[\substack{\text{rate-determining} \\ \text{step}}]{\text{slow}} CH_3-\underset{\underset{CH_3}{|}}{\overset{\overset{CH_3}{|}}{C}}^+ + :\ddot{B}r:^-$$

Step 2

$$CH_3-\underset{\underset{CH_3}{|}}{\overset{\overset{CH_3}{|}}{C}}^+ + :\ddot{O}H^- \xrightarrow{\text{fast}} CH_3-\underset{\underset{CH_3}{|}}{\overset{\overset{CH_3}{|}}{C}}-\ddot{O}H$$

What about the stereochemistry of the S_N1 reactions? Is our mechanism consistent with the observation that S_N1 reactions of optically active alkyl halides yield racemic products? The answer is yes. To see this, think about the geometry of the carbocation. It is planar with bond angles of 120°, and therefore not a chiral center. The nucleophile can attack the carbocation with equal probability from either face of the plane of the molecule, and therefore produce a racemized product (Figure 5.10).

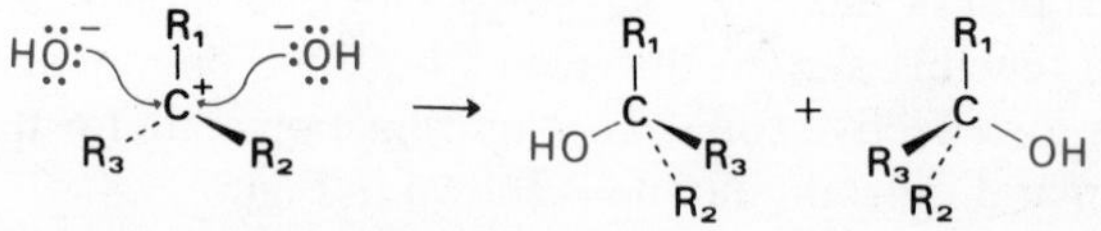

FIGURE 5.10 Attack of a nucleophile from either side of the planar carbocation to produce equal amounts of two enantiomers.

Of course, if the carbocation is bound in some manner in a chiral environment, for example, as on the surface of an enzyme, then it is very likely that the product will be entirely one enantiomer. However, in the usual laboratory reactions of carbocations, racemization is the rule.

One final comment on these nucleophilic substitutions. We have seen that the conversion of 2-bromobutane to 2-butanol proceeds by an S_N2 reaction mechanism, whereas the conversion of 2-bromo-2-methylpropane (*tert*-butyl bromide) to 2-methyl-2-propanol (*tert*-butyl alcohol) proceeds by an S_N1 process. What about other substitution reactions? By what mechanisms do they proceed? The answer is that the reaction path depends largely on the ease of forming the carbocation. If the carbocation is relatively easy to form, as in the case of the tertiary carbocation, then the reaction will proceed by the two-step process. If the carbocation is formed with difficulty, as in the case of the primary halides, then the reaction will proceed by the one-step process, the S_N2 mechanism. Further, it is possible to define experimental conditions that will favor one or the other of the two mechanisms. A high concentration of nucleophile will favor the S_N2 path; a low concentration will favor the S_N1 path. A solvent of high polarity favors the formation of the carbocation by the S_N1 path; a solvent of low polarity favors the S_N2 path. In addition, the nature of the nucleophile will affect the reaction.

5.7 Reactions of Alcohols

Alcohols of all kinds are available from a variety of sources including the hydration of alkenes (Chapter 3), reduction of aldehydes and ketones (Chapter 7), the hydrolysis of esters, and so on. Furthermore, alcohols themselves undergo a variety of important reactions including nucleophilic displacement of the hydroxyl group, dehydration, and oxidation to aldehydes, ketones and carboxylic acids, depending on the alcohol and the experimental conditions. Thus alcohols are not only readily available but they are also valuable starting materials for the synthesis of other compounds.

5.8 Conversion to Alkyl Halides

An alcohol can be converted to an <u>alkyl halide</u> by any one of several general methods. Perhaps the most common means for converting the simpler alcohols to alkyl halides is to use the hydrogen halides, HCl, HBr, or HI.

$$CH_3CH_2CH_2CH_2OH + HBr \xrightarrow[H_2O]{H_2SO_4} CH_3CH_2CH_2CH_2Br + H_2O$$

1-butanol   1-bromobutane
n-butyl alcohol   *n*-butyl bromide

$$CH_3CH_2CH_2OH + HCl \xrightarrow[reflux]{ZnCl_2} CH_3CH_2CH_2Cl + H_2O$$

1-propanol   1-chloropropane
n-propyl alcohol   *n*-propyl chloride

$$
\underset{\substack{\text{2-methyl-2-propanol}\\ \textit{tert}\text{-butyl alcohol}}}{\overset{\overset{\displaystyle CH_3}{|}}{\underset{\underset{\displaystyle CH_3}{|}}{CH_3COH}}} + HCl \xrightarrow[\text{room temp.}]{CaCl_2} \underset{\substack{\text{2-chloro-2-methylpropane}\\ \textit{tert}\text{-butyl chloride}}}{\overset{\overset{\displaystyle CH_3}{|}}{\underset{\underset{\displaystyle CH_3}{|}}{CH_3CCl}}} + H_2O
$$

$3° > 2° > 1°$

The differences in ease of conversion of <u>primary, secondary, and tertiary</u> <u>alcohols to their respective halides are dramatic.</u> The preparation of *tert*-butyl chloride is complete on shaking *tert*-butyl alcohol with concentrated HCl at room temperature for several minutes. The preparation of *n*-butyl chloride requires refluxing *n*-butyl alcohol with concentrated HCl and zinc chloride as a catalyst for several hours.

    <u>These nucleophilic displacements are the reverse of the hydrolysis of</u> <u>alkyl halides (except that the leaving group is H_2O and not OH^-) and we</u> <u>can discuss them in terms of S_N1 and S_N2 mechanisms as well.</u> With most primary alcohols, the displacement of the hydroxyl group proceeds by an S_N2 mechanism. The strong acid, HX, first converts the alcohol to an alkoxonium ion which is then displaced by the halide ion.

$$:\ddot{C}l:^- + CH_2\!-\!\overset{+}{\underset{\underset{\displaystyle CH_3}{|}}{\overset{\displaystyle H}{\ddot{O}}}}\!\!\diagdown_{\displaystyle H} \longrightarrow CH_3CH_2\!-\!\ddot{C}l: + H_2\ddot{O}:$$

In this displacement, halide ion is the nucleophile and water is the leaving group.

With tertiary alcohols, displacement of the hydroxyl group proceeds by an S_N1 mechanism. The strong acid first protonates the alcohol, which then ionizes to a tertiary carbocation and water. This ionization is the slow, rate-determining step and is immediately followed by reaction of the carbocation with halide ion.

$$CH_3-\underset{\underset{CH_3}{|}}{\overset{\overset{CH_3}{|}}{C}}-\ddot{O}-H \xrightarrow{H^+} CH_3-\underset{\underset{CH_3}{|}}{\overset{\overset{CH_3}{|}}{C}}-\overset{+}{\overset{/H}{\underset{\backslash H}{O}}} \xrightarrow[slow]{-H_2O} CH_3-\underset{\underset{CH_3}{|}}{\overset{\overset{CH_3}{|}}{\overset{+}{C}}} \xrightarrow[fast]{:\ddot{C}l:^-} CH_3-\underset{\underset{CH_3}{|}}{\overset{\overset{CH_3}{|}}{C}}-\ddot{C}l:$$

While this reaction can be used for converting alcohols to alkyl halides, it has a major disadvantage. It involves treating an alcohol with concentrated aqueous acid, often at reflux temperatures. If there are other functional groups in the same molecule that are at all sensitive to strong acid, then this method cannot be used.

Thionyl chloride, $SOCl_2$, is a particularly useful reagent for the synthesis of alkyl chlorides because the by-products, HCl and SO_2, are driven off as gases.

$$R-OH + Cl-\overset{\overset{O}{\|}}{S}-Cl \xrightarrow{-HCl} R-O-\overset{\overset{O}{\|}}{S}-Cl \xrightarrow{heat} R-Cl + SO_2$$

thionyl   an alkyl
chloride   chlorosulfite

Addition of one mole of thionyl chloride per mole of alcohol gives an alkyl chlorosulfite which generally decomposes on mild heating to give the alkyl chloride in good yield.

The phosphorus halides, PCl_5, PBr_3, and PI_3, are also used extensively to convert alcohols to alkyl halides. The by-product of the reaction with phosphorus tribromide is the high-boiling phosphorus acid.

$$3CH_3\underset{\underset{OH}{|}}{CH}CH_3 + PBr_3 \longrightarrow 3CH_3\underset{\underset{Br}{|}}{CH}CH_3 + H_3PO_3$$

phosphorus   phosphorus
tribromide   acid

5.9 Dehydration of Alcohols

An alcohol may be converted to an alkene by dehydration—that is, by elimination of a molecule of water. This dehydration requires the presence of an acid catalyst. In practice, dehydration reactions are usually carried out by heating the alcohol and sulfuric or phosphoric acid at temperatures of from 100° to 200°C. Often the water is removed from the reaction mixture by distillation. For example, the dehydration of ethanol yields ethylene.

$$CH_3CH_2OH \xrightarrow[180°C]{H_2SO_4} CH_2{=}CH_2 + H_2O$$

In the first step of such a dehydration, the hydroxyl group is protonated to form an <u>oxonium ion</u>. In the second step, the C—O bond is broken and a <u>carbocation</u> is formed. Note that this bond cleavage is made possible through the prior protonation of the hydroxyl group. Finally, the carbocation loses a proton to form the alkene.

Step 1 $H-\overset{\displaystyle H}{\underset{\displaystyle H}{C}}-\overset{\displaystyle H}{\underset{\displaystyle H}{C}}-\overset{..}{\underset{..}{O}}-H + H^+ \longrightarrow H-\overset{\displaystyle H}{\underset{\displaystyle H}{C}}-\overset{\displaystyle H}{\underset{\displaystyle H}{C}}-\overset{+}{\underset{..}{O}}-H$

Step 2 $H-\overset{\displaystyle H}{\underset{\displaystyle H}{C}}-\overset{\displaystyle H}{\underset{\displaystyle H}{C}}-\overset{+}{\underset{..}{O}}-H \longrightarrow H-\overset{\displaystyle H}{C}-\overset{+}{C}-H + H-\overset{..}{\underset{..}{O}}-H$

Step 3 $H-\overset{\displaystyle H}{\underset{\displaystyle H}{C}}-\overset{+}{C}\overset{H}{} \longrightarrow \overset{H}{}C=C\overset{H}{} + H^+$

Since the dehydration of an alcohol involves an intermediate carbocation, we would predict that the relative ease of dehydration of alcohols should be the same as the ease of formation of carbocations (Section 3.10). This prediction is borne out in experimental observations. The ease of dehydration of alcohols is of the order

$$3° > 2° > 1°$$

In dehydrations where more than one alkene is possible, it is generally the more substituted alkene that is the major product.

$$CH_3\overset{\displaystyle CH_3}{\underset{\displaystyle OH}{CHCHCH_3}} \xrightarrow[-H_2O]{H^+} \left[CH_3-\overset{\displaystyle CH_3}{CH}-\overset{+}{CH}-CH_3 \right]$$

3-methyl-2-butanol

$$\xrightarrow{-H^+} \overset{\displaystyle CH_3}{\underset{\displaystyle CH_3}{C}}=CH-CH_3$$

2-methyl-2-butene
(major product)

$$\xrightarrow{-H^+} CH_3-\overset{\displaystyle CH_3}{CH}-CH=CH_2$$

3-methyl-1-butene
(minor product)

In Section 3.11 we discussed the acid-catalyzed hydration of alkenes to yield alcohols. In this section we have discussed the dehydration of alcohols to yield alkenes. In fact, the <u>hydration-dehydration</u> reactions are reversible, and we must consider the amounts of alkene and alcohol present at equilibrium.

$$\overset{}{}C=C\overset{}{} + H_2O \rightleftharpoons -\overset{|}{\underset{|}{C}}-\overset{|}{\underset{|}{C}}- \\ \qquad\qquad\qquad H \quad OH$$

Large amounts of water favor alcohol formation, whereas operation under experimental conditions where water is removed favors formation of alkene. Depending on the conditions, it is possible to use the hydration-dehydration equilibrium to prepare either alcohols or alkenes, each in quite high yields.

5.10 Oxidation of Alcohols

Alcohols can be oxidized to aldehydes, ketones, or carboxylic acids depending on the number of hydrogen atoms attached to the carbon bearing the —OH group. The most commonly used oxidizing agents for this purpose are $K_2Cr_2O_7$ and CrO_3 (each containing Cr^{6+}) and $KMnO_4$ (containing Mn^{7+}).

Primary alcohols can also be oxidized to aldehydes. In general, aldehydes are more easily oxidized than are primary alcohols and therefore conditions that will suffice to oxidize a given primary alcohol will more than suffice to oxidize the desired aldehyde directly to the carboxylic acid. However, by the proper choice of oxidizing agent, solvent, and experimental conditions, it is possible to obtain aldehydes in good yield.

$$R-CH_2OH + Cr_2O_7^{2-} \longrightarrow R-\overset{\overset{\displaystyle O}{\|}}{C}-H + Cr^{3+}$$

Primary alcohols can be oxidized to carboxylic acids using either potassium permanganate, $KMnO_4$, or potassium dichromate, $K_2Cr_2O_7$. In this oxidation the aldehyde is an intermediate.

$$R-CH_2OH + Cr_2O_7^{2-} \longrightarrow \left[R-\overset{\overset{\displaystyle O}{\|}}{C}-H\right] \longrightarrow R-\overset{\overset{\displaystyle O}{\|}}{C}-OH + Cr^{3+}$$

primary alcohol (orange-red)   aldehyde   carboxylic acid (green)

$$CH_3CH_2OH + Cr_2O_7^{2-} \longrightarrow \left[CH_3-\overset{\overset{\displaystyle O}{\|}}{C}-H\right] \longrightarrow CH_3-\overset{\overset{\displaystyle O}{\|}}{C}-OH + Cr^{3+}$$

ethanol   acetic acid

The oxidation of secondary alcohols yields ketones. Ketones are resistant to further oxidation and normally no special precautions are needed in their preparation.

$$\underset{\text{secondary alcohol}}{R-\overset{\overset{\displaystyle OH}{|}}{C}H-R'} \xrightarrow{K_2Cr_2O_7} \underset{\text{ketone}}{R-\overset{\overset{\displaystyle O}{\|}}{C}-R'}$$

menthol $\xrightarrow[\text{heat}]{K_2Cr_2O_7}$ menthone

Tertiary alcohols are stable to oxidation except that if the oxidizing agent is acidic, the alcohol may be dehydrated (Section 5.9); the resulting alkene will be susceptible to oxidative attack.

There is a large number of reagents that may be used for the laboratory oxidation of alcohols. We have seen two of these, $K_2Cr_2O_7$ and $KMnO_4$. It should come as no surprise that neither of these reagents is involved in the very important oxidation reactions carried out in biological systems. Here the most important agents for enzyme catalyzed oxidations are <u>molecular oxygen</u> and <u>nicotinamide adenine dinucleotide</u>, abbreviated NAD^+

The reactive group of NAD^+ in biological oxidation is the pyridine ring (Section 6.8). This ring can accept two electrons and one proton to form the reduced structure NADH

NAD^+ is involved in a great many biological oxidation-reduction reactions. Three examples will show how it is involved in the oxidation of a primary alcohol, a secondary alcohol, and an aldehyde. In each of these oxidations, the reaction is catalyzed by a specific enzyme.

5.11 Synthesis and Reactions of Ethers

Most of the commercially available ethers are synthesized by the <u>acid-catalyzed dehydration of alcohols</u>. Typical is the formation of diethyl ether from ethanol.

Recall from Section 5.9 that acid-catalyzed dehydration of alcohols can also yield alkenes. Generally, it is possible to maximize the formation of either ether or alkene by the proper choice of experimental conditions. For example, at 140°C the formation of ethylene from ethanol is minimized and diethyl ether is the major product. At 180°C, however, ethylene becomes the major product.

Since, a single alcohol is used as the starting material, intermolecular dehydration yields ethers having identical alkyl groups. Such ethers are called symmetrical ethers. Furthermore, the preparation of ethers by acid-catalyzed dehydration of alcohols is most useful for the synthesis of ethers from primary alcohols. Secondary and tertiary alcohols are more readily converted to carbocations under the conditions of the reaction and alkene formation competes even more strongly with ether formation. This difficulty can be avoided by using another, very versatile method for ether formation, the so-called Williamson ether synthesis. This scheme can be used for the synthesis of symmetrical as well as unsymmetrical ethers.

The Williamson synthesis involves nucleophilic displacement (S_N2) of an alkyl halide by a metal alkoxide as, for example, the synthesis of methyl isopropyl ether from methyl iodide and sodium isopropoxide.

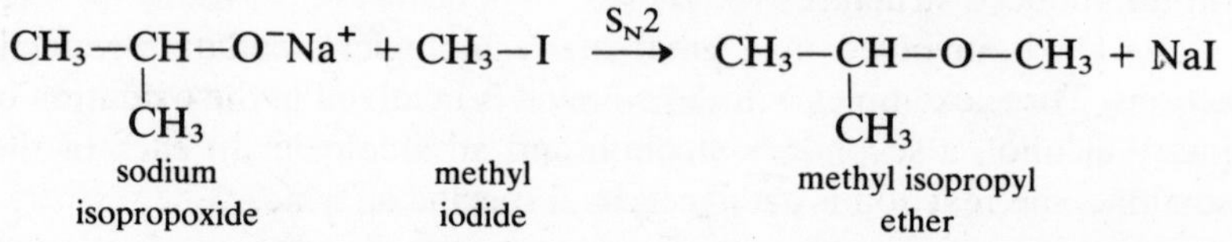

$$CH_3-\underset{\underset{CH_3}{|}}{CH}-O^-Na^+ + CH_3-I \xrightarrow{S_N2} CH_3-\underset{\underset{CH_3}{|}}{CH}-O-CH_3 + NaI$$

| sodium | methyl | methyl isopropyl |
| isopropoxide | iodide | ether |

Recall from Section 5.4 that the desired metal alkoxides are readily available by the reaction of alcohols with metallic sodium or potassium.

Obviously, the Williamson synthesis can be used for the preparation of any number of unsymmetrical ethers. There is, however, a side reaction that must be considered in planning any ether synthesis by this method. In addition to being good nucleophiles, metal alkoxides are also good bases and can bring about dehydrohalogenation of alkyl halides to form alkenes. Therefore, in planning a Williamson synthesis it is best to use a combination of alkyl halide and metal alkoxide that will maximize the desired S_N2 reaction and thereby minimize the formation of alkene side-products. Recall from Section 5.6 that S_N2 reactions are most favorable when the halide is displaced from a primary rather than a secondary or tertiary carbon atom. With this in mind, let us plan a synthesis for ethyl *tert*-butyl ether. There are two possible choices of starting materials, either an ethyl halide and a metal *tert*-butoxide, or a *tert*-butyl halide and a metal ethoxide. Each path is illustrated below.

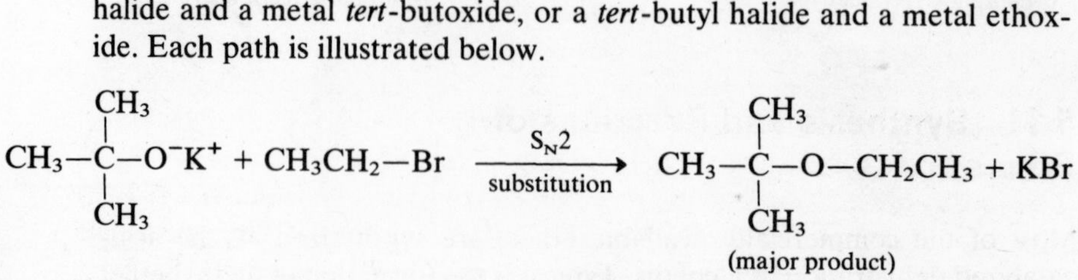

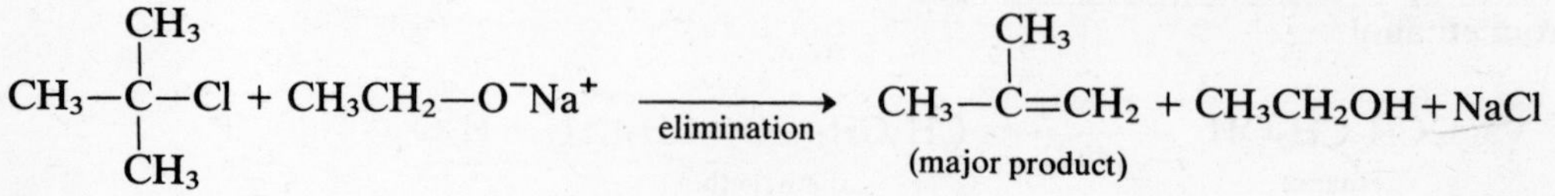

$$CH_3-\underset{\underset{CH_3}{|}}{\overset{\overset{CH_3}{|}}{C}}-O^-K^+ + CH_3CH_2-Br \xrightarrow[\text{substitution}]{S_N2} CH_3-\underset{\underset{CH_3}{|}}{\overset{\overset{CH_3}{|}}{C}}-O-CH_2CH_3 + KBr$$

(major product)

$$CH_3-\underset{\underset{CH_3}{|}}{\overset{\overset{CH_3}{|}}{C}}-Cl + CH_3CH_2-O^-Na^+ \xrightarrow[\text{elimination}]{} CH_3-\underset{}{\overset{\overset{CH_3}{|}}{C}}=CH_2 + CH_3CH_2OH + NaCl$$

(major product)

The yield of the desired ethyl *tert*-butyl ether is greatest using a primary halide. Since yields of substitution product over elimination product are greatest with primary halides, the Williamson synthesis of ethers should always use primary halides whenever possible.

As intermediates in chemical synthesis, ethers are far less important than alcohols. In fact ethers resemble the hydrocarbons in their resistance to other types of chemical reaction. They do not react with oxidizing agents such as potassium dichromate, potassium permanganate, or ozone. They are not affected by either strong acids or bases at moderate temperatures. It is precisely this general inertness to chemical reaction on the one hand and good solvent characteristics on the other which make ethers excellent solvents in which to carry out many organic reactions.

Two hazards must be avoided when working with ethers. First, ethers of low molecular weight are highly flammable. Consequently sparks or open flames must be avoided in any area where such an ether is being used. In addition, ether fires cannot be extinguished with water because they float on top of water. Fortunately, carbon dioxide extinguishers are effective. Second, ethers react with oxygen of the air to form peroxides. These peroxides are of higher molecular weight than the parent ethers and accordingly are less volatile. They tend to become concentrated as ether is distilled or evaporated. In concentrated or solid form these peroxides are dangerous because they are highly explosive. Commonly used ethers such as diethyl ether and tetrahydrofuran often become contaminated with peroxides on prolonged storage and exposure to air and light. For this reason purification of ethers by treatment with a reducing agent such as alkaline ferrous sulfate is frequently necessary before use.

5.12 Ether and Anesthesia*

Prior to the middle of the 19th century, surgery was performed only when absolutely necessary because there were no truly effective general anesthetics. More often than not, patients were drugged with certain alkaloids, hypnotized, or simply tied down. In 1772 Joseph Priestly isolated nitrous oxide. In 1799 Sir Humphrey Davy demonstrated its anesthetic effect and named it "laughing gas." In 1844 an American dentist, Horace Wells, administered nitrous oxide while extracting teeth. His first anesthetic trials were successful and he soon introduced nitrous oxide into general dental practice. However, one patient awakened prematurely, screaming with pain, and another died. Wells was forced to withdraw from practice, became embittered, depressed, and finally insane. He committed suicide at the age of 33. In the same period, a Boston chemist, Charles Jackson, etherized himself with diethyl ether, and persuaded a dentist, William Morton, to use it. Subsequently they persuaded a surgeon, John Warren, to give a public demonstration of surgery under anesthesia. The operation was completed painlessly, and soon general anesthesia by diethyl ether was widely adopted for surgical operations. Morton patented his preparation,

---

* Adapted from Lloyd N. Ferguson, *Organic Chemistry, A Science and An Art*, (Boston: Willard Grant Press, 1972).

but soon a great "ether controversy" arose. A Georgia physician, Crawford Long, had used diethyl ether anesthesia in surgery and obstetrics several years prior to the Morton-Warren experiment. However, he did not publish his observations or try to bring diethyl ether into general anesthesia until 1848, when he gave a report to the Georgia Medical Society. Long, Morton, Jackson, and the descendents of Wells all made claims for the $100,000 prize being offered by the U.S. Congress for development of a safe, effective anesthetic. The controversy continued for years and the award was never made.

Diethyl ether is still the most widely used anesthetic because it is easy to administer and causes excellent muscle relaxation. If any other agent had been used with the careless methods of earlier times, anesthetic mortality would have been tremendous. Blood pressure and rate of pulse and respiration are usually only slightly affected by ethyl ether. Diethyl ether's chief drawbacks are its irritating effect on the respiratory passages and its aftereffect of nausea. Nitrous oxide is pleasant to inhale and rapid in its action, and it is used frequently to induce unconsciousness and pave the way for more potent drugs such as diethyl ether.

Divinyl ether does not have a nauseous aftereffect and it is a rapid-acting, short-term anesthetic. It is favored by doctors for home and office use, but must be used with caution to prevent too deep a level of unconsciousness. Methylpropyl ether is claimed to be less irritating and more potent than ethyl ether. Cyclopropane, which produces deep anesthesia, would be about the best anesthetic if it were not explosive and did not require so much skill in its administration to avoid deleterious effects on the heart and lungs. Another recent, widely used anesthetic is Halothane, $C_2HBrClF_3$. It is nonflammable, nonexplosive, and causes minimum discomfort to the patient.

The mechanism of action of anesthetics is unknown. Correlations have been observed between anesthetic potency and certain physical properties such as oil/water distribution coefficients, water solubility, vapor pressure, adsorbability, and stability of clathrate hydrates. These generalizations have led to various theories about the mode of action of anesthetics, but correlations do not prove a theory, so that no one knows yet how these agents produce the state of unconsciousness.

5.13 Epoxides

The synthesis and reactions of ethylene oxide deserve special mention. Ethylene oxide and other three-membered ring ethers are classed together as epoxides and they show chemical reactivities quite different from those of other ethers.

$$H_2C-CH_2$$

epoxyethane
ethylene oxide

1,2-epoxycyclopentane
cyclopentene oxide

$$Cl-CH_2CH-CH_2$$

3-chloro-1,2-epoxypropane
epichlorohydrin

We have already noted one preparation of ethylene oxide by the silver-catalyzed air oxidation of ethylene (Section 5.5). Epoxides can also be synthesized in a two-step process beginning with the addition of hypochlorous acid to an alkene to form a chlorohydrin.

$$CH_2{=}CH_2 + HOCl \longrightarrow \underset{\underset{OH \quad Cl}{|\qquad|}}{CH_2{-}CH_2}$$

ethylene
chlorohydrin

Treatment of the chlorohydrin with strong base results in the elimination of the elements of HCl and formation of an epoxide.

$$\underset{\underset{Cl}{|}}{\overset{\overset{HÖ:}{|}}{CH_2{-}CH_2}} \xrightarrow{OH^-} \underset{\underset{:\ddot{C}l:}{\wedge}}{\overset{\overset{:\ddot{O}:^-}{|}}{CH_2{-}CH_2}} \longrightarrow \underset{O}{\overset{\cdot\ddot{O}\cdot}{CH_2{-}CH_2}} + :\ddot{C}l:^-$$

ethylene oxide

This second step in the sequence is an internal S_N2 reaction with the alkoxide ion displacing chloride ion.

The value of epoxides in laboratory and industrial chemistry lies in the fact that they can be prepared easily from alkenes and in turn undergo facile ring openings with a wide variety of nucleophiles. This versatility is well illustrated by ethylene oxide. With the abundant supplies of ethylene available from thermal and catalytic cracking processes in the petroleum industry and the development of the technology for the direct conversion of ethylene to ethylene oxide, this substance has grown from a laboratory curiosity to a 4 billion pound-per-year intermediate used in the production of thousands of consumer products. The largest use of ethylene oxide (50%) is ethylene glycol whose major uses are in automobile antifreeze and polyester fibers and films. Ethylene oxide is also the starting point for the synthesis of many other materials including ethanolamines, glycol ethers, and polyethylene glycols. These derivatives are used in the production of synthetic rubbers, synthetic fibers, resins, paints, adhesives, molded articles, solvents, brake fluids, and cosmetics.

$$CH_2{-}CH_2 \quad\begin{cases} \xrightarrow{\;H_2O\;} & HOCH_2CH_2OH \quad (antifreeze) \\ & \text{ethylene glycol} \\[4pt] \xrightarrow{\;NH_3\;} & H_2NCH_2CH_2OH \\ & \text{ethanolamine} \\[4pt] \xrightarrow{\;HOCH_2CH_2OH\;} & HOCH_2CH_2OCH_2CH_2OH \\ & \text{diethylene glycol} \\[4pt] \xrightarrow{\;CH_3OH\;} & CH_3OCH_2CH_2OH \\ & \text{2-methoxyethanol} \end{cases}$$

5.14 Thiols

Sulfur analogs of alcohols are known as thiols, thioalcohols, or mercaptans. The —SH group is called a sulfhydryl group. These compounds may also be considered as alkyl derivatives of hydrogen sulfide. Like H_2S, they have atrocious odors, they are weakly acidic, and they are easily oxidized.

Mercaptans may be prepared by heating an alkyl halide with sodium hydrosulfide.

$$CH_3CH_2Cl + NaSH \longrightarrow CH_3CH_2SH + NaCl$$

ethanethiol
ethyl mercaptan
bp 35°C

This reaction is an example of nucleophilic substitution at saturated carbon.

Since they are weakly acidic, thiols yield salts with bases and form insoluble precipitates with lead, mercury, and other heavy metal cations.

$$2RSH + HgCl_2 \longrightarrow (RS)_2Hg + 2HCl$$

Many enzymes contain sulfhydryl groups and can be precipitated or inactivated through the mercury salt.

When treated with mild oxidizing agents such as dilute hydrogen peroxide or cupric salts, thiols are converted into disulfides:

$$2CH_3CH_2SH + H_2O_2 \longrightarrow CH_3CH_2S-SCH_2CH_3 + 2H_2O$$

ethyl disulfide

Disulfides are analogous to peroxides but are much more stable. They are less volatile than thiols and have less offensive odors.

5.15 Spectroscopic Properties of Alcohols and Ethers

Alcohols show characteristic absorption of infrared radiation associated with stretching of the O—H and C—O bonds. The position of the O—H stretching frequency depends on whether the hydroxyl group is free or hydrogen-bonded. In dilute solutions of alcohols in carbon tetrachloride (a convenient solvent for recording infrared spectra) the unassociated O—H bond absorbs at 3600 to 3650 cm^{-1}. However, in more concentrated solutions in which the hydroxyl group is more extensively hydrogen bonded, the position of the O—H stretching frequency is shifted and appears as a broad peak in the region 3200 to 3400 cm^{-1}. Stretching of the C—O bond appears between 1000 cm^{-1} and 1300 cm^{-1}. In the infrared spectrum of 1-hexanol (Figure 5.11) notice the broad peak at 3330 cm^{-1} associated with stretching of the hydrogen-bonded O—H and the broad peak at 1050 cm^{-1} associated with C—O stretching. The peaks between 2900 cm^{-1} and 3000 cm^{-1} are associated with stretching of the various C—H bonds within the molecule.

Ethers do not contain an O—H functional group and hence show no absorption between 3200 to 3600 cm^{-1}. The presence or absence of absorption in this region of the infrared spectrum can be used to distinguish alcohols from isomeric ethers. Ethers do, however, show absorption in the region 1000 to 1300 cm^{-1} due to C—O stretching.

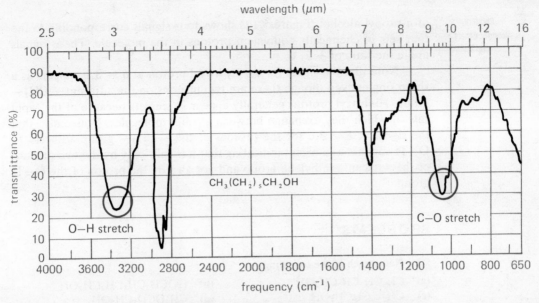

FIGURE 5.11 An infrared spectrum of 1-hexanol.

Neither saturated alcohols nor ethers absorb ultraviolet radiation. In fact, ethyl alcohol and diethyl ether are commonly used solvents for the recording of ultraviolet spectra.

In the NMR spectrum, the proton of the O—H group generally appears as a singlet anywhere within the range $\delta = 1$ to 5. Often the singlet is broad and difficult to see. The exact position and sharpness of the NMR signal depends on the extent of hydrogen bonding, temperature, and structure of the particular alcohol under examination. The NMR spectrum

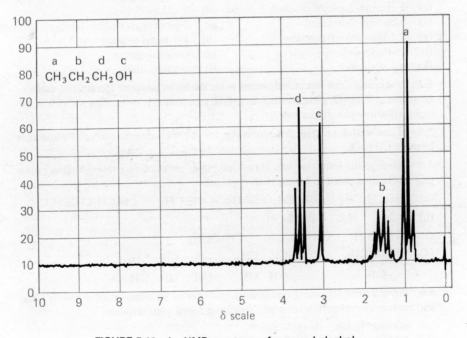

FIGURE 5.12 An NMR spectrum of n-propyl alcohol.

of *n*-propyl alcohol (Figure 5.12) shows four signals corresponding to the four sets of chemically equivalent protons in the molecule. These signals are in the ratio of $3:2:1:2$.

Note that the signal for the hydroxyl proton at $\delta = 3.1$ appears as a sharp singlet even though there are two protons on the adjacent $-CH_2-$ group. Hydroxyl protons generally appear as singlets because of the rapid rate at which they exchange between alcohol molecules. Notice also that the signal at $\delta = 3.6$ for the protons of the $-CH_2-$ group adjacent to O$-$H appears as a triplet indicating that it is split only by the two protons of the adjacent methylene group and not also by the proton of the O$-$H group.

PROBLEMS

5.1 Name each of the following:

(a) $CH_3CH_2CH_2CH_2OH$

(b) $HOCH_2CH_2CH_2CH_2OH$

(c) $CH_2{=}CHCH_2OH$

(d) $CH_3OCH_2CH_2OH$

(e)

(f)

(g) $CH_3CH_2CH_2OCHCH_3$ with CH_3 below

(h) $H_2NCH_2CH_2OH$

5.2 Write structural formulas for each of the following:

(a) methylisopropyl ether

(b) propylene glycol

(c) 2-methyl-2-propylpropane-1,3-diol

(d) 1-chloro-2-hexanol

(e) 5-methyl-2-hexanol

(f) 2-isopropyl-5-methyl-1-cyclohexanol

(g) 2,2-dimethyl-1-propanol

(h) *tert*-butyl alcohol

(i) methylcyclopropyl ether

(j) ethylene glycol

(k) wood alcohol

(l) grain alcohol

5.3 Name and draw structural formulas for the eight isomeric alcohols of molecular formula $C_5H_{12}O$. Classify each as primary, secondary, or tertiary. Which of the eight will show enantiomerism?

5.4 Name and draw structural formulas for the six isomeric ethers of molecular formula $C_5H_{12}O$.

5.5 Arrange the following sets of compounds in order of decreasing boiling points (highest boiling point to lowest boiling point).

(a) $CH_3CH_2CH_3$ $CH_3CH_2CH_2CH_2CH_2CH_2CH_3$ $CH_3CH_2CH_2CH_2CH_3$

(b) N_2H_4 H_2O_2 CH_3CH_3

(c) CH_3CO_2H CH_3CH_2OH CH_3OCH_3

(d) CH_3CHCH_3 with OH below; $CH_3{-}CH{-}CH_2$ with OH OH below; $CH_2{-}CH{-}CH_2$ with OH OH OH below

5.6 Arrange the following compounds in order of decreasing solubility in water and explain the principles on which you have based your answers.

(a) ethanol; butane; diethyl ether

(b) 1-hexanol; 1,2-hexanediol; hexane

5.7 Diethyl ether has a much lower boiling point than *n*-butyl alcohol, yet each of these compounds shows about the same solubility in water, 8 grams per 100 ml of water. How do you account for this observation?

$$CH_3CH_2CH_2CH_2OH \qquad CH_3CH_2OCH_2CH_3$$

n-butyl alcohol
bp 117°C

diethyl ether
bp 35°C

5.8 How would you account for the fact that ethane CH_3CH_3 is a gas (bp −88°C), whereas hydrazine, N_2H_4, is a liquid (bp 113.5°C)?

5.9 Both propanoic acid and methyl acetate have the molecular formula $C_3H_6O_2$. One of these compounds is a liquid, boiling point 141°C. The other is also a liquid, boiling point 57°C.

$$CH_3-CH_2-\overset{\overset{\textstyle O}{\|}}{C}-OH \qquad CH_3-\overset{\overset{\textstyle O}{\|}}{C}-O-CH_3$$

propanoic acid

methyl acetate

(a) Which of these two compounds would you predict to have the boiling point of 141°C; the boiling point of 57°C? Explain the basis for your prediction.

(b) Which of these two compounds would you predict to be the more soluble in water? Explain.

5.10 Compounds that contain the N—H bond also show considerable evidence of association. Would you expect this association to be stronger or weaker than that in compounds containing O—H groups? (Hint: remember the table of relative electronegativities.)

5.11 What is "absolute" ethanol? How is absolute ethanol prepared from the more readily available 95% ethanol?

5.12 Show the products of the reaction (if any) of 1-propanol (*n*-propyl alcohol) with:

(a) Na

(b) $SOCl_2$

(c) $Na_2Cr_2O_7$ (H^+, heat)

(d) PBr_3

(e) H_2SO_4 (0°)

(f) H_3PO_4 (heat)

(g) O_3, then Zn/H_2O

(h) HCl ($ZnCl_2$)

5.13 Show the products of the reaction (if any) of cyclohexanol with:

(a) Na

(b) HBr (heat)

(c) $Na_2Cr_2O_7$ (H^+, heat)

(d) H_2SO_4 (0°)

(e) H_3PO_4 (heat)

(f) H_2/Pt

(g) $SOCl_2$

(h) PI_3

5.14 What is meant by the terms S_N1 and S_N2? Compare and contrast S_N1 and S_N2 reactions in terms of (a) rate-determining step, (b) the stereochemistry of the product as related to that of the starting material.

5.15 Hydrolysis of 2-bromooctane yields 2-octanol:

$$CH_3(CH_2)_5\underset{\underset{\textstyle Br}{|}}{C}HCH_3 + H_2O \longrightarrow CH_3(CH_2)_5\underset{\underset{\textstyle OH}{|}}{C}HCH_3 + HBr$$

2-bromooctane

2-octanol

Assume that the starting material is R-2-bromooctane. Use suitable stereorepresentations to show the course of this reaction and the stereochemistry of the product if the reaction takes place by an S_N1 mechanism; by an S_N2 mechanism.

5.16 Predict the relative ease with which the alcohols below will undergo acid-catalyzed dehydration. Draw a structural formula for the major product of each dehydration.

$$CH_3-\underset{\underset{\textstyle |}{\overset{\overset{\textstyle OH}{|}}{C}}H}-\underset{\underset{\textstyle CH_3}{|}}{C}H-CH_2-CH_3$$

$$CH_3-\underset{\overset{\textstyle H_3C}{|}}{C}H-\underset{\underset{\textstyle CH_3}{|}}{\overset{\overset{\textstyle OH}{|}}{C}}-CH_2-CH_3$$

$$CH_2-CH_2OH$$

5.17 Complete the following substitution reactions showing major organic products.

(a) 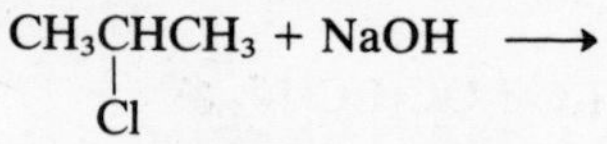 CH$_3$CHCH$_3$ + NaOH $\longrightarrow$
|
Cl

(b) 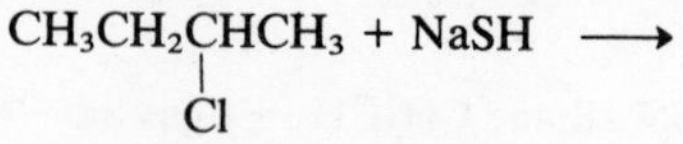 CH$_3$CH$_2$CHCH$_3$ + NaSH $\longrightarrow$
|
Cl

(c) 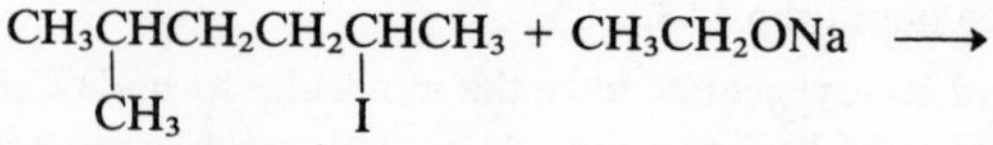 CH$_3$CHCH$_2$CH$_2$CHCH$_3$ + CH$_3$CH$_2$ONa $\longrightarrow$
| |
CH$_3$ I

(d) 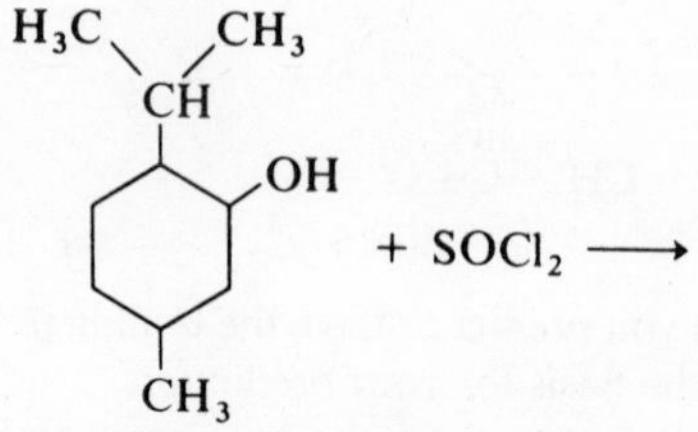 H$_3$C, CH$_3$
   \ /
    CH
     OH
          + SOCl$_2$ $\longrightarrow$

    CH$_3$

(e) CH$_2$=CHCH$_2$OH + PBr$_3$ $\longrightarrow$

(f) CH$_2$=CHCH$_2$Cl + CH$_3$ONa $\longrightarrow$

(g) cyclohexanol + HCl $\xrightarrow[\text{reflux}]{\text{CaCl}_2}$

(h) 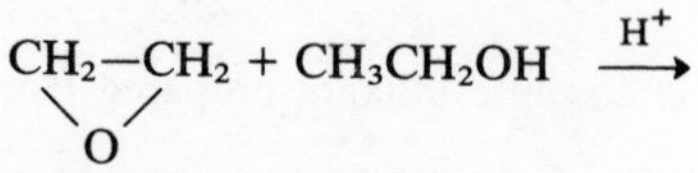 CH$_2$—CH$_2$ + CH$_3$CH$_2$OH $\xrightarrow{\text{H}^+}$
     \   /
       O

(i) 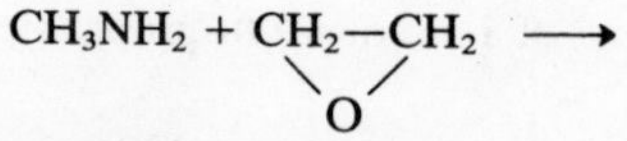 CH$_3$NH$_2$ + CH$_2$—CH$_2$ $\longrightarrow$
                   \   /
                     O

(j) CH$_3$OCH$_2$CH$_2$OH + PBr$_3$ $\longrightarrow$

5.18 Upon treatment with sulfuric acid, an equimolar mixture of ethyl alcohol and isopropyl alcohol yields not only ethyl isopropyl ether but two other ethers as well. Draw structural formulas for these two other ethers.

5.19 Draw structural formulas for the major products of each reaction:

        CH$_3$
         |
(a) 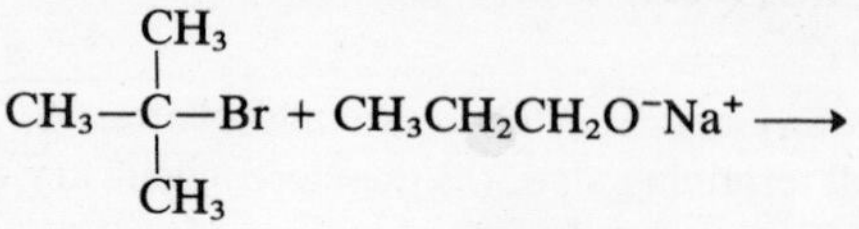 CH$_3$—C—Br + CH$_3$CH$_2$CH$_2$O$^-$Na$^+$ $\longrightarrow$
         |
        CH$_3$

        CH$_3$
         |
(b) 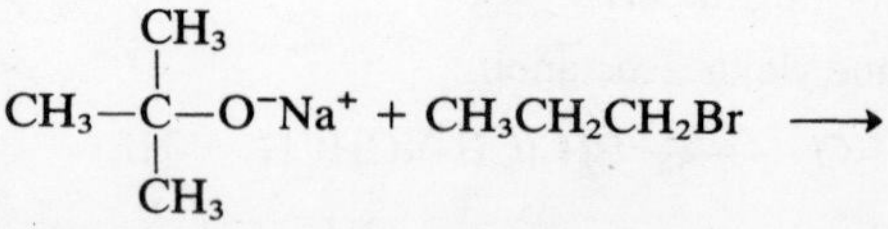 CH$_3$—C—O$^-$Na$^+$ + CH$_3$CH$_2$CH$_2$Br $\longrightarrow$
         |
        CH$_3$

5.20 Write equations to show the best combination of reactants to prepare the following ethers by the Williamson ether synthesis.

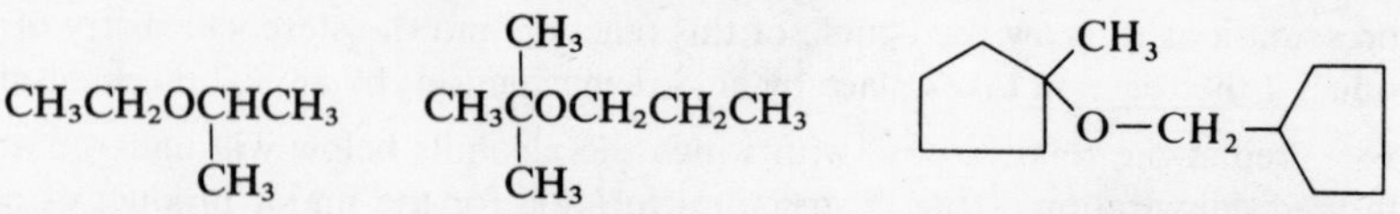

                          CH$_3$
                           |
CH$_3$CH$_2$OCHCH$_3$ CH$_3$COCH$_2$CH$_2$CH$_3$
         | |
        CH$_3$ CH$_3$

        CH$_3$

        O—CH$_2$

5.21 Starting with 1-butene and 2-methylpropene, show how you might synthesize each of the four isomeric alcohols for molecular formula C$_4$H$_{10}$O.

5.22 Each of the following conversions can be carried out in three steps or fewer. Show the reagents you would use, and draw structural formulas for intermediate compounds formed in each conversion.

(a)

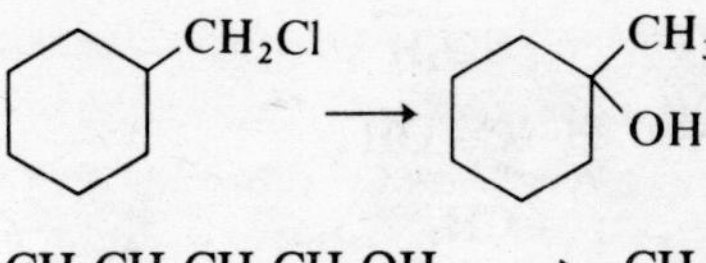

(b) $CH_3CH_2CH_2CH_2OH \longrightarrow CH_3CH_2CH_2CH_2OCH_2CH_2CH_2CH_3$

(c)

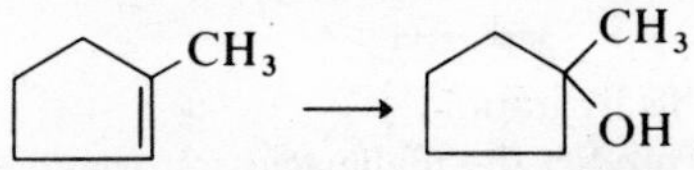

(d)

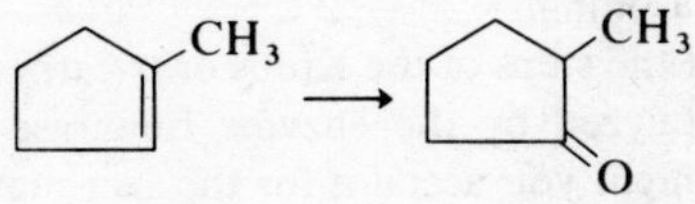

(e)

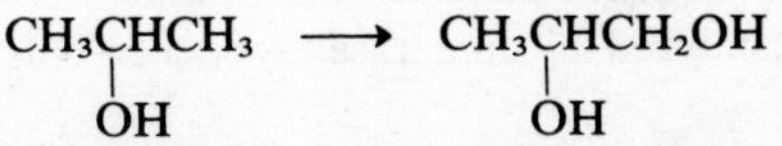

$$\underset{\underset{OH}{|}}{CH_3CHCH_3} \longrightarrow \underset{\underset{OH}{|}}{CH_3CHCH_2OH}$$

(f) $CH_2{=}CH_2 \longrightarrow CH_3CH_2OCH_2CH_2OH$

(g) 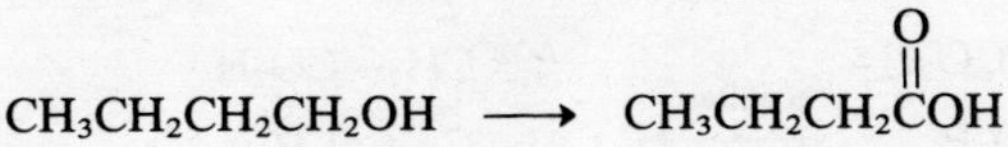 $CH_3CH_2CH_2CH_2OH \longrightarrow CH_3CH_2CH_2\overset{\overset{O}{\|}}{C}OH$

(h) 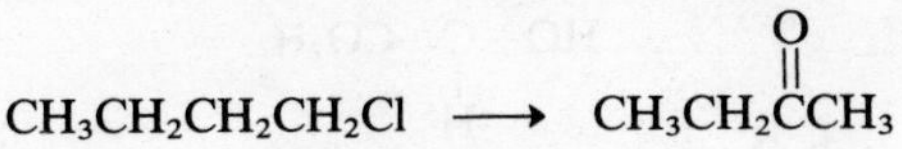 $CH_3CH_2CH_2CH_2Cl \longrightarrow CH_3CH_2\overset{\overset{O}{\|}}{C}CH_3$

(i) 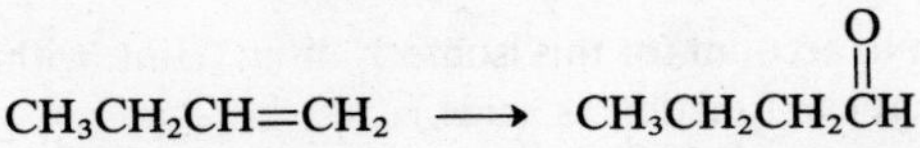 $CH_3CH_2CH{=}CH_2 \longrightarrow CH_3CH_2CH_2\overset{\overset{O}{\|}}{C}H$

(j) 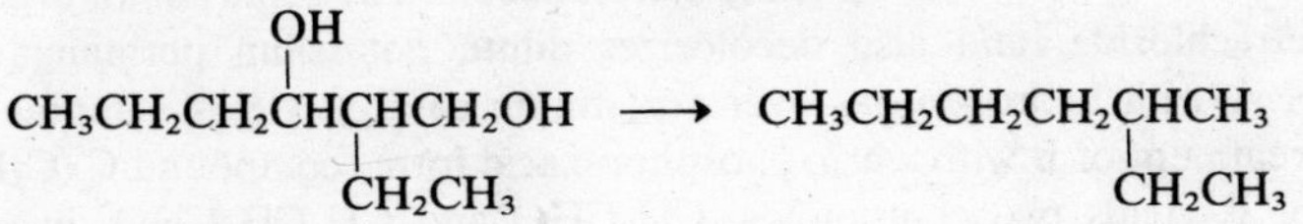 $\underset{\underset{CH_2CH_3}{|}}{CH_3CH_2CH_2\overset{\overset{OH}{|}}{C}HCHCH_2OH} \longrightarrow \underset{\underset{CH_2CH_3}{|}}{CH_3CH_2CH_2CH_2CHCH_3}$

(k)

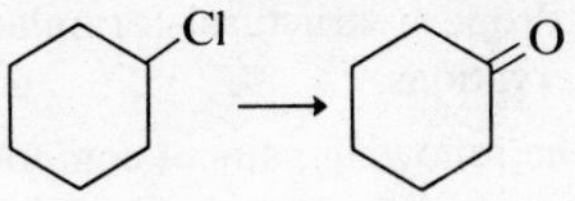

(l)

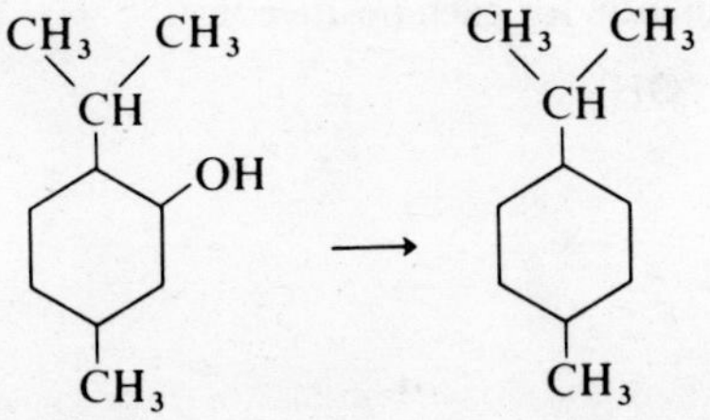

(m) 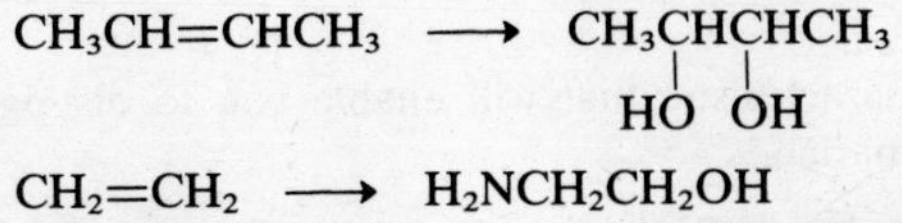 $CH_3CH{=}CHCH_3 \longrightarrow \underset{\underset{HO \quad OH}{|\quad\;|}}{CH_3CHCHCH_3}$

(n) $CH_2{=}CH_2 \longrightarrow H_2NCH_2CH_2OH$
(o) $CH_3CH_2CH_2CH_2Br \longrightarrow CH_3CH_2CH_2CH_2SH$

5.23 Propose a mechanism for the acid-catalyzed dehydration of cyclohexanol to give cyclohexene and water. One of the by-products of this reaction is dicyclohexyl ether. How might you account for the formation of this by-product?

5.24 Hydration of <u>fumaric acid</u>, catalyzed by aqueous sulfuric acid, produces <u>malic acid</u>.

$$
\begin{array}{ccc}
\underset{|}{CO_2H} & & \underset{|}{CO_2H} \\
CH & \xrightarrow{H_2SO_4} & H-C-OH \\
\| & +H_2O & | \\
CH & & CH_2 \\
| & & | \\
CO_2H & & CO_2H
\end{array}
$$

fumaric acid   malic acid

(a) Propose a reasonable mechanism for this hydration.

(b) Based on your mechanism, would you predict the malic acid so formed to be optically active or racemic? Explain your reasoning.

(c) The hydration of fumaric acid is one of the steps of the Krebs or tricarboxylic acid cycle. The biological hydration is catalyzed by the enzyme fumarase and produces optically active malic acid. How might you account for the fact that the hydration of fumaric acid catalyzed by aqueous sulfuric acid produces racemic malic acid while the same reaction, catalyzed by fumarase, produces only one enantiomer?

5.25 One of the reactions in the metabolism of glucose is the isomerization of citric acid to isocitric acid. The isomerization is catalyzed by the enzyme <u>aconitase</u>.

$$
\begin{array}{ccc}
CH_2-CO_2H & & CH_2-CO_2H \\
| & \xrightarrow[\text{aconitase}]{} & | \\
HO-C-CO_2H & \rightleftharpoons & H-C-CO_2H \\
| & & | \\
CH_2-CO_2H & & HO-C-CO_2H \\
& & | \\
& & H
\end{array}
$$

citric acid   isocitric acid

Propose a reasonable mechanism to account for this isomerization. (Hint: within its structure aconitase has groups that can function as acids.)

5.26 Compound A ($C_5H_{10}O$) is optically active, decolorizes a solution of bromine in carbon tetrachloride, and also decolorizes dilute potassium permanganate. Treatment of A with hydrogen gas over a platinum catalyst yields compound B ($C_5H_{12}O$). Treatment of B with warm phosphoric acid forms compound C (C_5H_{10}). Ozonolysis of C yields two compounds, CH_3CHO and CH_3CH_2CHO, in equal amounts. Only compound A is optically active. Propose structural formulas for compounds A, B, and C consistent with these observations.

5.27 Show how you might distinguish between the following pairs of compounds by a simple chemical test. In each case, tell what test you would perform, what you would expect to observe, and write an equation for each positive test.

(a)

and

(b)

and

(c) CH_3CH_2OH and $CH_3OCH_2CH_2OCH_3$

5.28 List one major spectral characteristic that will enable you to distinguish between the following pairs of compounds.

(a)

and   (NMR, IR)

(b)

and   (NMR, IR)

(c) CH_3CH_2OH and $CH_3OCH_2CH_2OCH_3$ (NMR, IR)

5.29 Predict the number of signals and the splitting pattern of each signal in the NMR spectrum of 2,2-dimethyl-1-propanol (neopentyl alcohol). Also predict the ratio of the areas under the signals.

5.30 An alcohol of molecular formula $C_5H_{12}O$ shows signals in the NMR spectrum at $\delta = 0.9$, 1.2, 1.7, and 5.0. The areas of these signals are in the ratio $3:6:2:1$.

(a) Draw a structural formula for this alcohol.

(b) Predict the splitting pattern of each signal.

6

Benzene and the Concept of Aromaticity

6.1 Introduction

<u>Benzene</u> is a liquid with a boiling point of 80°C. It was first isolated by Michael Faraday in 1825 from the oily liquid that collected in the illuminating gas lines of London. Its molecular formula, C_6H_6, suggests a high degree of unsaturation—remember that a saturated alkane would have the formula C_6H_{14}, and a saturated cycloalkane would have the formula C_6H_{12}. With this high degree of unsaturation you might expect that benzene would be highly reactive and show reactions characteristic of alkenes and alkynes. Surprisingly, benzene does not undergo characteristic alkene reactions. For example, benzene does not react readily with bromine, hydrogen chloride, hydrogen bromide, or other reagents that usually add to double and triple bonds. Furthermore, it is unaffected by the usual oxidizing agents. When it does react, benzene typically does so by substitution. In the presence of bromine and ferric bromide, benzene forms <u>bromobenzene</u> and hydrogen bromide. With fuming sulfuric acid (sulfuric acid containing dissolved SO_3) at room temperature or with concentrated sulfuric acid at elevated temperature, benzene undergoes a substitution reaction producing <u>benzenesulfonic acid</u>. The reaction is known as sulfonation.

$$C_6H_6 + Br_2 \xrightarrow{\text{FeBr}_3} \underset{\text{bromobenzene}}{C_6H_5Br} + HBr$$

$$C_6H_6 + SO_3 \xrightarrow[\text{room temp.}]{\text{H}_2\text{SO}_4\text{(conc)}} \underset{\text{benzenesulfonic acid}}{C_6H_5-SO_3H}$$

139

The terms aromatic and aromatic compounds were used by Kekulé to classify benzene and a number of its derivatives because many of them have rather pleasant odors. However, after a time it became clear that a sounder classification for these compounds should be based not on aroma but rather on chemical reactivities. Currently, the term aromatic is used to refer to the unusual chemical stability of benzene and its derivatives. This chemical inertness of benzene and other aromatic hydrocarbons—a resistance to uncatalyzed halogenation, a resistance to oxidation, and a tendency to react by substitution rather than addition—was both strikingly evident and truly puzzling to Kekulé and his contemporaries.

Let us put ourselves in the mid-19th century and examine the evidence on which chemists attempted to build an adequate model for the structure of benzene. First, it was clear that the molecular formula of benzene is C_6H_6. In the light of Kekulé's theory of a characteristic valence of four for carbon, it seemed evident that the benzene molecule should be highly unsaturated. Yet as we have already demonstrated, benzene does not show the chemical reactivity of the unsaturated compounds known at that time, namely the alkenes. Benzene does undergo reactions, but substitution rather than addition as is characteristic of alkenes. When the monosubstituted benzenes, for example, bromobenzene, were examined, it was found that no isomers were formed. There is one and only one bromobenzene. All efforts to isolate and identify additional isomers were unsuccessful. From this type of evidence, chemists concluded that all six of the hydrogens of benzene are equivalent. Finally, when a monosubstituted benzene was made to undergo further substitution, three different isomers of the disubstituted benzene were isolated. For example, in the case of the further bromination of bromobenzene, three isomers of dibromobenzene can be isolated. Two of the isomers are produced in major amounts, and the other one in only minor amounts, but the fact remains that three, and only three dibromoisomers can be isolated.

For Kekulé and his contemporaries, the problem was to incorporate these observations, along with the accepted tetravalence of carbon, into a structural formulation of the benzene molecule. Before we examine the structural formula for benzene proposed by Kekulé, we should note that the problem of an adequate description of the structure of benzene and other aromatic hydrocarbons has occupied the efforts of chemists for over a century. Only since the 1930s has a general understanding of this problem been realized, and this has required the best efforts of chemists and mathematicians to develop new mathematical tools and physical concepts equal to the problem.

6.2 The Structure of Benzene

In 1865, Kekulé proposed that the six carbons of benzene are arranged in a six-membered ring with one hydrogen attached to each carbon. To maintain the then-established tetravalence of carbon, he further proposed that the ring contains three double bonds which shift back and forth so rapidly that the two forms, Ia and Ib, cannot be separated.

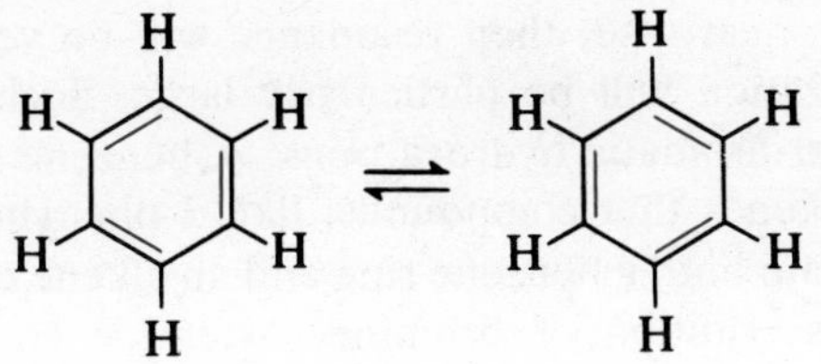

Ia ⇌ Ib

This proposal accounted nicely for the observation that there is only one bromobenzene but three dibromobenzenes.

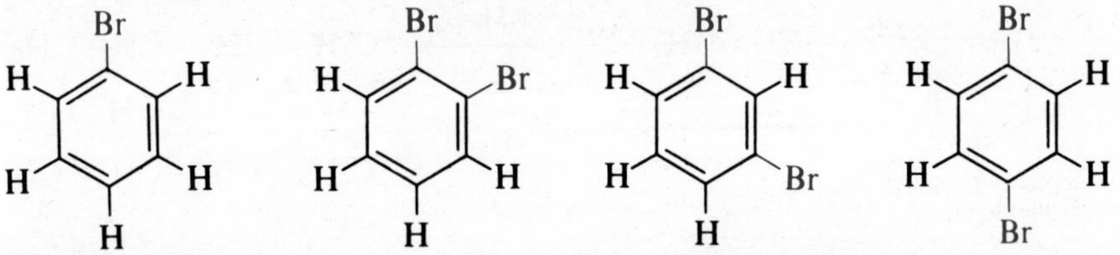

bromobenzene   three isomeric dibromobenzenes

Although Kekulé's proposal was consistent with many of the experimental observations, it did not totally solve the problem and was contested for years. The major objection was that it did not account for the unusual chemical behavior of benzene. For example, if benzene contained three double bonds, as Kekulé proposed, then, his critics argued, it should react with Br_2 in much the same manner as do other alkenes. Yet, benzene is relatively inert to these reagents. Therefore, they argued, benzene cannot have the three double bonds as Kekulé had suggested. The controversy over the structural formula of benzene continued from Kekulé's time well into the 20th century.

6.3   The Structure of Benzene—The Resonance Model

The resonance theory of Linus Pauling provided the first adequate description of the structure and unusual reactivities of benzene. Recall from Section 1.16 that according to the resonance theory, when a substance can have two or more equivalent or nearly equivalent structures that are interconvertible simply by the redistribution of valence electrons, the actual molecule does not conform to any one of the contributing structures but exists as a resonance hybrid of them all. The two major contributors to the benzene hybrid are shown below.

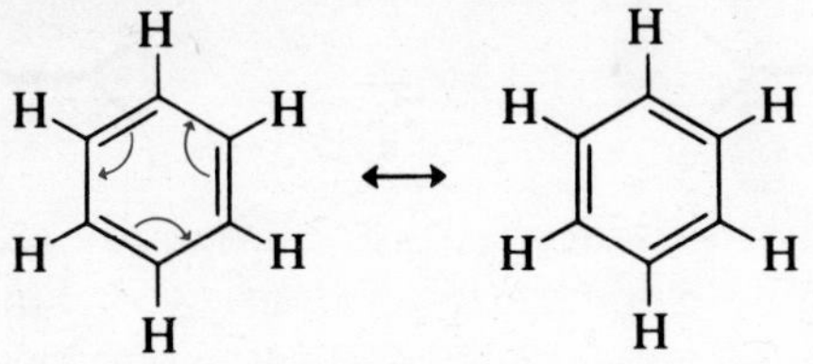

One consequence of resonance is a marked increase in stability of the hybrid over that of any contributing structure. If the contributing structures

141

are equivalent or nearly so, then resonance will be very important and resonance stabilization will be particularly large. Such is the case with benzene and other aromatic hydrocarbons. A benzene ring is so inert by comparison to alkenes that compounds, like 3-phenylpropene (allylbenzene), which contain both a benzene ring and an alkene can be put through reactions such as addition of bromine, oxidation by dilute potassium permanganate, or hydrogenation without alteration of the benzene ring.

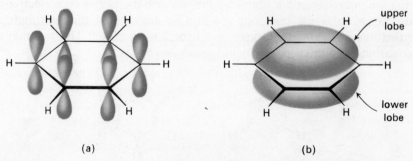

6.4 The Structure of Benzene—The Molecular Orbital Model

Let us also consider the structure of benzene in terms of the <u>molecular orbital model</u>, for just as this approach gave us a clearer understanding of the carbon-carbon double bond (as one sigma bond and one pi bond), so too will it give us a clearer understanding of the bonding in benzene. Benzene is a regular hexagon with bond angles of 120°. For this type of bonding, each carbon uses sp^2 hybrid orbitals. The six carbon atoms are joined together in a regular hexagon by single bonds formed by the overlap of sp^2 hybrid orbitals. Each carbon is further joined to one hydrogen by a sigma bond formed from the overlap of sp^2-$1s$ orbitals. These twelve C—C and C—H sigma bonds form the skeletal framework of the ring. Each carbon also has a single $2p$ orbital containing one electron. Overlap of these six $2p$ orbitals forms three bonding pi molecular orbitals. Because of the symmetry of the molecule, each $2p$ orbital can achieve bond formation with both of its neighbors, and consequently the pi electron density is completely symmetrical around the ring with half in the "upper lobe" and half in the "lower lobe" (Figure 6.1).

FIGURE 6.1 (a) The bonds of benzene formed by overlap of sp^2-sp^2 and sp^2-$1s$ orbitals. The six $2p$ orbitals are uncombined. (b) Overlap of six $2p$ orbitals to form the pi cloud.

Because each pi electron is associated with more than two nuclei, we refer to the pi cloud as being delocalized. Recall from our earlier discussions of the stability of covalent bonds that localization of an electron pair between two nuclei leads to a more stable arrangement than when the two atoms are separated. The association of two electrons between more than two nuclei in delocalized bonds leads to even greater stabilization. Specifically, molecular orbital calculations clearly show that the energy released when six $2p$ orbitals combine to form the pi system of benzene is greater than the energy released when six $2p$ orbitals combine to form three ordinary pi bonds like those in ethylene. In the molecular orbital model, the extra energy stabilizing the molecule is called delocalization or stabilization energy, while in the resonance model it is called resonance energy. One symbol used to represent benzene in the molecular orbital approach is a regular hexagon with a circle inside to indicate the symmetrical distribution of the six pi electrons.

We have now seen two approaches to the structure and bonding in benzene and each of these approaches uses a particular symbol to represent benzene. Two Kekulé structures connected by a double-headed arrow suggests to us the resonance approach. The regular hexagon with the inscribed circle suggests the molecular orbital approach. How should we, in the remainder of this text, represent the structure of benzene? As you might expect, there are very strongly held preferences among various groups of chemists. We shall find it most convenient to represent benzene by drawing just one Kekulé structure. We realize of course that this structural formula is not an adequate representation of the benzene molecule, but to the extent that the symbol serves us in communication, it is useful. Furthermore, this type of representation is more convenient since it allows us to see at a glance and count all valence electrons and to draw easily the contributing structures.

6.5 The Resonance Energy of Benzene

It has been possible to measure the resonance or delocalization energy of benzene. The most direct method is to compare the heats of hydrogenation of the actual benzene molecule with its hypothetical counterpart 1,3,5-cyclohexatriene, a six-membered ring with three separate, non-interacting pi bonds. Since cyclohexatriene does not exist, we can only estimate its heat of hydrogenation. Hydrogenation of an alkene is an exothermic reaction. The conversion of cyclohexene to cyclohexane liberates 28.6 kcal/mole.

$$+ \ H_2 \longrightarrow \qquad \Delta H = -28.6 \ \text{kcal/mole}$$

From this value we can estimate that hydrogenation of the hypothetical cyclohexatriene should release $3 \times 28.6 = 85.8$ kcal/mole. When benzene is reduced to cyclohexane, the heat released is 49.8 kcal/mole.

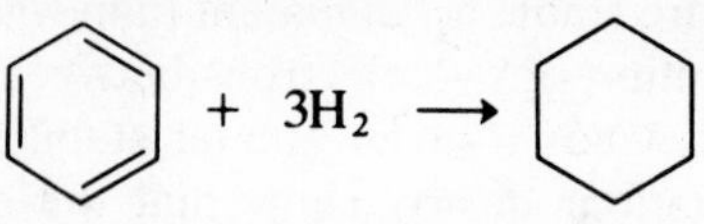

$$\text{benzene} + 3H_2 \longrightarrow \text{cyclohexane} \qquad \Delta H = -49.8 \text{ kcal/mole}$$

The difference in energy between the real molecule, benzene, and the hypothetical molecule, cyclohexatriene, is $(85.8 - 49.8) = 36$ kcal/mole. In other words the resonance or delocalization energy of benzene is 36 kcal/mole.

6.6 Nomenclature of Aromatic Hydrocarbons

In the naming of benzene derivatives, as with other classes of organic compounds, we encounter a blend of systematic and common names. As a matter of fact, the use of common names is particularly prevalent and there are many with which you should be familiar. For the IUPAC names, no new rules are necessary.

Many monosubstituted alkyl benzenes are named according to their alkyl substituents or by common names.

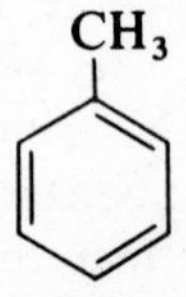

methylbenzene
toluene

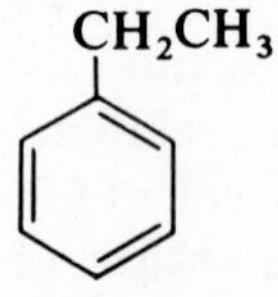

ethylbenzene

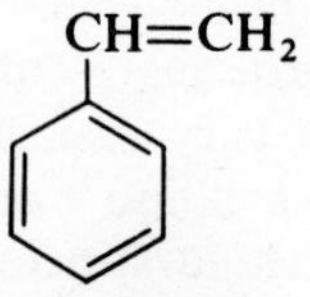

phenylethene
styrene

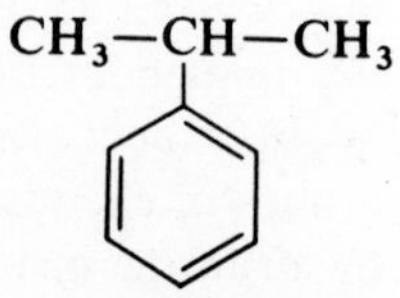

isopropylbenzene
cumene

When there are two substituents on the ring, three structural isomers are possible. The prefixes *ortho* (*o*), *meta* (*m*), and *para* (*p*) are used to locate the substituents.

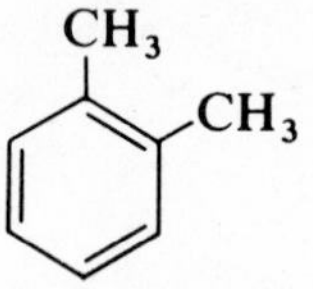

ortho-xylene

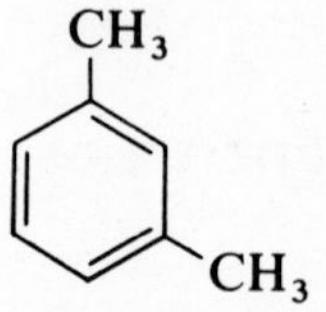

meta-xylene

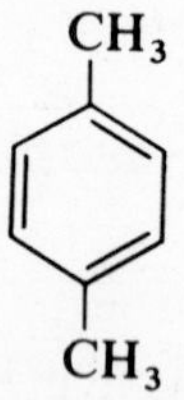

para-xylene

With three of more substituents, a numbering system is used.

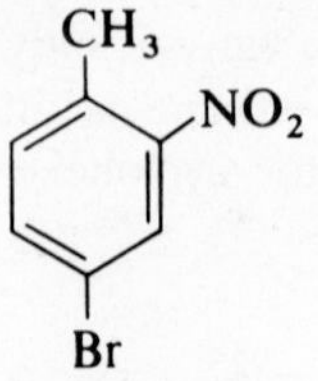

4-bromo-2-
nitrotoluene

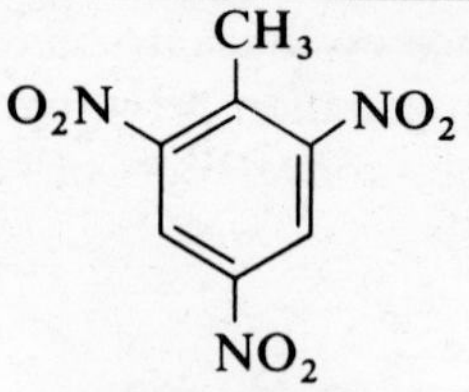

2,4,6-trinitrotoluene
(TNT)

144

In naming substituted benzenes as derivatives of alkanes, two names are particularly common.

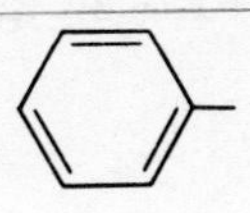

phenyl group 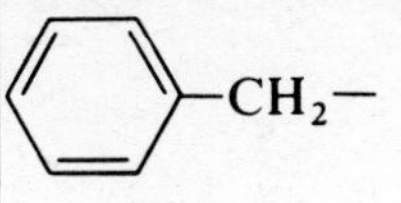   benzyl group

The hydrocarbon group C_6H_5- is called a phenyl group and is sometimes abbreviated by the symbol ϕ. The hydrocarbon group $C_6H_5CH_2-$ is called a benzyl group.

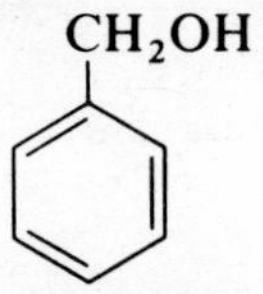

CH_2OH CH_2CH_2OH CH_2CO_2H

benzyl alcohol 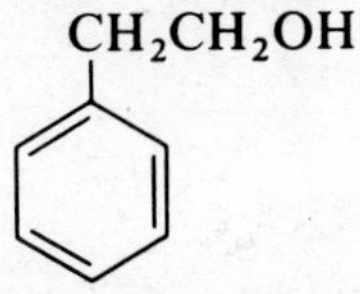   2-phenylethanol 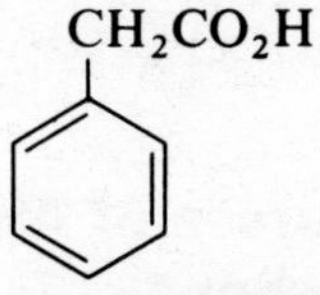   phenylacetic acid

Compounds containing an hydroxyl group on a benzene ring are known as phenols. The structure of phenol itself is shown below. Other phenols are named either as derivatives of the parent hydrocarbon or by common names.

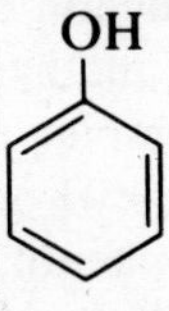

OH   OH   OH   OH

phenol 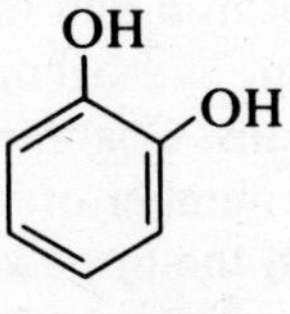   catechol 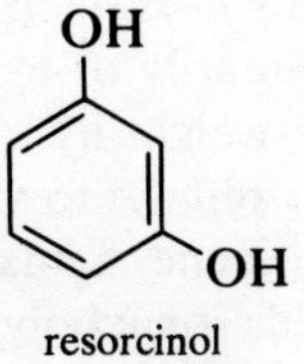   resorcinol   OH
                                    hydroquinone

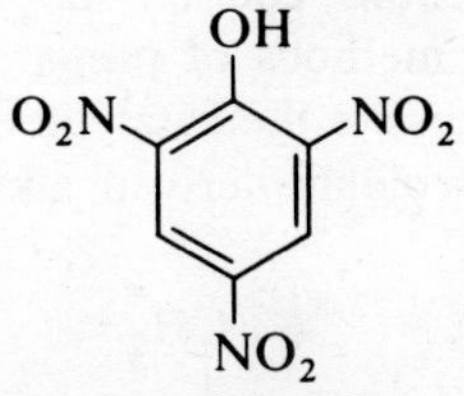

OH   OH

O_2N NO_2

NO_2 CH_3
2,4,6-trinitrophenol   m-cresol
picric acid

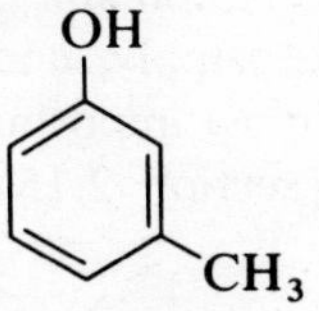

Compounds containing an $-NH_2$ group on a benzene ring are known as aromatic amines and are usually named as derivatives of aniline. Many of the simple substitution derivatives are also known by common names, as for example anisidine (methoxyaniline) and toluidine (methylaniline).

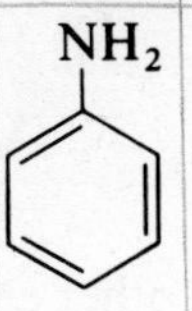

NH_2 NH_2 NH_2 NH_2

                              CH_3

                                              NO_2
            OCH_3

aniline 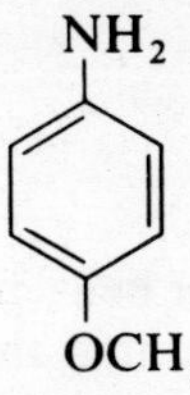   p-methoxyaniline   o-methylaniline   m-nitroaniline
           p-anisidine   o-toluidine

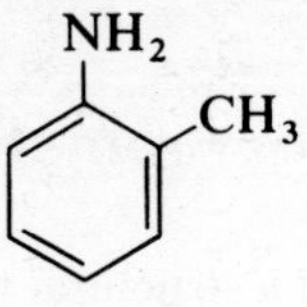

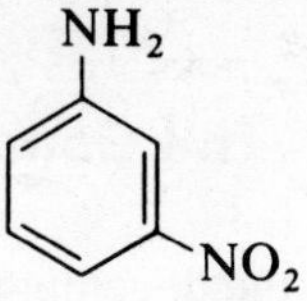

145

Closely related to benzene are numerous polynuclear aromatic hydrocarbons having one or more six-membered rings fused together. For each an IUPAC numbering system is used to locate substituents.

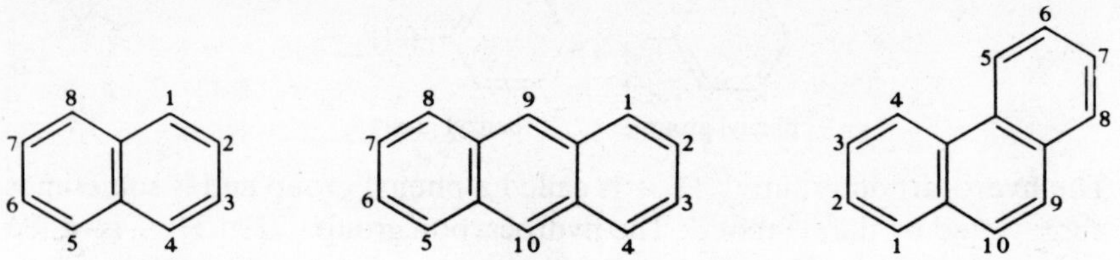

naphthalene
mp 80°C

anthracene
mp 217°C

phenanthrene
mp 99°C

6.7 Sources of Aromatic Hydrocarbons

Until quite recently aromatic hydrocarbons were obtained mainly from coal tar (and aliphatic hydrocarbons were obtained mainly from petroleum). Destructive distillation of coal—that is, heating coal in the absence of air—produces coke, coal gas, and coal tar. Coke is used in tremendous quantities in the refining of iron and the production of steel. Coal gas consists primarily of hydrogen, hydrogen sulfide, methane, and other low-molecular-weight hydrocarbons. Coal tar is a complex mixture which can be further refined to yield a number of aromatic hydrocarbons including benzene, toluene, a mixture of the three xylenes, phenol, and so on. Coal tar also yields some polynuclear aromatic hydrocarbons, the most important of which are naphthalene and anthracene.

Because of the tremendous demand by the chemical industry for benzene and other aromatic hydrocarbons, coal tar simply does not provide an adequate supply, and other methods of preparation have been developed. The most important of these is the catalytic dehydrogenation and reforming (Section 2.15) of petroleum-derived alkanes and cyclo-alkanes.

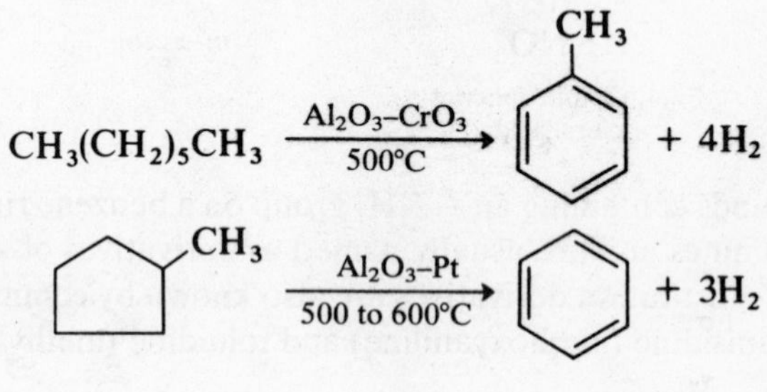

$$CH_3(CH_2)_5CH_3 \xrightarrow[500°C]{Al_2O_3-CrO_3} \text{(toluene)} + 4H_2$$

$$\text{(methylcyclopentane)} \xrightarrow[\text{500 to 600°C}]{Al_2O_3-Pt} \text{(benzene)} + 3H_2$$

6.8 The Concept of Aromaticity

With the development of a structural theory for benzene and other aromatic hydrocarbons, it seemed logical to suppose that the unusual chemical and physical properties of benzene were due to the presence of three double bonds in a cyclic, fully conjugated system. The question that followed was: are there other cyclic, fully conjugated molecules that also have aromatic properties? The answer is yes. We shall look at the structural formulas of several such molecules, each of which (1) has a substantial resonance energy and (2) tends to undergo substitution reactions like benzene rather than addition reactions like the alkenes.

Pyridine and pyrimidine are heterocyclic analogs of benzene in which first one CH group and then two are replaced by nitrogen atoms. These molecules are shown below along with the numbering system used to specify the location of substituents.

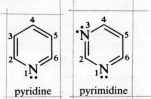

In these molecules, the nitrogen atoms are sp^2 hybridized and each contributes one of the six pi electrons in the ring. The unshared pair of electrons on the nitrogen atoms lies in an sp^2 orbital in the plane of the ring and is not a part of either six pi electron system. We can represent each molecule as a hybrid of two benzene-like contributing structures.

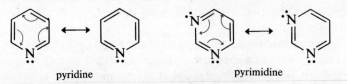

The resonance energy of pyridine is about 32 kcal/mole, just slightly less than that of benzene itself. The resonance energy of pyrimidine is 26 kcal/mole.

The five-membered rings pyrrole, furan, and imidazole also show aromatic character.

In each of these molecules, the heteroatoms are sp^2 hybridized. In pyrrole and furan, the six pi electrons are derived from four $2p$ electrons of the ring carbon atoms and two $2p$ electrons of the heteroatom. If you compare the origin of the six pi electrons in pyrrole and pyridine, you will note an important difference. In pyridine, the unshared pair of electrons on nitrogen is not a part of the six pi electrons; in pyrrole the unshared pair on nitrogen is very much a part of the aromatic sextet. The resonance energy of pyrrole is about 21 kcal/mole; that of furan is about 16 kcal/mole. In later chapters we will discuss the occurrence in nature of all three of these heterocyclic aromatic substances.

Nature abounds in polynuclear heterocyclic aromatic compounds, that is, compounds having two or more heterocyclic aromatic rings fused together. Two such substances especially important in the biological world are purine and indole.

purine

indole

Purine contains two fused rings, a six-membered pyrimidine ring, and a five-membered imidazole ring. Substances derived from the purine and pyrimidine bases appear in the molecules that comprise the genetic units of heredity, namely deoxyribonucleic acid (DNA) and ribonucleic acid (RNA). The chemistry and function of DNA and RNA are discussed in Chapter 14. Indole contains two fused rings, a six-membered benzene ring, and a five-membered pyrrole ring. Substances derived from indole include the essential amino acid L-tryptophan (Chapter 13), the neurotransmitter serotonin, and many other substances of plant and animal origin.

Thus, we can see that benzene is by no means unique in its aromatic properties. Rather, it is but one of a group of cyclic substances which have stabilities and chemical reactivities quite different from those of alkenes and alkynes.

In this section we have examples of five- and six-membered rings and fused ring systems that show aromatic properties. The feature common to all of them is the presence of six pi electrons in a cyclic, fully conjugated ring. The question we should now ask is: do compounds of other ring sizes and different numbers of pi electrons also show aromaticity? For example, does cyclooctatetraene, an eight-membered ring with eight pi electrons, show aromatic behavior?

cyclooctatetraene

To answer this question, cyclooctatetraene was synthesized in the first decade of this century and was found to undergo reactions typical of alkenes. For example, it readily reacts with bromine by addition and is easily oxidized by potassium permanganate. It certainly does not show the properties and reactivities typical of aromatic compounds. It was then suggested that the special aromaticity of benzene and its derivatives was peculiar to compounds containing $(4n + 2)\pi$ electrons in a cyclic, fully conjugated system. Here, n = any integer. This was supported by the fact that other compounds displaying aromatic characteristics were known in which the size of the ring was other than six. Without going into detail, it was the application of the molecular orbital theory which gave chemists a more fundamental understanding of aromaticity and also the ability to make very good predictions about which types of unsaturated rings should or should not show aromatic character.

6.9 Phenols

Both phenols and alcohols contain the hydroxyl group, —OH. However, phenols are grouped together as a separate class because they are significantly more acidic than alcohols and because the reactions of phenols are those typical of aromatic compounds.

Phenol, or carbolic acid as it was once called, is a low-melting solid only slightly soluble in water. In sufficiently high concentrations, it is corrosive to all kinds of cells. In dilute solutions, it has some antiseptic properties and was used for the first time in the 19th century by Joseph Lister for antiseptic surgery. Its medical use is now limited as it has been replaced by antiseptics which are both more powerful and have fewer undesirable side effects. Among these are o-phenylphenol (Lysol) and n-hexylresorcinol (Sucrets and mouthwashes), both widely used in household preparations as mild antiseptics and disinfectants.

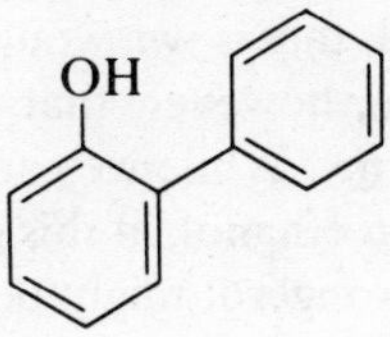

o-phenylphenol

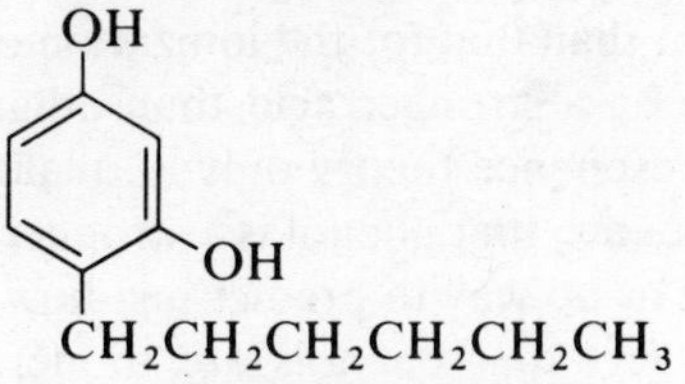

n-hexylresorcinol

Phenols are widely distributed in nature. Phenol itself and the isomeric cresols (ortho-, meta-, and para-cresol) are found in coal tar and petroleum. Thymol and vanillin are important constituents of thyme and vanilla beans.

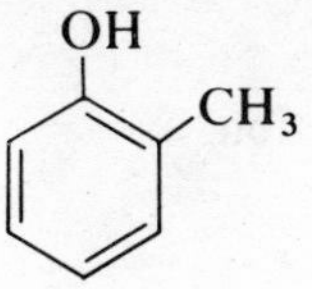

o-cresol 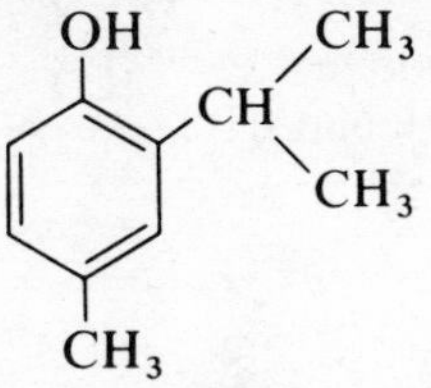   thymol 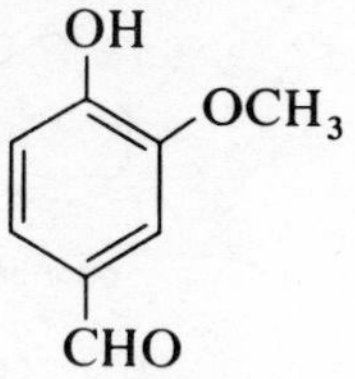   vanillin

As already noted, phenol is a much stronger acid than typical aliphatic alcohols. Just how much stronger it is can be seen by comparing the acid dissociation constants of phenol and ethanol.

$$CH_3CH_2\ddot{O}H \rightleftharpoons CH_3CH_2\ddot{O}:^- + H^+ \qquad K_a = 1 \times 10^{-16}$$
ethanol   ethoxide ion

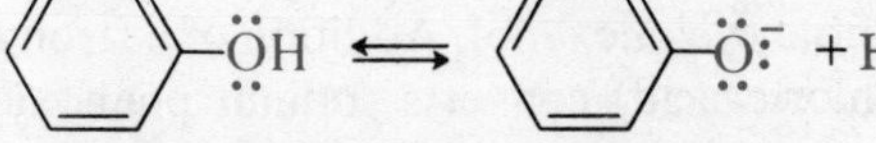

$$\text{phenol} \rightleftharpoons \text{phenoxide ion} + H^+ \qquad K_a = 1 \times 10^{-10}$$

phenol   phenoxide ion

Notice that phenol is about 1,000,000 times stronger an acid than ethanol. We can account for this enhanced acidity by using the resonance model and looking at the relative stabilities of the ethoxide ion and the phenoxide ion. There is no possibility for resonance or resonance stabilization in the ethoxide ion. However, for the phenoxide ion we can draw five contributing structures, two Kekulé structures with the negative charge on oxygen plus three additional ones delocalizing the negative charge to the *ortho* and *para* positions of the aromatic ring (Figure 6.2).

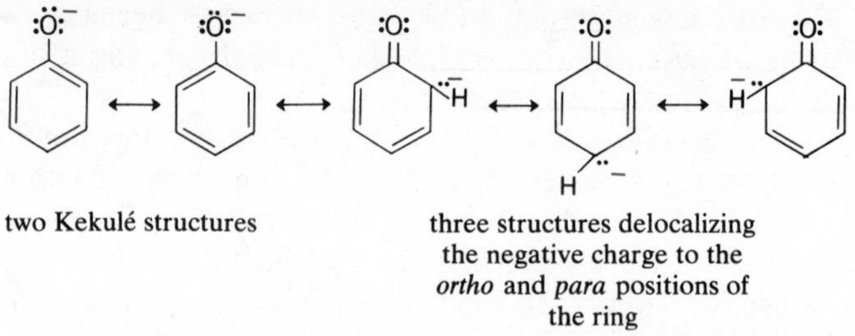

two Kekulé structures   three structures delocalizing
the negative charge to the
ortho and *para* positions of
the ring

FIGURE 6.2   Five contributing structures for the phenoxide ion.

Since phenoxide ion is more stabilized by resonance than is ethoxide ion, we would expect the equilibrium for the ionization of phenol to lie farther to the right than that for the ionization of ethanol, that is, we would expect phenol to be a stronger acid than ethanol. Note, however, that we have used the resonance theory only in qualitative terms. While we can predict with confidence that phenol is a stronger acid than ethanol, at this point we have little or no way to predict just how much stronger it might be.

The fact that phenols are moderately acidic whereas alcohols are essentially neutral provides a very convenient way to separate phenols from water-insoluble alcohols. As an example suppose we want to separate phenol from cyclohexanol. Each is quite insoluble in water so they cannot be separated by washing the mixture to dissolve one and not the other. We can however take advantage of differences in acidity.

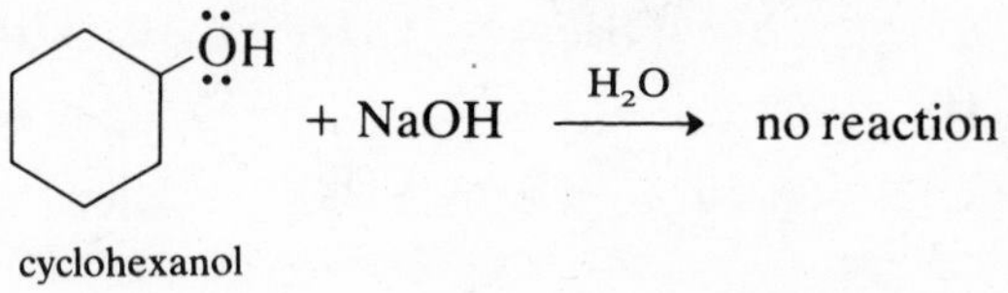

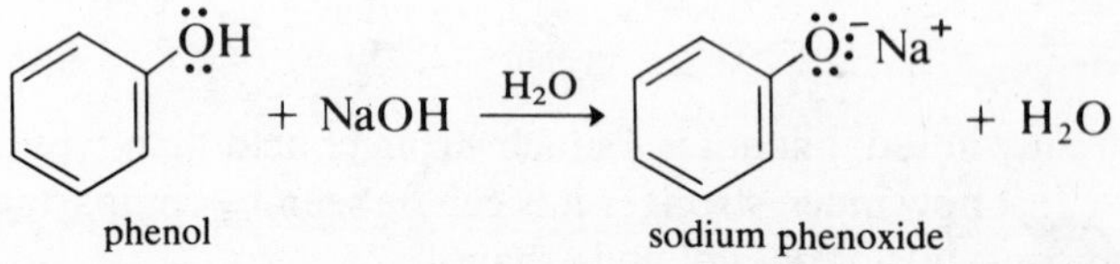

cyclohexanol

phenol   sodium phenoxide

Phenol reacts with aqueous sodium hydroxide to form the water-soluble salt, sodium phenoxide. Cyclohexanol remains as a water-insoluble layer. At this point the water layer containing the sodium phenoxide can be separated from the water-insoluble cyclohexanol. Addition of a strong acid (for example aqueous hydrochloric acid) converts sodium phenoxide to phenol.

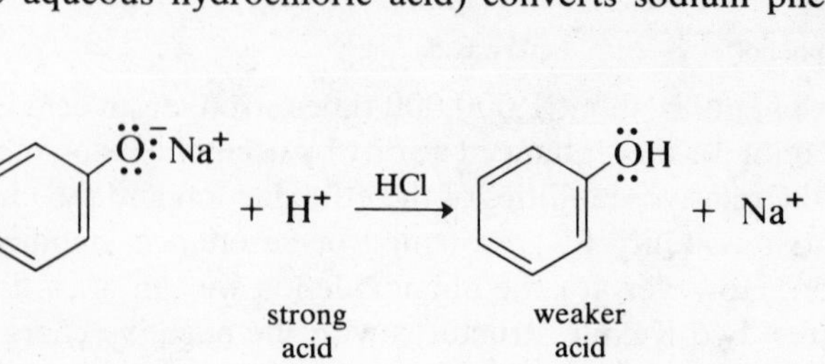

strong   weaker
acid   acid

6.10 Electrophilic Aromatic Substitution

Perhaps the most striking characteristic of aromatic hydrocarbons is their tendency to react by substitution rather than by addition. The following are a few of the more common substitution reactions.

$$+ HNO_3 \xrightarrow{H_2SO_4} \quad NO_2 \quad + H_2O$$

$$+ H_2SO_4 \longrightarrow \quad SO_3H \quad + H_2O$$

$$+ Br_2 \xrightarrow{FeBr_3} \quad Br \quad + HBr$$

$$+ Cl_2 \xrightarrow{AlCl_3} \quad Cl \quad + HCl$$

$$+ CH_3CH_2Cl \xrightarrow{AlCl_3} \quad CH_2CH_3 \quad + HCl$$

$$+ R-Cl \xrightarrow{AlCl_3} \quad R \quad + HCl$$

In studying these reactions we will deal with the questions of how they occur and why benzene, in contrast to alkenes, typically undergoes substitutions rather than additions.

A wealth of experimental evidence suggests that in nearly all substitution reactions, the benzene ring itself is attacked by an electrophilic reagent. We can illustrate this by the bromination of benzene. As we discuss a mechanism for the bromination of benzene (a substitution reaction) let us compare and contrast it to the bromination of an alkene (an addition reaction). Recall from Section 3.12 that the addition of bromine to ethylene can be formulated as a two-step process.

Step 1 $CH_2{=}CH_2 + :\ddot{B}r{-}\ddot{B}r: \longrightarrow \ ^+CH_2{-}CH_2{-}\ddot{B}r: + :\ddot{B}r:^-$

Step 2 $:\ddot{B}r:^- + \ ^+CH_2{-}CH_2{-}\ddot{B}r: \longrightarrow \ :\ddot{B}r{-}CH_2{-}CH_2{-}\ddot{B}r:$

Depending on how we look at it, the first step is either a nucleophilic attack of ethylene on bromine or an electrophilic attack of bromine on ethylene. Of course we can also look at this reaction in a greatly oversimplified way as the attack of a bromonium ion on ethylene.

$$CH_2{=}CH_2 + \ddot{B}r:^+ \longrightarrow \ ^+CH_2{-}CH_2{-}\ddot{B}r:$$

151

This simplification points out clearly the attack of the electrophile on the double bond. However, because there is no firm evidence that the free bromonium ion exists, it is therefore more accurate to show the reaction as beginning with the interaction of ethylene and the bromine molecule.

Benzene and most substituted benzenes react very slowly with bromine. However, in the presence of a Lewis acid catalyst such as $AlCl_3$, $AlBr_3$, $FeCl_3$, or $FeBr_3$, the reaction proceeds quite rapidly. The function of the catalyst is to polarize the Br—Br bond and develop positive character on one of the bromine atoms.

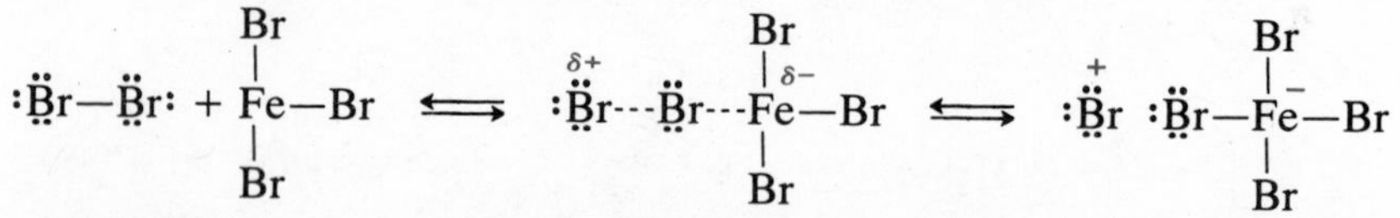

One of the bromine atoms now bears at least a partial positive charge and is therefore a much better electrophile. Reaction between the electrophilic bromine atom (shown as Br^+) and the pi electrons of the benzene ring produces a resonance-stabilized carbocation. Loss of a proton then regenerates the aromatic ring.

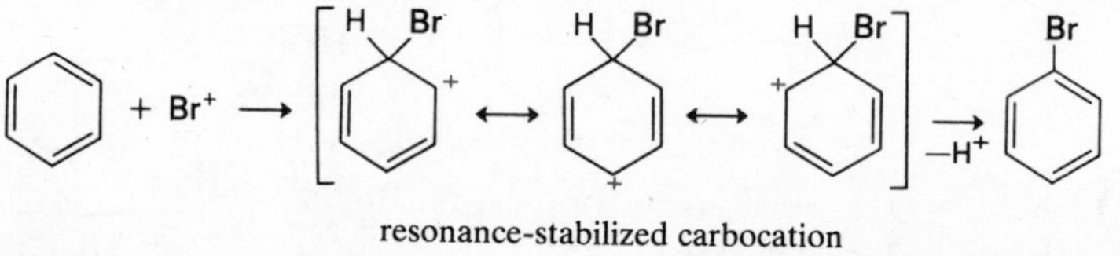

resonance-stabilized carbocation

FIGURE 6.3 A mechanism for the bromination of benzene. Electrophilic aromatic substitution.

Combining these two steps gives

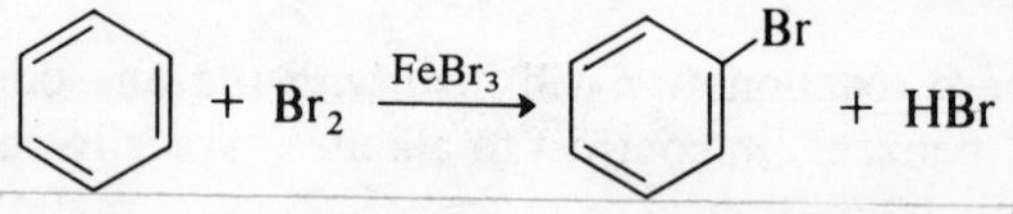

The major difference between halogen addition to an alkene and halogen substitution on an aromatic ring centers on the fate of the positively charged intermediate formed in the first step. In the case of an alkene, this intermediate reacts with a nucleophile to complete the addition. In the case of an aromatic hydrocarbon, the intermediate loses a proton to regenerate the aromatic ring and regain the large resonance stabilization. There is no such resonance stabilization to be regained in the case of an alkene.

Chlorination, nitration, sulfonation, and alkylation of benzene can be formulated in much the same way. In nitration, it is assumed that the added sulfuric acid facilitates the formation of the nitronium ion. The nitronium ion is the electrophile attacking the aromatic ring.

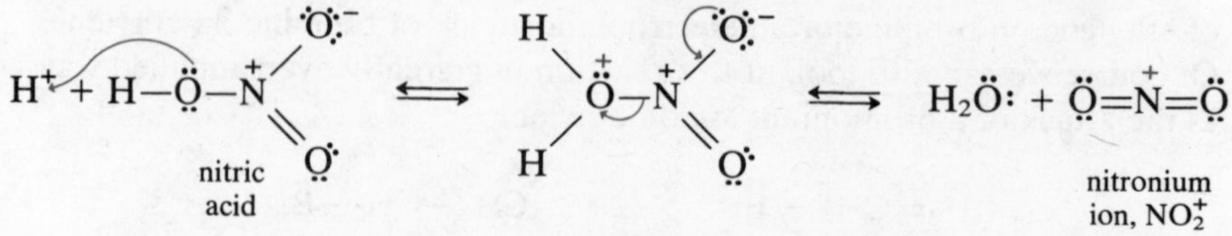

nitric
acid

nitronium
ion, NO_2^+

Sulfonation of aromatic compounds can be brought about with either concentrated or fuming sulfuric acid. The latter reagent contains sulfur trioxide, SO_3, dissolved in concentrated sulfuric acid and is very much more reactive. In either case, the electrophile appears to be sulfur trioxide itself.

benzenesulfonic
acid

Benzenesulfonic acid is an acid comparable in strength to sulfuric acid.

$$C_6H_5-SO_3H + H_2O \rightleftharpoons C_6H_5-SO_3^- + H_3O^+ \qquad K_a = 2 \times 10^{-1}$$

The reaction of an alkyl halide with benzene in the presence of a Lewis acid catalyst is known as the Friedel-Crafts reaction.

The function of the aluminum chloride is to polarize the C—Cl bond of the alkyl halide and generate carbocation character on the alkyl group.

An alkyl carbocation, or something close to it in structure and polarity, is the attacking electrophile.

While the Friedel-Crafts alkylation is useful in the synthesis of certain alkyl benzenes, the reaction has one serious drawback. The carbon skeleton of the alkyl group sometimes rearranges during the course of the alkylation process. For example, reaction of benzene with propyl chloride in the presence of aluminum chloride produces mostly isopropyl benzene rather than the expected n-propyl benzene.

I think due to
H-Shift.

It is the alkyl carbocation which undergoes rearrangement prior to its attack on the aromatic ring. If a primary carbocation can undergo molecular rearrangement to a more stable secondary or tertiary carbocation, it is likely to do so under the conditions of the Friedel-Crafts alkylation.

6.11 Disubstituted Derivatives of Benzene

Substitution reactions of benzene itself are relatively few in number; we have examined nitration, sulfonation, halogenation, and alkylation. Of course there are other substitution reactions, but the number of products that can be synthesized by direct substitution is limited. As we move from monosubstitution to di-, tri-, and polysubstitution, the number of possible derivatives that can be synthesized from benzene becomes much greater and at the same time the chemistry becomes more complex and more challenging. Our concern in this section is to examine the influence of a substituent already on the ring and then to account for variations in structure and reactivity in terms of modern organic theory.

Let us begin by looking at a series of experimental observations. As we do so it will become clear that a substituent group on benzene (1) directs the position at which further substitution takes place and, at the same time (2) alters the reactivity of the ring toward further substitution.

The directing effect can be seen very readily by comparing the products of nitration of anisole and nitrobenzene. Nitration of anisole yields a mixture of p-nitroanisole and o-nitroanisole. Of these two products, the para isomer predominates. More important, there is virtually no m-nitroanisole formed.

anisole   o-nitroanisole   p-nitroanisole

Because it directs the entering nitro group (or any other entering group for that matter) to the ortho and para positions, methoxyl is said to be an ortho,para-directing group.

Nitration of nitrobenzene yields a mixture consisting of approximately 93% of the meta isomer and less than 7% of the ortho and para isomers combined.

nitrobenzene   m-dinitrobenzene

Because it directs the entering group to the meta position, nitro is said to be a meta-directing group.

Substitution reactions such as these have been done for a wide variety of monosubstituted derivatives of benzene and the directing influences of various functional groups determined. While these directing influences are not absolute, for most functional groups and in particular for the ortho,para-directing groups, there is a very high directional specificity;

generally no more than 5 to 10% of the "unpredicted" isomer is formed. Listed in Table 6.1 are the directing influences for many of the important functional groups presented in this text.

TABLE 6.1 Directing effects of some common functional groups.

[handwritten annotations: activating; deactivating = sub. accourro slower than w/ benzene alone; deactivating -e⁻ withdrawing; activating - e⁻ releasing]

ortho,para-directing	meta-directing
—F, —Cl, —Br, —I	$-\overset{+}{N}\underset{O^-}{\overset{O}{\|}}$ $-\overset{O}{\underset{O}{\overset{\|}{S}}}-OH$
—CH₃, —C₂H₅, —R	
—OH, —OCH₃, —OR	
—NH₂, —NHR, —NR₂, —NHCR (O)	—COH, —COR, —CNH₂ (each C=O)
—OCR (O)	—CH, —CR (each C=O)
(benzene ring)	—N⁺(CH₃)₃ with CH₃ groups

If we examine these *ortho-*, *para-* and *meta-*directing groups for structural similarities and differences, we can make two empirical generalizations. Notice that if the atom attached directly to the aromatic ring bears a multiple bond, the group is a *meta-*director; otherwise it is an *ortho,para-*director. Of the groups listed in Table 6.1, the only two exceptions to these generalizations are phenyl and the trimethylammonium group, —N(CH₃)₃⁺.

Now let us turn to the activating/deactivating influence of a substituent group. We can determine whether a group activates the ring for further substitution or deactivates it in either of two ways. We can compare the relative rates of substitution under identical experimental conditions or we can compare the conditions necessary to produce equal rates of substitution. Whatever the method used, it is found that anisole undergoes nitration considerably more readily than does benzene itself, while nitrobenzene undergoes nitration considerably less readily. Accordingly, methoxyl is said to be an activating group and nitro is said to be a deactivating group. Similar studies may be carried out for each of the functional groups listed in Table 6.1, and from observations of this type we can arrive at two additional empirical generalizations. First, the presence of a *meta-*directing group makes further substitution slower, often considerably slower, than in benzene itself. Second, the presence of an *ortho,para-*directing group makes substitution faster. The presence of hydroxyl, alkoxyl, or amino groups can make substitution faster than in benzene by several powers of ten. The one notable exception to these generalizations is the halogens; they deactivate the ring and yet are *ortho,para-*directors.

We can illustrate the usefulness of these generalizations by considering the synthesis of two different disubstituted derivatives of benzene. Suppose we wish to prepare *m-*bromonitrobenzene from benzene. Such a conversion can be done in two steps, nitration and bromination. If the steps

155

are carried out in just this order, the product is indeed the desired *m*-bromonitrobenzene. The nitro group is a *meta*-director and will therefore direct the incoming bromine atom to the *meta* position.

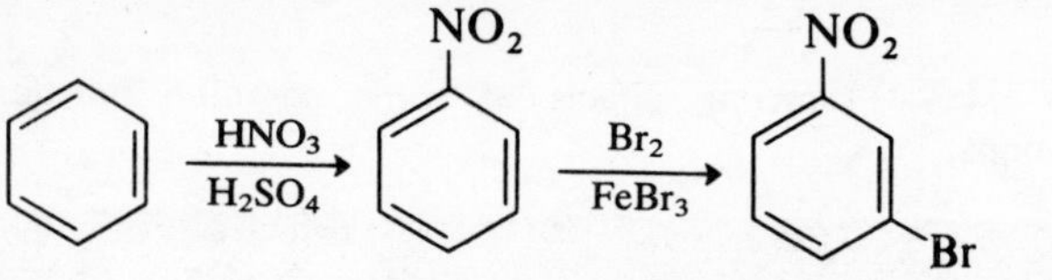

m-bromonitrobenzene

But what if we reverse the order of these two steps? Bromination produces bromobenzene. Bromine is an *ortho,para*-directing group and will therefore direct the incoming nitro group to the *ortho* and *para* positions. Therefore bromination followed by nitration yields a mixture of the *ortho* and *para* isomers and not the desired *meta* isomer.

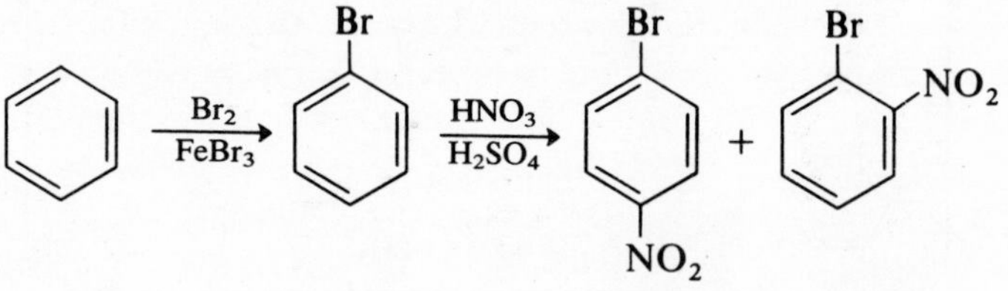

p-bromo-
nitrobenzene

o-bromo-
nitrobenzene

Clearly the order in which electrophilic aromatic substitution reactions are carried out is critical. As another example, consider the conversion of toluene into *p*-nitrobenzoic acid. The —NO$_2$ group can be introduced using a nitrating mixture of nitric and sulfuric acids. The —COOH group can be produced by oxidation of the —CH$_3$ group (Section 8.5).

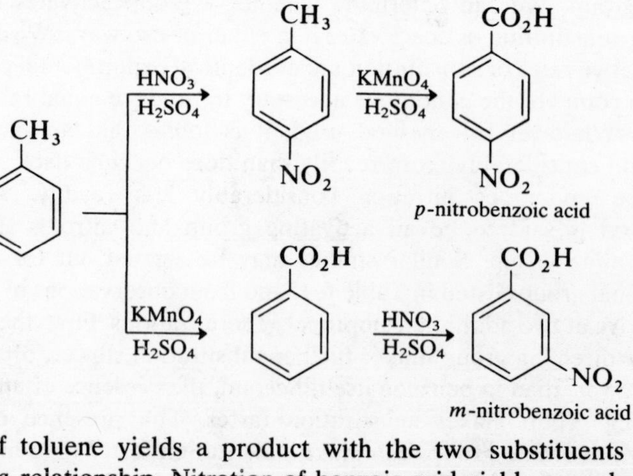

p-nitrobenzoic acid

m-nitrobenzoic acid

Nitration of toluene yields a product with the two substituents in the desired *para* relationship. Nitration of benzoic acid yields a product with the substituents *meta* to each other. Again we see that the order in which the steps are performed is critical. Note in this last example that we have shown the nitration of toluene producing only the *para* isomer. Since —CH$_3$ is an *ortho,para*-directing group, both *ortho* and *para* isomers will be formed. However, in problems of this type where you are asked to prepare one or the other of these isomers, we will assume that there are chemical or physical methods which can be used to separate the desired isomer in pure form.

6.12 Theory of Directing and Activating/Deactivating Influences

We have seen ample evidence that substituent groups on a benzene ring exert both a directing and an activating/deactivating influence, and we have used certain empirical generalizations about these influences to plan syntheses. Now let us turn to the task of discovering a relationship between structure and reactivity in terms of modern organic theory. As we do so, we shall attempt to understand the underlying basis for the generalizations themselves and also for the apparent exceptions to these generalizations. We shall proceed as we have done in other situations of this kind, namely, by proposing a mechanism for each particular reaction and then examining the relative stabilities or reactivities of the various intermediates.

First consider the nitration of anisole. The rate of electrophilic aromatic substitution is determined by the slowest step in the mechanism. For nitration of anisole, and in fact for every substitution reaction we shall consider, the slow step is the attack by the electrophile on the aromatic ring. Electrophilic attack by the nitronium ion, NO_2^+, produces a resonance-stabilized cation. Shown in Figure 6.4 are the cations formed by *meta* and *para* attack of the electrophile.

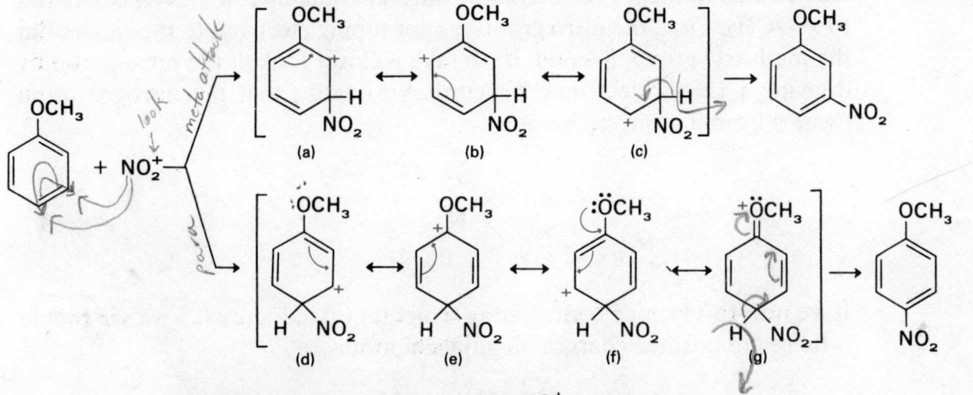

FIGURE 6.4   Electrophilic attack by NO_2^+ on anisole.

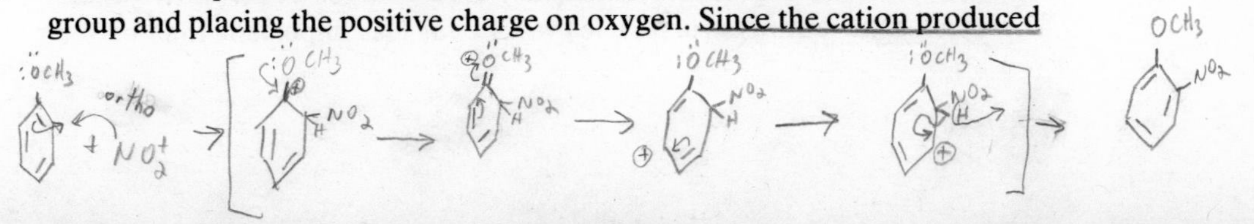

Attack of the nitronium ion in the position *meta* to the methoxyl group yields the resonance-stabilized cation (a) to (c). Attack in the *para* position produces the resonance-stabilized cation (d) to (g). (For the purposes of this discussion we will ignore for the moment the fact that attack can also be in the *ortho* position.) Note that for each mode of attack we can draw three contributing structures that place the positive charge on carbon atoms of the ring. These three structures are the only important ones that can be drawn for *meta* attack. In the case of *para* attack, there is a fourth important contributing structure, one involving the methoxyl group and placing the positive charge on oxygen. Since the cation produced in *para* attack is stabilized more by resonance (a greater number of important contributing structures and therefore a greater delocalization of the positive charge) than is the cation produced by *meta* attack, anisole undergoes nitration more readily in the *para* position than in the *meta* position.

Now let us examine the *meta*-directing influence of the nitro group in much the same manner as we have done for the *ortho,para*-directing influence of the methoxyl group. Shown in Figure 6.5 are the cations formed by *meta* and *para* attack of the nitronium ion on nitrobenzene.

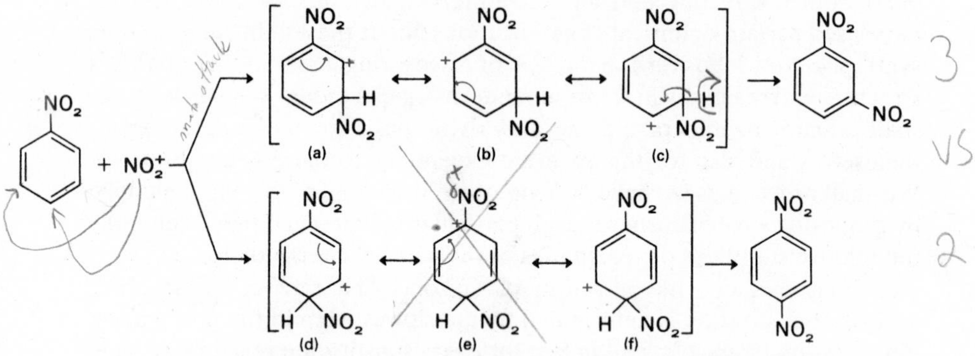

FIGURE 6.5 Electrophilic attack by NO_2^+ on nitrobenzene.

Each cation is a hybrid of three contributing structures and there are no additional important ones that can be drawn. There are two significant factors accounting for the markedly different influence of $-NO_2$ compared to $-OCH_3$. First, the nitro group cannot supply electrons to the ring as did the methoxyl group. Second, if we take a closer look at the nitro group by drawing a Lewis electronic structure, we realize that the nitrogen atom bears a formal positive charge.

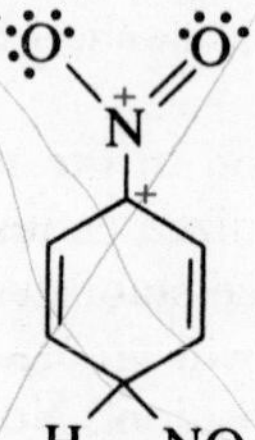

If we now re-examine contributing structure (e) in Figure 6.5 we see that in it there are positive charges on adjacent atoms.

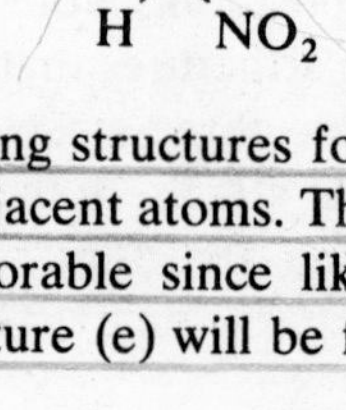

None of the other contributing structures for either *meta* or *para* attack places positive charges on adjacent atoms. This build-up of positive charge on adjacent atoms is unfavorable since like charges repel each other. Therefore contributing structure (e) will be far less important than either

(d) or (f) in stabilizing the cation produced in *para* attack. As a consequence, resonance stabilization for *meta* attack is greater (three important contributing structures) than for *para* attack (only two important contributing structures) and therefore the reaction will proceed by *meta* rather than by *para* attack.

The same type of argument can be applied to each of the other *meta*-directing groups listed in Table 6.1. In each instance, the atom attached directly to the aromatic ring bears either a formal positive charge or at least a partial positive charge because of bond polarization. Typical is the polarization of the carbonyl group in benzaldehyde and in acetophenone.

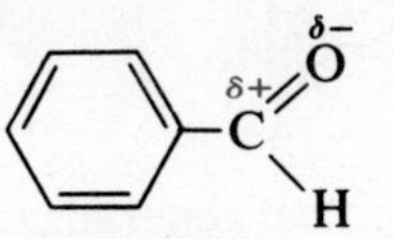

benzaldehyde 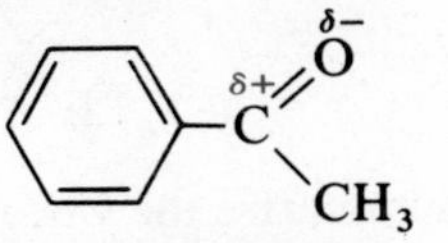   acetophenone

Next, let us turn to the activating or deactivating influences of substituent groups. Below are listed several derivatives in order of increasing ease of substitution.

$$C_6H_5NO_2 < C_6H_5Cl < C_6H_6 < C_6H_5CH_3 < C_6H_5OCH_3$$

Anisole undergoes substitution more readily than benzene because the cation derived from anisole is stabilized more by resonance than is the cation derived from benzene. Toluene and other alkyl derivatives of benzene undergo substitution more readily than benzene itself because methyl and other alkyl groups are able to stabilize the resulting cation. In this regard, recall the discussion of the relative ease of formation of primary, secondary, and tertiary carbocations in Section 3.10. Nitrobenzene undergoes substitution far less readily than benzene because of the presence of a formal positive charge on the nitrogen atom. This formal charge causes a polarization of electron density toward the nitro group and away from the aromatic ring itself. This electron-withdrawing effect is even more apparent if we consider nitrobenzene as best represented as a hybrid of structures such as the following:

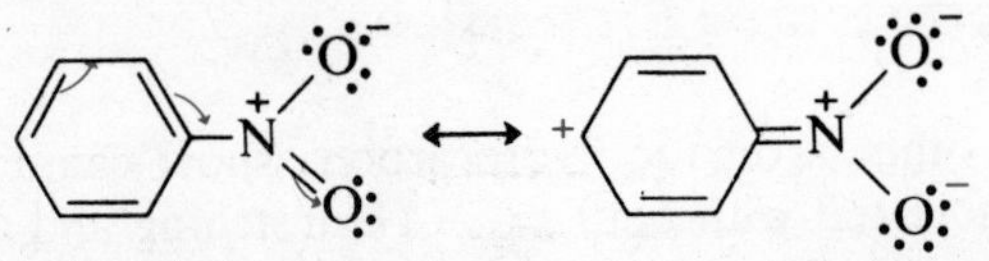

While such structures may not make much of a contribution to the hybrid because of the extreme separation of unlike charge, to the extent that they make any contribution, they serve to reduce the electron density on the ring itself. As a consequence of this reduction in electron density, the aromatic ring is less able to supply electrons to the approaching electrophile and hence the rate of substitution is reduced. It is for this reason that all the *meta*-directing groups listed in Table 6.1 are also deactivating.

The one apparent exception to this close correlation between *meta*-directing influence and deactivation of the aromatic ring is the halogens. They deactivate the ring and yet direct *ortho,para*. Deactivation by halogen can be accounted for by the greater electronegativity of halogen compared to carbon, and therefore, the ability of halogen to decrease the electron density on the ring.

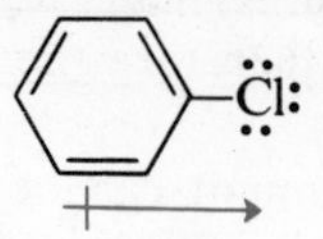

The *ortho,para*-directing influence of halogen can be accounted for by the ability of halogen to participate in stabilization of the cation produced on attack of an electrophile, E^+.

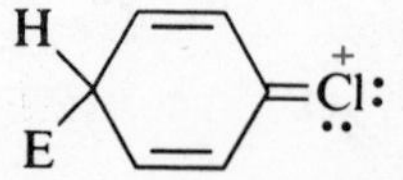

We can summarize the various empirical generalizations presented in these two sections and the underlying theory as follows:

1. The position of substitution is controlled by the group already present on the ring, not by the entering group.
2. If the atom attached to the aromatic ring is part of a multiple-bond, the group is *meta*-directing; otherwise it is *ortho,para*-directing. While the directing influence of a particular group is not 100% effective, generally less than 5 to 10% of the "unpredicted" product is formed. Of the groups listed in Table 6.1, the only exceptions to these generalizations are phenyl which directs *ortho,para* and trimethylammonium which directs *meta*. All of these directing influences can be accounted for in terms of resonance stabilization of the cation formed on attack of the electrophile.
3. The presence of a *meta*-directing group makes substitution slower than in benzene itself; the presence of an *ortho,para*-directing group makes substitution more rapid than in benzene. Hydroxyl, methoxyl, and amino groups make substitution faster by several powers of ten. The halogens are the only *ortho,para*-directing groups which deactivate the ring toward further substitution.

6.13   Spectroscopic Properties

Benzene and other aromatic hydrocarbons show characteristic infrared absorption associated with both $=$C—H stretching and C$=$C stretching. The C—H stretching frequencies occur between 3000 cm^{-1} and 3100 cm^{-1} and generally consist of weak to medium intensity bands. Recall from Section 3.16 that this is the same region of the spectrum in which alkene $=$C—H stretching frequencies occur. In addition, aromatic hydrocarbons also show a number of bands between 1430 cm^{-1} and 1665 cm^{-1} corresponding to stretching of carbon-carbon bonds of the aromatic ring. These two sets of stretching frequencies can be seen in the infrared spectrum of *n*-propyl benzene (Figure 6.6).

The conjugated system of benzene is responsible for several absorption maxima in the ultraviolet-visible region of the spectrum. The most important of these absorptions occur at 200 nm, the second at 257 nm. Toluene shows ultraviolet absorption at 265 nm, a value close to that of

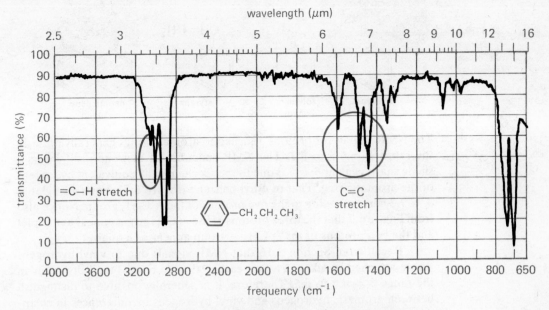

FIGURE 6.6 An infrared spectrum of n-propyl benzene.

benzene itself. Groups attached to the benzene ring, particularly those which contain pi electrons and can thereby extend the conjugated system, cause a shift of the absorption maxima toward the visible region of the spectrum. Styrene in which an additional carbon-carbon double bond is conjugated with the aromatic ring absorbs at 282 nm. Naphthalene, containing two fused aromatic rings, absorbs at 314 nm.

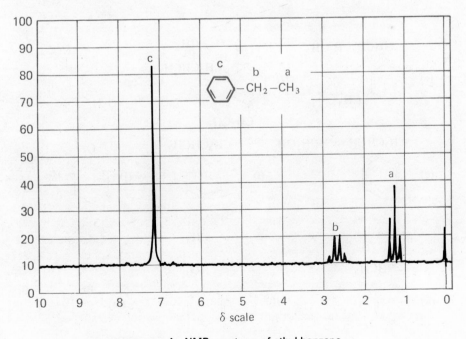

FIGURE 6.7 An NMR spectrum of ethyl benzene.

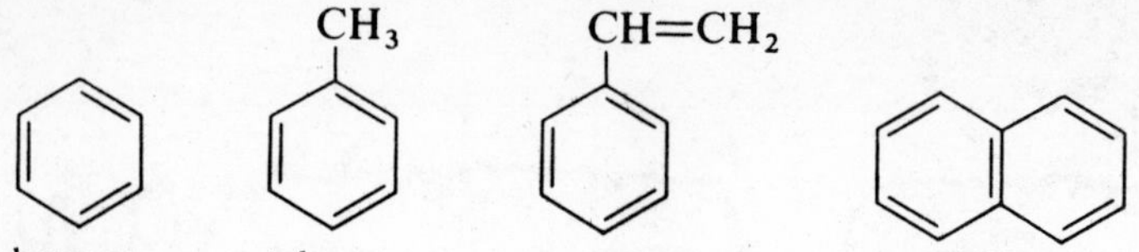

CH$_3$

CH=CH$_2$

benzene
257 nm

toluene
265 nm

styrene
282 nm

naphthalene
314 nm

The NMR signals for protons attached to aromatic rings generally occur in the range δ = 6.5 to 8.3. The NMR spectrum of benzene itself shows a single sharp peak at δ = 7.3 due to the six chemically equivalent hydrogens of the aromatic ring. That of ethyl benzene (Figure 6.7) shows a singlet at δ = 7.3 corresponding to the five protons of the —C$_6$H$_5$ group. Notice also from Figure 6.7 that the three protons of the —CH$_3$ group appear as a triplet and the two protons of the —CH$_2$— group appear as a quartet.

Recall from Section 3.16 that NMR signals due to vinyl hydrogens (those attached to alkene carbon-carbon double bonds) typically occur in the range δ = 4.5 to 6.5. Therefore, it is generally possible to distinguish between aromatic hydrogens and vinyl hydrogens by differences in chemical shifts.

PROBLEMS

6.1 Name the following molecules by using the IUPAC system.

(a)  NO$_2$ / Cl

(b) 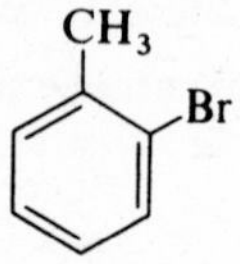 CH$_3$ / Br

(c) 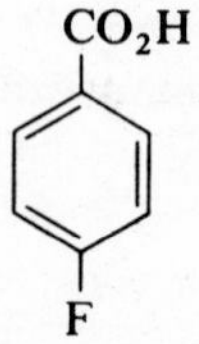 CO$_2$H / F

(d) 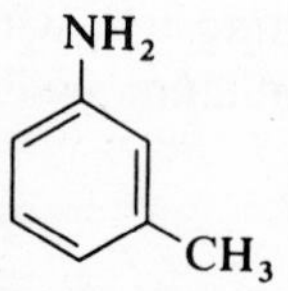 NH$_2$ / CH$_3$

(e) 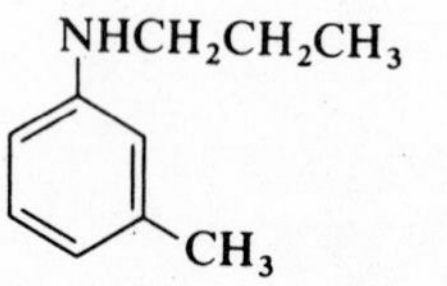 NHCH$_2$CH$_2$CH$_3$ / CH$_3$

(f) OH / CH$_2$CHCH$_3$ with CH$_3$ / CH$_2$CH$_3$

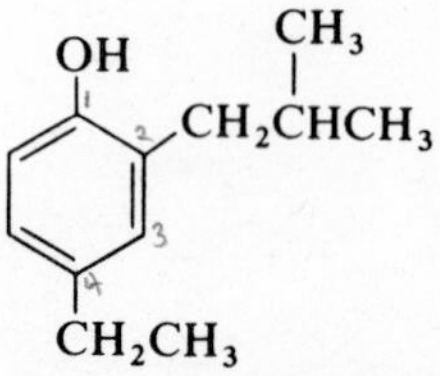

(g) 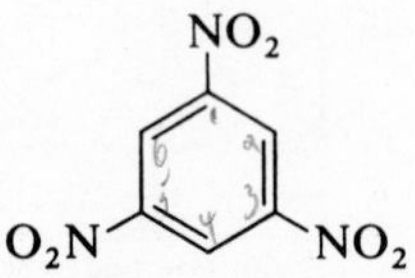 NO$_2$ / O$_2$N / NO$_2$

(h) CH$_3$CHCH=CHCH$_2$OH

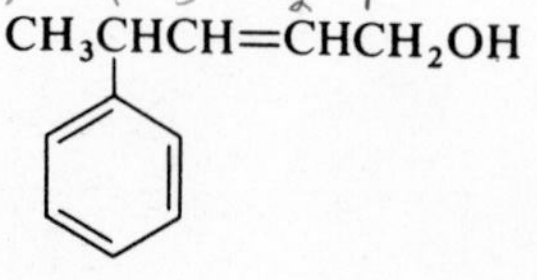

(i) 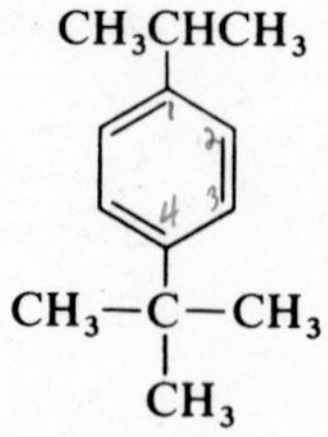 CH$_3$CHCH$_3$ / CH$_3$—C—CH$_3$ with CH$_3$

(j) 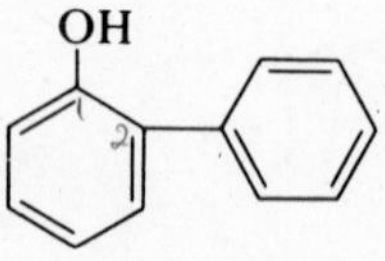 OH

(k) 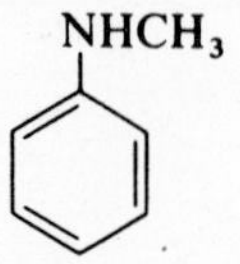 NHCH$_3$

(l) 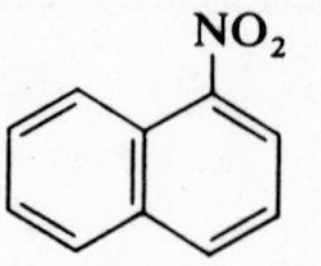 NO$_2$

(m) 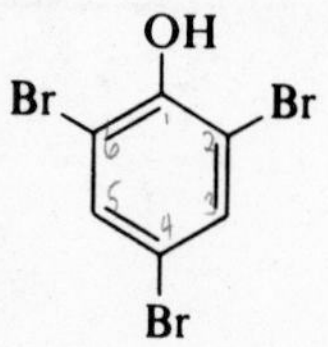 OH / Br / Br / Br

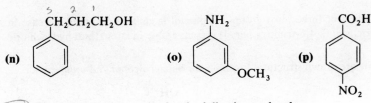

(n) $CH_2CH_2CH_2OH$

(o) NH_2 / OCH_3

(p) CO_2H / NO_2

6.2 Draw structural formulas for the following molecules.

(a) *m*-dibromobenzene
(b) *p*-aminophenol
(c) *p*-chloroiodobenzene
(d) 2-hydroxy-4-isopropyltoluene
(e) phenol
(f) benzyl alcohol
(g) *m*-xylene
(h) *p*-cresol
(i) 2-ethylnaphthalene
(j) *o*-chlorophenol
(k) *p*-diiodobenzene
(l) 2-phenyl-2-pentene
(m) *p*-methoxy-N-ethylaniline
(n) N,N-dimethylaniline
(o) anthracene
(p) isopropylbenzene (cumene)

6.3 Name and draw structural formulas for all derivatives of benzene having the following molecular formulas.

(a) $C_6H_3Br_3$ **(b)** C_8H_{10} **(c)** C_8H_9Cl **(d)** C_9H_{12}

6.4 One of the earliest methods for determining the relative orientations of substituent groups on an aromatic ring was that devised by Wilhelm Körner. To this day it is known as the Körner method of absolute orientation. This method was first applied to determine the structural formulas of the three isomeric dibromobenzenes, $C_6H_4Br_2$. Let us call them isomer *A* (mp +87°C), isomer *B* (mp +7°C), and isomer *C* (mp −7°C). Each of these isomers can be converted into one or more mononitroderivatives, $C_6H_3Br_2NO_2$. Isomer *A* yields one mononitroderivative; isomer *B* yields two mononitroderivatives; and isomer *C* yields three mononitroderivatives. From this information assign structural formulas to *A*, *B*, and *C*.

6.5 There are three isomeric tribromobenzenes, melting points 44°, 88°, and 122°C. Show how the Körner method of absolute orientation might be applied to determine the structural formulas of these isomers.

6.6 Draw five principle resonance contributing structures for the following:

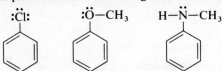

$:\ddot{C}l:$ $:\ddot{O}-CH_3$ $H-\ddot{N}-CH_3$

6.7 Draw the three Kekulé structures for naphthalene and the four Kekulé structures for anthracene.

6.8 Using the resonance theory, account for the fact that the dipole moment of nitrobenzene is larger than that of nitromethane.

nitrobenzene
4.22 D

nitromethane
3.40 D

6.9 Using the resonance theory, account for the fact that *p*-nitrophenol is a stronger acid than phenol itself.

p-nitrophenol
$K_a = 7 \times 10^{-8}$

phenol
$K_a = 1 \times 10^{-10}$

163

The substitution of three nitro groups on phenol leads to a dramatic increase in acidity. Picric acid (2,4,6-trinitrophenol) is comparable in strength to hydrochloric acid.

6.10 Below are drawn structural formulas for benzyl alcohol and *ortho*-cresol.

benzyl alcohol
bp 205°C

o-cresol
bp 191°C

Describe a procedure you might use to separate these two compounds and recover each in pure form.

6.11 Complete the following electrophilic aromatic substitution reactions:

(a) $C_6H_5OCH_3 + H_2SO_4 \xrightarrow{heat}$

(b) $C_6H_5\overset{O}{\overset{\|}{C}}CH_3 + HNO_3 \xrightarrow{H_2SO_4}$

(c) $C_6H_5\overset{O}{\overset{\|}{C}}OCH_3 + Br_2 \xrightarrow{FeBr_3}$

(d) $C_6H_5O\overset{O}{\overset{\|}{C}}CH_3 + Br_2 \xrightarrow{FeBr_3}$

(e) $C_6H_5NH\overset{O}{\overset{\|}{C}}CH_3 + Cl_2 \xrightarrow{FeCl_3}$

(f) $C_6H_5\overset{O}{\overset{\|}{C}}NHCH_3 + Cl_2 \xrightarrow{FeCl_3}$

(g) $C_6H_5-C_6H_5 + HNO_3 \xrightarrow{H_2SO_4}$

(h) $C_6H_5SO_3H + HNO_3 \xrightarrow{H_2SO_4}$

(i) $C_6H_5Br + H_2SO_4 \xrightarrow{heat}$

6.12 Show the reagents and conditions you would use to convert benzene into the following:

(a) toluene
(b) *p*-bromotoluene
(c) *o*-bromobenzoic acid
(d) *m*-bromobenzoic acid
(e) 2,4,6-trinitrotoluene (TNT)
(f) *p*-dichlorobenzene
(g) *p*-chlorobenzenesulfonic acid
    (a moth repellant)
(i) *m*-nitrobenzoic acid
(h) *o*-nitrobenzoic acid

6.13 Show reagents and conditions you would use for the following conversions:

(a) phenol to 2,4,6-tribromophenol
(b) benzyl bromide to methyl benzyl ether
(c) styrene to 1-phenyl-1-ethanol

6.14 Arrange the following in order of increasing reactivity to bromination: nitrobenzene, toluene, aniline, benzene.

6.15 Each of the following compounds undergoes electrophilic aromatic substitution in one ring in preference to the other. For each, indicate which ring you would expect to be attacked in bromination, and draw the structural formula of the principal product.

(a)

(b)

6.16 According to modern organic theory, the factor favoring *ortho,para* substitution is the ability of the atom attached directly to the ring to participate in stabilizing the intermediate cation. This participation by the oxygen of methoxyl is shown in contributing structure (g) in Figure 6.4. Draw comparable contributing structures showing resonance stabilization by the following *ortho,para*-directing groups.

(a) —OH   **(b)** —NHCH₃   **(c)** —Cl

$$\text{(d)} \quad -\overset{\overset{\textstyle O}{\textstyle \|}}{N}HCCH_3 \qquad \text{(e)}$$

6.17 Show how you might distinguish between the following pairs of compounds by a simple chemical test. In each case, tell what test you would perform, what you would expect to observe, and write an equation for each positive test.

(a) benzene and cyclohexene
(b) ethylbenzene and styrene

6.18 List one major spectral characteristic that will enable you to distinguish between the following compounds.

(a) benzene and cyclohexene (UV, NMR)
(b) methylcyclopentane and benzene (IR, UV, NMR)
(c) toluene and styrene (UV, NMR)
(d) *p*-xylene and ethylbenzene (NMR)
(e) *p*-chlorotoluene and benzyl chloride (NMR)
(f) toluene and methylcylohexane (IR, UV, NMR)
(g) ethylbenzene and isopropylbenzene (NMR)
(h) *p*-diethylbenzene and 1,3,5-trimethylbenzene (NMR)

Aldehydes and Ketones

7.1 Introduction

Without doubt, the carbonyl group ($C=O$) is one of the most important functional groups in all of organic chemistry. It is the central feature of aldehydes and ketones, carboxylic acids, esters, amides, and acid halides and anhydrides. Reactions of the carbonyl group are essentially very simple and an understanding of its few basic reaction themes will lead quickly to an understanding of a wide variety of synthetic and biochemical transformations. We will discuss the chemistry of the carbonyl group in two separate chapters, first in this chapter on aldehydes and ketones, and then again in Chapter 9 on the functional derivatives of carboxylic acids.

7.2 Structure and Nomenclature

Aldehydes and ketones contain the carbonyl group, and are referred to as carbonyl compounds. Aldehydes are compounds of the general formula RCHO (or ArCHO); ketones have the general formula R—CO—R′ (either or both R or R′ may be aliphatic or aromatic).

The IUPAC nomenclature of aldehydes follows the familiar pattern of selecting as the name the longest continuous chain that contains the functional group and changing the ending of that hydrocarbon from -e to -al. Common names of aldehydes are very often derived from the common names of the corresponding carboxylic acids by dropping the suffix -ic and adding -aldehyde. Since we will not study the nomenclature of carboxylic acids until the next chapter (Section 8.3), at this point you will not be able to propose common names for aldehydes. However, you should learn the few common names given in this chapter.

We can illustrate how the common system of naming aldehydes works by referring to a carboxylic acid that you are familiar with, namely, acetic acid, CH_3CO_2H. The two-carbon aldehyde, CH_3CHO, is related to acetic acid and therefore is named acetaldehyde.

$$H-\overset{\overset{\displaystyle O}{\|}}{C}-H \qquad CH_3-\overset{\overset{\displaystyle O}{\|}}{C}-H \qquad CH_3CH_2-\overset{\overset{\displaystyle O}{\|}}{C}-H$$

methanal   ethanal   propanal
formaldehyde acetaldehyde propionaldehyde

$$CH_3-\underset{\underset{\displaystyle CH_3}{|}}{CH}-CH_2-\overset{\overset{\displaystyle O}{\|}}{C}-H \qquad \overset{\overset{\displaystyle O}{\|}}{C}-H$$

3-methylbutanal
β-methylbutyraldehyde

2,3-dihydroxypropanal
glyceraldehyde

benzaldehyde   2-hydroxybenzaldehyde
salicylaldehyde

Formaldehyde, the simplest of the aldehydes, derives its name from that of an oxidation product, formic acid. Formaldehyde is a gas (bp −21°C), but is not usually handled in this form. It is commonly produced as a 37% aqueous solution (formalin) which is used for the preservation of biological specimens. Alternatively, formaldehyde is produced in solid form as paraformaldehyde (a polymer) or as trioxymethylene (a trimer). Formaldehyde is regenerated from either solid form on heating. Candles of paraformaldehyde are sometimes used as fumigating agents.

$$HO(CH_2OCH_2OCH_2O)_xH$$

paraformaldehyde   trioxymethylene

The IUPAC system names ketones by using the ending -one to indicate the carbonyl group. As usual, the compound is named as a derivative of the longest continuous carbon chain that contains the functional group. In the case of ketones, a number must be used to designate the position of the carbonyl group.

Simple ketones are often named by indicating the names of the two hydrocarbon groups attached directly to the carbonyl.

$$CH_3-\overset{\overset{\displaystyle O}{\|}}{C}-CH_2-CH_3$$

butanone
methyl ethyl ketone

$$CH_3-\overset{\overset{\displaystyle O}{\|}}{C}-CH_3$$

propanone
acetone

$$CH_3-\overset{\overset{\displaystyle O}{\|}}{C}-\overset{\overset{\displaystyle CH_3}{|}}{CH}-CH_3$$

3-methyl-2-butanone
methyl isopropyl ketone

cyclohexanone

acetophenone
methyl phenyl ketone

ethyl cyclohexyl ketone

Acetone is an example of a compound whose name is part common or trivial and part derived. Acetone can be prepared by the pyrolysis of the barium salt and certain other heavy metal salts of acetic acid. The name acetone is derived from acetic acid, with the ending -one used to designate that the compound is a ketone.

In naming more complicated aldehydes and ketones, the carbonyl group generally takes precedence over alkene, hydroxyl, and most other functional groups.

$$CH_3\overset{\overset{\displaystyle OH}{|}}{C}HCH_2\overset{\overset{\displaystyle O}{\|}}{C}CH_3$$

4-hydroxy-2-pentanone

$$CH_3CH_2\overset{\overset{\displaystyle O}{\|}}{C}CH=CHCH_2CH_3$$

4-hepten-3-one

3-methyl-2-cyclohexen-1-one

Note that in the first example the —OH group is indicated by -hydroxy. Also note that in the name 4-hepten-3-one, the heptane root is changed to 4-heptene to indicate the presence of an alkene between carbons 4 and 5 and that the terminal e is converted to -3-one to indicate the presence of a carbonyl group at carbon 3.

A great variety of aldehydes and ketones have been isolated from natural sources, and are best known by their common or trivial names. Such names are usually derived from a source from which the compound can be isolated, or they may refer to a characteristic property of the compound. Figure 7.1 shows structural formulas of several aldehydes and ketones of natural occurrence together with their common names.

Natural or dextrorotatory camphor is characterized by a fragrant penetrating odor. It is obtained from the bark of the camphor tree (*Cinnamomum camphora*) indigenous to the island of Formosa. Camphor has some medicinal uses as a weak antiseptic and as an analgesic; it is found in certain liniments. Aldehydes and ketones are in particular abundance in plant essential oils. Irone, the odoriferous principal of the violet, is isolated readily from the rhizomes of iris. Citral is the major constituent (70%) of lemon grass oil. Because of its presence in almond seed, benzaldehyde has been known as oil of bitter almond. Benzaldehyde is also the chief constituent of oils expressed from kernels of the peach, cherry, laurel, and other fruits. Vanillin is the fragrant constituent of the vanilla bean. Cinnamaldehyde is the chief constituent of the oil of cinnamon.

Testosterone (Section 12.11), one of the most prominent members of the male sex hormone group, is produced in the testes and is involved in

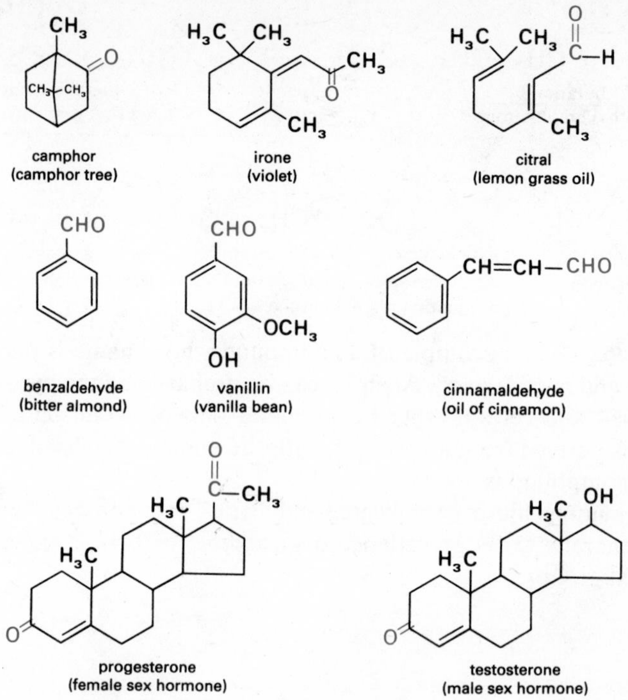

FIGURE 7.1 Common names of some aldehydes and ketones of natural occurrence.

the development of accessory sex functions in the male. Progesterone (Section 12.11), secreted by ovarian tissue (corpus luteum), is one of several sex hormones involved in stimulating growth of the uterine mucosa in preparation for implantation of the fertilized ovum.

In terms of use, acetone is by far the most important ketone. Because of its complete miscibility with water (Table 7.1) and many nonpolar organic solvents, it is used in large volumes as a solvent. Further it is used as an intermediate in the synthesis of many more complex organic molecules. Acetone is formed in the body during fat metabolism. Normally its concentration in the blood is very low, but under conditions where fat metabolism is accelerated, as in the disease diabetes mellitus, there is an accumulation of acetone in the blood. The odor of acetone on the breath of a patient is evidence that he may be a diabetic.

Formaldehyde is used for the preservation of biological materials and in the manufacture of silvered mirrors (Section 7.4). Probably the largest commercial use of formaldehyde is in the base-catalyzed condensation with phenol to yield resins known as Bakelites.

7.3 Physical Properties

The carbonyl group makes aldehydes and ketones polar molecules. Hence they have higher boiling points than nonpolar compounds of comparable molecular weight. Since they do not have a hydrogen atom directly bonded to oxygen, they do not show the same degree of interaction and association as alcohols and carboxylic acids of comparable molecular weight. The

TABLE 7.1 Physical properties of aldehydes and ketones.

Structure	Name Common	Name IUPAC	bp (°C)	Solubility (g/100 g H₂O)
HCHO	formaldehyde	methanal	−21	very
CH₃CHO	acetaldehyde	ethanal	20	∞
CH₃CH₂CHO	propionaldehyde	propanal	49	16
CH₃(CH₂)₂CHO	butyraldehyde	butanal	76	7
CH₃(CH₂)₃CHO	valeraldehyde	pentanal	103	slightly
CH₃(CH₂)₄CHO	caproaldehyde	hexanal	129	slightly
CH₃COCH₃	acetone	propanone	56	∞
CH₃COCH₂CH₃	methyl ethyl ketone	butanone	80	26
CH₃COCH₂CH₂CH₃	methyl propyl ketone	2-pentanone	102	6
CH₃CH₂COCH₂CH₃	diethyl ketone	3-pentanone	101	5

lower aldehydes and ketones are appreciably soluble in water because of hydrogen bonding between the carbonyl group and the solvent, water.

7.4 Oxidation of Aldehydes

As we pointed out in Chapter 5, ketones are resistant to oxidation except under the most vigorous conditions. Aldehydes, on the other hand, are readily oxidized to carboxylic acids even by such weak oxidizing agents as silver ion, Ag^+. In Tollens' test, a solution of silver nitrate in ammonium hydroxide is added to the substance suspected of being an aldehyde. Within a few minutes at room temperature, silver ion is reduced to metallic silver and the aldehyde is oxidized to the acid. This test requires an alkaline medium, and to prevent the precipitation of the insoluble silver oxide, the silver ion is converted with ammonia into the soluble silver-ammonia complex ion, $Ag(NH_3)_2^+$.

$$RCHO + 2Ag(NH_3)_2^+ + 3OH^- \longrightarrow RCOO^- + 2Ag + 4NH_3 + 2H_2O$$

If the test is done properly, the metallic silver will deposit as a mirror on a glass surface. It is for this reason that the test is commonly called the silver mirror test. Oxidation by silver ion is a convenient test to distinguish aldehydes from ketones (which do not react). The reduction of silver ion by formaldehyde is used in a commercial process to silver mirrors.

7.5 Polarity of the Carbonyl Group

In Section 1.3 we drew a Lewis electronic structure for formaldehyde, the simplest of the organic molecules containing the carbonyl group, and later in Section 1.10 we described the bonding in the carbonyl group in terms of the overlap of hybridized orbitals.

171

In terms of electronic structure and bonding, the carbon-oxygen double bond closely resembles the carbon-carbon double bond. However, there is also a major difference between the two. The bond between the two unlike atoms in the carbonyl group is highly polarized due to the large difference in electronegativity between carbon and oxygen. This polarization (partial ionic character) of the colvalent bond is indicated below using both arrow and delta notations.

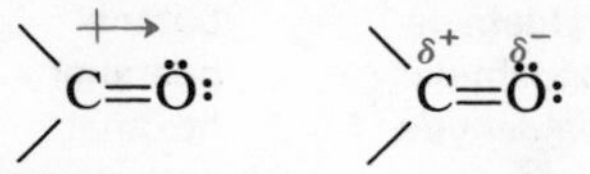

$$\begin{array}{cc} \diagdown \\ \diagup C=\ddot{O}\colon \end{array} \qquad \begin{array}{cc} \diagdown \\ \diagup C\overset{\delta^+}{=}\overset{\delta^-}{\ddot{O}}\colon \end{array}$$

Alternatively, the carbonyl group may be pictured as a hybrid of two major contributing structures.

$$\begin{array}{cc} \diagdown \\ \diagup C=\ddot{O}\colon \end{array} \longleftrightarrow \begin{array}{cc} \diagdown \\ \diagup \overset{+}{C}-\ddot{O}\colon^- \end{array}$$
(a)   (b)

Structure (b) makes a significant contribution to the hybrid for its places the negative charge on the more electronegative oxygen atom and the positive charge on the less electronegative carbon atom.

7.6 Reactions of the Carbonyl Group

The typical reactions of aldehydes and ketones may be divided into two classes: (1) nucleophilic addition to the carbonyl group, and (2) reactions at the carbon atom alpha to the carbonyl group. These reactions are among the most important in organic and biochemistry for the formation of new carbon-carbon bonds and for this reason they deserve special attention. In this chapter we shall develop the chemistry of nucleophilic addition to the carbonyl group of aldehydes and ketones, and in Chapter 9 we shall discuss nucleophilic addition to the carbonyl groups of carboxylic acids, esters, amides, acid anhydrides, and acid halides. In all of these nucleophilic additions, the same generalization applies: reaction with a nucleophile produces a tetrahedral carbonyl addition compound. In the following illustrations, the nucleophilic reagent is indicated by the symbol H—Nu: to emphasize the presence of an unshared pair of electrons on the nucleophile.

$$H-\ddot{Nu}\colon + \begin{array}{c}\diagdown\\\diagup\end{array}C=\ddot{O}\colon \rightleftharpoons H-\overset{+}{\ddot{Nu}}-\overset{|}{\underset{|}{C}}-\ddot{O}\colon^- \rightleftharpoons Nu-\overset{|}{\underset{|}{C}}-\ddot{O}H$$

tetrahedral
carbonyl addition
compound

Example:

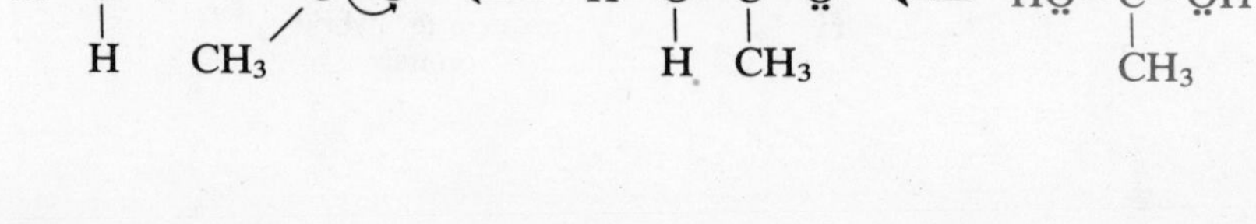

$$H-\ddot{O}\colon + \begin{array}{c}CH_3\\\diagup\end{array}C=\ddot{O}\colon \rightleftharpoons H-\overset{CH_3}{\underset{H}{\overset{+}{\ddot{O}}-\underset{CH_3}{C}}}-\ddot{O}\colon^- \rightleftharpoons H\ddot{O}-\overset{CH_3}{\underset{CH_3}{C}}-\ddot{O}H$$

It is because of the polarity of the carbonyl group that the carbonyl carbon is susceptible to attack by nucleophiles. Another consequence of this polarity and of the presence of two unshared pairs of electrons is that the carbonyl oxygen is susceptible to attack by electrophiles. The most common electrophile in carbonyl addition reactions is the proton H^+.

$$\ce{\underset{}{C=\overset{..}{\underset{..}{O}}:} + H^+ \rightleftharpoons \overset{+}{C}=\overset{..}{O} \longleftrightarrow \overset{+}{C}-\overset{..}{\underset{..}{O}}}$$

resonance-stabilized cation

Proton transfer to a carbonyl oxygen produces a resonance-stabilized cation and by further increasing the electron deficiency of the carbonyl carbon, makes it even more susceptible to attack by a nucleophile.

$$H-Nu: + \overset{+}{C}-\overset{..}{\underset{..}{O}} \rightleftharpoons \overset{+}{Nu}-C-\overset{..}{O}H \overset{-H^+}{\rightleftharpoons} Nu-C-OH$$

Example:

$$H_2\overset{..}{O}: + \overset{CH_3}{\underset{CH_3}{\overset{+}{C}-\overset{..}{O}}} \rightleftharpoons H-\overset{+}{\underset{H}{O}}-\overset{CH_3}{\underset{CH_3}{C}}-\overset{..}{O}H \overset{-H^+}{\rightleftharpoons} H\overset{..}{O}-\overset{CH_3}{\underset{CH_3}{C}}-\overset{..}{O}H$$

Notice that these reactions, whether initiated by electrophilic attack of H^+ or by nucleophilic attack of Nu:, correspond to addition of Nu:H (HO—H in the particular example chosen for illustration) to the carbon-oxygen double bond. These pathways for addition to the carbonyl group are two of the basic themes referred to in the introduction to this chapter.

7.7 Addition of Water—Hydration

As suggested by the two examples in the last section, aldehydes and ketones add water reversibly to form hydrated carbonyl compounds. In pure water at 20°C, formaldehyde is about 99.99% hydrated.

$$\underset{H}{\overset{H}{C}}=O + H_2O \rightleftharpoons \underset{H}{\overset{H}{C}}\underset{OH}{\overset{OH}{}}$$

formaldehyde   99.99%

Under the same conditions acetaldehyde is about 58% hydrated. The extent of hydration of ketones at equilibrium is very small. While it is not possible to observe directly acetone hydrate in aqueous solution, it is possible to infer indirectly its existence using $H_2{}^{18}O$ and to demonstrate that there is oxygen exchange between water and acetone. This exchange involves the formation of hydrated acetone.

$$\underset{H_3C}{\overset{H_3C}{C}}=O + H_2{}^{18}O \rightleftharpoons \underset{H_3C}{\overset{H_3C}{C}}\underset{OH}{\overset{{}^{18}OH}{}} \rightleftharpoons \underset{H_3C}{\overset{H_3C}{C}}={}^{18}O + H_2O$$

173

Exchange is slow in pure water, but is accelerated by the addition of small amounts of HCl or other strong acids. Under these conditions protonation of the carbonyl oxygen forms a resonance-stabilized cation which reacts with water more readily than does acetone itself.

$$
\underset{H_3C}{\overset{H_3C}{>}}C=\overset{..}{O}: + H^+ \rightleftharpoons \left[\underset{H_3C}{\overset{H_3C}{>}}C \overset{+}{=} \overset{..}{O}H \longleftrightarrow \underset{H_3C}{\overset{H_3C}{>}}\overset{+}{C}-\overset{..}{O}H \right] \overset{H_2O}{\rightleftharpoons}
$$

resonance stabilized
cation

$$
\underset{H_3C}{\overset{H_3C}{>}}\underset{\overset{|}{H}}{\overset{OH}{\underset{\overset{+}{O}-H}{C}}} \rightleftharpoons \underset{H_3C}{\overset{H_3C}{>}}\underset{OH}{\overset{OH}{C}} + H^+
$$

Generally, the hydrated forms of aldehydes and ketones cannot be isolated because water is readily lost and the carbonyl group is reformed.

7.8 Addition of Alcohols— Formation of Acetals and Ketals

Alcohols add to aldehydes and ketones in the same manner as described for the addition of water.

$$
CH_3-CHO + CH_3OH \rightleftharpoons CH_3-\underset{OCH_3}{\overset{OH}{\underset{|}{\overset{|}{C}}}}-H
$$

a hemiacetal
(acetaldehyde hemiacetal)

Addition of one molecule of alcohol to an aldehyde results in the formation of a hemiacetal. The comparable reaction with a ketone forms a hemiketal. Hemiacetals are only minor constituents of an equilibrium mixture of alcohol and aldehyde, except in one very important case. When the hydroxyl group is a part of the same molecule that contains the carbonyl group and a five- or six-membered ring can form, the cyclic hemiacetal structure is frequently the normal form in which the compound exists at equilibrium.

$$
\underset{\overset{|}{OH}}{CH_3CHCH_2CH_2}\overset{\overset{O}{\|}}{CH} \rightleftharpoons
$$

4-hydroxypentanal   a cyclic hemiacetal

Hemiacetals and hemiketals react further with alcohol, with the formation of compounds known as acetals and ketals. The reaction is catalyzed by acids.

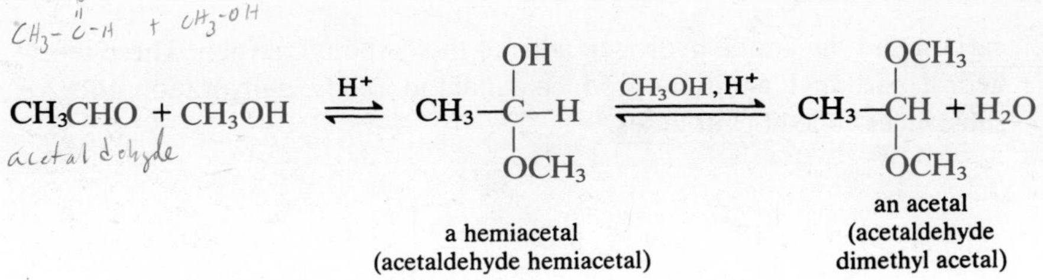

CH₃–C(=O)–H + CH₃–OH *(handwritten)*

$$CH_3CHO + CH_3OH \rightleftharpoons CH_3-\underset{\underset{OCH_3}{|}}{\overset{\overset{OH}{|}}{C}}-H \underset{}{\overset{CH_3OH,\,H^+}{\rightleftharpoons}} CH_3-\underset{\underset{OCH_3}{|}}{CH} + H_2O$$

acetaldehyde *(handwritten)*

a hemiacetal
(acetaldehyde hemiacetal)

an acetal
(acetaldehyde
dimethyl acetal)

A mechanism for the acid-catalyzed conversion of acetaldehyde hemiacetal into acetaldehyde dimethyl acetal is shown below. It involves the protonation of —OH to form an oxonium ion, loss of a molecule of water to form a resonance-stabilized cation, condensation with a second molecule of alcohol, and finally loss of a proton to give the acetal.

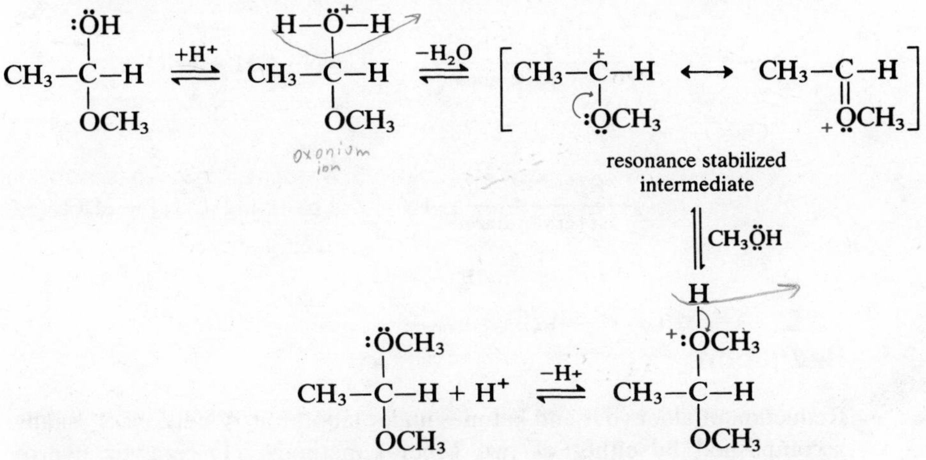

$$CH_3-\underset{\underset{OCH_3}{|}}{\overset{\overset{:\ddot{O}H}{|}}{C}}-H \overset{+H^+}{\rightleftharpoons} CH_3-\underset{\underset{OCH_3}{|}}{\overset{\overset{H-\overset{+}{\ddot{O}}-H}{|}}{C}}-H \overset{-H_2O}{\rightleftharpoons} \left[CH_3-\overset{+}{\underset{\underset{:\ddot{O}CH_3}{|}}{C}}-H \longleftrightarrow CH_3-\underset{\underset{+\ddot{O}CH_3}{||}}{C}-H \right]$$

oxonium ion *(handwritten)*

resonance stabilized intermediate

$$\Big\Vert CH_3\ddot{O}H$$

$$CH_3-\underset{\underset{OCH_3}{|}}{\overset{\overset{:\ddot{O}CH_3}{|}}{C}}-H + H^+ \overset{-H+}{\rightleftharpoons} CH_3-\underset{\underset{OCH_3}{|}}{\overset{\overset{+:\overset{H}{\ddot{O}}CH_3}{|}}{C}}-H$$

Acetal formation is an equilibrium reaction and in order to obtain high yields of the acetal, it is necessary to remove water from the reaction mixture and thus favor the formation of the product acetal. Ketones in the same type of reaction form ketals. The addition mechanism is essentially the same as that described for acetal formation but the equilibrium is less favorable to the formation of ketals than it is of acetals.

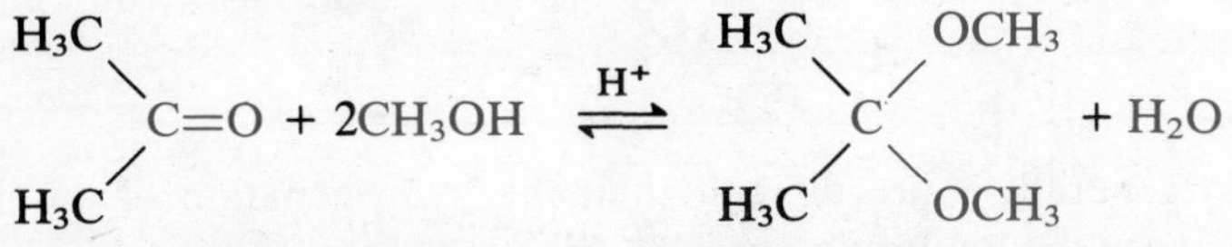

$$\underset{H_3C}{\overset{H_3C}{>}}C=O + 2CH_3OH \overset{H^+}{\rightleftharpoons} \underset{H_3C}{\overset{H_3C}{>}}C\underset{OCH_3}{\overset{OCH_3}{<}} + H_2O$$

Acetal and ketal formation is catalyzed by acid. Their hydrolysis in water is also catalyzed by acid. However, acetals and ketals are stable and unreactive to aqueous base.

7.9 Addition of Ammonia and Its Derivatives

Ammonia, primary amines (R—NH₂), and certain monosubstituted amines such as hydroxylamine (HO—NH₂) are strong enough nucleophiles to add directly to aldehydes and ketones. The amine nitrogen adds to the carbonyl carbon and the amine hydrogen adds to the carbonyl oxygen. These tetra-hedral carbonyl addition products undergo ready dehydration to give substances called <u>Schiff bases</u>.

$$\diagdown\!\!\!C{=}O + H_2NR \longrightarrow \left[\begin{array}{c} OH \\ \diagup \\ \diagdown\!\!\!C \\ \diagup \quad \diagdown \\ NHR \end{array} \right] \longrightarrow \diagdown\!\!\!C{=}NR + H_2O$$

a Schiff
base

In the case of certain derivatives of ammonia, the reaction products are often crystalline solids with sharp melting points whose physical properties can be used to identify and characterize unknown aldehydes and ketones.

$$\diagdown\!\!\!C{=}O \quad\begin{array}{c} \xrightarrow[\text{hydroxylamine}]{H_2N-OH} \diagdown\!\!\!C{=}N-OH + H_2O \\ \\ \xrightarrow[\text{phenylhydrazine}]{H_2N-NH-C_6H_5} \diagdown\!\!\!C{=}N-NH-C_6H_5 + H_2O \end{array}$$

an oxime

a phenylhydrazone

7.10   Addition of Hydrogen— Reduction

Reduction of aldehydes and ketones under laboratory conditions is usually accomplished by either of two general methods: (1) catalytic hydrogenation or (2) chemical reduction using a complex metal hydride.

<u>Catalytic hydrogenation</u> is completely analogous to the reduction of carbon-carbon double bonds. The carbonyl compound in an inert solvent and hydrogen are shaken under pressure and in the presence of a heavy metal catalyst. The most commonly used catalysts are palladium, platinum, nickel, or copper chromite. Catalytic hydrogenation of an aldehyde or

$$\text{(cyclohexanone)} + H_2 \xrightarrow{Ni} \text{(cyclohexanol, OH)}$$

ketone is generally more difficult than the hydrogenation of an alkene. Consequently, if a molecule contains both a carbon-carbon double bond and an aldehyde or ketone, the conditions necessary to reduce the carbonyl group may also saturate the alkene.

The second and more common laboratory method involves the use of inorganic hydrides such as lithium aluminum hydride, $LiAlH_4$, and sodium borohydride, $NaBH_4$. <u>Lithium aluminum hydride</u> reacts almost violently with even such weak acids (proton donors) as water or alcohols and consequently these reductions must be carried out in dry ether or some other nonhydroxylic solvent.

$$4CH_3CH_2OH + LiAlH_4 \longrightarrow LiOCH_2CH_3 + Al(OCH_2CH_3)_3 + 4H_2$$

lithium
ethoxide

aluminum
ethoxide

Lithium aluminum hydride reduction of an aldehyde or ketone is most often carried out in ether and yields the lithium and aluminum salts of the alcohol. Hydrolysis of these salts (metal alkoxides) in water or dilute aqueous acid produces the alcohol.

$$4CH_3CHO + LiAlH_4 \xrightarrow{\text{ether}} (CH_3CH_2O)_3Al + CH_3CH_2OLi$$

aluminum   lithium
ethoxide   ethoxide

$$\downarrow H_3O^+$$

$$4CH_3CH_2OH + Al^{3+} + Li^+$$

This reagent appears to reduce carbonyl groups by transferring a hydride ion ($H:^-$) to the carbonyl carbon. Hydride ion is not only an exceedingly strong base (hence the reaction with alcohols to produce hydrogen gas) but it is also a good nucleophile. The transfer of hydride ion and the formation of an aluminum (or lithium) alkoxide occur simultaneously.

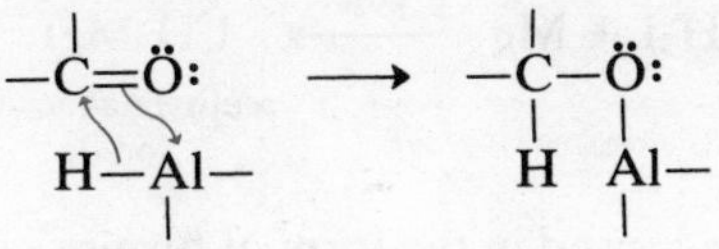

$$-\overset{|}{\underset{|}{C}}=\ddot{O}: \longrightarrow -\overset{|}{\underset{|}{C}}-\ddot{O}:$$
$$H-Al- \qquad\qquad H \quad Al-$$

Actually, lithium aluminum hydride is an extremely versatile carbonyl reducing agent for it will reduce not only aldehydes and ketones but also esters, amides, and anhydrides. However, it is also selective in that it will not reduce carbon-carbon double or triple bonds since they are not subject to nucleophilic attack to nearly the same degree as are the highly polarized carbon-oxygen double bonds.

$$CH_3CH=CH-\overset{\overset{O}{\|}}{C}H \xrightarrow[\text{ether}]{LiAlH_4} \xrightarrow{H_2O} CH_3CH=CHCH_2OH$$

Sodium borohydride, $NaBH_4$, is also useful for reducing aldehydes and ketones to alcohols. This reagent does not react as readily with alcohols or water to liberate hydrogen and therefore it can be used in aqueous solutions of methanol or ethanol. It is also more selective than $LiAlH_4$ in

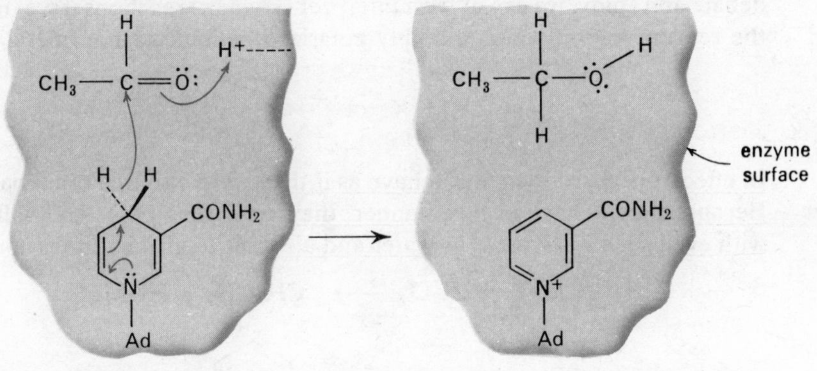

FIGURE 7.2 Schematic diagram for the enzyme-catalyzed reduction of acetaldehyde by NADH showing the transfer of a hydride ion.

that under most conditions it will reduce aldehydes and ketones, but few other functional groups.

The enzyme-catalyzed reduction of an aldehyde or ketone by <u>NADH</u> can also be formulated as a transfer of a hydride ion from NADH to the biomolecule. In the schematic diagram (Figure 7.2) the enzyme is shown binding NADH, acetaldehyde, and the proton needed to complete formation of the primary alcohol.

7.11 Addition of Organometallics— The Grignard Reaction

In 1901 <u>Victor Grignard</u> discovered that magnesium metal reacts with alkyl or aryl halides to give a reagent containing a carbon-metal bond. For example, methyl iodide reacts with magnesium in diethyl ether to give an ether-soluble substance called methylmagnesium iodide.

$$CH_3I + Mg \xrightarrow{\text{ether}} CH_3MgI$$

methylmagnesium
iodide

Generally magnesium is used in the form of turnings or shavings. The ether must be pure and absolutely free from traces of moisture or ethanol. When this reaction is initiated under proper conditions, the magnesium metal first becomes etched and then gradually disintegrates in a highly exothermic reaction to give a cloudy solution of an alkylmagnesium halide, known as a "Grignard reagent."

$$R-X + Mg \xrightarrow{\text{ether}} R-MgX$$

aryl- or alkyl-
magnesium halide
"Grignard reagent"

<u>Grignard reagents</u> are highly versatile substances and participate in a wide range of important reactions. For the discovery and study of this class of organometallic compounds, Victor Grignard was awarded the Nobel Prize in Chemistry in 1912.

Although the real structure of the Grignard reagent is still a matter of debate and study, we can at least interpret Grignard reactions if we think of the reagent as containing a highly polarized or ionic carbon-metal bond.

$$-\overset{|}{\underset{|}{C}}\overset{\delta^-}{-}\overset{\delta^+}{MgX} \quad \text{or} \quad -\overset{|}{\underset{|}{C}}{:}^- MgX^+$$

In effect, Grignard reagents behave as if they were a source of carbanions. Because they behave in this manner, they are strong bases and will react with even such weak acids as water and alcohols to give hydrocarbons.

$$CH_3CH_2MgBr + H_2O \longrightarrow CH_3CH_3 + HOMgBr$$

$$+ \ CH_3CH_2OH \longrightarrow \qquad + \ CH_3CH_2OMgBr$$

It is for this reason that the diethyl ether used as a solvent for Grignard reactions must be rigorously dried and free from traces of ethanol. This reaction with even weak acids like water does have at least one practical value. If D$_2$O is used in place of H$_2$O, then deuterium-labeled hydrocarbons can be prepared.

$$\text{(m-tolyl)—MgCl} + D_2O \longrightarrow \text{(m-tolyl)—D} + DOMgCl$$

For our purposes, the most important use of Grignard reagents is in reaction with carbonyl compounds. <u>The reagent adds to the carbonyl group of aldehydes and ketones to produce metal alkoxide salts. Hydrolysis of these salts in aqueous acid yields alcohols. For organic chemistry, this reaction sequence is extremely important since a new carbon-carbon bond is formed in the process.</u>

Hydrolysis of salt in aqueous acid

$$R—MgX + H—\overset{O}{\underset{||}{C}}—H \longrightarrow R—\underset{OMgX}{\overset{|}{CH_2}} \xrightarrow{H_3O^+} R—\underset{OH}{\overset{|}{CH_2}}$$

formaldehyde   *an alkoxide salt*   a primary alcohol

$$R—MgX + R'—\overset{O}{\underset{||}{C}}—H \longrightarrow R—\underset{OMgX}{\overset{|}{CH}}—R' \xrightarrow{H_3O^+} R—\underset{OH}{\overset{|}{CH}}—R'$$

an aldehyde   a secondary alcohol

$$R—MgX + R'—\overset{O}{\underset{||}{C}}—R'' \longrightarrow R—\underset{\underset{R''}{|}}{\overset{OMgX}{\overset{|}{C}}}—R' \xrightarrow{H_3O^+} R—\underset{\underset{R''}{|}}{\overset{OH}{\overset{|}{C}}}—R'$$

a ketone   a tertiary alcohol

<u>Reaction of a Grignard reagent with formaldehyde gives a primary alcohol; reaction with any other aldehyde gives a secondary alcohol. Reaction with any ketone gives a tertiary alcohol.</u> Some specific examples of these reactions are shown in Figure 7.3. In each example, the intermediate organomagnesium compound is hydrolyzed in aqueous acid and only the final alcohol is shown.

<u>Grignard reagents also add to ethylene oxide and other epoxides.</u>

$$RMgBr + H_2C\overset{}{\underset{O}{\diagdown\diagup}}CH_2 \longrightarrow R—CH_2CH_2OH$$

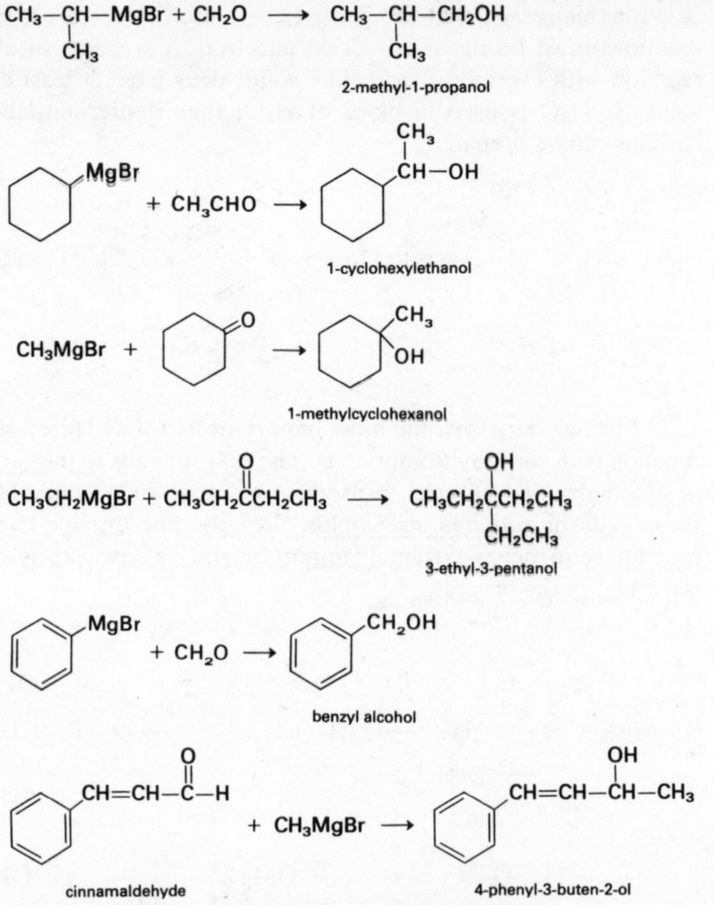

$$CH_3-\underset{\underset{CH_3}{|}}{CH}-MgBr + CH_2O \longrightarrow CH_3-\underset{\underset{CH_3}{|}}{CH}-CH_2OH$$

2-methyl-1-propanol

+ CH_3CHO →

1-cyclohexylethanol

CH_3MgBr + →

1-methylcyclohexanol

$$CH_3CH_2MgBr + CH_3CH_2\overset{\overset{O}{||}}{C}CH_2CH_3 \longrightarrow CH_3CH_2\underset{\underset{CH_2CH_3}{|}}{\overset{\overset{OH}{|}}{C}}CH_2CH_3$$

3-ethyl-3-pentanol

+ CH_2O →

benzyl alcohol

+ CH_3MgBr →

cinnamaldehyde

4-phenyl-3-buten-2-ol

FIGURE 7.3 Examples of the use of Grignard reactions for the synthesis of primary, secondary, and tertiary alcohols. Assume that magnesium alkoxide salts are hydrolyzed in aqueous acid.

This is a very convenient method for lengthening a carbon chain by two atoms

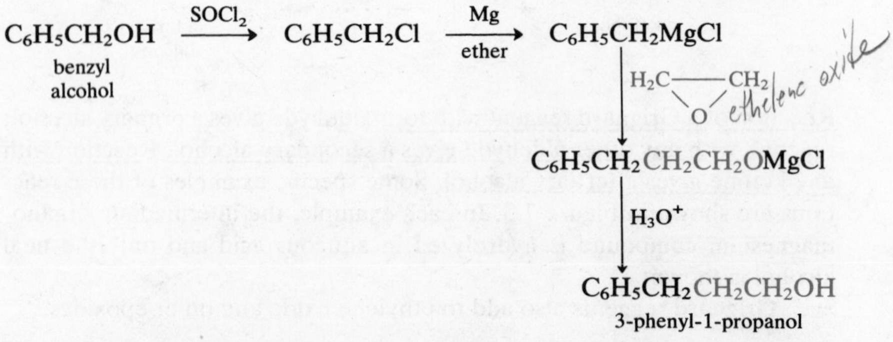

$$C_6H_5CH_2OH \xrightarrow{SOCl_2} C_6H_5CH_2Cl \xrightarrow[\text{ether}]{Mg} C_6H_5CH_2MgCl$$

benzyl
alcohol

$$\underset{O}{\overset{H_2C\underset{\diagdown}{\quad}CH_2}{}} \text{ ethelene oxite}$$

$$\downarrow$$

$$C_6H_5CH_2CH_2CH_2OMgCl$$

$$\downarrow H_3O^+$$

$$C_6H_5CH_2CH_2CH_2OH$$
3-phenyl-1-propanol

Finally, Grignard reagents add to carbon dioxide. The initial product, the magnesium salt of the carboxylic acid, is then treated with dilute mineral acid to yield the free carboxylic acid.

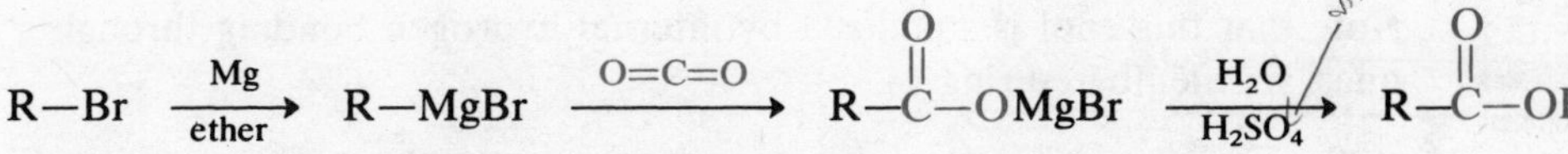

$$R-Br \xrightarrow[\text{ether}]{Mg} R-MgBr \xrightarrow{O=C=O} R-\overset{\overset{\displaystyle O}{\|}}{C}-OMgBr \xrightarrow[\text{H}_2\text{SO}_4]{\text{H}_2\text{O}} R-\overset{\overset{\displaystyle O}{\|}}{C}-OH$$

This is a very convenient method for converting an alkyl or aryl halide to a carboxylic acid. Shown below is the conversion of bromobenzene into benzoic acid.

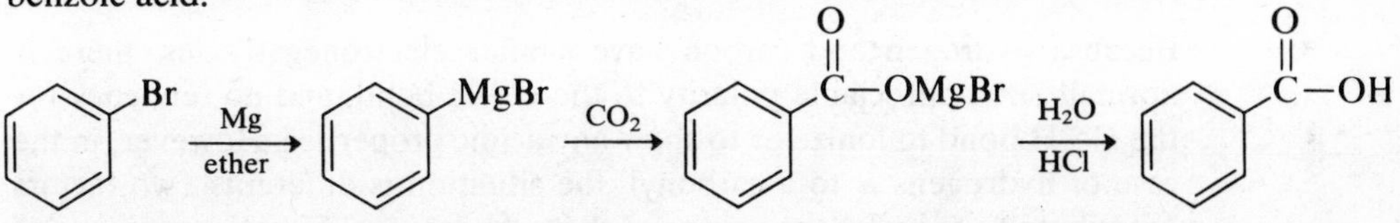

7.12 Tautomerism

Aldehydes, ketones, and other carbonyl compounds (for example, esters) with an α-hydrogen are in rapid and reversible equilibrium with an isomer formed by the migration of a proton from the α-carbon to oxygen. This new isomer is called an enol, a name derived from the IUPAC designation of it as both alkene and an alcohol.

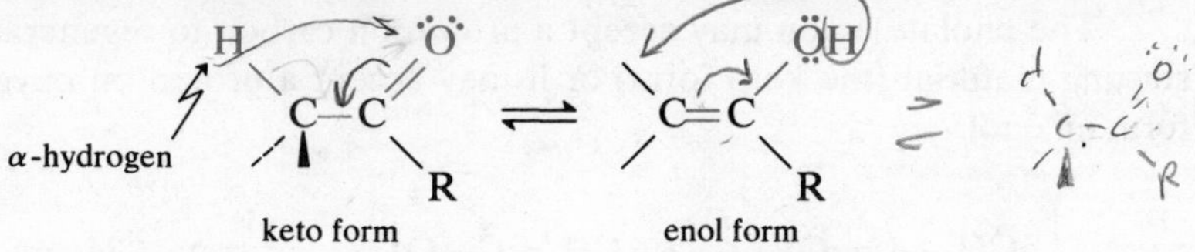

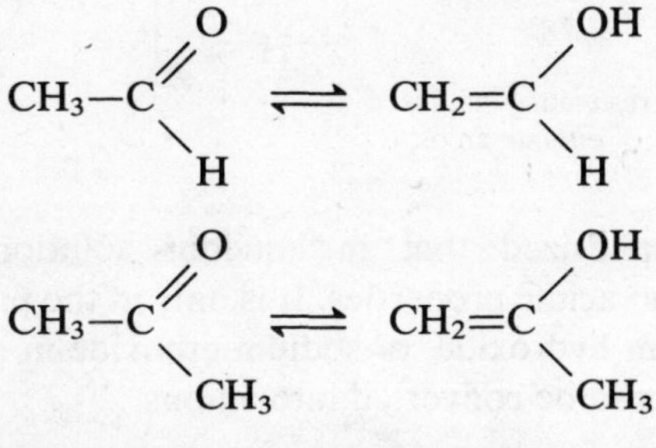

$$CH_3-\overset{\overset{\displaystyle O}{\|}}{C}\diagdown_H \;\rightleftharpoons\; CH_2=C\overset{\diagup OH}{\diagdown_H}$$

$$CH_3-\overset{\overset{\displaystyle O}{\|}}{C}\diagdown_{CH_3} \;\rightleftharpoons\; CH_2=C\overset{\diagup OH}{\diagdown_{CH_3}}$$

The term tautomerism is used to designate this rearrangement. Most generally tautomeric shifts are observed with migration of a proton between two or more sites within a molecule. In the case of simple aldehydes and ketones, the keto form predominates over the enol form at equilibrium by factors of well over 1000 to 1, and no one has yet isolated a pure enol form of a simple aldehyde or ketone because the enol form converts to the more stable keto forms.

When the α-hydrogen is flanked by a second carbonyl group, as in ethyl acetoacetate, the equilibrium shifts toward the enol form. Liquid ethyl acetoacetate consists of an equilibrium mixture containing approximately 75% of the enol form.

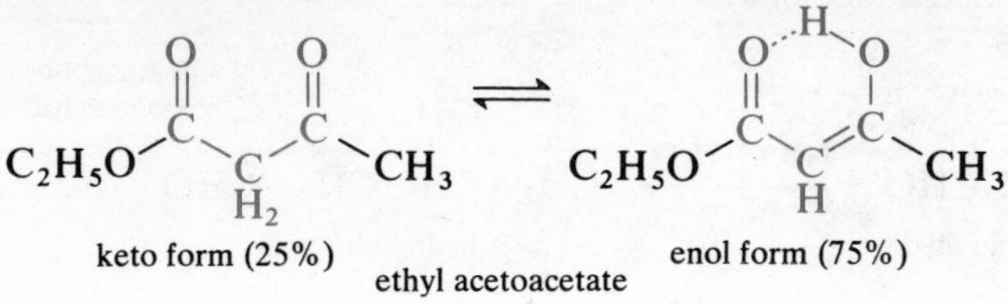

keto form (25%)   enol form (75%)

ethyl acetoacetate

Note that this enol is stabilized by internal hydrogen bonding through a quasi six-membered ring.

7.13   Acidity of α-Hydrogens

Because hydrogen and carbon have similar electronegativities, there is normally no appreciable polarity to the C—H bond, and no tendency for the C—H bond to ionize or to show any acidic properties. However, in the case of hydrogens α to a carbonyl, the situation is different. Two factors contribute to enhance the acidity of the α-hydrogens. First the presence of the adjacent carbon-oxygen dipole polarizes the electron pair of the C—H bond to such an extent that the hydrogen may be removed as a proton by a strong base. Ionization of the α-hydrogen results in the formation of an anion. The second and perhaps the more important factor in the acidity of the α-hydrogens is the fact that the resultant anion can be stabilized by resonance. The enolate anion is a hybrid of two important structures, one with the negative charge on carbon (a carbanion) and the other with the negative charge on oxygen.

The enolate anion may accept a proton on carbon to regenerate the starting material (the keto form) or it may accept a proton on oxygen to form an enol.

$$\underset{\substack{\mid\\ \text{H}}}{\text{CH}_3-\text{C}=\text{O}} \quad \underset{+\text{H}^+}{\overset{-\text{H}^+}{\rightleftharpoons}} \quad \left[\underset{\substack{\mid\\ \text{H}}}{{}^-\ddot{\text{C}}\text{H}_2-\text{C}=\ddot{\text{O}}:} \quad \longleftrightarrow \quad \underset{\substack{\mid\\ \text{H}}}{\text{CH}_2=\text{C}-\ddot{\text{O}}:^-}\right] \quad \underset{-\text{H}^+}{\overset{+\text{H}^+}{\rightleftharpoons}} \quad \underset{\substack{\mid\\ \text{H}}}{\text{CH}_2=\text{C}-\text{OH}}$$

keto form   resonance-stabilized enolate anion   enol form

It should be emphasized that in aqueous solution, most carbonyl compounds show no acidic properties. It is only in the presence of a strong base such as sodium hydroxide, or sodium ethoxide in ethanol, that aldehydes and ketones can be converted into anions.

7.14   Addition of Carbanions—The Aldol Condensation

The most important type of reaction of an anion derived from an aldehyde or ketone is the addition to the carbonyl of another aldehyde or ketone. For example, the condensation of two molecules of acetaldehyde in aqueous alkali yields 3-hydroxybutanal, commonly known as aldol. The name aldol is derived from the structural features of the compound; it is an aldehyde and an alcohol.

a new carbon-carbon bond is formed

$$2\text{CH}_3\text{CHO} \xrightarrow{\text{NaOH, H}_2\text{O}} \underset{\substack{\mid\\ \text{OH}}}{\text{CH}_3-\text{CH}-\text{CH}_2-\text{CHO}}$$

acetaldehyde   3-hydroxybutanal aldol

Step 1 in the mechanism for this reaction uses a molecule of base to form the anion of acetaldehyde. Actually, the resonance-stabilized enolate anion is produced, but for convenience we focus our attention on the carbanion contributing structure. Nucleophilic addition of the carbanion to another molecule of acetaldehyde in step 2 gives a carbonyl addition intermediate, which is turn reacts with water to give the final product and to regenerate a molecule of base.

Step 1 $CH_3CHO + OH^- \rightleftharpoons :\bar{C}H_2-CHO + HOH$

Step 2 $CH_3-\overset{:\overset{..}{O}:}{\overset{\|}{C}}-H + :\bar{C}H_2-CHO \rightleftharpoons CH_3-\overset{:\overset{..}{O}:^-}{\underset{H}{\overset{|}{C}}}-CH_2-CHO$

Step 3 $CH_3-\overset{:\overset{..}{O}:^-}{\underset{H}{\overset{|}{C}}}-CH_2-CHO + HOH \rightleftharpoons CH_3-\overset{:\overset{..}{O}H}{\underset{H}{\overset{|}{C}}}-CH_2-CHO + OH^-$

Ketones also undergo the aldol condensation as illustrated by the condensation of acetone in the presence of a barium hydroxide catalyst.

$$2CH_3-\overset{O}{\overset{\|}{C}}-CH_3 \underset{Ba(OH)_2}{\rightleftharpoons} CH_3-\overset{OH}{\underset{CH_3}{\overset{|}{C}}}-CH_2-\overset{O}{\overset{\|}{C}}-CH_3$$

4-hydroxy-4-methyl-2-pentanone
diacetone alcohol

The ingredients in the key step of the aldol condensation are an anion and a carbonyl group. In the self-condensation both roles are played by one kind of molecule, but there is no reason why this should be a necessary condition for the reaction. Many kinds of mixed aldol condensations are possible. Consider the aldol condensation of acetone and formaldehyde. Formaldehyde cannot function as an enolate anion because it contains no α-hydrogen, but it can function as a particularly good enolate anion acceptor because of the freedom from steric crowding around the carbonyl group. Acetone forms an enolate anion easily but its carbonyl group is a relatively poor enolate anion acceptor compared to that of formaldehyde. Consequently the aldol condensation of acetone and formaldehyde occurs readily.

$$CH_3-\overset{O}{\overset{\|}{C}}-CH_3 + H-\overset{O}{\overset{\|}{C}}-H \xrightarrow{OH^-} CH_3-\overset{O}{\overset{\|}{C}}-CH_2-\overset{OH}{\overset{|}{C}}H_2$$

4-hydroxy-2-butanone

In the case of mixed aldol condensations where there is not an appreciable difference in reactivity between the two compounds, mixtures of products result. For example, in the condensation of equimolar quantities of propanal and butanal, both α carbons are alike and the carbonyls are alike. As a consequence, the aldol condensation in which a mixture of these two aldehydes is used, results in a mixture of four possible products.

183

β-Hydroxyaldehydes and β-hydroxyketones are very easily dehydrated. Often the conditions necessary to bring about the condensation itself will suffice to cause dehydration. Alternatively, warming the aldol product in dilute mineral acid leads to dehydration. The major product of the loss of water is one with the carbon-carbon double bond α-β to the carbonyl.

$$\underset{\text{aldol}}{CH_3-\overset{\overset{\displaystyle OH}{|}}{CH}-CH_2-CHO} \xrightarrow[\text{warm}]{\text{dilute HCl}} \underset{\text{2-butenal}\atop\text{crotonaldehyde}}{CH_3-\overset{\beta}{CH}=\overset{\alpha}{CH}-CHO} + H_2O$$

The double bonds of alkenes, aldehydes, and ketones are readily reduced by catalytic hydrogenation or by chemical means. Hence the aldol condensation is often used for the preparation of saturated alcohols. For example, acetaldehyde may be converted into 1-butanol by first making 2-butenal.

$$2CH_3CHO \xrightarrow[\text{then dehydration}]{\text{aldol condensation}} \underset{\text{2-butenal}}{CH_3-CH=CH-CHO} \xrightarrow{2H_2,\, Pt} \underset{\text{1-butanol}}{CH_3CH_2CH_2CH_2OH}$$

Alternatively, if the β-hydroxyaldehyde is isolated, selective oxidation of the aldehyde with silver nitrate in ammonia solution will produce a β-hydroxycarboxylic acid.

$$\underset{\text{aldol}}{CH_3-\overset{\overset{\displaystyle OH}{|}}{CH}-CH_2-CHO} + 2Ag^+ \xrightarrow{NH_4OH} \underset{\text{3-hydroxybutanoic acid}\atop\beta\text{-hydroxybutyric acid}}{CH_3-\overset{\overset{\displaystyle OH}{|}}{CH}-CH_2-COOH} + 2Ag$$

An important example of a biochemical process that may be formulated as an aldol condensation is the reaction, catalyzed by the enzyme aldolase, which forms fructose 1,6-diphosphate from dihydroxyacetone phosphate and glyceraldehyde 3-phosphate. Note that at physiological pH, near 7, the phosphate groups are ionized.

$$
\begin{array}{c}
CH_2OPO_3^{2-} \\
| \\
C=O \\
| \\
CH_2OH \\
+ \\
H-C=O \\
| \\
H-C-OH \\
| \\
CH_2OPO_3^{2-}
\end{array}
\quad \underset{\longleftarrow}{\overset{\text{aldolase}}{\longrightarrow}} \quad
\begin{array}{c}
CH_2OPO_3^{2-} \\
| \\
C=O \\
| \\
HO-C-H \\
| \\
H-C-OH \\
| \\
H-C-OH \\
| \\
CH_2OPO_3^{2-}
\end{array}
$$

dihydroxyacetone phosphate + glyceraldehyde 3-phosphate   fructose 1,6-diphosphate

This reaction occurs in the biosynthesis of glucose in green plants. The reverse reaction occurs during the metabolism of glucose to produce energy (glycolysis).

Aldehydes and ketones show characteristic infrared absorption between 1710 and 1740 cm^{-1} associated with stretching of the carbonyl group. This absorption is generally very sharp and very intense as can be seen in the spectrum of menthone (Figure 7.4).

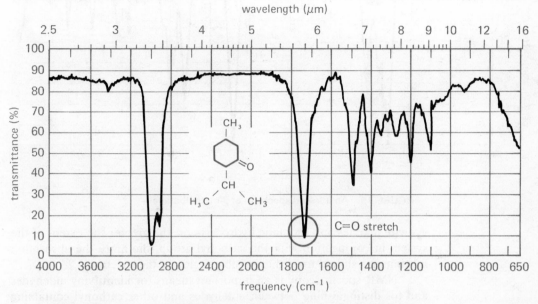

FIGURE 7.4   An infrared spectrum of menthone.

Because few other bond vibrations absorb energy in this region of the spectrum, absorption between 1680 and 1750 cm^{-1} is a very reliable means for confirming the presence of a carbonyl group. However, since several different functional groups contain the carbonyl group, it is not possible to tell from absorption in this region alone whether the carbonyl-containing substance is an aldehyde, ketone, or as we shall see in the next two chapters, a carboxylic acid, ester, or amide. Aldehydes show absorption (generally, two closely spaced peaks) between 2720 and 2830 cm^{-1} due to the stretching of the aldehyde C—H bond and can be distinguished from ketones by careful examination of this region of the infrared spectrum. For example, compare the spectrum of 3-methylbutanal (valeraldehyde, Figure 7.5) with that of menthone (Figure 7.4) and note the presence of the aldehyde C—H stretching frequencies at 2740 cm^{-1} and 2855 cm^{-1}.

Simple aldehydes and ketones show only weak absorption in the ultraviolet region of the spectrum. This absorption is due entirely to electronic excitation in the carbonyl group. If, however, the carbonyl group is conjugated with a carbon-carbon double bond as in the α,β-unsaturated ketone 3-pentene-2-one, the intensity of the absorption is sharply increased. 3-Pentene-2-one itself shows absorption at 224 nm.

$$CH_3-CH=CH-\overset{\displaystyle O}{\overset{\displaystyle \|}{C}}-CH_3$$

3-pentene-2-one
224 nm

185

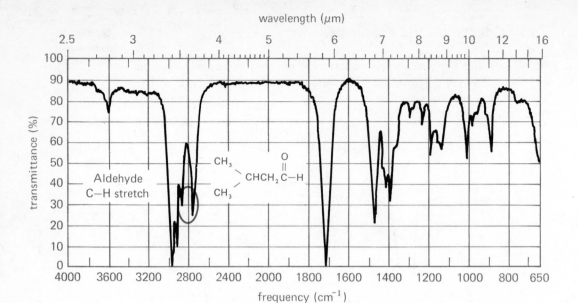

FIGURE 7.5   An infrared spectrum of 3-methylbutanal.

As with alkenes and aromatic hydrocarbons, the greater the extent of the system in conjugation with the carbonyl group, the more the absorption maximum is shifted toward the visible region of the spectrum.

NMR spectroscopy is an important means for identifying aldehydes and for distinguishing between aldehydes and other carbonyl-containing compounds, that is, ketones, carboxylic acids, esters, and amides. For example, the NMR spectrum of butanal shows a very closely spaced triplet due to the aldehyde proton at $\delta = 9.5$ (Figure 7.6). Ketones, of course, do

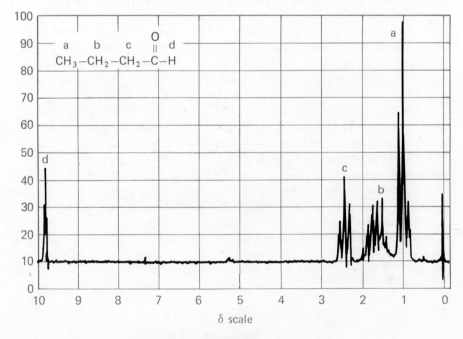

FIGURE 7.6   An NMR spectrum of butanal.

186

not contain a hydrogen bonded to the carbonyl group, and therefore do not give rise to a signal in this region. Notice also in the NMR spectrum of butanal that the signal at $\delta = 2.4$ due to the protons on the $-CH_2-$ group adjacent to the carbonyl group is split into a complex pattern. It is split into a triplet by the two protons of the adjacent $-CH_2-$ group, and each peak of the triplet is in turn split into a doublet by the single aldehyde proton. In general, we will not attempt to analyze splitting of this complexity but will instead simply refer to such signals as multiplets.

PROBLEMS

7.1 Name each of the following compounds.

(a) CH₃—CH—CHO
        |
        CH₃

(b) CH₃CH₂CH₂CCH₂CH₂CH₃ (with =O on the central C)

(c) 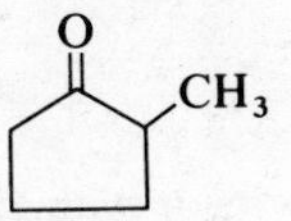 (cyclopentanone with CH₃ substituent)

(d) 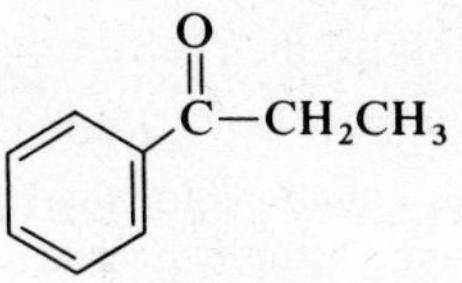 (phenyl C(=O)—CH₂CH₃)

(e) CH₃—CH—C—CH₂—CH₃ (C=O, CH with OH)

(f) CH₃—CH=CH—CHO

(g) 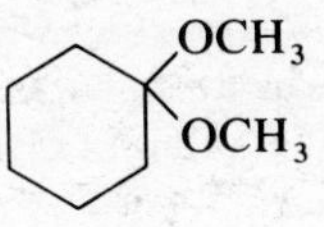 (cyclohexane with OCH₃ and OCH₃)

(h) 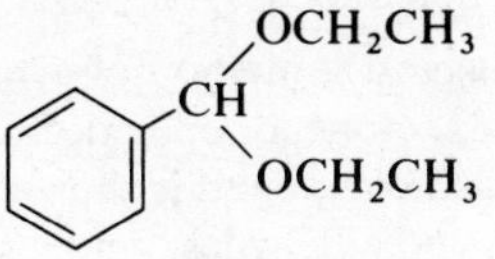 (phenyl CH with OCH₂CH₃ and OCH₂CH₃)

(i) CH₃—O—CH₂—CH₂—CHO

(j) 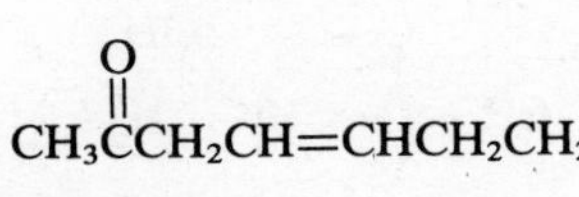 CH₃O—⟨benzene⟩—C—CH₃ (C=O)

(k) CH₃—C—CH₂—C—CH₃ (C with OH and CH₃; second C=O)

(l) CH₃CH₂CHCH₂CHCH₂OH (with OH and CH₃ substituents)

(m) (cyclopentane-1,3-dione)

(n) CH₃CCH₂CH=CHCH₂CH₂C—H (two C=O groups)

7.2 Write structural formulas for the following compounds.
(a) cycloheptanone
(b) propanal
(c) 2-methylpropanal
(d) benzaldehyde
(e) methyl *tert*-butyl ketone
(f) 1,1-dimethoxycyclohexane
(g) the phenylhydrazone of acetophenone
(h) the oxime of cyclopentanone
(i) propenal

187

(j) p-bromocinnamaldehyde
(k) 3-methoxy-4-hydroxybenzaldehyde (vanillin from the vanilla bean)
(l) 3-phenyl-2-propenal (from oil of cinnamon)
(m) 3,7-dimethyl-2,6-octadienal (from orange blossom oil)

7.3 Draw the structures for all aldehydes of formula C_4H_8O; of formula $C_5H_{10}O$. Name each by a common name and an IUPAC name.

7.4 Draw the structures of all ketones of formula C_4H_8O; of formula $C_5H_{10}O$. Name each by a common name and an IUPAC name.

7.5 What is the structure of the simplest aldehyde that can show enantiomerism? Name the aldehyde.

7.6 What is the structure of the simplest ketone that can show enantiomerism? Name it.

7.7 Write equations for the reactions (if any) of cyclopentanone with each of the following:

(a) $LiAlH_4$, followed by H_2O (b) $NaBH_4$, followed by H_2O
(c) H_2/Pt (d) CH_3OH, H^+
(e) $Ag(NH_3)_2^+$ (f) $K_2Cr_2O_7$, H^+
(g) $C_6H_5NHNH_2$ (h) NH_2OH
(i) CH_3MgBr, then H_3O^+

7.8 Repeat problem 7.7 using butanal.

7.9 Write equations for the reaction of propylmagnesium bromide with each of the following. Assume that the magnesium alkoxide salts are hydrolyzed in aqueous acid.

(a) formaldehyde   (b) cyclohexanone
(c) ethylene oxide   (d) diethyl ketone
(e) carbon dioxide

7.10 Suggest a method of synthesis for the following alcohols starting with any aldehyde, ketone, or epoxide and an appropriate Grignard reagent. In parentheses below each compound is shown the number of different combinations of Grignard reagents and aldehydes, ketones, or epoxides that might be used.

(a)
$$\begin{array}{c} OH \\ | \\ CH_3-C-CH_2CH_2CH_3 \\ | \\ CH_2CH_3 \end{array}$$
(3 different ways)

(b)
$$\begin{array}{c} OH \\ | \\ CH_3-CH-CH_2CH_2CH_3 \end{array}$$
(2 different ways)

(c)
$$\begin{array}{c} CH_3 \quad\;\; OH \\ | \qquad\; | \\ CH_3-C-CH_2-CH-CH_2-CH_2-CH_3 \\ | \\ CH_3 \end{array}$$
(2 different ways)

(d) 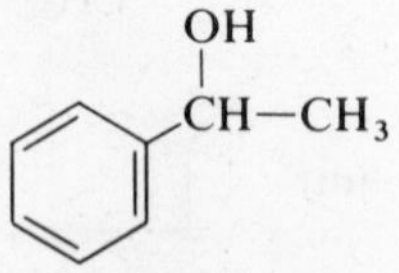
(2 different ways)

(e) 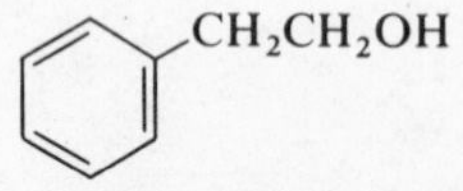
(2 different ways)

(f) 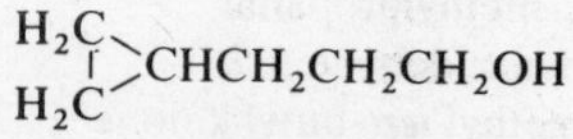
$$\begin{array}{c} H_2C \\ | \quad\; >CHCH_2CH_2CH_2OH \\ H_2C \end{array}$$
(2 different ways)

7.11 How would you account for the fact that Grignard reagents add readily to carbon-oxygen double bonds but do not add to carbon-carbon double bonds?

7.12 Draw structural formulas for the major products from the aldol condensations of the following:

(a) $\underset{\underset{\text{O}}{\overset{\overset{\text{O}}{\|}}{}}{CH_3CH_2CH}$

(b) phenyl-$\underset{\underset{\text{O}}{\overset{\overset{\text{O}}{\|}}{}}{C}$-$CH_3$

(c) cyclohexanone

(d) $CH_3CH_2\overset{\overset{\text{O}}{\|}}{C}CH_2CH_3$

7.13 Draw structural formulas for the major products from the mixed aldol condensations of the following:

(a) $CH_3-\overset{\overset{\text{O}}{\|}}{C}-H + CH_3-\overset{\overset{\text{O}}{\|}}{C}-CH_3$

(b) phenyl-$\overset{\overset{\text{O}}{\|}}{C}-H$ + $CH_3-\overset{\overset{\text{O}}{\|}}{C}$-phenyl

(c) cyclohexanone + $H-\overset{\overset{\text{O}}{\|}}{C}-H$

7.14 Draw structural formulas for the four possible aldol condensation products from a mixture of propanal and butanal.

7.15 Show reagents and conditions to illustrate how the following products can be synthesized from the indicated starting materials by way of aldol condensation reactions.

(a) $CH_3CHO \longrightarrow CH_3-CH_2-CH_2-CH_2OH$

(b) $CH_3CHO \longrightarrow CH_3-CH=CH-CO_2H$

(c) $CH_3-\overset{\overset{\text{O}}{\|}}{C}-CH_3 \longrightarrow CH_3-\underset{\underset{\text{CH}_3}{|}}{\overset{\overset{\text{OH}}{|}}{C}}-CH_2-\overset{\overset{\text{OH}}{|}}{CH}-CH_3$

(d) $CH_3-\overset{\overset{\text{O}}{\|}}{C}-CH_3 \longrightarrow CH_3-\underset{\underset{\text{CH}_3}{|}}{CH}-CH_2-CH_2-CH_3$

(e) $CH_3-CH_2-CH_2-CHO \longrightarrow CH_3-CH_2-CH_2-CH_2-\underset{\underset{\text{CH}_3-\text{CH}_2}{|}}{CH}-CH_2OH$

7.16 Show reagents and conditions by which you could prepare each of the following substances from cyclopentanone. In addition to cyclopentanone, use any additional organic or inorganic reagents necessary.

(a) cyclopentane with OH

(b) cyclopentene

(c) cyclopentane with Cl

(d) cyclopentane with CH_2OH

(e) cyclopentane with CO_2H

(f) cyclopentane with CH_2-CH_2OH

(g) cyclopentane with $\underset{\underset{\text{CH}_2-\text{CH}_3}{|}}{\overset{\overset{\text{CH}_3}{|}}{C}}-OH$

(h) $H-\overset{\overset{\text{O}}{\|}}{C}-CH_2-CH_2-CH_2-\overset{\overset{\text{O}}{\|}}{C}-H$

(i) cyclopentane with $\underset{\underset{\text{CH}_2-\text{CH}_3}{}}{\overset{\overset{\text{OH}}{}}{}}$

189

(j)

(k)

(l)

7.17 5-Hydroxyhexanal readily forms a six-membered cyclic hemiacetal. Draw a chair conformation for this hemiacetal. Label substituent groups axial or equatorial.

7.18 Propose a reasonable mechanism to account for the formation of cyclic acetal from 4-hydroxypentanal and one molecule of methyl alcohol.

$$CH_3CHCH_2CH_2\overset{\overset{O}{\|}}{CH} + CH_3OH \underset{}{\overset{H^+}{\rightleftharpoons}} H_3C\underset{}{\overset{}{\diagdown}}O\diagup OCH_3 + H_2O$$
$$\underset{OH}{|}$$

If the carbonyl oxygen of 4-hydroxypentanal were enriched with oxygen-18, would you predict that the oxygen-18 would appear in the cyclic ketal or in the water?

7.19 Acetaldehyde reacts with ethylene glycol in the presence of a trace of sulfuric acid to give a cyclic acetal of formula $C_4H_8O_2$. Draw the structural formula of this acetal and propose a mechanism for its formation.

7.20 What is meant by the term *tautomerism*? Illustrate tautomerism by drawing keto and enol tautomers for acetaldehyde, acetone, butanal, and 2-methyl-cyclohexanone (two enol forms).

7.21 Describe how resonance and tautomerism differ.

7.22 How would you account for the fact that in dilute aqueous alkali glyceraldehyde is converted into an equilibrium mixture of glyceraldehyde and dihydroxyacetone?

$$\underset{\substack{\text{glyceraldehyde}}}{\overset{\text{CHO}}{\underset{\text{CH}_2\text{OH}}{\overset{|}{\text{CHOH}}}}} \xrightarrow[\text{base}]{\text{dilute}} \underset{\substack{\text{glyceraldehyde}}}{\overset{\text{CHO}}{\underset{\text{CH}_2\text{OH}}{\overset{|}{\text{CHOH}}}}} + \underset{\substack{\text{dihydroxyacetone}}}{\overset{\text{CH}_2\text{OH}}{\underset{\text{CH}_2\text{OH}}{\overset{|}{\text{C}=\text{O}}}}}$$

7.23 In dilute-aqueous base, the α-hydrogens of butanal show acidity, but neither the β- nor γ-hydrogens show any acidity whatsoever. Account for this difference in acidities.

7.24 Cyclohexene can be converted into cyclopentene carboxaldehyde by the following series of steps. Ozonolysis of cyclohexene followed by work up in the presence of zinc and acetic acid forms a compound $C_6H_{10}O_2$. Treatment of this compound with dilute base transforms it into an isomer also of formula $C_6H_{10}O_2$. Warming this isomer in dilute acid yields cyclopentene carboxaldehyde.

cyclohexene $C_6H_{10}O_2$ $C_6H_{10}O_2$ cyclopentene carboxaldehyde

Propose structural formulas for the isomeric compounds of formula $C_6H_{10}O_2$ and account for the conversion of one isomer into the other in the presence of dilute base.

7.25 A compound A ($C_5H_{12}O$) does not give a precipitate with phenylhydrazine. Oxidation of A with potassium dichromate gives B ($C_5H_{10}O$). Compound B reacts readily with phenylhydrazine but does not give a precipitate with silver nitrate in ammonia. The original compound, A, can be dehydrated with sulfuric acid to give a hydrocarbon C (C_5H_{10}). Ozonolysis of the hydrocarbon C gives acetone and acetaldehyde. Propose structural formulas for compounds A, B, and C.

7.26 Compare and contrast the acid-catalyzed hydrations of 2-methylpropene (isobutylene) and acetone in terms of

(a) the role of the acid catalyst
(b) the structure of the cation intermediate
(c) resonance stabilization of the cation intermediate
(d) the relative yields of the hydration products in each case

7.27 In Section 5.11, we stated that ethers are quite resistant to the action of strong acids, strong bases, and oxidizing and reducing agents. There are two general exceptions to this rule: (1) epoxides in which the ether oxygen is a part of a three-membered ring, and (2) acetals and ketals in which there are two ether linkages to the same carbon. How might you account for the fact that acetals and ketals undergo ready cleavage in dilute aqueous acid whereas ethers like diethyl ether and tetrahydrofuran do not?

7.28 Make a list of the reagents or types of reagents we have seen so far that add to the carbon-oxygen double bonds of aldehydes and ketones. Do the same for reagents that add to the carbon-carbon double bond.

(a) Which of these reagents will add to both carbon-carbon and carbon-oxygen double bonds?
(b) Which of these reagents will add to carbon-carbon double bonds but not to carbon-oxygen double bonds? How might you account for these differences in reactivity?
(c) Which of these reagents will add to carbon-oxygen double bonds but not to carbon-carbon double bonds? How might you account for these differences in reactivity?

7.29 Make a list of the reactions we have examined so far that lead to the formation of new

(a) carbon-carbon bonds
(b) carbon-oxygen bonds
(c) carbon-nitrogen bonds

Do not include reactions that merely transform a single bond into a double bond (as for example an alcohol into an aldehyde or ketone), or vice versa.

7.30 Show how you might distinguish between the following pairs of compounds by a simple chemical test. In each case, tell what test you would perform, what you would expect to observe, and write an equation for each positive test.

(a) cyclohexanone and cyclohexanol

(b) benzaldehyde and acetophenone $(C_6H_5-\overset{\overset{\textstyle O}{\|}}{C}-CH_3)$
(c) benzaldehyde and benzyl alcohol

7.31 List a major spectral characteristic that will enable you to distinguish between the following pairs of compounds.

(a) cyclohexanone and cyclohexanol (IR)
(b) benzaldehyde and acetophenone (IR, NMR)
(c) benzaldehyde and benzyl alcohol (IR, NMR)
(d) cyclohexanone and 1,1-dimethoxycyclohexane (IR, NMR)
(e) progesterone (Figure 7.1) and testosterone (IR)
(f) cyclohexane carboxaldehyde $(C_6H_{11}-CHO)$ and benzaldehyde (IR, NMR)

(g) and (IR)

(h) and (NMR)

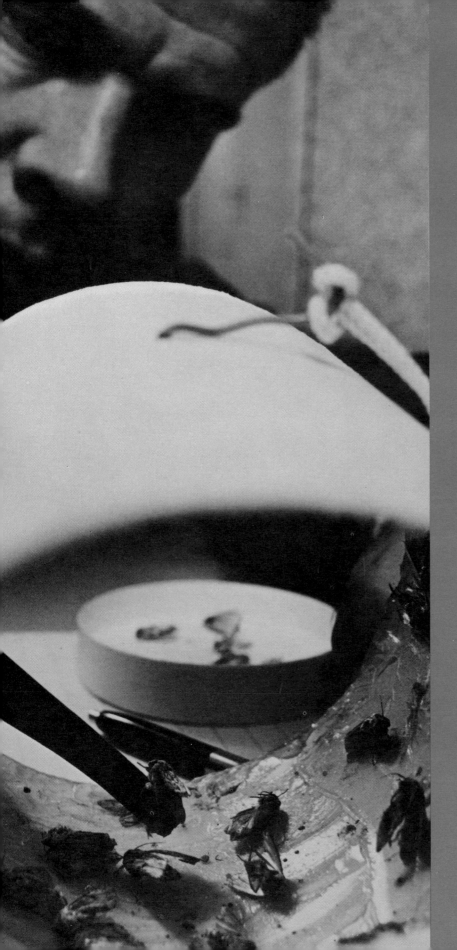

II

Mini-Essays

Insect Juvenile Hormone

Insect Pheromones

Insect Juvenile Hormone

Present knowledge of insect juvenile hormone stems quite logically from basic research on insect physiology and on the chemical events that control insect development. In the four decades since recognizing the existence of insect juvenile hormones, scientists have isolated and identified such hormones from insects, discovered extremely potent and selective hormone-mimicking substances in the plant world, synthesized hundreds of hormone analogs in the laboratory, and discovered substances in the plant world with anti-juvenile hormone activity. As a result we may now be on the verge of realizing a new generation of pesticides, compounds so selective in their action that they will attack only certain insect pests without presenting a hazard to other organisms.

To put these results in perspective, we must first look at insect development itself. Insect growth and metamorphosis generally proceed through four stages—egg to larva to pupa to adult. Internal glands secrete certain hormones which control these various stages. One such hormone, known as juvenile hormone, is secreted by the corpora allata, two tiny glands in the head of the insect. At certain stages in development this hormone must be present; at certain other stages it must be absent. For example, juvenile hormone must be present for the immature larva to progress through the usual stages of larval growth. Then, for the mature larva to undergo metamorphosis into a mature adult, the secretion must stop. If juvenile hormone is supplied at this critical time, either by implantation of active corpora allata or by application of the pure hormone itself, then the pupa does not form a viable, mature adult. Juvenile hormone must also be absent from insect eggs for them to undergo normal embryonic development. If the hormone is applied to the eggs, either they fail to hatch or the immature insects die without reproducing.

Although the existence of juvenile hormone was recognized as early as 1939 and its site of production in the corpora allata was established, all efforts to extract and isolate it from living insects failed. Then in 1956, Carroll Williams of Harvard University made the fortuitous discovery that the abdomen of the adult male Cecropia moth contains a rich storage depot of the hormone. In retrospect, the discovery of this depot seems most remarkable for even today the only other known insect from which juvenile hormone can be extracted is the closely related male Cynthia moth.

The first crude preparations of Cecropia juvenile hormone were obtained by ether extraction of the excised abdomens. Evaporation of the ether extracts leaves a golden colored oil, about 0.2 ml per abdomen.

Closeup of three yellow mealworms shows a normal pupa (left); an abnormal adult that has been sprayed with synthetic juvenile hormone which caused it to develop an adult head and thorax, but kept it from developing an adult abdomen (center); and a normal adult (right). (USDA photo)

Injection of this crude oil produced all the effects achieved by implanting active corpora allata. In fact, injection of the hormone is not even necessary to produce these effects. Merely placing the oily extract on unbroken skin has the same results—derangement of growth and the formation of nonviable adults. It is just this type of disrupted development coupled with the simplicity of application that first suggested juvenile hormone as a potential insecticide. Although its activity varies from family to family, natural Cecropia hormone affects such diverse insects as representatives of Coleoptera, Lepidoptera, Hemiptera, and Orthoptera.

Before the structure of Cecropia juvenile hormone was established, scientists had observed some slight degree of juvenile hormone activity in farnesol, a naturally occurring plant sesquiterpene alcohol (Figure 1). Following this lead, William Bowers of the United States Department of Agriculture (USDA) experimental station at Beltsville, Md., began to make systematic structural modifications of farnesol hoping to synthesize new and even more active compounds. In 1965, he prepared methyl 10,11-epoxyfarnesoate from farnesol by forming an epoxide between carbons 10 and 11 and oxidizing the alcohol on carbon 1 to a carboxylic acid, followed by esterification.

FIGURE 1 Farnesol, a naturally occurring alcohol, and methyl 10,11-epoxyfarnesoate, a juvenile hormone analog synthesized in 1965.

farnesol

methyl 10,11-epoxyfarnesoate

The new substance was 1600 times more active than farnesol, and even though it had only about 0.02% of the Cecropia hormone activity, this suggested that, when juvenile hormone was identified, it would bear at least some structural resemblance to farnesol. Bowers speculated that "...it is quite possible that the juvenile hormone will be synthesized before it is structurally identified from natural sources."

Then in 1964, another juvenile hormone analog was discovered by what most certainly must be regarded as serendipity smiling and chance favoring the prepared mind. Karel Sláma, a young Czech entomologist, arrived at Harvard University to study with Carroll Williams. Sláma brought with him species of the European bug *Pyrrhocoris apterus*, a species he had reared and studied for many years in Prague. To his considerable surprise and mystification, these bugs, when reared in the Harvard environment, failed to metamorphose into sexually mature adults. Instead, they continued to grow as larvae or molted into adultlike forms while retaining many larvalike characteristics. All died without attaining maturity. Behavior of this type had previously been observed only upon application of juvenile hormone, and it appeared that the bugs had access to some unknown source of the hormone. Finally it was established that the source was none other than the paper towels placed in each rearing jar to give the bug a surface to walk on. Almost all American paper had the same effect, but surprisingly, paper of European or Japanese manufacture had no effect. For want of a better name, the substance was termed paper factor. The origin of paper factor was traced to balsam fir (*Abies balsamea*), a principle source of American paper pulp. Balsam fir synthesizes the active material, which stays with the paper pulp through the

Closeup of three linden bugs shows the insect in the last stage of the nymph phase (left), an overgrown nymph that was treated with juvabione (center), and a normal adult (right). The treated nymph will not develop into an adult, and thus will not reproduce. (USDA photo)

197

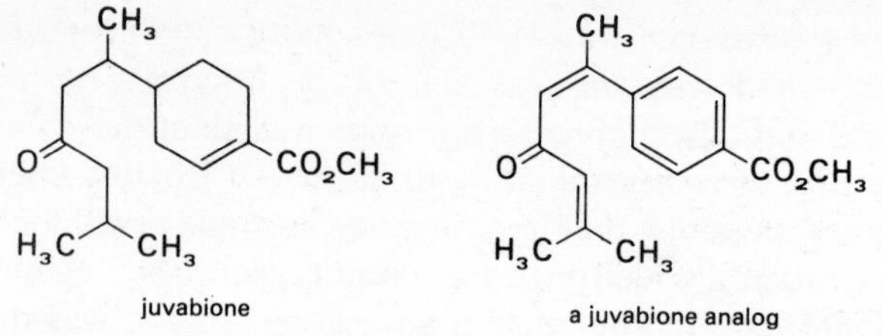

juvabione   a juvabione analog

FIGURE 2 Juvabione, a juvenile hormone analog of the balsam fir, active only on insects of the family Pyrrhocoridae. A juvenile hormone analog derived from p-(1,5-dimethylhexyl) benzoic acid.

entire manufacturing process. "Paper factor" eventually became known as juvabione (Figure 2), suggesting its relation to juvenile hormone.

With the structure of juvabione determined, Sláma and his associates in Czechoslovakia prepared a number of compounds structurally related to juvabione but incorporating a benzene rather than a cyclohexene ring. One such analog is shown in Figure 2. Some of these derivatives are about 100 times more active than juvabione itself, and all retain specific action only for Pyrrhocoid bugs.

Unlike the juvenile hormone of the male Cecropia moth, juvabione is active only in the Pyrrhocoridae, an insect family containing some of the most destructive pests of cotton. Closely related families appear to be totally unaffected. This exciting discovery was the first evidence for the existence of juvenile hormonelike material with highly selective action on a particular family of insects. This type of specificity is essential if we hope to tailor-make insecticides against only certain predetermined pests.

Research in this area came full cycle when in 1965 Herbert Röller of the University of Wisconsin isolated the male Cecropia hormone itself and, in 1967, using less than 0.3 mg of pure material, determined its structure.

Bower's prediction in 1963 that Cecropia hormone would bear some structural similarity to farnesol is amply confirmed. Furthermore, Cecropia hormone is remarkably similar to the farnesol derivative synthesized by Bowers in 1965. The difference is only two carbon atoms; the alkyl groups at carbons 7 and 11 in Cecropia hormone are $-CH_2CH_3$ rather than $-CH_3$. Cecropia hormone contains two *trans* double bonds and a *cis* epoxide. The *trans* configuration of both double bonds appears crucial for biological activity. In contrast, the stereochemistry of the epoxide ring is of secondary importance (Figure 3).

Since Roller's isolation and determination of the structure of JH1 in 1967, two additional juvenile hormones have been discovered, and both bear obvious structural similarities to the Cecropia hormone. JH2 has a methyl rather than an ethyl group at carbon 7, while JH3 has methyl rather than ethyl groups at carbons 7 and 11. Note that of the three hormones,

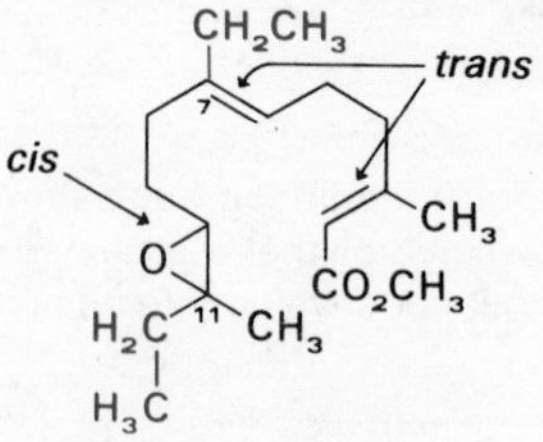

FIGURE 3 Juvenile hormone (JH1) of the male Cecropia moth.

198

only JH3 is a true terpene, for only it can be divided into three isoprene units.

Clearly, juvenile hormones and hormone analogs offer promise of a new approach in pest control. In contrast to DDT and other persistent chemical pesticides, these new, third-generation pesticides are extremely potent, and so highly selective in their action that they scarcely affect the surrounding biosphere. The trick is to use them at critical times in the insects' life cycle to so derange their normal growth and development that they fail to mate, reproduce, and multiply. Both laboratory and field studies are currently underway to test these substances in insect control programs.

But the story of insect juvenile hormones has not yet ended. Rather, it has taken a new and intriguing turn. Bowers and others reasoned that if plants contain substances with insect juvenile hormone activity, might they not also contain substances with anti-juvenile hormone activity? Bowers began a screening program by extracting plants with nonpolar solvents and testing the effects of the extracts on immature insects. He discovered that an extract from the common ornamental bedding plant, *Ageratum houstonianum*, did possess anti-juvenile hormone activity. Isolation and structure determination showed that there were two active components of the extract, precocene I and precocene II.

precocene I   precocene II

Notice the similarity in structure between the two; they differ only in the presence of a second methoxyl group in precocene II. Clearly, the discovery of these substances adds a new dimension to the study of insect juvenile hormone activity. Furthermore, they offer even greater potential for pest control, for while juvenile hormones and hormone analogs are effective only during a brief period in the insect's life cycle, precocene and precocene analogs may be effective over much of the insect's life.

REFERENCES

Bowers, W. S., O. Tomishisa, J. S. Cleere, and P. A. Marsella. *Science*, vol. 93:542, August 13, 1976.

Judy, K. J., et al. *Proceedings of the National Academy of Sciences*, USA, 70, 1509 (1973).

Roller, H., K. H. Dahm, C. C. Sweely, and B. M. Frost. *Angewant Chemie. International Edition (English)*, vol. 6, 179, (1967).

Williams, C. "Third-Generation Pesticides," *Scientific American*, vol. 217 (July 1967), p. 13.

———. in *Chemical Ecology*, E. Sondheimer and J. Simeone, eds. Academic Press, New York, 1970.

Insect Pheromones

Chemical communication abounds in nature: the clinging, penetrating odor of the skunk's defensive spray; the hound, nose to the ground, in pursuit of prey; the female cat advertising her sexual availability; the female moth attracting males from over great distances for mating. As biologists and chemists cooperate to extend our knowledge to other animals, it is becoming increasingly clear that chemical communication is not only a widespread biological phenomenon, but it is probably the primary mode of communication in the vast majority of species of the animal world. Prior to 1950, isolation of enough of the biologically active material to permit identification seemed an insurmountable task preventing us from deciphering any of these chemical communications. However, rapid progress in instrumental techniques, particularly in chromatography and spectroscopy, has now made it possible to isolate and carry out structural determinations on as little as a few milligrams and even a few micrograms of material. Yet, even with these advances, the isolation and identification of such chemicals remains a major challenge of technical and experimental expertise. For example, obtaining a mere 12 milligrams of gypsy moth sex attractant required the processing of some 500,000 virgin female moths, each yielding only 0.02 micrograms of attractant. In other insect species, it is not uncommon to process at least 20,000 insects to obtain enough material for chemical identification.

The term *pheromone* (from the Greek *pherein*, "to carry" and *horman* "to excite") has become accepted as the name for chemicals secreted by an organism of one species to evoke a specific response in another member of the same species. In this essay, we will look only at insect pheromones for these have been the most widely studied. Table 1 lists eight insect pheromones along with the name of the insect species by which it is used.

Pheromones are generally divided into two classes—releaser and primer pheromones—depending on their mode of action. Releaser pheromones produce a rapid, reversible change in behavior such as sexual attraction and stimulation, assembly, territorial and home range marking, recruiting for foraging efficiency. Primer pheromones produce more subtle effects. They cause important physiological changes that affect the animal's development and later behavior. The most clearly understood primer pheromones regulate caste systems in social insects (bees, ants, and termites). A typical colony of honey bees (*Apis mellifera*) consists of one queen, several hundred drones (males), and thousands of workers (underdeveloped females). The queen bee, the only fully developed female in the colony, secretes a "queen substance" which prevents the development of worker's ovaries and the construction of royal colony cells for the rearing of new queens. One of the primer pheromones in the queen substance has

TABLE 1 Insect pheromones.

Primer pheromones

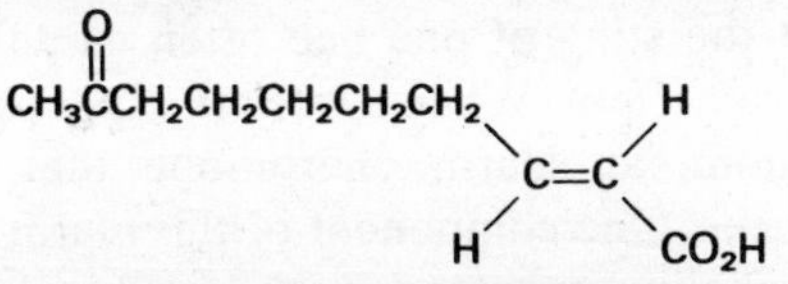

CH₃CCH₂CH₂CH₂CH₂CH₂ ... C=C ... H / CO₂H   **Honey bee**

9-keto-*trans*-2-decenoic acid

Alarm pheromones

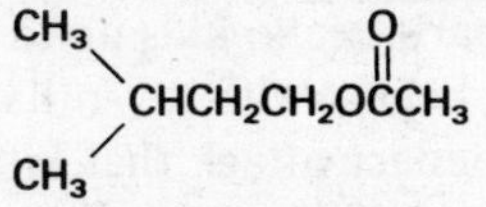

CH₃ / CH₃ CHCH₂CH₂OCCH₃   **Honey bee**

isoamyl acetate

Aggregating pheromones

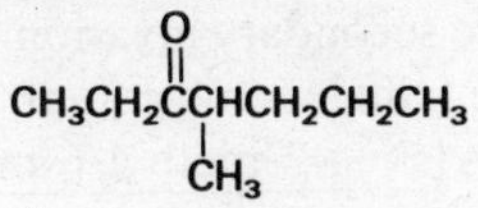

CH₃CH₂CCHCH₂CH₂CH₃ / CH₃   **European elm beetle**

4-methyl-3-heptanone

CH₂ ... CH₂ / HO ...   **Bark beetle**

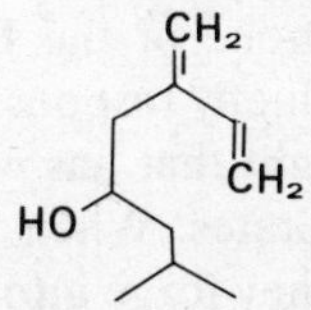

2-methyl-6-methylene-7-octen-4-ol

Sex pheromones

CH₃(CH₂)₃ ... C=C ... (CH₂)₅CH₂OCCH₃   **Cabbage looper**

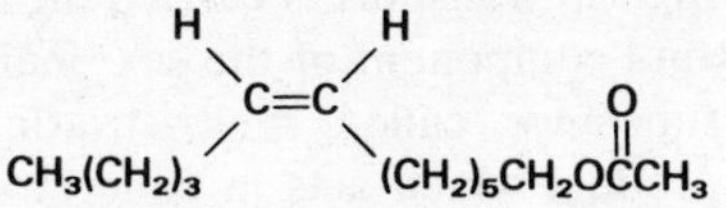

cis-7-dodecenyl acetate

CH₃(CH₂)₃ ... C=C ... C=C ... (CH₂)₆OCCH₃   **Pink bollworm moth**

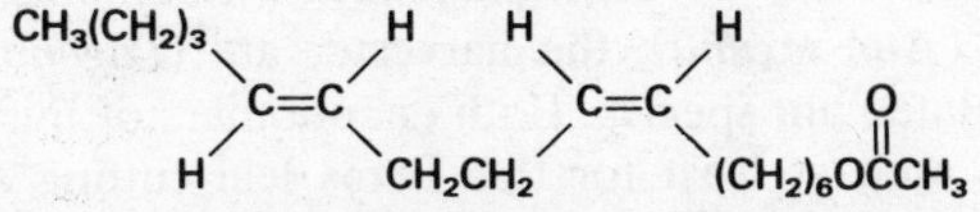

cis,trans-7,11-hexadecadienyl acetate
and the *cis,cis*-7,11- isomer (not shown)

CH₃ / CH₂—CH₂OH / C—CH₃ / CH₂   **Boll weevil**

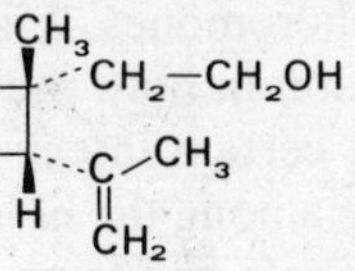

grandisol

CH₃CH₂ ... C=C ... H / (CH₂)₁₀OCCH₃   **European corn borer**

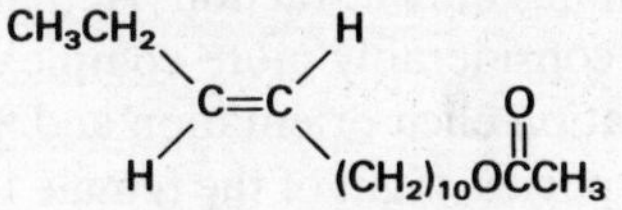

trans-11-tetradecenyl acetate
and the *cis*-11- isomer (not shown)

been identified as 9-keto-*trans*-2-decenoic acid. In addition, this same compound serves as a sex pheromone, attracting drones to the queen during her mating flight.

Some of the earliest observations on alarm pheromones have been recorded on the honey bee. Bee-keepers, and perhaps some of the rest of us too, are well aware that the sting of one bee often causes swarms of angry workers to attack the same spot. When a worker stings an intruder, it discharges, along with venom, an alarm pheromone that evokes the aggressive attack by other bees. One component of this alarm pheromone is isoamyl acetate, a sweet-smelling substance with an odor similar to that of banana oil.

One of the few aggregating pheromones thus far identified is that of the bark beetle, *Ips confusus*. The bark beetle kills up to five billion board feet of timber each year. Invasion begins with an initial attack by a few beetles, followed by a massive secondary attack that kills the tree. In the case of *Ips confusus*, a few males initially bore into the tree to construct nuptial chambers. During this process they excrete frass, a mixture of fecal pellets and wood fragments. These fecal pellets contain the aggregating pheromone that triggers the massive secondary invasion by both males and females. One male mates with three females, and colonization is on its way. The aggregating pheromone of frass is (−)-2-methyl-6-methylene-7-octen-4-ol.

The ketone, 4-methyl-3-heptanone, is one of the three known components of the aggregation pheromone of the European elm beetle (*Scolytus multristriatus*). This beetle is chiefly responsible for the spread of Dutch Elm disease, an insect-borne blight that has wiped out Dutch Elm trees in about one-half of the United States. When virgin females find a tree suitable for breeding and feeding, they forge into the elm. As they do, they release an aggregating pheromone which very soon triggers a mass attack on the tree by other European elm beetles. The fungus that actually destroys the tree by interrupting sap transport is carried on the bodies of the attacking beetles. The second component of the sex pheromone mixture is a beetle-produced substance called multristriatin. The third component is a natural tree product which acts in concert with the two insect-produced chemicals. Interestingly enough, the same ketone, 4-methyl-3-heptanone, is one of the principle alarm pheromones of the Texas leaf-cutting ant (*Atta texana*), the harvester ant (*Pogonomyrmex barbatus*), and other related ant species. Both enantiomers of this ketone have been synthesized and, at least for the Texas leaf-cutting ant, the dextrorotatory isomer is almost 400 times more effective in eliciting the alarm response than is the levorotatory isomer.

Of all of the various classes of pheromones, it is the sex pheromones that have received the greatest attention in both the scientific community and the popular press. The larvae of certain insects which release them, particularly the moths and beetles, are among the most serious agricultural pests. Sex pheromones are commonly referred to as "sex attractants" but this term is misleading for it implies only attraction. Actually, the behavior elicited by these phermones is considerably more complex than that. Low levels of sex pheromone stimulation elicit orientation and flight of the male toward the female (or in some species, flight of the female to the male) and, if the level of stimulation is high enough, copulation follows. At least in the

Male cockroaches responding to synthetic sex pheromone on a piece of filter paper. (USDA photo)

case of the cabbage looper (*Trichoplusia mi*), none of these behavioral responses requires the presence of the female. A spot of female cabbage looper sex pheromone, *cis*-7-dodecenyl acetate, on a piece of filter paper elicits orientation, flight, and even copulatory behavior of the male, all directed toward the spot of the evaporating pheromone.

Grandisol is one of four components of the sex pheromone of the male cotton boll weevil (*Anthonomus grandis*). Work at the U.S. Department of Agriculture's Boll Weevil Research Laboratory showed that at any given time a male contains only about 200 nanograms (200×10^{-9} gram) of the attractant, and that about 1.5 micrograms of attractant are released daily. The female of the species does not produce any of the attractant. Another sex attractant that has received a great deal of attention is that of the pink bollworm moth (*Pectinophora gossypiella*) which, like the boll weevil, is a destructive pest in cotton-growing areas. The isolation of this sex attractant was reported in 1966 by scientists at the U.S. Department of Agriculture (USDA) Laboratory, Beltsville, Md. Efforts to obtain sufficient material for structure determination required the isolation of sex pheromones from 850,000 virgin female moths. In 1973, the attractive material, active in both laboratory and field assays, was identified as a $1:1$ mixture of *cis,cis*-7,11 and *cis,trans*-7,11-hexadecadienyl acetate. This attractant was named "gossyplure," a name obviously related to the scientific name of this troublesome pest. The synthesis of this material proved an exciting challenge to the organic chemist, and within a few years there were published syntheses of all four stereoisomers, that is, of the four possible combinations of *cis*- and *trans*-configurations at the two carbon-carbon double bonds. Some interesting findings emerged from the laboratory and field tests of these isomers. First, a $1:1$ ratio of the *cis,cis*-7,11 and *cis,trans*-7,11 isomers is essential for full attractiveness. Second, neither of the *trans*-7 isomers is at all attractive. In fact, not only are the 7-*trans* isomers not attractive, but as little as 10% contamination by either markedly reduces the attractiveness of the natural pheromone mixture.

Several groups of scientists have studied the components of the sex pheromone of both Iowa and New York strains of the European corn borer. Females of each closely related species appear to secrete the same sex attractant, 11-tetradecenyl acetate. Males of the Iowa strain show maximum response to a mixture containing about 96% of the *cis* isomer and 4% of the *trans* isomer. In fact, when the pure *cis* isomer is used alone, males are only weakly attracted. The response of males of the New York strain shows an entirely different pattern in that they respond maximally to a mixture containing only 3% of the *cis* isomer and 97% of the *trans* isomer. There is now more and more evidence that this type of optimum response to a narrow range of stereoisomers may be very widespread in nature and an important means of preventing cross-attraction of species. In other words, some species of insects may be able to maintain species isolation (at least for the purposes of mating and reproduction) by the very nature of the stereochemistry of their pheromones.

Finally, let us consider the structural characteristics of insect pheromone molecules themselves. Specifically, let us consider how the molecular architecture of these substances is suited for the tasks they must perform. Any ordinary community of insects may contain hundreds to thousands of different species, each set with its own characteristic pheromones to communicate its needs for food, protection, reproduction, and so on. Such an odor environment must be enormously complex, yet each species of insect in this environment must be able to select only those signals from its own species. An insect does this by means of chemoreceptors highly sensitive to relevant stimuli and far less sensitive to irrelevant ones. What kinds of structural and physical properties might we expect of the organic molecules that meet these demanding requirements of species specificity?

First, these molecules must be ones that can be synthesized and stored easily by the insect. It is probably for this reason that the majority of insect pheromones so far discovered are related in structure to either fatty acids or terpenes, molecules for whose production there are already elaborate biochemical pathways. Second, since insect pheromones are released into the air and transported as gasses, they must be volatile enough so that the concentration of pheromone in a given volume of air is high enough to activate the chemoreceptors of males (or females, as the case may be) of the species. Molecules of very low molecular weight would be too volatile and disperse too rapidly. Molecules of very high molecular weight would not be volatile enough. It is probably for these reasons that most airborne pheromones have molecular weights in the range of 40 to 300. Third, a molecule that is to serve as a pheromone must have a distinctive enough molecular architecture to insure that there is very little or no possibility for cross-attraction between even closely related species. Clearly, there is an astronomical range of diversity and, therefore, of biological individuality available with only a relatively few carbon atoms through variations in structure, functional groups, and configuration about chiral carbons and carbon-carbon double bonds. This potential for biological individuality is further compounded by the use of pheromones containing blends of substances as, for example, blends of *cis-trans* isomers, enantiomers, functional group isomers, and other species-specific substances.

We have discussed insect pheromones only. It is now certain that pheromones are also used by other animals as well. The other animal groups in which pheromone communication has been intensively studied are the rodents, some primates, and a few other mammals. Here pheromones have been found to regulate social life and reproduction. There is also some information on fright and alarm pheromones in fish and amphibians. And what of man?

REFERENCES

Birch, M. C., ed. "Pheromones," *Frontiers of Biology*, vol. 32, American, Elsevier (1974).

Hummel, et al. "Clarification of the Chemical Status of the Pink Bollworm Sex Pheromone," *Science*, vol. 181, 873 (1973).

Klun, J. A., et al. "Insect Sex Pheromones: Minor Amounts of Opposite Geometrical Isomer Critical to Attraction," *Science*, vol. 181, 661 (1973).

Sanders, Howard J. "New Weapons Against Insects: A Special Report," *Chemical and Engineering News*, July 28, 1975.

Shorey, H. H. *Animal Communication by Pheromones*. Academic Press, New York, 1976.

Sondheimer, E. and J. B. Simeone, eds. *Chemical Ecology*. Academic Press, New York, 1970.

Wilson, E. O. *Bio-Organic Chemistry*, M. Calvin and M. Jorgenson eds. W. H. Freeman & Co., San Francisco, 1968.

Wood, D. L., R. M. Silverstein, and M. Nakajama, eds. *Control of Insect Behavior by Natural Products*. Academic Press, New York, 1970.

Carboxylic Acids

8.1 Introduction

Of the various classes of organic compounds that show acidity, carboxylic acids are by far the most important. These acids and their functional derivatives are ubiquitous both in the biological world and, thanks to the blend of research and technology, in the world of man-made materials (for example, see the mini-essay "Nylon and Dacron").

In this chapter we will discuss the structure and acidity of carboxylic acids. In addition, we will describe a special group of carboxylic acids known as fatty acids. This in turn will lead us into the chemistry of soaps and detergents. In Chapter 9, we will discuss several functional derivatives of carboxylic acids.

8.2 Structure

Substances containing a carbonyl group, (C=O), bonded directly to a hydroxyl group (—OH) are called carboxylic acids. We have already described (Chapter 1) the Lewis electronic structure for the carboxylic acid functional group. Based on our discussions of bonding, we predict bond angles of 120° about the carbonyl carbon, and 109.5° about the hydroxyl oxygen (see Figure 8.1). In addition the three atoms attached to the carbonyl carbon are in the same plane.

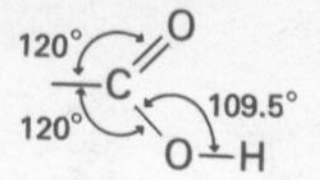

FIGURE 8.1 Structure of
the COOH group.

$$120° \quad 120° \quad C \overset{O}{\underset{O-H}{\big\langle}} \quad 109.5°$$

8.3 Nomenclature

Carboxylic acids are easily isolated from natural sources and quite a few have been known since the early days of alchemy. As is often the case with organic nomenclature, the common names refer to the natural sources of these acids, and the names have no relation to any systematic nomenclature. For example formic acid adds to the sting of the bite of an ant (Latin, *formica*, "ant"); butyric acid gives rancid butter its characteristic smell (Latin, *butyrum*, "butter"); caproic acid is found in goat fat (Latin, *caper*, "goat"); lauric acid from laurel.

Table 8.1 shows the structural formulas and names of some representative aliphatic mono- and dicarboxylic acids.

TABLE 8.1 Carboxylic acids.

Formula	Name (IUPAC)	Name (Common)
HCOOH	methanoic	formic
CH_3COOH	ethanoic	acetic
CH_3CH_2COOH	propanoic	propionic
$CH_3(CH_2)_2COOH$	butanoic	butyric
$CH_3(CH_2)_3COOH$	pentanoic	valeric
$CH_3(CH_2)_4COOH$	hexanoic	caproic
$CH_3(CH_2)_{10}COOH$	dodecanoic	lauric
$CH_3(CH_2)_{14}COOH$	hexadecanoic	palmitic
$CH_3(CH_2)_{16}COOH$	octadecanoic	stearic
HOOC—COOH		oxalic
$HOOC-CH_2COOH$		malonic
$HOOC-CH_2CH_2COOH$		succinic
$HOOC-CH_2CH_2CH_2COOH$		glutaric

The IUPAC system of nomenclature employs the name of the longest continuous carbon chain that contains the —COOH group, and indicates the acid function by using the suffix -oic acid. All of the other rules for naming organic compounds apply. Since the —COOH group must be on the terminal carbon of the chain, it is usually not necessary to indicate its position by the number 1-. When using common names, Greek letters (α, β, γ, δ, etc.) are often used to locate substituents. Note that the α-carbon is the one next to the carboxyl group and, therefore, an α-substituent is equivalent to a 2-substituent.

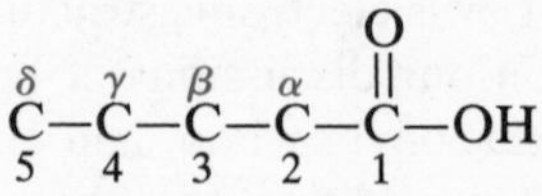

$$\overset{\delta}{\underset{5}{C}}-\overset{\gamma}{\underset{4}{C}}-\overset{\beta}{\underset{3}{C}}-\overset{\alpha}{\underset{2}{C}}-\overset{\overset{O}{\|}}{\underset{1}{C}}-OH$$

The aliphatic dicarboxylic acids are known almost exclusively by their common names. Aromatic acids are usually named by common names or as derivatives of benzoic acid.

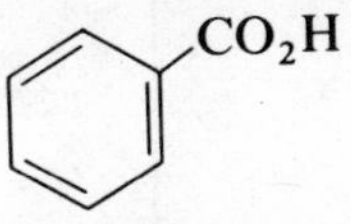

SECTION 8.4
Physical Properties

benzoic acid

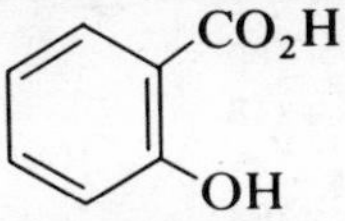

salicylic acid
o-hydroxybenzoic acid

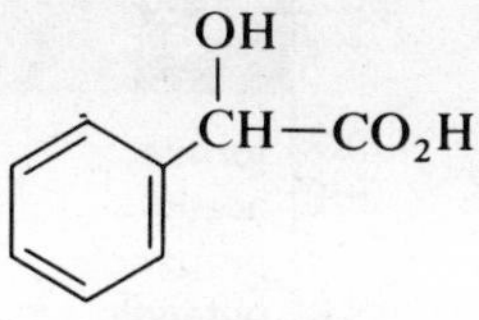

mandelic acid
α-hydroxyphenylacetic
acid

In more complex structural formulas, the carboxyl group may be named as a substituent, a carboxy group, as in 2-carboxycyclohexanone shown below. Alternatively, the —COOH group may be named by adding the words -carboxylic acid to the name of the parent hydrocarbon structure.

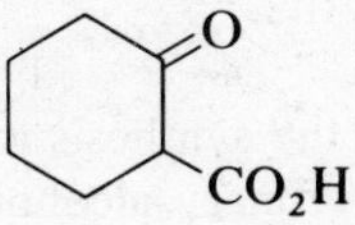

2-carboxycyclohexanone

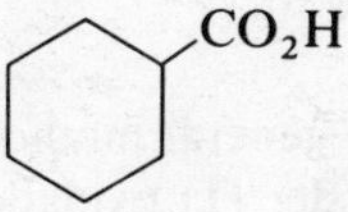

cyclohexanecarboxylic
acid

Salts of organic acids are named in much the same manner as inorganic salts; the cation is named first and then the carboxylate anion. The anions are named by dropping the terminal -ic from the name of the acid and adding -ate.

$$CH_3CO_2^-Na^+$$

sodium acetate

$$C_6H_5CO_2^-NH_4^+$$

ammonium benzoate

$$(CH_3CH_2CO_2^-)_2Ca^{2+}$$

calcium propanoate

8.4 Physical Properties

As we would expect from their structures, the carboxylic acids are polar compounds. They show evidence of strong hydrogen bonding. Even the higher members of the aliphatic series show a solubility in water considerably greater than that of alkanes, alkyl halides, and ethers of comparable molecular weight. Boiling points of carboxylic acids indicate that they are more highly associated in the liquid state than even the alcohols. For example, propanoic acid (mw 74, bp 141°C) boils over 20° higher than 1-butanol (mw 74, bp 117°C).

The first four members of the series (formic through butanoic) are colorless liquids, are completely miscible with water, and have sharp, disagreeable odors. Vinegar is a 4 to 5% solution of acetic acid in water. It is the acetic acid that gives vinegar its characteristic odor and flavor. Pure acetic acid (mp 16°C) solidifies to an icy-looking mass when cooled slightly below normal room temperature and for this reason is often called glacial acetic acid. Table 8.2 shows physical properties of some aliphatic carboxylic acids.

TABLE 8.2 Physical properties of carboxylic acids.

Name	mp (°C)	bp (°C)	Solubility grams/100 grams H_2O
formic	8	100	∞
acetic	16	118	∞
propanoic	−22	141	∞
butanoic	−6	164	∞
hexanoic	−3	205	1.0
palmitic	63	390	insoluble
oxalic	189	dec.*	9
malonic	136	dec.*	74
succinic	185	dec.*	6

* dec. indicates decomposes before boiling.

8.5 Preparation

There are several general methods for the synthesis of simple carboxylic acids. These include: (1) oxidation of primary alcohols, (2) oxidation of alkenes of the type RCH=CHR, (3) oxidation of alkyl substituents on benzene rings, and (4) carbonation of Grignard reagents. Let us look briefly at each of these methods.

First, vigorous oxidation of primary alcohols (Section 5.10) yields carboxylic acids. In the laboratory, <u>potassium dichromate and potassium permanganate</u> are the most commonly used oxidizing agents.

1) Oxidation of 1° Alcohol.

$$CH_3(CH_2)_5CH_2OH + Cr_2O_7^{2-} \xrightarrow{H_3O^+} CH_3(CH_2)_5CO_2H + Cr^{3+}$$

1-heptanol   heptanoic acid

$$HOCH_2CH_2CH_2CH_2OH + MnO_4^- \xrightarrow{H_3O^+} HO_2CCH_2CH_2CO_2H + Mn^{2+}$$

1,4-butanediol   succinic acid

2) Second, <u>ozonolysis of disubstituted alkenes followed by treatment with hydrogen peroxide results in cleavage of the carbon skeleton and formation of two carboxylic acids. The same type of oxidative cleavage can be done using KMnO₄ or K₂Cr₂O₇.</u>

$$CH_3CH_2CH_2CH=CHCH_2CHCH_3 + Cr_2O_7^{2-}$$
                         |
                         CH_3

2-methyl-4-octene

$\xrightarrow{H_3O^+}$

$$CH_3CH_2CH_2\overset{O}{\overset{\|}{C}}OH + HO\overset{O}{\overset{\|}{C}}CH_2CHCH_3 + Cr^{3+}$$
                                              |
                                              CH_3

butanoic acid   3-methylbutanoic acid

$$\bigcirc\!= + Cr_2O_7^{2-} \xrightarrow{H_3O^+} \text{(cyclohexane ring with } CO_2H, CO_2H) + Cr^{3+}$$

cyclohexene   adipic acid

Note that oxidation of a cyclic alkene also results in cleavage of a carbon-carbon double bond but yields a single molecule of a dicarboxylic acid.

Third, carboxylic acids can be prepared by the oxidation of alkyl substituents on an aromatic ring. Recall from Chapter 2 that most alkanes and cycloalkanes are quite resistant to oxidation by $K_2Cr_2O_7$ or $KMnO_4$. However, when the alkyl group has attached to it a benzene ring and at least one hydrogen atom, the alkyl group is more susceptible to vigorous oxidation. No matter how long the alkyl group is, it is degraded to a substituted benzoic acid.

Finally, carboxylic acids can be prepared by the reaction of Grignard reagents with carbon dioxide (Section 7.11). This reaction gives good yields and is widely applicable to the preparation of both aliphatic and aromatic carboxylic acids.

8.6 Acidity of Carboxylic Acids

All carboxylic acids, whether soluble or insoluble in water, react quantitatively with aqueous solutions of sodium hydroxide to form acid salts. They also react quantitatively with solutions of sodium bicarbonate and sodium carbonate to liberate carbon dioxide. These latter reactions can be used as a simple qualitative test to distinguish carboxylic acids from most other organic compounds. Crystalline salts may be obtained by reaction of equivalent amounts of acid and base and evaporation of the solution to dryness.

211

In addition to reacting quantitatively with strong inorganic bases, carboxylic acids also ionize in water to give acidic solutions, and at least in this operational sense are like the common inorganic acids. However, quantitatively they are quite different from inorganic acids such as HCl, HBr, HNO_3, and H_2SO_4. These inorganic acids are strong acids, that is, they are completely ionized in water.

$$HCl \longrightarrow H^+ + Cl^-$$

Carboxylic acids, on the other hand, are only incompletely ionized in dilute aqueous solution and an equilibrium is established between the carboxylic acid, the carboxylate anion, and H^+.

$$CH_3CO_2H \rightleftharpoons CH_3CO_2^- + H^+$$

The equilibrium constant for this ionization is called K_a, the acid dissociation or ionization constant, and has the form

$$K_a = \frac{[CH_3CO_2^-][H^+]}{[CH_3CO_2H]} = 1.8 \times 10^{-5}$$

Values of K_a for some representative carboxylic acids are given in Table 8.3.

TABLE 8.3 K_a and pK_a for some carboxylic acids.

Name	Structure	K_a	pK_a
formic	HCOOH	1.8×10^{-4}	3.74
acetic	CH_3COOH	1.8×10^{-5}	4.74
propanoic	CH_3CH_2COOH	1.4×10^{-5}	4.85
butanoic	$CH_3CH_2CH_2COOH$	1.6×10^{-5}	4.80
chloroacetic	$ClCH_2COOH$	1.4×10^{-3}	2.85
dichloroacetic	$Cl_2CHCOOH$	3.3×10^{-2}	1.48
trichloroacetic	Cl_3CCOOH	2.3×10^{-1}	0.64
methoxyacetic	CH_3OCH_2COOH	3.3×10^{-4}	3.48
benzoic	C_6H_5COOH	6.5×10^{-5}	4.19

It should be obvious from the expression for K_a that strong (highly dissociated) acids will have large values of K_a and, conversely, that weak (incompletely dissociated) acids will have considerably smaller values of K_a. From the K_a values in Table 8.3 we know that formic acid is a stronger acid than acetic acid. Other than the K_a for formic acid, the values for unsubstituted aliphatic carboxylic acids are essentially the same as that for acetic acid. Therefore, the value of K_a for acetic acid is a useful number to remember.

You might wonder why carboxylic acids are so much more acidic than water or alcohols, compounds that also contain the —OH functional group. The acidity of ethanol is undetectable in water but is estimated to be about 10^{-16}. The value of K_a for acetic acid is over ten trillion (10^{10}) times larger than that of ethanol.

$$CH_3CH_2OH \rightleftharpoons CH_3CH_2O^- + H^+ \qquad K_a = 10^{-16}$$

$$CH_3-\overset{\overset{\displaystyle O}{\|}}{C}-OH \rightleftharpoons CH_3-\overset{\overset{\displaystyle O}{\|}}{C}-O^- + H^+ \qquad K_a = 1.8 \times 10^{-5}$$

The question may be stated another way: Why is acetic acid so much more extensively ionized in water than ethanol? We can account for this enhanced acidity (enhanced ionization) by using the resonance model and looking at the relative stabilities of the acetate ion and the ethoxide ion. Recall that this is the approach we used in Section 6.9 to account for the greater acidity of phenol compared to ethanol.

Resonance has no effect on the dissociation of ethanol for there is no possibility for resonance or resonance stabilization in the ethoxide ion. For the acetate ion, however, we can write two equivalent contributing structures.

$$CH_3CH_2-\ddot{O}-H \;\rightleftharpoons\; CH_3CH_2-\ddot{O}\colon^- + H^+$$

$$CH_3-\overset{\displaystyle :O:}{\overset{\|}{C}}-\ddot{O}-H \;\rightleftharpoons\; CH_3-\overset{\displaystyle :\ddot{O}}{\overset{\|}{C}}-\ddot{O}\colon^- \;\longleftrightarrow\; CH_3-\overset{\displaystyle :\ddot{O}\colon^-}{\overset{|}{C}}=\ddot{O}\colon + H^+$$

resonance-stabilized
anion

Since acetate is stabilized by resonance compared to ethoxide, we can expect the equilibrium for the ionization of acetic acid to lie farther to the right than that for the ionization of ethanol, that is, we can expect acetic acid to be a stronger acid than ethanol.

Substituents, particularly on the α-carbon, can exert a marked effect on the acidity of a carboxylic acid. Compare for example the K_a value given in Table 8.3 for acetic acid with those of its mono-, di-, and trichloro-derivatives. You will see that chloroacetic acid is almost 80 times stronger an acid than acetic acid. Trichloroacetic acid is a very much stronger acid, and in fact in dilute aqueous solutions is about as strong as sulfuric acid. The most usual explanation for these facts is that, because of its electronegativity, the chlorine atom of chloroacetic acid attracts electron density, thereby polarizing the electrons of the Cl—C bond and inducing a partial positive charge on the carbon atom. This in turn induces a polarization of electrons in the C—C bond and in the C—O bond. The hydroxyl oxygen of chloroacetic acid is slightly more positive than the corresponding oxygen of acetic acid and consequently the proton is more easily removed. This polarization of electrons through sigma bonds is called the inductive effect and is indicated by an arrow on the sigma bonds themselves with the head of the arrow pointing toward the more electronegative atom.

$$\overset{\textstyle H}{\underset{\textstyle H}{Cl\leftarrow C\leftarrow C}}\overset{\textstyle O}{\underset{\textstyle O\leftarrow H}{\diagup\!\!\diagup}}$$

Such inductive effects are largest when the electronegative substituent is close to the carboxylic acid group, and becomes progressively weaker as the chlorine substituent is moved farther away. Whereas chloroacetic acid is about 80 times stronger than acetic, 3-chloropropanoic acid is only about six times stronger, and 4-chlorobutanoic acid is only about twice as strong as acetic acid.

chloroacetic acid	$Cl-CH_2-CO_2H$	$K_a = 140 \times 10^{-5}$
3-chloropropanoic acid	$Cl-CH_2-CH_2-CO_2H$	$K_a = 10 \times 10^{-5}$
4-chlorobutanoic acid	$Cl-CH_2-CH_2-CH_2-CO_2H$	$K_a = 3 \times 10^{-5}$

Before we leave this discussion of the acidity of carboxylic acids, we should compare this class of organic acids with another class of acids which we discussed in Section 6.9, namely, phenols which are also weakly acidic. With a K_a of 1.2×10^{-10}, phenol is sufficiently acidic to react with an aqueous solution of sodium hydroxide to form the sodium salt, sodium phenoxide. Furthermore, we saw how the difference in acidity between a phenol and a water-insoluble alcohol can be used to separate a mixture of two such substances. Most phenols, however, are considerably weaker acids than carboxylic acids and will not dissolve in aqueous sodium bicarbonate because the bicarbonate ion is too weakly basic to remove a proton from the phenol molecule. This fact permits the ready separation of phenols from the more strongly acidic carboxylic acids. An aqueous solution of sodium bicarbonate will dissolve carboxylic acids as their sodium salts but will not dissolve phenols. The water-soluble carboxylate salt can then be physically separated from the water-insoluble phenol.

8.7 Reduction of Carboxylic Acids

Reduction of carboxylic acids is generally quite difficult. While they are inert to normal conditions of hydrogen in the presence of a metal catalyst, catalytic hydrogenation at high pressures (100 atm) in the presence of a copper chromite ($CuCrO_2$) catalyst does reduce carboxylic acids to alcohols. Lithium aluminum hydride also reduces carboxylic acids smoothly to alcohols.

$$R-CO_2H \quad \xrightarrow{\text{LiAlH}_4} \quad R-CH_2OH$$

$$\xrightarrow[\text{high pressure}]{\text{H}_2/\text{CuCrO}_2} \quad R-CH_2OH$$

8.8 Fatty Acids

Fatty acids are long-chain aliphatic carboxylic acids, so named because they are readily obtained by acid, alkaline, or enzyme-catalyzed hydrolysis of fats. The common neutral fats are triesters of glycerol and their general formula is shown in Figure 8.2.

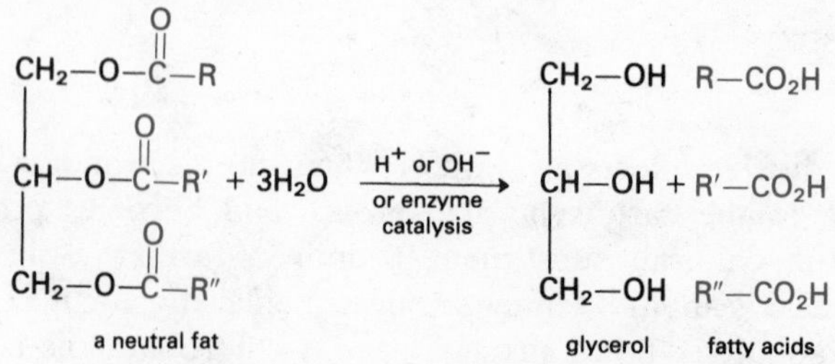

$$
\begin{array}{l}
CH_2-O-\overset{\overset{\displaystyle O}{\|}}{C}-R \\
CH-O-\overset{\overset{\displaystyle O}{\|}}{C}-R' \;+\; 3H_2O \quad \xrightarrow[\substack{\text{or enzyme}\\ \text{catalysis}}]{H^+ \text{ or } OH^-} \quad CH-OH \;+\; R'-CO_2H \\
CH_2-O-\overset{\overset{\displaystyle O}{\|}}{C}-R''
\end{array}
$$

a neutral fat   glycerol   fatty acids

FIGURE 8.2 The hydrolysis of a neutral fat.

TABLE 8.4 Some important naturally occurring fatty acids.

Carbon Atoms	Structure	Common Name	mp (°C)
Saturated fatty acids			
12	$CH_3(CH_2)_{10}COOH$	lauric	44
14	$CH_3(CH_2)_{12}COOH$	myristic	58
16	$CH_3(CH_2)_{14}COOH$	palmitic	63
18	$CH_3(CH_2)_{16}COOH$	stearic	70
20	$CH_3(CH_2)_{18}COOH$	arachidic	77
Unsaturated fatty acids			
16	$CH_3(CH_2)_5CH{=}CH(CH_2)_7COOH$	palmitoleic	−1
18	$CH_3(CH_2)_7CH{=}CH(CH_2)_7COOH$	oleic	16
18	$CH_3(CH_2)_4CH{=}CHCH_2CH{=}CH(CH_2)_7COOH$	linoleic	−5
18	$CH_3CH_2(CH{=}CHCH_2)_3(CH_2)_6COOH$	linolenic	−11
20	$CH_3(CH_2)_3(CH_2CH{=}CH)_4(CH_2)_3COOH$	arachidonic	−49

We shall discuss the structure and properties of fats in Chapter 12. For the moment let us look at the fatty acids themselves. Over 70 fatty acids have been isolated from various cells and tissues. Table 8.4 gives structural formulas of some important fatty acids.

We can make certain generalizations about the more abundant fatty acid components of higher plants and animals.

1. Nearly all have an even number of carbon atoms, usually between 14 to 22 carbons in an unbranched chain. Those having 16 or 18 carbon atoms are by far the most abundant in nature.
2. Unsaturated fatty acids have lower melting points than their saturated counterparts. The physical properties of the particular fatty acid components also affect the fats into which they are incorporated and, as we shall see in Section 12.2, fats rich in unsaturated fatty acids are lower melting than those rich in saturated fatty acids.
3. In most of the unsaturated fatty acids of higher organisms there is a double bond between carbon atoms 9 and 10. In these unsaturated fatty acids, *cis* isomers predominate; the *trans* configuration is very rare.

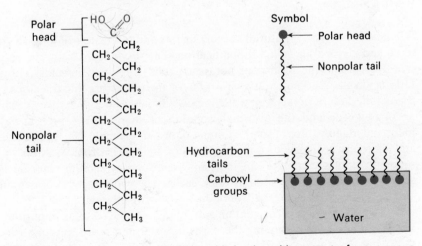

FIGURE 8.3 The interaction of a fatty acid droplet with water to form a mono-molecular layer.

Because of their long hydrocarbon chains, fatty acids are essentially insoluble in water. However, they do interact with water in a particular manner. If a droplet of fatty acid is placed on the surface of water, it will spread out to form a thin film one molecule thick (a monomolecular layer) with the polar carboxyl groups dissolved in the water and the nonpolar hydrocarbon chains forming a hydrocarbon layer on the surface of the water. This is illustrated in Figure 8.3.

8.9 Essential Fatty Acids

If fatty acids (in the form of fats) are entirely withheld from the diet of rats, the rats soon begin to suffer from retarded growth, scaly skin, kidney damage, and eventually, early death. Addition of the unsaturated fatty acids—linoleic, linolenic, and arachidonic—will cure this condition. Strictly speaking, linoleic is the critical fatty acid for it can be converted within the cell to linolenic and arachidonic acids. Because linoleic acid must be obtained in the diet for normal growth and well-being of high animals and man, it is classified as an essential fatty acid. Most animal fats are relatively rich in saturated fatty acids, and though the percentage of unsaturated fatty acids is also high, this unsaturation is due mostly to oleic acid. Vegetable fats, on the other hand, generally have a lower content of saturated fatty acids and a relatively higher content of unsaturated fatty acids including linoleic acid. Corn, cottonseed, soybean, and wheat-germ oils are especially rich in linoleic acid.

There is no set minimum requirement for this essential fatty acid, but the Food and Nutrition Board states that for adults a linoleic acid intake of about 6 grams per day in a diet of 2700 calories should be sufficient. For infants and premature babies, the requirements are higher. Human milk and commercially prepared infant formulas provide a generous allowance of linoleic acid.

8.10 Soaps

Alkaline hydrolysis of naturally occurring fats and oils (esters of long-chain acids and glycerol) is called saponification. The products of saponification are glycerol and the sodium or potassium salts of carboxylic acids. These fatty acid salts are known as soaps. One of the oldest organic reactions is the boiling of lard (a fat) with a slight excess of soda (sodium hydroxide) in an open kettle and the eventual isolation of soap. In the present-day industrial manufacture of soap, molten tallow (for example, the fat of cattle and sheep) is heated with a slight excess of sodium hydroxide. After the saponification is complete, an inorganic salt such as sodium chloride is added to precipitate the soap as thick curds. The water layer is drawn off and the glycerol is recovered from it by vacuum distillation.

The crude soap curds contain salt, alkali, and glycerol as impurities. These are removed by boiling the curds in water and reprecipitating with salt. After several such purifications, the soap may be used without further processing as an inexpensive industrial soap. Fillers such as sand or pumice may be added to make a scouring soap. Other treatments transform the crude soap into laundry soaps, medicated soaps, cosmetic soaps, liquid soaps, and so on.

Soap owes its remarkable cleansing properties to its ability to act as an emulsifying agent. Let us consider the sodium salt of stearic acid as a specific example of a soap. Regarded from one end, sodium stearate is a highly ionic salt and, therefore, is strongly attracted to water (it is <u>hydrophilic</u>). Regarded from the other end, sodium stearate is a long hydrocarbon chain and is said to be <u>hydrophobic</u> (repelling water) and <u>lyophilic</u> (attracting) toward nonpolar organic solvents. Because its long hydrocarbon chain is intrinsically insoluble in water, there is little tendency for sodium stearate to dissolve in water to form a true solution. However, it readily disperses in water to form <u>micelles</u> in which the charged carboxylate groups form a negatively charged surface and the nonpolar, water-insoluble hydrocarbon chains lie buried within the center (Figure 8.4). Such micelles have a net negative charge and remain suspended or dispersed because of mutual repulsion of one for another.

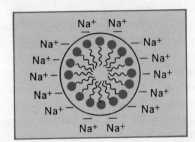

FIGURE 8.4 Sodium stearate micelle.

Soaps also seek out the interface between water and fats, oils, or grease—substances that by themselves are quite insoluble in water. (Most dirt is held to clothes by a thin film of grease or oil.) If the oil is dispersed into tiny droplets throughout the water by shaking, the soap again forms micelles, now with the oil droplet at the center (Figure 8.5). In this way, the oil, grease, or dirt is then emulsified and may be carried away in the wash water.

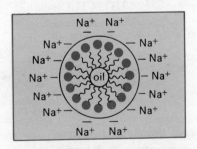

FIGURE 8.5 Sodium stearate micelle with a "dissolved" oil droplet.

Yet soaps are not without their disadvantages. First, soaps are sodium or potassium salts of weak acids and are converted by strong mineral acid into the free fatty acids.

$$CH_3(CH_2)_{16}CO_2^-Na^+ + H^+ \longrightarrow CH_3(CH_2)_{16}CO_2H \downarrow + Na^+$$

The free acids are far less soluble than their potassium or sodium salts and they precipitate, forming a scum. Therefore, soaps cannot be used in acidic solution. Second, soaps form insoluble salts when used in water containing calcium, magnesium, or ferric ions ("hard" water).

$$2CH_3(CH_2)_{16}CO_2^- + Ca^{2+} \longrightarrow [CH_3(CH_2)_{16}CO_2^-]_2Ca^{2+} \downarrow$$

This precipitate or scum formation creates problems including rings around the bathtub, the films that spoil the luster of hair, and the grayness and harshness of feel that build up on textiles after repeated washing.

Given these limitations on the use of the natural soaps, the problem for the chemist is to create a new type of cleansing agent that will be readily soluble in both acidic and alkaline solutions, and will not form insoluble precipitates when used in hard water. Despite considerable effort, there was no significant progress until late in the 1940s with the introduction of synthetic detergents.

8.11 Synthetic Detergents

One of the most useful innovations in cleansing has been the development in recent years of synthetic detergents (often called syndets). These synthetic products have cleansing power as good or better than ordinary soaps, and at the same time they avoid the two major difficulties already listed for soaps. Given an understanding of the mechanism of action of the soaps, the design criteria for a synthetic detergent are as follows: a molecule with a long hydrocarbon chain (preferably 12 to 18 carbon atoms) and a highly polar group or groups at one end of the molecule. It was recognized that the essential characteristics of a soap could be produced in a molecule containing a sulfate group rather than a carboxylate group. Such compounds, known as alkyl acid sulfate esters, are strong acids comparable in strength to sulfuric acid. Furthermore, the calcium, magnesium, and ferric salts of alkyl acid sulfate esters are soluble in water.

In the earliest method of syndet production, a long-chain alcohol is allowed to react with sulfuric acid and form an alkyl acid sulfate ester. Neutralization with sodium hydroxide forms a synthetic detergent.

$$CH_3(CH_2)_{10}CH_2OH + H_2SO_4 \longrightarrow CH_3(CH_2)_{10}CH_2O{-}SO_3H + H_2O.$$

1-dodecanol
lauryl alcohol

dodecyl hydrogen sulfate
lauryl hydrogen sulfate

$$\Big\downarrow NaOH$$

$$CH_3(CH_2)_{10}CH_2O{-}SO_3^-Na^+ + H_2O$$

sodium dodecyl sulfate (SDS)
sodium lauryl sulfate

The physical resemblances between this synthetic detergent and the ordinary soaps are obvious: a long nonpolar hydrocarbon chain and a highly polar end.

Yet the major advance in detergents came in the late 1940s when it became technologically feasible to make the so-called alkylbenzene sulfonate detergents. The essential raw materials for this synthesis, propylene and benzene, had become readily available from the petroleum refining industry. Propylene was polymerized to a tetramer and reacted with benzene. This product was then sulfonated and reacted with sodium hydroxide to yield an alkylbenzene sulfonate salt of the type shown below.

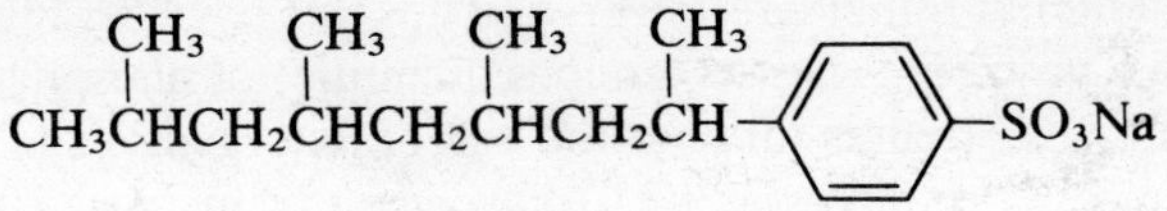

$$CH_3CHCH_2CHCH_2CHCH_2CH \text{—} \langle\text{benzene}\rangle \text{—} SO_3Na$$

with CH_3 groups on each of the four branch carbons

a sodium alkylbenzene sulfonate

These alkylbenzene sulfonate detergents were introduced in the 1950s and were accepted very rapidly. Within a decade U.S. production of synthetic detergents increased twenty-fold. Today they command close to 90% of the market once held by soaps.

The cleansing power of synthetic detergents of this type can be enhanced enormously by certain additives. Sodium tripolyphosphate ($Na_5P_3O_{10}$), or STPP, is added as a "builder"; it has the ability to coordinate with and suspend calcium, magnesium, copper, iron, and many other ions. Thus STPP is able to break up and suspend certain clays and pigments by forming complexes with the metal ions, thereby facilitating their removal. Phosphates were introduced into cleansing agents in 1948 when Proctor & Gamble Company introduced Tide. By 1953, STPP was used in more than half of the detergents sold in the United States and by 1970 almost all detergents contained phosphates, sometimes as much as 60% by weight. Other common additives are whitening agents (optical brighteners), sudsing enhancers and repressors, and granular salts to create a satisfactory consistency for the commercial product.

Yet as useful as the synthetic detergents proved to be, they have created two major problems. One of the problems, that of disposability and biodegradability, has been solved. The other, the phosphate additives, is now being brought under control.

The first of these problems began to appear as excessive foaming in natural waters and sewage treatment plants. Most of it was found to be caused by the alkylbenzene sulfonate detergents. Soaps are removed from sewage waters by precipitation or by degradation in the treatment plants by microorganisms that are able to metabolize the linear alkyl hydrocarbon chains of the natural soaps derived from fats and oils. Such soaps are said to be biodegradable. It was discovered the the first alkylbenzene sulfonate detergents marketed could not be removed in either of these ways; they could not be precipitated and they could not be degraded by the microorganisms in sewage treatment plants. Instead they remained in suspension, causing sudsing and foaming, polluting streams, and in some cases finding their way into municipal drinking supplies. The solution to the problem seemed to be in replacing the nonbiodegradable branched-chain part of the alkylbenzene sulfonate by a biodegradable linear-chain hydrocarbon. Fortunately just such linear hydrocarbons had become available through advances in petroleum refining, and in particular the use of molecular sieves to separate the desired detergent-range hydrocarbons. In 1965 the detergent industry converted entirely to the new linear alkylbenzene sulfonates of the type shown below.

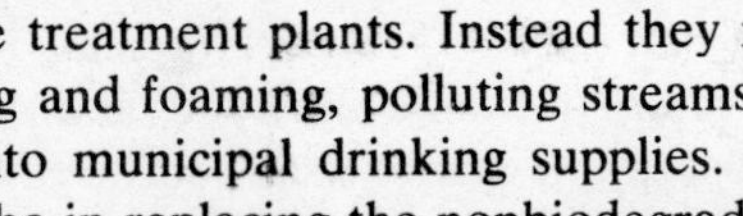

$$CH_3(CH_2)_{11} \text{—} \langle\text{benzene}\rangle \text{—} SO_3^-Na^+$$

sodium dodecylbenzenesulfonate (a syndet)

The second problem is that of the phosphate additives themselves and their contribution to water pollution. Strictly speaking, phosphate is not a pollutant, but rather a fertilizer, and it is as a fertilizer that phosphate has created a problem. The tremendous quantities of phosphates added to lakes, rivers, and streams through the use of detergents (and agricultural phosphate-based fertilizers as well) have greatly increased the nutrient quality of the water. According to the House of Representatives' Subcommittee on Conservation and Natural Resources, when a rich stream of fertilizer flows steadily into a lake:

> Overstimulated, the water plants grow to excess. Seasonally they die off and rot. . . . In the process of decay they exhaust the dissolved oxygen of the water and produce the rotten-egg stench of hydrogen sulfide. . . . The game fish die of oxygen deficiency. . . . Intake filters for potable water become clogged, and boat propellers fouled with algae. The lake loses its value as a water supply, as an esthetic and recreational asset, and as an avenue of commerce. Finally the water itself is displaced by the accumulating masses of living and dead vegetation and their decay products and the lake becomes a bog, and eventually, dry land.

This evolution of a lake is, of course, a natural process, but one which has been greatly accelerated. Lake Erie is the most notorious example of a dying American lake. It is said to have "aged" 15,000 years in the last 50. Since phosphate-containing detergents were introduced in 1948, the phosphate content of the lake's western basin has more than tripled. Fortunately, this situation has not been allowed to continue. Public pressure and intelligent legislation have combined to require much lower phosphate levels in all detergents and, in many instances, to ban the use of phosphate-containing detergents. There are a number of nonphosphate-containing detergents on the market at the present time.

8.12 Spectroscopic Properties

The carboxyl group of carboxylic acids gives rise to two characteristic absorptions in the infrared spectrum. One of these occurs in the region 1700 to 1725 cm^{-1} and is associated with the stretching vibration of the carbonyl group. Note that this is essentially the same range of absorption as that for the carbonyl group of simple aldehydes and ketones (Section 7.15). The other infrared absorption characteristic of the carboxyl group is a peak between 2500 to 3000 cm^{-1} due to the stretching vibration of the O—H group. This absorption is generally very broad due to hydrogen bonding between molecules of carboxylic acid. Both C=O and O—H stretching frequencies can be seen in the infrared spectrum of butanoic acid (Figure 8.6).

Like aldehydes and ketones (Section 7.15) simple unconjugated carboxylic acids show only weak absorption in the ultraviolet spectrum.

The hydrogen of the carboxyl group gives rise to a signal in the NMR spectrum in the range $\delta = 10$ to 13.0. Note that the chemical shift of the carboxyl hydrogen is even larger than that of the aldehyde hydrogen. In fact, this chemical shift is so large that it serves to distinguish carboxyl

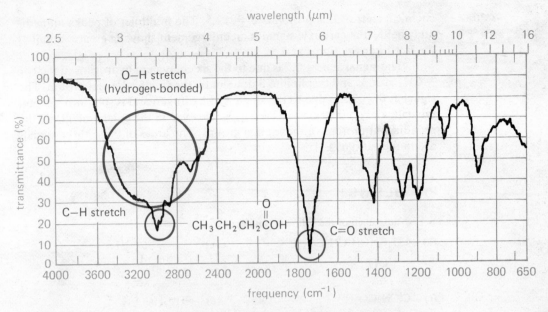

FIGURE 8.6 An IR spectrum of butanoic acid.

hydrogens from most other types of hydrogens. The NMR signal for carboxyl hydrogens generally falls off the scale of chart papers, most of which are calibrated from $\delta = 0$ to $\delta = 10$. Therefore, in order to record this signal, the pen is set back on the paper at some arbitrary point and a note is made of how much it has been displaced or offset. The signal for the carboxyl proton of 2-methylpropanoic acid appears in the top right corner of the NMR spectrum in Figure 8.7 and has been offset by 10δ. Thus, the

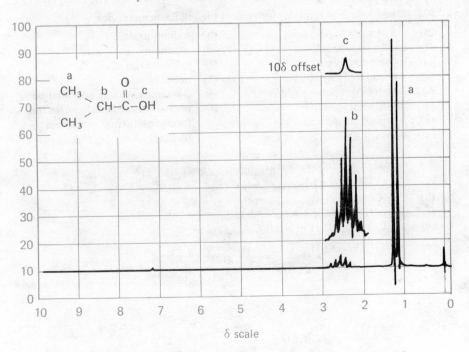

FIGURE 8.7 An NMR spectrum of 2-methylpropanoic acid (isobutyric acid).

221

chemical shift of this proton is $\delta = 12.5$. The multiplet of peaks immediately below the carboxyl signal is an enlargement showing greater detail of the low intensity signal at $\delta = 2.6$.

The signal at $\delta = 1.2$ is due to the six equivalent hydrogens of the two —CH_3 groups and is split into a doublet by the single adjacent proton. The signal at $\delta = 2.5$ is due to the single CH hydrogen and is split into a septet (actually only five peaks are clearly visible even in the enlargement) by the six adjacent protons of the methyl groups. The areas of these three signals are in the ratio $6:1:1$.

PROBLEMS

8.1 Name each of the following molecules.

(a) $CH_3CHCH_2CH_2CO_2H$
$\quad\quad\ \ |$
$\quad\quad\ \ OH$

(b) 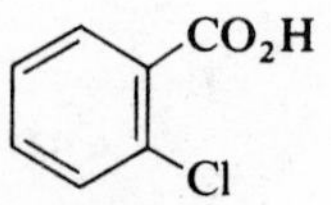 (structure: benzene ring with CO_2H and Cl ortho)

(c) $ClCH_2CO_2H$

(d) CH_3CHCO_2H
$\quad\quad\ |$
$\quad\quad\ CH_3$

(e) $C_6H_5CH_2CO_2H$

(f) 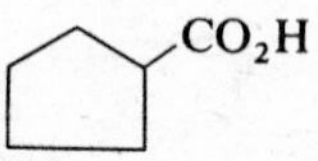 (cyclopentane ring with CO_2H)

(g) $C_6H_5CO_2Na$

(h) $CH_3CH_2CH_2CH_2CO_2NH_4$

(i) CH_3CHCO_2H
$\quad\quad\ |$
$\quad\quad\ CH_2CO_2H$

(j) $CF_3—CO_2H$

(k) $(CH_3CH_2CO_2)_2Ca$

(l) $CH_3\overset{O}{\overset{\|}{C}}CH_2CH_2CH_2CH_2CH_2CH{=}CHCO_2H$

8.2 Draw structural formulas for each of the following molecules.

(a) 3-hydroxybutanoic acid
(b) sodium oxalate
(c) trichloroacetic acid
(d) 4-aminobutanoic acid
(e) sodium hexadecanoate
(f) calcium stearate
(g) potassium phenylacetate
(h) octanoic acid
(i) 2-hydroxypropanoic acid (lactic acid)
(j) 2-aminopropanoic acid (alanine)
(k) p-aminobenzoic acid
(l) potassium 2,4-hexadienoate (the food preservative, potassium sorbate)

8.3 Complete the following reactions.

(a) $CH_3CO_2H + NaOH_{(aqueous)} \longrightarrow$
(b) $C_6H_5CO_2H + NH_{3(aqueous)} \longrightarrow$
(c) $(CH_3CH_2CO_2^-)_2Ca^{2+} + H_2SO_{4(aqueous)} \longrightarrow$
(d) $CH_3CH_2CH_2CO_2H + NaHCO_{3(aqueous)} \longrightarrow$

(e) 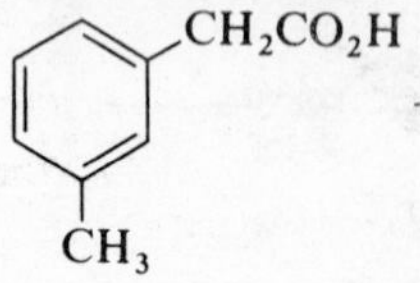 (benzene ring with CH_2CO_2H and CH_3) $+ Na_2Cr_2O_7 \xrightarrow[\text{heat}]{H_2O}$

(f) 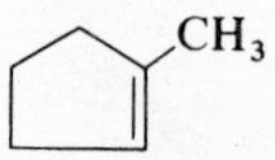 (cyclopentene ring with CH_3) $+ Na_2Cr_2O_7 \xrightarrow[\text{heat}]{H_2O}$

(g) $CH_3CH_2CHCH_2MgBr + CO_2 \longrightarrow$
            $|$
            CH_3

(h) product (g) + H_3O^+ $\longrightarrow$

(i) [ring structure with CH_2CH_2OH and CH_3 substituents] + $Na_2Cr_2O_7$ $\xrightarrow[\text{heat}]{H_2O}$

8.4 Show reagents and conditions you would use to carry out the following transformations.

(a) 1-octanol to octanoic acid
(b) 1-octanol to heptanoic acid
(c) 1-octanol to nonanoic acid
(d) cyclopentene to cyclopentanecarboxylic acid
(e) cyclopentanol to glutaric acid
(f) 1-butene to 2-methylbutanoic acid
(g) 2-methylpropene to 2,2-dimethylpropanoic acid
(h) toluene to terephthalic acid (1,4-benzenedicarboxylic acid)
(i) benzene to *p*-nitrobenzoic acid

8.5 Compound *A* ($C_5H_{10}O_3$) readily dissolves in water to give an acidic solution and can be titrated with aqueous sodium hydroxide. Compound *A* is also optically active and contains an alcohol group. Oxidation of compound *A* by potassium permanganate gives compound *B* ($C_5H_8O_4$). Compound *B* is a dicarboxylic acid and is optically inactive. Deduce structures for compounds *A* and *B* consistent with these observations.

8.6 There are four isomeric alcohols of molecular formula $C_4H_{10}O$. Draw and name each. Which of these alcohols is indicated by the following experimental observations? Compound *D* ($C_4H_{10}O$), an oxidation by potassium permanganate in acid solution, gives compound *E* ($C_4H_8O_2$), a carboxylic acid. Treatment of compound *D* with warm phosphoric acid brings about dehydration and yields compound *F* (C_4H_8). Treatment of compound *F* with warm aqueous sulfuric acid gives *G* ($C_4H_{10}O$), a new alcohol isomeric with compound *D*. Compound *G* is resistant to oxidation. Propose structures for compounds *D*, *E*, *F*, and *G* consistent with these observations.

8.7 Examine the structural formulas for lauric, palmitic, stearic, oleic, linoleic, and arachidonic acids. For each that will show *cis-trans* isomerism, state the total number of such isomers possible.

8.8 By using structural formulas illustrate how a molecule of fatty acid interacts with water to form a monomolecular layer on the surface of water.

8.9 By using structural formulas, show how a soap "dissolves" fats, oils, and grease.

8.10 Show by balanced equations the reaction of a soap with (a) hard water, and (b) acidic solution.

8.11 Characterize the structural features necessary to make a detergent of good cleansing ability. Illustrate by structural formulas two different classes of synthetic detergents. Name each example.

8.12 Below are given structural formulas for a cationic detergent and a nonionic detergent. How would you account for the detergent properties of each?

            CH_3
            $|$
$C_6H_5-CH_2-N^+-CH_3Cl^-$
            $|$
            C_8H_{17}

benzyldimethyloctylammonium
chloride

        O CH_2OH
        $||$ $|$
$CH_3(CH_2)_{14}C-O-CH_2-C-CH_2OH$
                      $|$
                      CH_2OH

pentaerythrityl palmitate

223

8.13   What does it mean to say that linoleic acid is an "essential" fatty acid? Name several dietary sources of linoleic acid.

8.14   Show how you might distinguish between the following pairs of compounds by a simple chemical test. In each case, tell what test you would perform, what you would expect to observe, and write an equation for each positive test.

(a)   acetic acid and acetaldehyde

(b)   hexanoic acid and 1-hexanol

(c)   benzoic acid and phenol

(d)   sodium salicylate and salicylic acid

(e)   oleic acid and stearic acid (see Table 8.4 for structural formulas of these fatty acids)

(f)   phenylacetic acid and acetophenone (methyl phenyl ketone)

(g)   4-oxohexanoic acid and hexanoic acid

(h)   sodium lauryl sulfate (a synthetic detergent, Section 8.11) and sodium stearate (a natural soap)

8.15   List one major spectral characteristic that will enable you to distinguish between the following compounds.

(a)   acetic acid and acetaldehyde (IR and NMR)

(b)   hexanoic acid and 1-hexanol (IR and NMR)

(c)   benzoic acid and phenol (IR)

(d)   phenylacetic acid and acetophenone (IR and NMR)

(e)   4-oxohexanoic acid and hexanoic acid (NMR)

(f)   benzoic acid and cyclohexanecarboxylic acid (UV and NMR)

(g)   2-methylpropanoic acid (isobutyric acid) and butanoic acid (n-butyric acid)

III

Mini-Essay

Prostaglandins

Prostaglandins

In 1930 Raphael Kurzrok and Charles Lieb, two gynecologists practicing in New York, reported that human seminal fluid stimulates the contraction of isolated uterine muscle. A few years later in Sweden, Ulf von Euler confirmed this report and noted that it also produces contraction of intestinal smooth muscle and lowers blood pressure when injected into the blood stream. Von Euler proposed the name *prostaglandin* for the mysterious substance or substances responsible for these diverse effects for at that time, it was thought that they originated in the prostate gland. Hence the name prostaglandin. We now know that prostaglandins are produced in many different types of cells and tissues within the body and that they are by no means limited to the prostate gland. However, the richest source of prostaglandins in man is seminal fluid where they are present at a total concentration of approximately 10^{-3} M.

Prostaglandins were first isolated in pure crystalline form in 1959 and the structural formulas of several were determined shortly thereafter. These substances contain 20 carbon atoms and have the same carbon skeleton as that of prostanoic acid.

prostanoic acid

They are abbreviated PG with an additional letter and numerical subscript to indicate the type and series. The various types differ in the functional groups present in the five-membered ring. Those of the A-type are α,β-unsaturated ketones; those of the E-type are β-hydroxyketones; and those of the F-type are 1,3-diols. The subscript α in those of the F-type indicates that the hydroxyl group at carbon-9 is below the plane of the five-membered ring and on the same side as the hydroxyl at carbon-11. The various series of prostaglandins differ in the number of double bonds on the two side chains. Those of the 1-series have only one double bond; those of the 2-series have two double bonds; and those of the 3-series have three double bonds. Shown in Figure 1 are structural formulas for PGA_2, PGE_2 and $PGF_{2\alpha}$, all synthesized within the body from the essential fatty acid, arachidonic acid.

Coincident with the investigations of the chemical structure of prostaglandins, clinical scientists began to study the biochemistry of these remarkable substances and their potential as drugs. Initially, research was hampered by the high cost and great difficulty in isolating and purifying

227

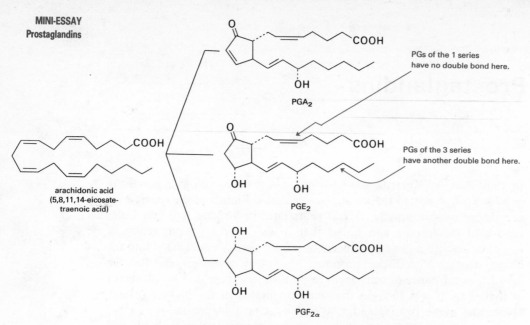

FIGURE 1 Prostaglandins PGA$_2$, PGE$_2$, and PGF$_{2\alpha}$. Each is synthesized within the body from the essential fatty acid, arachidonic acid.

these substances. If they could not be isolated easily, could they be synthesized instead? The first totally synthetic prostaglandins became available in 1968 when Dr. John Pike of the Upjohn Company and Professor E. J. Corey of Harvard University each announced laboratory synthesis of several prostaglandins and prostaglandin analogs. However, costs were still high. Then in 1969, the price of prostaglandins dropped dramatically with the discovery that the gorgonian sea whip or sea fan, *Plexaura homomalla*, which grows on the coral reefs off the coast of Florida and in the Caribbean, is a rich source of prostaglandin-like materials. The concentration of PG-like substances in this marine organism is about 100 times the normal concentration in most mammalian sources. The PG-like compounds were extracted and then transformed in the laboratory into prostaglandins and prostaglandin analogs. In effect, chemists took advantage of the ability of *P. homomalla* to perform steps which in the laboratory gave only poor yields. At the present time, however, there is little need to depend on even this natural source for chemists have succeeded in devising highly effective and stereospecific laboratory schemes for the synthesis of almost any prostaglandin or prostaglandin-like substance.

We now know that prostaglandins are intimately involved in a host of bodily processes. Clinically, it has been found that they are involved in almost every phase of reproductive physiology; they can act to regulate menstruation, to prevent conception, and to induce childbirth and abortion. Certain prostaglandins stimulate blood clotting by promoting blood platelet aggregation; others inhibit clotting by inhibiting platelet aggregation. Prostaglandins have been implicated in both the induction of the inflammatory response and in its relief. They are also involved in the allergic response. The medical significance of their involvement in the inflammatory and allergic responses is obvious when we realize that millions of Americans suffer from rheumatoid arthritis, an inflammatory disease, and from asthma and other allergic diseases. Certain prostaglandins also appear to stimulate the enzyme adenyl cyclase which in turn catalyzes the conversion of adenosine triphosphate (ATP) into cyclic adenosine monophosphate (cAMP). Cyclic-AMP is a compound of major importance in the regulation of cellular metabolism and it may well be that the ability of prostaglandins to influence the metabolism of this substance is the key to their wide range of physiological activities.

In this essay, we will discuss two aspects of the biochemistry and function of prostaglandins; their role in regulating the clotting of blood, and their recent introduction into clinical medicine to induce childbirth and second-trimester abortion.

First the participation of prostaglandins in blood clotting. There are three distinct phases to the physiological mechanisms which come into play within the body to stop bleeding from a ruptured blood vessel. In the first phase, initiated by agents such as thrombin, the blood platelets become sticky and form a platelet plug at the site of the injury. This phase is known as platelet aggregation. If the damage is minor and the blood vessel is small, this platelet plug may be sufficient to stop the loss of blood from the vessel. If it is not sufficient, the platelets are stimulated to release a group of substances (the platelet release reaction) which in turn promotes a second wave of platelet aggregation and the constriction of the injured vessel. The third phase is the triggering of the actual blood coagulation process.

We have learned within the past few years that among the substances released in the platelet release reaction are two prostaglandins known as PGG_2 and PGH_2. Both contain peroxide linkages between carbons 9 and 11 of the five-membered ring and, therefore, are referred to as cyclic endoperoxides. They are synthesized from arachidonic acid stored within the platelets in a series of reactions catalyzed by the enzyme cyclo-oxygenase.

arachidonic acid

$$\xrightarrow[\text{oxygenase}]{\underset{\text{cyclo-}}{2O_2}}$$

PGG_2

PGH_2

Both PGG_2 and PGH_2 are very potent stimulators of platelet aggregation and blood vessel constriction. Thus they play a key role in the second phase of blood clotting.

Just as we have known for some time that thrombin stimulates the second and irreversible phase of platelet aggregation, we have also known that aspirin and aspirin-like drugs such as indomethacin inhibit this second phase. How these drugs are able to do this has remained a mystery to clinical scientists. There is now good evidence that aspirin inhibits the clumping of platelets by inhibiting the activity of cyclooxygenase, the enzyme system that catalyzes the conversion of platelet-bound arachidonic acid to PGG_2 and PGH_2. Aspirin's ability to do other things as well may also be related to its ability to influence other aspects of prostaglandin metabolism. For example, aspirin is a potent fever-reducing agent. So is PGE_1. Therefore, aspirin may act to reduce fever by stimulating the biosynthesis of PGE_1. Aspirin also acts to reduce inflammation. It may be that its anti-inflammatory action is also related to its ability to influence prostaglandin metabolism. Clearly, the research generated by the discovery of prostaglandins has given us new insights into the long-standing problem of how aspirin inhibits blood clotting. Further research may help us understand more about inflammatory diseases such as rheumatoid arthritis, and asthma and other allergic responses.

As indicated in the introduction to this essay, the first recorded observations on the biological activity of the prostaglandins were those of gynecologists Kurzrok and Lieb. Coincidentally, the first widespread clinical application of these substances is by gynecologists and obstetricians. The observation that prostaglandins stimulate the contraction of uterine smooth muscle led quite naturally to a widespread interest in their use to induce labor and abortion and to stimulate menstruation. The first published study of the use of prostaglandins for the induction of labor was by Dr. Sultan Karim then at Uganda's Makerere University medical school. Between 1968 and 1971 he reported induction of labor at term in over 500 women by the use of PGE_2 and $PGF_{2\alpha}$. These studies coupled with those of M. P. Embrey at Oxford and M. Bygdeman and M. Wiqvist at Stockholm leave no doubt about the ability of PGE_2 and $PGF_{2\alpha}$ to induce labor at term.

A very serious clinical problem for the obstetrician is that of induction of labor after fetal death or missed abortion, and for abortion in the second trimester. For many years, the most commonly used method for the termination of second-trimester pregnancy was intra-amniotic administration of hypertonic saline solution (20% saline solution). A number of clinical observations suggested that prostaglandins might also be used for abortion and termination of second-trimester pregnancy. Yet, one problem with the use of naturally occurring prostaglandins for this purpose is that they are very rapidly degraded within the body. Therefore, their use required repeated administration over a period of hours. In the search for less rapidly degraded prostaglandins, a number of semisynthetic

an extra methyl group at
carbon 15

15-methyl prostaglandin $F_{2\alpha}$

prostaglandin analogs were synthesized. One of the most effective of these was 15-methyl prostaglandin $F_{2\alpha}$. This drug is longer acting and has 10 to 20 times the potency of $PGF_{2\alpha}$.

The potential clinical use of prostaglandins and prostaglandin analogs for abortion in the second trimester of pregnancy were explored in a study designed and conducted by the World Health Organization Task Force on the Use of Prostaglandins for the Regulation of Fertility. This impressive multicenter, multinational study entitled, *Prostaglandins and Abortion*, involved a total of 2969 patients and is described in the American Journal of Obstetrics and Gynecology (1977). The study compared the effectiveness of two prostaglandins, $PGF_{2\alpha}$ and 15-methyl $PGF_{2\alpha}$, given by intramuscular, extra-amniotic and intra-amniotic administration. Intramuscular injection of $PGF_{2\alpha}$ is the least effective of the means studied. After intramuscular injection, the prostaglandins reach the systemic circulation and stimulate the contraction of not only uterine smooth muscle but other smooth muscles as well, especially those of the gastrointestinal tract. This in turn produces an unacceptably high incidence of nausea, vomiting, and diarrhea in most patients. The most effective drug and method of administration was found to be a single intra-amniotic injection of either 50 mg of $PGF_{2\alpha}$ or 2.5 mg of 15-methyl $PGF_{2\alpha}$. The success rate with the single intra-amniotic injection of 15-methyl $PGF_{2\alpha}$ approached 96% at 48 hours, a rate which is reached by the hypertonic saline methodology only at about 72 hours. The study concludes that the single intra-amniotic method provides a safe and effective means for the termination of second-trimester pregnancy.

In this mini-essay we have looked at only two aspects of the importance of prostaglandins in human physiology. From even this brief encounter, it should be clear to you that our understanding of the chemistry and biochemistry of these remarkable substances is only in its infancy. And it should also be clear that the enormous prostaglandin research effort now under way offers great promise for even deeper insight into human physiology and for the development of new and highly effective drugs for use in clinical medicine.

REFERENCES

Bergstrom, S. "Prostaglandins: Members of a New Hormonal System," *Science*, vol. 157, 382 (1967).

———, Carlson, L. A., and Weeks, J. R., "The Prostaglandins: A Family of Biologically Active Lipids," *Pharmacological Review*, vol. 20, 1 (1968).

P. Needleman, et al. *Nature* (London) 261 (1976), pp. 558–60.

"Prostaglandins and Abortion," *The American Journal of Obstetrics and Gynecology*, vol. 129 (1977), pp. 593–606.

The Prostaglandins (Plenum Press, New York).

B. Samuellson, et al. "Prostaglandins," *Annual Review of Biochemistry*, vol. 44 (1975), pp. 669–95.

Functional Derivatives of Carboxylic Acids

9.1 Introduction

In addition to undergoing the reactions with bases discussed in Chapter 8, which involve the loss of a proton from the carboxyl group, acids can be transformed into a variety of derivatives in which the structural alteration of the carboxyl group is somewhat greater. In this chapter we shall examine the structure and chemical reactivity of organic esters, amides, anhydrides, and acid chlorides—all functional derivatives of carboxylic acids in which the —OH of the carboxyl group has been replaced by —OR, —NH$_2$, OCOR, or —Cl.

$$
\begin{array}{cccc}
\overset{\displaystyle O}{\overset{\|}{R-C-OR}} & \overset{\displaystyle O}{\overset{\|}{R-C-NH_2}} & \overset{\displaystyle O}{\overset{\|}{R-C-O}}\overset{\displaystyle O}{\overset{\|}{-C-R}} & \overset{\displaystyle O}{\overset{\|}{R-C-Cl}} \\
\text{ester} & \text{amide} & \text{anhydride} & \text{acid chloride}
\end{array}
$$

9.2 Nomenclature

Esters are named as derivatives of carboxylic acids by dropping the suffix -ic from the IUPAC or the common name of the acid and adding -ate. The alkyl or aryl group on oxygen is named first followed by the name of the acid from which the ester is derived.

233

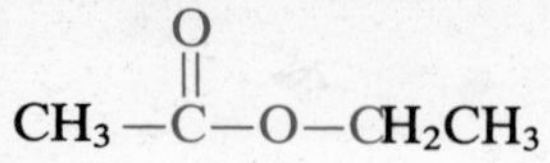

$$CH_3-\overset{\overset{\displaystyle O}{\|}}{C}-O-CH_2CH_3$$

ethyl acetate

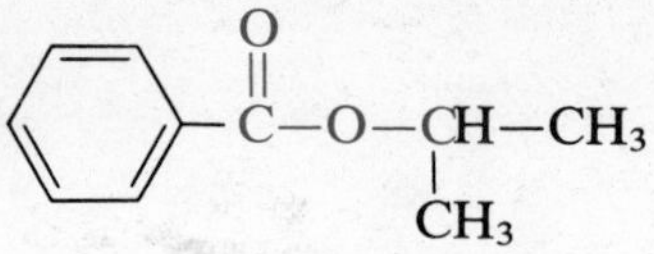

isopropyl benzoate

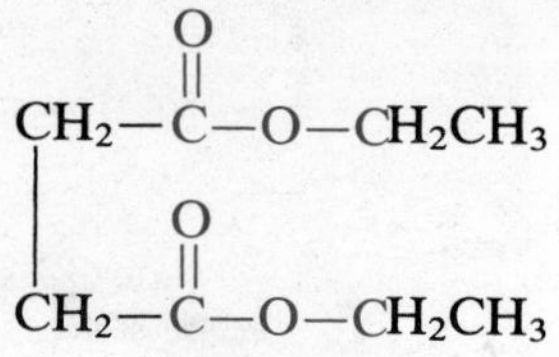

$$CH_2-\overset{\overset{\displaystyle O}{\|}}{C}-O-CH_2CH_3$$
$$CH_2-\overset{\overset{\displaystyle O}{\|}}{C}-O-CH_2CH_3$$

diethyl succinate

Esters of phosphoric acid are exceedingly important in the whole of biological chemistry. Shown below are the phosphate esters of glyceraldehyde and dihydroxyacetone, both key intermediates in the metabolism of glucose. Also shown is pyridoxal phosphate, one of the metabolically active forms of vitamin B_6.

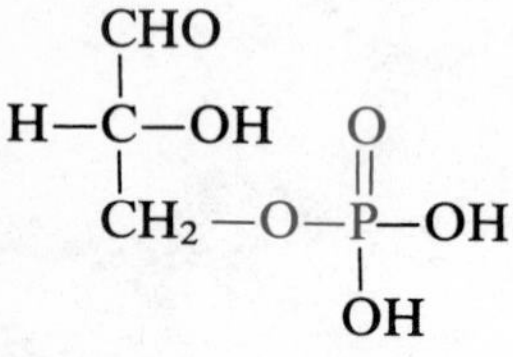

glyceraldehyde
3-phosphate

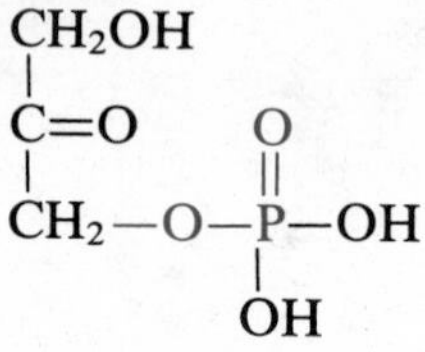

dihydroxyacetone
phosphate

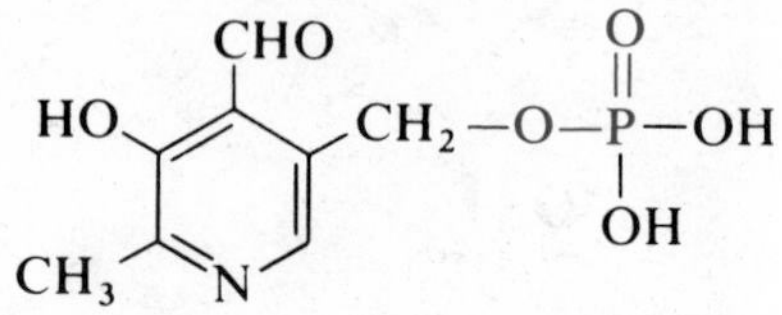

pyridoxal phosphate

Several organic esters of nitric acid and nitrous acid have been used as drugs for more than 100 years. Two of these are glyceryl trinitrate, or as it is more commonly known, nitroglycerine, and isoamyl nitrite.

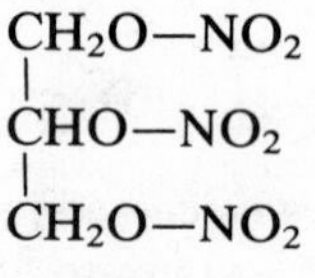

nitroglycerine

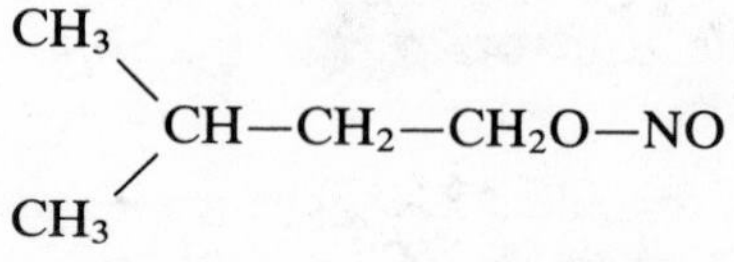

isoamyl nitrite

Both nitroglycerine and isoamyl nitrite produce rapid relaxation of most smooth muscles of the body. Medically, their most important action is relaxation of the smooth muscle of blood vessels and dilation of all large and small arteries of the heart, for which reason they are called vasodilators. Both esters are used extensively for the treatment of angina pectoris, a heart disease characterized by agonizing pain.

Amides are named as derivatives of carboxylic acids by dropping the suffix -oic from the IUPAC name of the acid, or the suffix -ic from the common name of the acid, and adding -amide

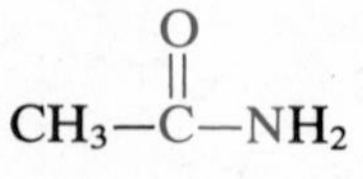

acetamide

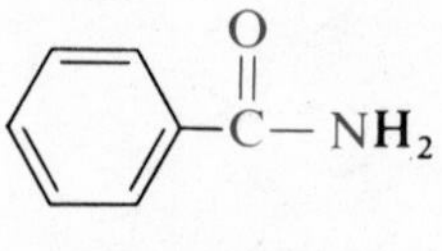

benzamide

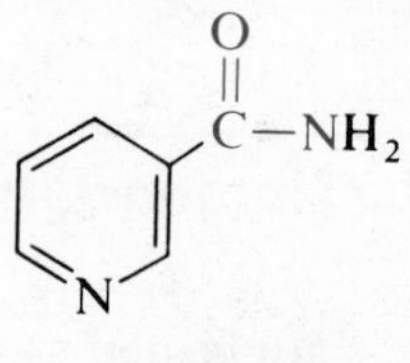

nicotinamide

If the nitrogen atom is substituted with an alkyl or aryl group, the substituent is named and its location on nitrogen is indicated by a capital N-.

234

$$
\begin{array}{cc}
\underset{\substack{\text{N,N-dimethylformamide}}}{\overset{\displaystyle\overset{O}{\underset{\|}{}}\quad\overset{CH_3}{\diagup}}{H-C-N}}_{\diagdown CH_3}
&
\underset{\substack{\text{N-phenylbutanamide}\\ \text{N-phenylbutyramide}}}{CH_3CH_2CH_2\overset{\displaystyle\overset{O}{\underset{\|}{}}}{C}-NH-\bigcirc}
\end{array}
$$

Acid halides are named as derivatives of carboxylic acids by dropping the suffix -ic acid from the IUPAC or common name and adding -yl and the name of the halogen atom. The most commonly encountered are acid chlorides and acid bromides.

$$
\underset{\text{acetyl chloride}}{CH_3-\overset{\displaystyle\overset{O}{\|}}{C}-Cl}
\qquad\qquad
\underset{\text{benzoyl bromide}}{\bigcirc-\overset{\displaystyle\overset{O}{\|}}{C}-Br}
$$

9.3 Nucleophilic Substitution at Unsaturated Carbon

The basic reaction theme common to the carbonyl group of aldehydes, ketones, carboxylic acids, esters, amides, and acid halides and anhydrides in nucleophilic addition to the carbonyl group to form a tetrahedral carbonyl addition intermediate. In the case of aldehydes and ketones, this addition product is either isolated as such or undergoes loss of water to give an unsaturated derivative of the original aldehyde or ketone. For example, recall that aldol condensation products are isolated as such (Section 7.14),

$$
\underset{}{CH_3\overset{\displaystyle\overset{O}{\|}}{C}H} + \underset{}{CH_3\overset{\displaystyle\overset{O}{\|}}{C}H} \xrightarrow{\text{base}} \underset{}{CH_3\overset{\displaystyle\overset{OH}{|}}{C}HCH_2\overset{\displaystyle\overset{O}{\|}}{C}H}
$$

while the reaction of an aldehyde or ketone with hydroxylamine (Section 7.9) yields an unsaturated derivative of the original aldehyde or ketone.

$$
CH_3\overset{\displaystyle\overset{O}{\|}}{C}CH_3 + H_2NOH \longrightarrow \left[CH_3-\underset{\underset{\displaystyle CH_3}{|}}{\overset{\overset{\displaystyle OH}{|}}{C}}-\underset{\underset{\displaystyle OH}{}}{\overset{\overset{\displaystyle H}{\diagup}}{N}} \right] \longrightarrow CH_3\overset{\displaystyle\overset{NOH}{\|}}{C}CH_3 + H_2O
$$

With the new functional groups to be studied in this chapter, the addition compounds undergo subsequent collapse to regenerate the carbonyl group.

$$
\underset{}{R-\overset{\displaystyle\overset{O}{\|}}{C}-Y} + H-Z \longrightarrow \underset{\substack{\text{tetrahedral carbonyl}\\ \text{addition intermediate}}}{\left[R-\underset{\underset{\displaystyle Y}{|}}{\overset{\overset{\displaystyle OH}{|}}{C}}-Z \right]} \longrightarrow R-\overset{\displaystyle\overset{O}{\|}}{C}-Z + H-Y
$$

It is for this reason that we characterize these reactions of the carbonyl group as nucleophilic substitution at unsaturated carbon. As an example, the reaction of methyl acetate, an ester, with ammonia follows.

$$CH_3\overset{O}{\overset{\|}{C}}-OCH_3 + NH_3 \longrightarrow \left[CH_3\overset{\overset{OH}{|}}{\underset{\underset{OCH_3}{|}}{C}}\overset{H}{\underset{H}{-N}} \right] \longrightarrow CH_3\overset{O}{\overset{\|}{C}}-NH_2 + CH_3OH$$

The result of this sequence of reactions is substitution of —NH$_2$ for —OCH$_3$.

9.4 Preparation of Esters

A carboxylic acid can be converted into an ester by heating with an alcohol in the presence of an acid catalyst, usually dry hydrogen chloride, concentrated sulfuric acid, or an ion-exchange resin in the hydrogen ion form.

Direct esterification of alcohols and acids in this manner is called Fischer esterification.

$$R-\overset{O}{\overset{\|}{C}}-OH + HO-R' \underset{}{\overset{H^+}{\rightleftharpoons}} R-\overset{O}{\overset{\|}{C}}-O-R' + H_2O$$

The reaction involves condensing a molecule of acid and a molecule of alcohol through the elimination of a molecule of water. This reaction is reversible, and generally at equilibrium there are appreciable quantities of both ester and alcohol present. For example, if 60.0 g (one mole) of acetic acid and 60.0 g (one mole) of n-propyl alcohol are refluxed for a short time in the presence of a few drops of concentrated sulfuric acid, the reaction mixture at equilibrium will contain about 68.0 g (0.67 mole) of ester, 12.0 g (0.67 mole) of water, and 20.0 g (0.33 mole) each of acetic acid and n-propyl alcohol. In other words, at equilibrium there is about 67% conversion of acid and alcohol into ester.

Direct esterification can be used to prepare esters in high yields. For example, if the alcohol is particularly inexpensive, we may use a large excess of it and achieve a high conversion of the acid into ester. Or we may take advantage of a situation in which the boiling points of the reactants and ester are higher than that of water. Heating the reaction mixture somewhat above 100°C will remove water as it is formed and shift the equilibrium toward the production of higher yields of ester.

Fischer esterification is but one of the general methods of preparing esters. From the standpoint of the organic chemist interested in the laboratory synthesis of esters, two much more important methods are the reaction of hydroxyl compounds (alcohols, phenols) with acid anhydrides or acid halides. We shall see these methods of ester preparation in Sections 9.12 and 9.13.

9.5 Acid Catalysis of Esterification-Mechanism

In considering the preparation of an ester by the Fischer method, one question we might ask is: Does the oxygen eliminated as water come from the acid or from the alcohol. A clear demonstration of the source of the water oxygen comes from an experiment in which one of the oxygen atoms is labeled. The use of a mass spectrometer shows that ordinary oxygen in nature is a mixture of three isotopes: 99.7% ^{16}O, 0.04% ^{17}O and 0.20% ^{18}O. Through the use of modern methods of separating isotopes, a variety of compounds with significant enrichment in ^{18}O are commercially available.

When methanol enriched with ^{18}O is allowed to react with acetic acid containing ordinary oxygen, the ester is found to contain the enriched oxygen. None is present in the product water. This demonstrates that esterification involves the rupture of the C—O bond of the acid and the H—O bond of the alcohol.

$$CH_3\overset{\overset{\displaystyle O}{\|}}{C}-OH + H-^{18}OCH_3 \rightleftharpoons CH_3\overset{\overset{\displaystyle O}{\|}}{C}-^{18}OCH_3 + H_2O$$

A mechanism consistent with this observation and with the fact that Fischer esterification is acid catalyzed postulates initial attack of a proton on the carbonyl oxygen. We can describe this as reaction between an acid and a base, or, alternatively, as reaction between an electrophile and a nucleophile. As a result of this attack on the carbonyl oxygen, the carbonyl carbon becomes more electron deficient and more susceptible to nucleophilic attack. Addition of an alcohol molecule to the protonated carbonyl group gives a _tetrahedral carbonyl addition intermediate_. This intermediate may then either (1) lose a molecule of alcohol to regenerate the starting material, or (2) lose a molecule of water to give an ester. Shown below is a step-by-step mechanism for acid-catalyzed esterification.

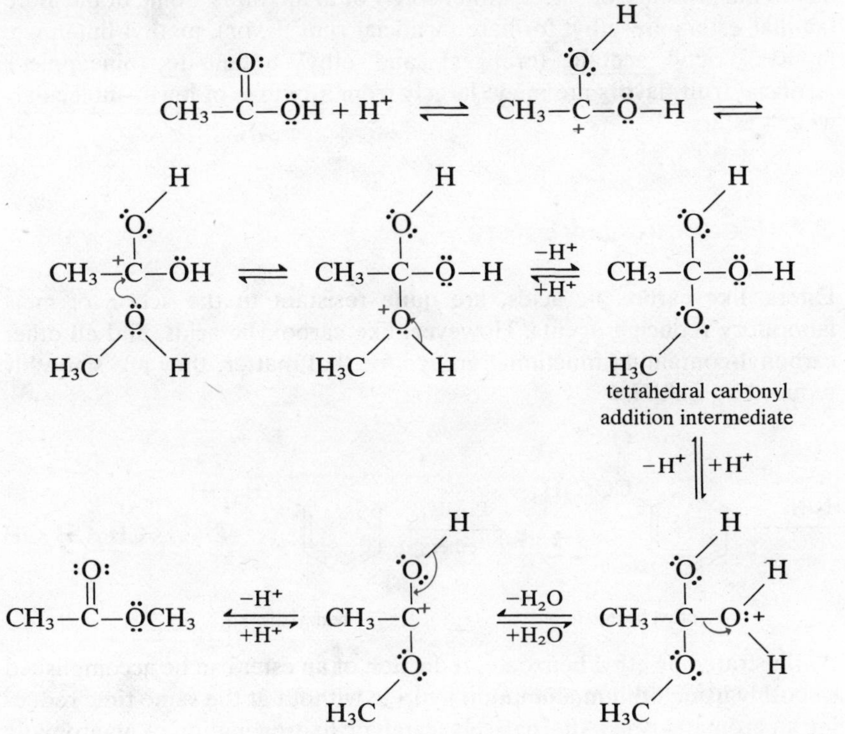

This may seem complicated because of the several steps in the pathway.

However you should be able to break the mechanism down into the following components:

1. four proton-transfer reactions;
2. formation of a carbon-oxygen single bond by the reaction of a nucleophile and an electrophile;
3. rupture of a carbon-oxygen single bond by the separation of a nucleophile and an electrophile.

If we omit all proton-transfer reactions, we see that Fischer esterification is a specific example of the general mechanism shown in the previous section for nucleophilic substitution at unsaturated carbon.

$$CH_3-\overset{\overset{\displaystyle O}{\|}}{C}-OH + HOCH_3 \rightleftharpoons \left[CH_3-\overset{\overset{\displaystyle OH}{|}}{\underset{\underset{\displaystyle OH}{|}}{C}}-OCH_3 \right] \rightleftharpoons CH_3-\overset{\overset{\displaystyle O}{\|}}{C}-OCH_3 + H_2O$$

tetrahedral carbonyl
addition intermediate

9.6 Physical Properties of Esters

Esters are neutral substances, less soluble in water and lower boiling than isomeric carboxylic acids. Unlike the carboxylic acids from which they are derived, the low-molecular-weight esters have rather pleasant odors. The characteristic fragrances of many flowers and fruits are in many instances due to the presence of esters, either singly or in mixtures. Some of the more familiar esters are ethyl formate (artificial rum flavor), methyl butanoate (apples), octyl acetate (oranges), and ethyl butanoate (pineapples). Artificial fruit flavors are made largely from mixtures of lower-molecular-weight esters.

9.7 Reduction of Esters

Esters, like carboxylic acids, are quite resistant to the action of most laboratory reducing agents. However, like carboxylic acids, and all other carbonyl-containing functional groups for that matter, they are smoothly reduced by LiAlH$_4$.

ethyl
cyclohexanecarboxylate   ethyl benzoate   benzyl alcohol

As illustrated by ethyl benzoate, reduction of an ester can be accomplished smoothly using lithium aluminum hydride without at the same time reducing an aromatic ring. Alternatively, catalytic hydrogenation of an aromatic ring can be accomplished without reducing an ester group.

238

9.8 Hydrolysis of Esters

Esters of carboxylic acids, as well as those of other acids such as phosphoric acid and sulfuric acid, may be reconverted to the corresponding acids and alcohols by hydrolysis in either aqueous acid or base.

$$CH_3CO_2CH_2CH_3 + H_2O \underset{}{\overset{H^+}{\rightleftharpoons}} CH_3CO_2H + CH_3CH_2OH$$

In the case of acid-catalyzed hydrolysis, the H^+ functions in the same manner as we encountered in esterification, namely, to make the carbonyl carbon more susceptible to nucleophilic attack. Since each step in the mechanism of acid-catalyzed esterification (Section 9.5) is reversible, this mechanism can equally well account for acid-catalyzed hydrolysis. By carrying out the reaction in a large excess of water, the position of the equilibrium is shifted to favor the formation of acid and alcohol.

In alkaline hydrolysis of esters, the OH^- is a very powerful nucleophile which readily adds to the carbonyl carbon to form a tetrahedral carbonyl addition intermediate. This intermediate then reacts to eliminate a molecule of alcohol and to generate the carboxylate anion, $RCOO^-$.

$$R-\overset{\overset{\displaystyle O}{\|}}{C}-OR' + OH^- \rightleftharpoons \left[R-\overset{\overset{\displaystyle O^-}{|}}{\underset{\underset{\displaystyle OR'}{|}}{C}}-OH \right] \longrightarrow R-\overset{\overset{\displaystyle O}{\|}}{C}-O^- + HOR'$$

tetrahedral carbonyl
addition intermediate

This mechanism, like that for acid-catalyzed esterification and hydrolysis, involves cleavage of the C—O bond of the carboxylic acid portion of the molecule rather than the C—O bond of the alcohol. Alkaline hydrolysis of an ester is not an equilibrium reaction because the carboxylate anion, the final product, shows no tendency to react with alcohol.

9.9 Ammonolysis of Esters

Treatment of esters with ammonia, often in a solvent such as ethanol, yields an amide. This reaction is similar to hydrolysis and is called ammonolysis. Ammonia is a strong nucleophile, and adds directly to the carbonyl without a catalyst being necessary.

$$CH_3\overset{\overset{\displaystyle O}{\|}}{C}-OCH_2CH_3 + NH_3 \longrightarrow \left[CH_3\overset{\overset{\displaystyle OH}{|}}{\underset{\underset{\displaystyle NH_2}{|}}{C}}-OCH_2CH_3 \right] \longrightarrow CH_3\overset{\overset{\displaystyle O}{\|}}{C}-NH_2 + CH_3CH_2OH$$

tetrahedral
carbonyl addition
intermediate

We should note here that although ammonolysis is an equilibrium reaction, the concentration of ester present at equilibrium is so small that for all practical purposes it may be regarded as zero. It is not possible to prepare an ester by treating an amide with alcohol. We shall see other evidence later in the chapter that amides are the least reactive of the four classes of carboxylic acid derivatives.

Another example of this ammonolysis reaction is the laboratory synthesis of barbituric acid and barbiturates. Heating urea and diethyl malonate at 110°C in the presence of sodium ethoxide yields <u>barbituric acid</u>.

$$
\underset{\text{diethyl malonate}}{
\begin{array}{c}
\text{O} \\
\parallel \\
\text{C-OCH}_2\text{CH}_3 \\
\diagup \\
\text{H}_2\text{C} \\
\diagdown \\
\text{C-OCH}_2\text{CH}_3 \\
\parallel \\
\text{O}
\end{array}}
\;+\;
\underset{\text{urea}}{
\begin{array}{c}
\text{H}_2\text{N} \\
\diagdown \\
\;\;\;\text{C=O} \\
\diagup \\
\text{H}_2\text{N}
\end{array}}
\;\longrightarrow\;
\underset{\text{barbituric acid}}{
\begin{array}{c}
\text{O} \\
\parallel \\
\text{C-NH} \\
\diagup\qquad\diagdown \\
\text{H}_2\text{C}\qquad\;\text{C=O} \\
\diagdown\qquad\diagup \\
\text{C-NH} \\
\parallel \\
\text{O}
\end{array}}
\;+\;2\text{CH}_3\text{CH}_2\text{OH}
$$

Mono- and disubstituted malonic esters yield substituted barbituric acids known as <u>barbiturates</u>.

$$
\underset{\substack{\text{phenobarbital} \\ \text{(Luminal)}}}{
\begin{array}{c}
\text{C}_6\text{H}_5 \qquad\quad \overset{\displaystyle\text{O}}{\overset{\parallel}{\text{C}}}\text{-NH} \\
\diagdown\qquad\diagup\qquad\quad\diagdown \\
\text{C}\qquad\qquad\text{C=O} \\
\diagup\qquad\diagdown\qquad\quad\diagup \\
\text{C}_2\text{H}_5 \qquad\quad \underset{\parallel}{\text{C}}\text{-NH} \\
\qquad\qquad\quad\text{O}
\end{array}}
\qquad\qquad
\underset{\substack{\text{pentobarbital} \\ \text{(Nembutal)}}}{
\begin{array}{c}
\text{CH}_3\text{CH}_2 \qquad \overset{\displaystyle\text{O}}{\overset{\parallel}{\text{C}}}\text{-NH} \\
\diagdown\qquad\diagup\qquad\quad\diagdown \\
\text{C}\qquad\qquad\text{C=O} \\
\diagup\qquad\diagdown\qquad\quad\diagup \\
\text{CH}_3\text{CH}_2\text{CH}_2\text{CH}\qquad \underset{\parallel}{\text{C}}\text{-NH} \\
\quad\mid\qquad\qquad\quad\text{O} \\
\quad\text{CH}_3
\end{array}}
$$

$$
\underset{\substack{\text{thiopental} \\ \text{(Pentothal)}}}{
\begin{array}{c}
\text{CH}_3\text{CH}_2 \qquad \overset{\displaystyle\text{O}}{\overset{\parallel}{\text{C}}}\text{-NH} \\
\diagdown\qquad\diagup\qquad\quad\diagdown \\
\text{C}\qquad\qquad\text{C=S} \\
\diagup\qquad\diagdown\qquad\quad\diagup \\
\text{CH}_3\text{CH}_2\text{CH}_2\text{CH}\qquad \underset{\parallel}{\text{C}}\text{-NH} \\
\quad\mid\qquad\qquad\quad\text{O} \\
\quad\text{CH}_3
\end{array}}
$$

The barbiturates produce effects ranging from mild sedation to deep anesthesia, and even death, depending on the dose and the particular barbiturate. The sedation, long or short acting, depends on the structure of the barbiturate. Phenobarbital is long acting while pentobarbital acts for a shorter time, about three hours. Thiopental is very fast acting and is used as an anesthetic for producing deep sedation quickly. With barbiturates in general, sleep can be produced with as little as 0.1 g (1 capsule) and toxic symptoms and even death can result from 1.5 g (15 capsules). Barbiturates produce both a general tranquilizing effect and a sleep-producing effect. Recently it has been possible to separate these effects with another class of drugs. Reserpine and the chloropromazine derivatives tranquilize without producing sleep at the same time. (Read the mini-essay "Inside the Molecules of the Mind.")

9.10 Reaction of Esters with Grignard Reagents

The reaction of an ester of a carboxylic acid with a Grignard reagent is a common synthetic method to prepare tertiary alcohols.

$$\underset{\text{ester}}{CH_3-\overset{\overset{\displaystyle O}{\|}}{C}-OCH_3} + 2RMgX \longrightarrow \underset{\substack{\text{tertiary} \\ \text{alcohol}}}{CH_3\overset{\overset{\displaystyle OH}{|}}{\underset{\underset{\displaystyle R}{|}}{C}}-R} + CH_3OMgX$$

Addition of one mole of Grignard reagent to the carbonyl group produces an unstable intermediate which breaks down during the course of the reaction to produce a ketone.

$$CH_3\overset{\overset{\displaystyle O}{\|}}{C}-OCH_3 + RMgBr \longrightarrow \left[CH_3\overset{\overset{\displaystyle OMgBr}{|}}{\underset{\underset{\displaystyle R}{|}}{C}}-OCH_3 \right] \longrightarrow CH_3\overset{\overset{\displaystyle O}{\|}}{C}-R + CH_3OMgBr$$

This new ketone then reacts with a second molecule of Grignard reagent to form a tertiary alcohol.

9.11 Preparation and Hydrolysis of Amides

Amides may be prepared by heating the ammonium salt of a carboxylic acid above its melting point.

$$\underset{\substack{\text{ammonium} \\ \text{acetate}}}{CH_3\overset{\overset{\displaystyle O}{\|}}{C}-O^-NH_4^+} \xrightarrow{\text{heat}} \underset{\text{acetamide}}{CH_3\overset{\overset{\displaystyle O}{\|}}{C}-NH_2} + H_2O$$

They are also prepared by the reaction of an ester with ammonia or a derivative of ammonia as illustrated in Section 9.9. Ammonolysis of ethyl nicotinate yields nicotinamide.

(ethyl nicotinate) $+ NH_3 \longrightarrow$ (nicotinamide) $+ CH_3CH_2OH$

Later in this chapter we will see two additional methods for amide preparation, namely, the reaction of acid halides (Section 9.12) or acid anhydrides (Section 9.13) with ammonia and derivatives of ammonia.

Amides are very resistant to hydrolysis even in boiling water. However, in the presence of moderately concentrated aqueous acid or base, hydrolysis does occur, though not as rapidly as in the case of esters. It is not unusual to reflux an amide for several hours in concentrated hydrochloric acid to effect hydrolysis.

$$\underset{\text{phenylacetamide}}{C_6H_5CH_2\overset{\displaystyle O}{\overset{\displaystyle \|}{C}}NH_2} + H_2O \xrightarrow[\text{reflux}]{\text{35\% HCl}} \underset{\substack{\text{phenylacetic}\\\text{acid}}}{C_6H_5CH_2CO_2H} + NH_4Cl$$

9.12 Preparation and Reactions of Acid Halides

Acid chlorides are most often prepared by reacting a carboxylic acid with either thionyl chloride or phosphorus pentachloride in much the same way that alkyl chlorides are prepared from alcohols (Section 5.8).

$$\underset{\substack{\text{acetic}\\\text{acid}}}{CH_3-\overset{\displaystyle O}{\overset{\displaystyle \|}{C}}-OH} + \underset{\substack{\text{thionyl}\\\text{chloride}}}{Cl-\overset{\displaystyle O}{\overset{\displaystyle \|}{S}}-Cl} \longrightarrow \underset{\substack{\text{acetyl}\\\text{chloride}}}{CH_3-\overset{\displaystyle O}{\overset{\displaystyle \|}{C}}-Cl} + HCl + \underset{\substack{\text{sulfur}\\\text{dioxide}}}{SO_2}$$

$$\underset{\text{benzoic acid}}{\text{C}_6\text{H}_5-\overset{\displaystyle O}{\overset{\displaystyle \|}{C}}-OH} + PCl_5 \longrightarrow \underset{\substack{\text{benzoyl}\\\text{chloride}}}{\text{C}_6\text{H}_5-\overset{\displaystyle O}{\overset{\displaystyle \|}{C}}-Cl} + POCl_3 + HCl$$

Acid chlorides undergo rapid reaction with a wide variety of nucleophiles including water, ammonia, amines, alcohols, and phenols. Reaction with water merely regenerates the parent carboxylic acid.

$$CH_3-\overset{\displaystyle O}{\overset{\displaystyle \|}{C}}-Cl + H_2O \longrightarrow CH_3-\overset{\displaystyle O}{\overset{\displaystyle \|}{C}}-OH + HCl$$

Acetyl chloride and many other low molecular weight aliphatic acid halides react so readily with water that they must be protected from atmospheric moisture during storage. Acid chlorides react with alcohols and phenols to yield esters. Such reactions are most often carried out in the presence of an organic base such as pyridine, which both catalyzes the reaction and neutralizes the HCl formed during esterification.

$$CH_3-\overset{\displaystyle O}{\overset{\displaystyle \|}{C}}-Cl + CH_3-\overset{\overset{\displaystyle CH_3}{|}}{\underset{\underset{\displaystyle CH_3}{|}}{C}}-OH \xrightarrow{\text{base}} CH_3-\overset{\displaystyle O}{\overset{\displaystyle \|}{C}}-O-\overset{\overset{\displaystyle CH_3}{|}}{\underset{\underset{\displaystyle CH_3}{|}}{C}}-CH_3 + HCl$$

tert-butyl
acetate

Recall that esters can also be prepared by the acid-catalyzed Fischer esterification (Section 9.4). Of these two methods for the preparation of esters, the use of acid chlorides is often more convenient primarily because both the preparation of the acid chlorides and their reaction with alcohols are rapid and irreversible reactions. In contrast, Fischer esterification is a slow, reversible reaction, and yields of ester depend on the position of the equilibrium.

Acid chlorides react with ammonia, primary, and secondary amines to give amides. In this reaction, it is necessary to use two moles of amine for each mole of acid chloride, the first to form the amide and the second to neutralize the HCl produced during the ammonolysis reaction

$$CH_3-\overset{\overset{\displaystyle O}{\|}}{C}-Cl + 2NH_3 \longrightarrow CH_3-\overset{\overset{\displaystyle O}{\|}}{C}-NH_2 + NH_4^+Cl^-$$

acetyl chloride   acetamide

benzoyl chloride   N-methylbenzamide   methylammonium chloride

9.13 Reactions of Anhydrides

Anhydrides also undergo rapid reaction with water, ammonia, alcohols, and phenols. In these reactions, anhydrides are equally as reactive as acid chlorides.

$$CH_3-\overset{\overset{\displaystyle O}{\|}}{C}-O-\overset{\overset{\displaystyle O}{\|}}{C}-CH_3$$

$\xrightarrow{H_2O}$ $CH_3COOH + CH_3COOH$

$\xrightarrow{NH_3}$ $CH_3CONH_2 + CH_3COOH$

$\xrightarrow{CH_3OH}$ $CH_3CO_2CH_3 + CH_3COOH$

$\xrightarrow{CH_3CH_2SH}$ $CH_3COSCH_2CH_3 + CH_3COOH$

Aspirin is prepared by the reaction of acetic anhydride and salicylic acid.

salicylic acid   acetylsalicylic acid
                           aspirin

9.14 Relative Reactivities of Acid Halides, Anhydrides, Esters, and Amides

The four common functional derivatives of carboxylic acid we have described in this chapter show marked differences in reactivities. For

TABLE 9.1 Interconversion of functional derivatives of carboxylic acids.

Acyl derivative	can be converted to	by reaction with	Name of Process
acid anhydride or acid chloride	carboxylic acid	water	hydrolysis
	ester	alcohol or phenol	alcoholysis
	amide	NH₃, primary or secondary amine	ammonolysis
ester	carboxylic acid	water	hydrolysis
	amide	NH₃, primary or secondary amine	ammonolysis
amide	carboxylic acid	water	hydrolysis

example, consider the ease of hydrolysis of an acid chloride, an anhydride, an ester, and an amide. Both acetyl chloride and acetic anhydride react so readily with water that they must be protected from atmospheric moisture during storage. Ethyl acetate reacts slowly with water at room temperature but hydrolyzes readily on heating in the presence of an acid or base catalyst. Acetamide is very resistant to hydrolysis except in the presence of moderately strong acid or base and heat. The reactivity of these functional derivatives of carboxylic acids decreases in the following order.

$$\underset{\substack{\text{acid}\\\text{chloride}}}{CH_3C-Cl} \approx \underset{\substack{\text{acid}\\\text{anhydride}}}{CH_3C-O-CCH_3} > \underset{\text{ester}}{CH_3C-OCH_2CH_3} > \underset{\text{amide}}{CH_3C-NH_2}$$

It follows directly from this order of reactivity that any less reactive derivative of a carboxylic acid may be prepared directly from a more reactive derivative, but not vice versa. For example, an ester can be synthesized from an acid anhydride or acid chloride plus an alcohol or phenol. However, an ester cannot be synthesized from an amide plus an alcohol. These interconversions are summarized in Table 9.1.

9.15 The Claisen Condensation: β-Ketoesters

The Claisen condensation is a carbonyl reaction closely related to the aldol condensation. The reaction involves a carbanion derived from an ester in a condensation with another molecule of ester. As an example, when ethyl acetate is treated with sodium ethoxide in ethanol and the resulting solution acidified, ethyl 3-oxobutanoate, commonly known as ethyl acetoacetate, is obtained. Note that in the IUPAC name for this substance, the prefix oxo- is used to indicate the presence of the ketone group on the butanoate chain.

$$\underset{\text{ethyl acetate}}{2CH_3C-OCH_2CH_3} \xrightarrow[CH_3CH_2OH]{CH_3CH_2O^-Na^+} \underset{\substack{\text{ethyl 3-oxobutanoate}\\\text{ethyl acetoacetate}\\\text{(a }\beta\text{-ketoester)}}}{CH_3C-CH_2COCH_2CH_3} + CH_3CH_2OH$$

The net reaction is one of substitution at one carbonyl by the α-carbon of another. Recall our discussion, in Section 7.13, of the acidity of hydrogens located on a carbon α to a carbonyl group. Hydrogen atoms on a carbonyl α to an ester also show acidity in the presence of a strong base such as sodium ethoxide. The first step in the Claisen condensation is the formation of a resonance-stabilized anion from ethyl acetate. This anion is a powerful nucleophile and attacks the carbonyl carbon of a second molecule of ethyl acetate forming a tetrahedral carbonyl addition intermediate. Elimination of ethoxide ion then leads to ethyl acetoacetate, a β-ketoester.

Step 1
$$CH_3CH_2\ddot{O}:^- + H{-}CH_2{-}\overset{\displaystyle O}{\overset{\|}{C}}{-}OC_2H_5 \rightleftharpoons CH_3CH_2OH + {}^-:CH_2{-}\overset{\displaystyle O}{\overset{\|}{C}}{-}OC_2H_5$$

Step 2
$$CH_3{-}\overset{\displaystyle :\ddot{O}:}{\underset{\displaystyle OC_2H_5}{C}} + {}^-:CH_2{-}\overset{\displaystyle O}{\overset{\|}{C}}OC_2H_5 \rightleftharpoons CH_3{-}\overset{\displaystyle :\ddot{O}:^-}{\underset{\displaystyle OC_2H_5}{C}}{-}CH_2{-}\overset{\displaystyle O}{\overset{\|}{C}}{-}OC_2H_5$$

tetrahedral carbonyl
addition intermediate

Step 3
$$CH_3{-}\overset{\displaystyle :\ddot{O}:^-}{\underset{\displaystyle :\ddot{O}C_2H_5}{C}}{-}CH_2{-}\overset{\displaystyle O}{\overset{\|}{C}}{-}OC_2H_5 \rightleftharpoons CH_3{-}\overset{\displaystyle :O:}{\overset{\|}{C}}{-}CH_2{-}\overset{\displaystyle O}{\overset{\|}{C}}OC_2H_5 + CH_3CH_2\ddot{O}:^-$$

This reaction is readily reversible and in fact the equilibrium is quite unfavorable for the formation of ethyl acetoacetate. However, the yield of the β-ketoester can be increased in either of two ways. (1) Ethanol, one of the products, can be removed by distillation as it is formed. (2) A more usual method is to add at least one mole of sodium ethoxide per mole of product formed. Ethyl acetoacetate is a stronger acid than ethanol. In fact the K_a of ethyl acetoacetate is 10^{-10} which makes it about as strong an acid as phenol. Ethyl acetoacetate will react with sodium ethoxide to precipitate as the sodium salt and thus be removed from the equilibrium.

$$CH_3CH_2\ddot{O}:^- + CH_3\overset{\displaystyle O}{\overset{\|}{C}}{-}CH_2{-}\overset{\displaystyle O}{\overset{\|}{C}}OCH_2CH_3 \rightleftharpoons CH_3\overset{\displaystyle O}{\overset{\|}{C}}{-}\overset{\displaystyle ..}{C}H{-}\overset{\displaystyle O}{\overset{\|}{C}}OCH_2CH_3 + CH_3CH_2OH$$
ethyl acetoacetate

Claisen condensation can be carried out under these conditions with any ester providing it has at least two α-hydrogens. Removal of the first α-hydrogen forms the reactive carbanion for condensation; removal of the second α-hydrogen shifts the equilibrium to the right through conversion of the β-ketoester to its insoluble salt.

In the case of a <u>mixed Claisen condensation</u>, that is, a condensation between two different esters, a mixture of four possible products will result unless there is an appreciable difference in reactivity between one ester and the other. One such difference is if one of the esters has no α-hydrogens and, therefore, cannot serve as a carbanion. Examples of such esters are ethyl formate, ethyl benzoate, and diethyl carbonate.

$$\underset{\text{ethyl benzoate}}{C_6H_5\overset{\overset{\displaystyle O}{\|}}{C}OCH_2CH_3} + \underset{\text{ethyl propanoate}}{CH_3CH_2\overset{\overset{\displaystyle O}{\|}}{C}OCH_2CH_3} \longrightarrow C_6H_5\overset{\overset{\displaystyle O}{\|}}{C}\underset{\underset{\displaystyle CH_3}{|}}{CH}\overset{\overset{\displaystyle O}{\|}}{C}OCH_2CH_3 + CH_3CH_2OH$$

ethyl 2-methyl-3-oxo-
3-phenylpropanoate

Diesters of six- or seven-carbon diacids undergo intramolecular Claisen condensation to form five- and six-membered rings.

diethyl adipate → ethyl 2-oxocyclopentane carboxylate (1. C$_2$H$_5$ONa 2. H$_2$O)

Ester condensations are of great importance in biosynthesis. One example is the biosynthesis of acetoacetyl coenzyme A, a key intermediate in synthesis of a variety of compounds of plant and animal origin.

$$2CH_3-\overset{\overset{\displaystyle O}{\|}}{C}-SCoA \xrightarrow{\text{enzyme}} CH_3-\overset{\overset{\displaystyle O}{\|}}{C}-CH_2-\overset{\overset{\displaystyle O}{\|}}{C}-SCoA + CoA-SH$$

acetyl coenzyme A   acetoacetyl coenzyme A   coenzyme A

The starting material for this reaction is acetyl coenzyme A, an ester of acetic acid and a complex molecule ($C_{21}H_{36}O_{16}N_7P_3S$) called coenzyme A. A major structural feature of coenzyme A is the presence of a sulfhydryl group (—SH). This molecule is generally abbreviated CoA—SH to emphasize the importance of the sulfhydryl group, or often it is written more simply as CoA. An enzyme-catalyzed Claisen-type ester condensation of two molecules of acetyl coenzyme A leads to the formation of acetoacetyl coenzyme A. This substance is a key intermediate in the synthesis of terpenes (Section 3.5) and steroids (Section 12.7). Hydrolysis of β-ketoesters in warm acid or base yields the corresponding β-ketoacids.

$$CH_3\overset{\overset{\displaystyle O}{\|}}{C}CH_2\overset{\overset{\displaystyle O}{\|}}{C}-OCH_2CH_3 + H_2O \xrightarrow[\text{heat}]{H^+} CH_3\overset{\overset{\displaystyle O}{\|}}{C}CH_2\overset{\overset{\displaystyle O}{\|}}{C}-OH + CH_3CH_2OH$$

3-oxobutanoic acid
β-ketobutyric acid

Such β-ketoacids readily lose carbon dioxide on heating to yield ketones.

$$CH_3\overset{\overset{\displaystyle O}{\|}}{C}CH_2-\overset{\overset{\displaystyle O}{\|}}{C}-OH \xrightarrow{\text{heat}} CH_3\overset{\overset{\displaystyle O}{\|}}{C}CH_3 + CO_2$$

In the more usual reaction, ester hydrolysis and decarboxylation occur together and only the resulting ketone is isolated.

$$\xrightarrow[\text{heat}]{H_3O^+} + CH_3CH_2OH + CO_2$$

Such decarboxylation is a unique property of β-ketoacids and is not observed with α, γ or higher ketoacids. The reaction involves a cyclic six-membered ring transition state and, by rearrangement of electrons, leads directly to the enol form of the product ketone and carbon dioxide. Equilibration of the keto and enol forms generates the ketone.

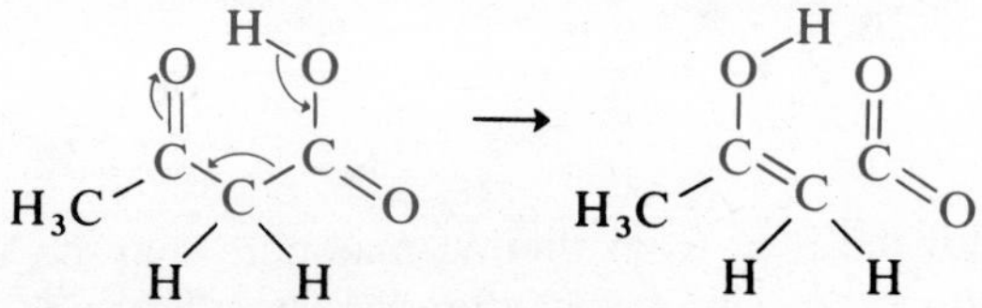

9.16 Alkylation of Acetoacetic and Malonic Esters

We have already seen that ethyl acetoacetate is readily converted to a resonance-stabilized enolate anion by strong base. Other 1,3-dicarbonyl compounds show comparable acidity and are also readily converted to resonance-stabilized carbanions. One such compound is diethyl malonate.

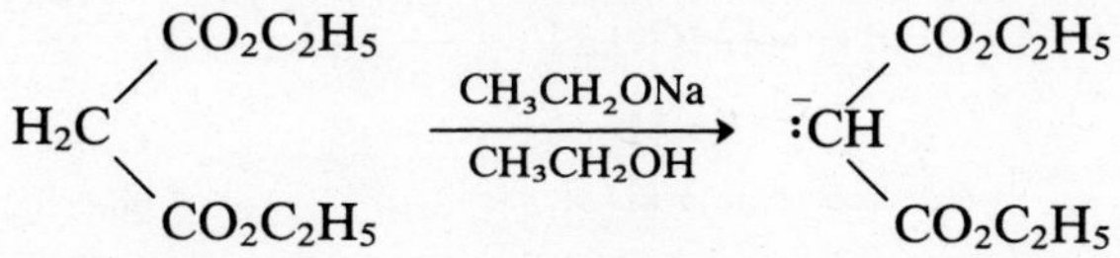

The anions derived from ethyl acetoacetate and diethyl malonate are powerful nucleophiles and will displace halide from primary and secondary alkyl halides in typical S_N2 fashion. This process involves substitution of an alkyl group for a hydrogen atom and is known as alkylation. Acid- or base-catalyzed ester hydrolysis followed by decarboxylation of the resulting β-ketoacid yields a ketone. This method for ketone synthesis is illustrated below with the conversion of 1-bromopentane to 2-octanone.

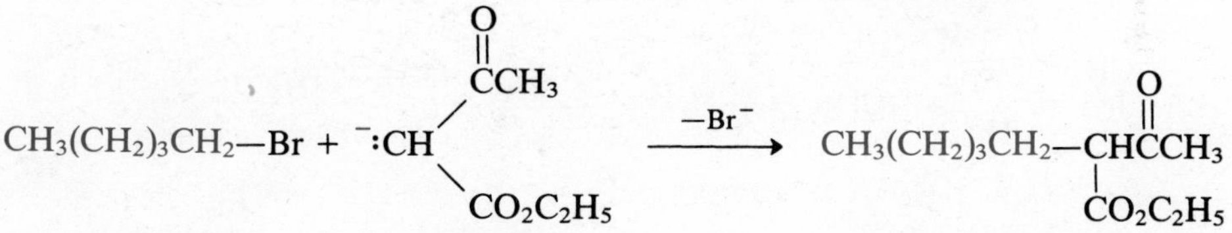

1-bromopentane — anion from ethyl acetoacetate

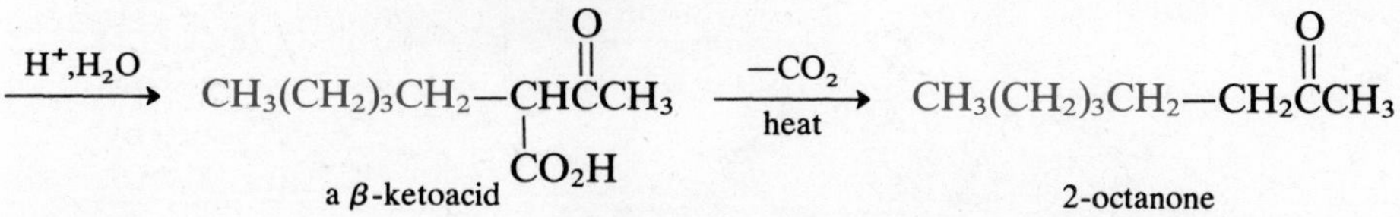

a β-ketoacid — 2-octanone

This scheme is known as the acetoacetic ester synthesis, and it is useful for the synthesis of ketones of the following type:

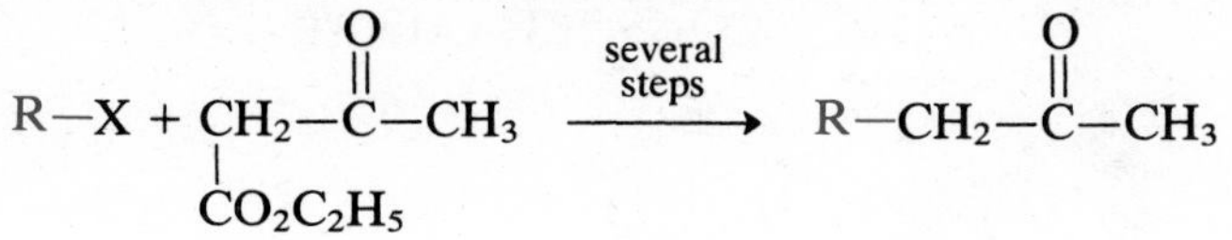

This same series of reactions, that is, conversion to an anion followed by alkylation, may be repeated, and the two α-hydrogens of ethyl acetoacetate may be replaced by alkyl groups. Subsequent ester hydrolysis and decarboxylation produces a more complex ketone.

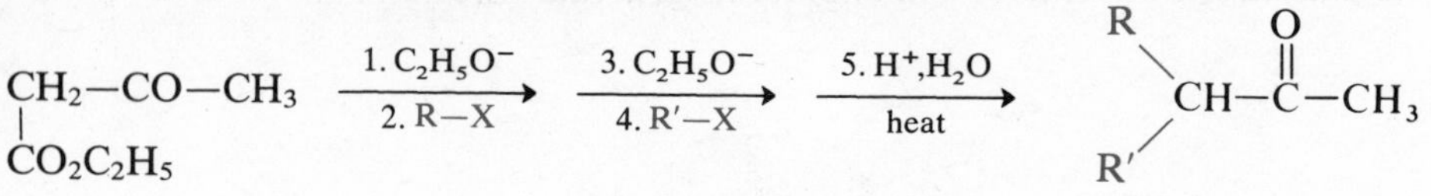

$$CH_2-CO-CH_3 \quad \xrightarrow[\text{2. R}-\text{X}]{\text{1. C}_2\text{H}_5\text{O}^-} \quad \xrightarrow[\text{4. R}'-\text{X}]{\text{3. C}_2\text{H}_5\text{O}^-} \quad \xrightarrow[\text{heat}]{\text{5. H}^+,\text{H}_2\text{O}} \quad \begin{array}{c} R \\ \diagdown \\ CH-C-CH_3 \\ \diagup \quad \parallel \\ R' \quad\quad O \end{array}$$
$$|$$
$$CO_2C_2H_5$$

The anion derived from diethyl malonate may be used in similar fashion for the synthesis of substituted carboxylic acids. This series of reactions is illustrated below with the conversion of benzyl chloride to 3-phenylpropanoic acid.

$$C_6H_5CH_2-Cl + \text{}^-\text{:}CH\begin{array}{c} CO_2C_2H_5 \\ \diagup \\ \diagdown \\ CO_2C_2H_5 \end{array} \quad \xrightarrow{-Cl^-} \quad C_6H_5CH_2-CHCO_2C_2H_5$$
$$|$$
$$CO_2C_2H_5$$

benzyl   anion from
chloride   diethyl malonate

$$\xrightarrow{H^+,H_2O} \quad C_6H_5CH_2-CHCO_2H \quad \xrightarrow[\text{heat}]{-CO_2} \quad C_6H_5CH_2-CH_2CO_2H$$
$$|$$
$$CO_2H$$

a β-dicarboxylic acid   3-phenylpropanoic acid

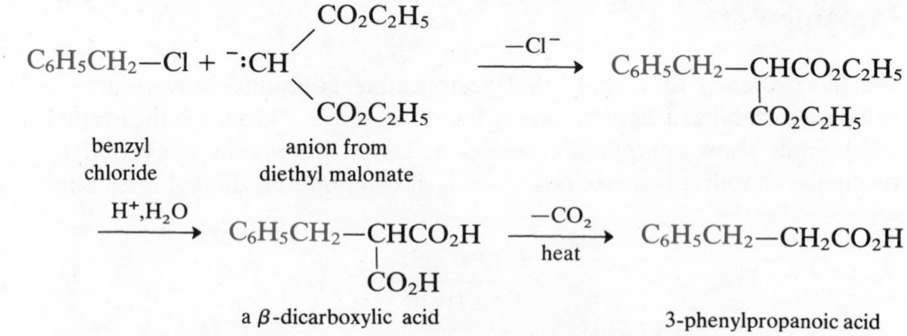

This scheme is known as the <u>malonic ester synthesis</u> and can be used for the synthesis of carboxylic acids of the following type.

$$R-CH_2-CO_2H \quad\quad \text{and} \quad\quad R-CH-CO_2H$$
$$|$$
$$R$$

Both the malonic ester and acetoacetic ester syntheses are very versatile and the selective double alkylation scheme is a major laboratory

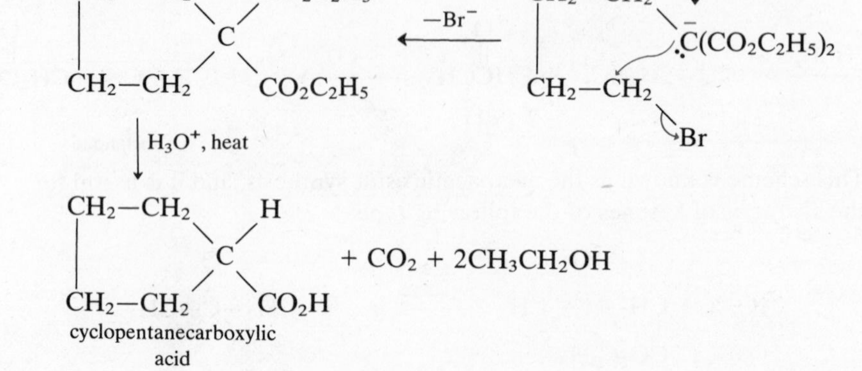

$$\begin{array}{c} CH_2-CH_2 \\ | \quad\quad\quad\quad \diagdown^{Br} \\ CH_2-CH_2 \\ \diagdown \\ Br \end{array} \quad \text{:CH}(CO_2C_2H_5)_2 \quad \xrightarrow{-Br^-} \quad \begin{array}{c} CH_2-CH_2 \\ | \quad\quad\quad\quad \diagdown \\ CH_2-CH_2 \quad CH(CO_2C_2H_5)_2 \\ \diagdown \\ Br \end{array}$$

1,4-dibromobutane

$$\Big\downarrow C_2H_5ONa$$

$$\begin{array}{c} CH_2-CH_2 \quad CO_2C_2H_5 \\ | \quad\quad\quad \diagup \\ \quad\quad C \\ | \quad\quad\quad \diagdown \\ CH_2-CH_2 \quad CO_2C_2H_5 \end{array} \quad \xleftarrow{-Br^-} \quad \begin{array}{c} CH_2-CH_2 \\ | \quad\quad\quad\quad \diagdown \\ CH_2-CH_2 \quad \text{:}\ddot{C}(CO_2C_2H_5)_2 \\ \diagdown_{Br} \end{array}$$

$$\Big\downarrow H_3O^+, \text{heat}$$

$$\begin{array}{c} CH_2-CH_2 \quad H \\ | \quad\quad\quad \diagup \\ \quad\quad C \\ | \quad\quad\quad \diagdown \\ CH_2-CH_2 \quad CO_2H \end{array} \quad + CO_2 + 2CH_3CH_2OH$$

cyclopentanecarboxylic
acid

method for building up branched-chain carbon skeletons. The same double alkylation scheme can also be used for constructing three-, four-, five-, and six-membered rings as illustrated by the synthesis of cyclopentanecarboxylic acid from 1,4-dibromobutane and diethyl malonate.

9.17 Spectroscopic Properties

All carbonyl-containing organic molecules show characteristic absorption in the infrared spectrum due to the stretching vibration of the carbonyl group. These absorption ranges for unconjugated aldehydes, ketones, carboxylic acids, esters, and amides are summarized below in Table 9.2

TABLE 9.2 Characteristic infrared absorptions of carbonyl-containing groups.

Functional Group	Carbonyl Stretching Frequency (cm^{-1})	Other Characteristic Stretching Frequency (cm^{-1})	
$R-\overset{\overset{\displaystyle O}{\|\|}}{C}-H$	1710–1740	2720–2830	C—H
$R-\overset{\overset{\displaystyle O}{\|\|}}{C}-R$	1710–1740		
$R-\overset{\overset{\displaystyle O}{\|\|}}{C}-OH$	1700–1725	2500–3000	O—H
$R-\overset{\overset{\displaystyle O}{\|\|}}{C}-OR$	1725–1750	1000–1300	O—R
$R-\overset{\overset{\displaystyle O}{\|\|}}{C}-NH_2$	1680–1690	3400–3500	N—H

Notice that while these carbonyl stretching frequencies fall within a relatively narrow range, there is some variation and it is sometimes possible to distinguish between these functional groups by the position of the carbonyl stretching absorption. Of greater use, however, in distinguishing between carbonyl-containing functional groups is the presence or absence of peaks in the spectrum corresponding to aldehyde C—H stretching (Section 7.15), carboxyl O—H stretching (Section 8.12), or amide N—H stretching (Section 10.8).

The most characteristic feature of the NMR spectrum of esters is the chemical shift of hydrogens on the carbon atom attached to the carbonyl group and to the ester oxygen. The presence of the carbonyl group and the oxygen shifts the signals of adjacent protons to larger delta values. For example, the NMR spectrum of methyl propanoate (Figure 9.1) shows three signals; a triplet at $\delta = 1.1$, a quartet at $\delta = 2.3$, and a singlet at $\delta = 3.6$. These signals are in the ratio $3:2:3$.

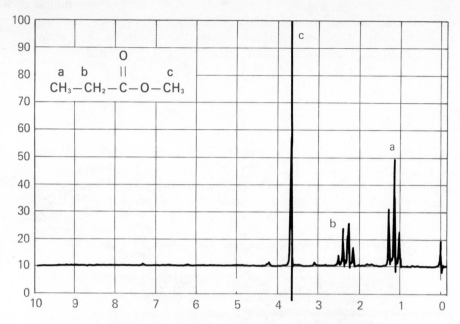

FIGURE 9.1 An NMR spectrum of methyl propanoate.

Compare the chemical shifts of the protons on the two methyl groups in this molecule. The methyl protons of CH_3CH_2- appear as a triplet at $\delta = 1.1$, while the methyl protons of CH_3O- appear as a singlet at $\delta = 3.6$. Note that this same shift to larger delta values is seen in the protons attached to oxygen atoms in alcohols and ethers. (Compare Table 15.3.)

PROBLEMS

9.1 Give an acceptable name for the following:

(a) $CH_3CH_2\overset{\overset{\displaystyle O}{\|}}{C}O\underset{\underset{\displaystyle CH_3}{|}}{C}HCH_3$

(b) $CH_3\overset{\overset{\displaystyle O}{\|}}{C}NH_2$

(c)

(d)

(e) $CH_3CH_2CH_2CH_2\overset{\overset{\displaystyle O}{\|}}{C}-N\overset{\diagup H}{\diagdown CH_3}$

(f) $CH_3CH_2O\overset{\overset{\displaystyle O}{\|}}{C}CH_2CH_2\overset{\overset{\displaystyle O}{\|}}{C}OCH_2CH_3$

9.2 Draw a structural formula for each of the following.
(a) phenyl acetate
(b) diethyl carbonate
(c) benzamide
(d) cyclobutyl butanoate

(e) methyl 3-butenoate
(g) diethyl oxalate
(i) acetamide
(k) propanoyl bromide
(m) diethyl malonate

(f) isopropyl 3-methyhexanoate
(h) ethyl *cis*-2-pentenoate
(j) *p*-chlorophenyl acetate
(l) N-phenylbutanamide
(n) formamide

9.3 Draw structural formulas for the nine isomeric esters of molecular formula $C_5H_{10}O_2$. Name each.

9.4 Arrange the following compounds in order of increasing boiling points.

(a) $CH_3CH_2\overset{\overset{\displaystyle O}{\|}}{C}OH$ (b) $CH_3CH_2CH_2CH_2OH$ (c) $CH_3CH_2OCH_2CH_3$

9.5 Both acetic acid and methyl formate have the same molecular formula, $C_2H_4O_2$. One of these compounds is a liquid of boiling point 32°C. The other is a liquid of boiling point 118°C.

$$CH_3\overset{\overset{\displaystyle O}{\|}}{-}C-OH \qquad H-\overset{\overset{\displaystyle O}{\|}}{C}-O-CH_3$$

acetic acid   methyl formate

Which of the two compounds would you predict to have the boiling point of 118°C; the boiling point of 32°C? Explain your reasoning.

9.6 Arrange the following compounds in order of increasing acidity.

(a) 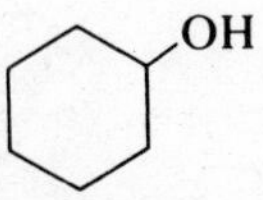 (b) 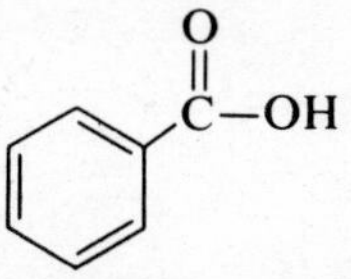 (c)

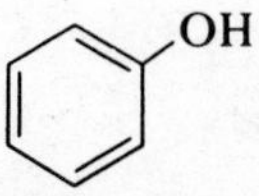

9.7 Write structural formulas for the products of hydrolysis of the following esters, amides, and anhydrides.

(a) $CH_3\overset{\overset{\displaystyle O}{\|}}{C}OCH_2CH_3 + H_2O \longrightarrow$

(b) $\begin{matrix} CH_2OCOCH_3 \\ | \\ CHOCOCH_3 \\ | \\ CH_2OCOCH_3 \end{matrix} + 3H_2O \longrightarrow$

(c) $\begin{matrix} CHO \\ | \\ H-C-OH \\ | \\ CH_2OPO_3^{2-} \end{matrix} + H_2O \longrightarrow$

(d) $CH_3\overset{\overset{\displaystyle O}{\|}}{C}OCH_2CH_2O\overset{\overset{\displaystyle O}{\|}}{C}CH_3 + 2H_2O \longrightarrow$

(e) $\begin{matrix} CH_2ONO_2 \\ | \\ CHONO_2 \\ | \\ CH_2ONO_2 \end{matrix} + 3H_2O \longrightarrow$

(f) $\begin{matrix} H_3C \\ \quad\diagdown \\ \quad\quad CHCH_2CH_2ONO \\ \quad\diagup \\ H_3C \end{matrix} + H_2O \longrightarrow$

(g) $CH_3(CH_2)_{10}CH_2OSO_3H + H_2O \longrightarrow$

(h) $CH_3\overset{\overset{\displaystyle O}{\|}}{C}SCH_2CH_3 + H_2O \longrightarrow$

(i) $H_2N\overset{\overset{\displaystyle O}{\|}}{C}NH_2 + 2H_2O \longrightarrow$

(j) 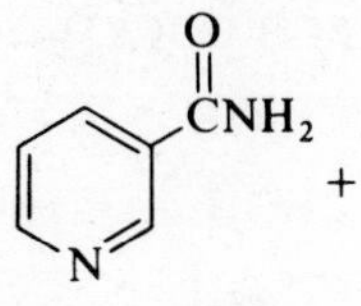 $+ H_2O \longrightarrow$

9.8 Propose structural formulas for the tetrahedral carbonyl addition intermediates formed during the following reactions.

(a) $CH_3CO_2H + HOCH_3 \overset{H^+}{\longrightarrow} CH_3CO_2CH_3 + H_2O$

251

(b) $CH_3CO_2CH_2CH_3 + NH_3 \longrightarrow CH_3CONH_2 + HOCH_2CH_3$

(c) $CH_3CH_2CONH_2 + H_2O \xrightarrow{H^+} CH_3CH_2CO_2H + NH_4^+$

(d) $CH_3\overset{O}{\overset{||}{C}}O\overset{O}{\overset{||}{C}}CH_3 + H_2O \longrightarrow 2CH_3CO_2H$

(e) $CH_3\overset{O}{\overset{||}{C}}O\overset{O}{\overset{||}{C}}CH_3 + NH_3 \longrightarrow CH_3CONH_2 + CH_3CO_2^-NH_4^+$

(f) $CH_3\overset{O}{\overset{||}{C}}O\overset{O}{\overset{||}{C}}CH_3 + CH_3OH \longrightarrow CH_3CO_2CH_3 + CH_3CO_2H$

(g) $CH_3\overset{O}{\overset{||}{C}}O\overset{O}{\overset{||}{C}}CH_3 + HSCH_2CH_3 \longrightarrow CH_3\overset{O}{\overset{||}{C}}SCH_2CH_3 + CH_3CO_2H$

(h) $2CH_3\overset{O}{\overset{||}{C}}S{-}CoA \xrightarrow{\text{enzyme}} CH_3\overset{O}{\overset{||}{C}}CH_2\overset{O}{\overset{||}{C}}S{-}CoA + CoA{-}SH$

(i) 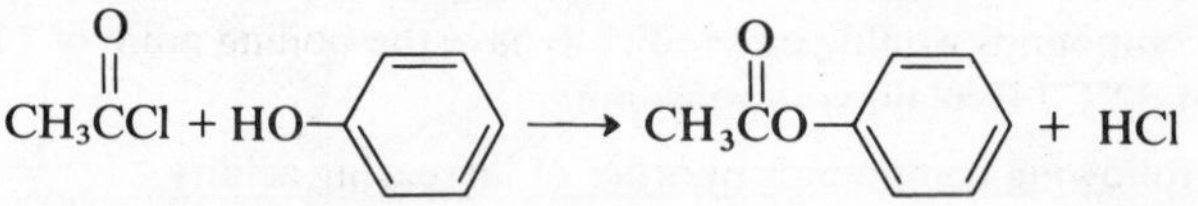 $CH_3\overset{O}{\overset{||}{C}}CCl + HO{-}\bigcirc \longrightarrow CH_3\overset{O}{\overset{||}{C}}O{-}\bigcirc + HCl$

9.9 Complete the following reactions.

(a) $CH_3CH_2CO_2CH_3 + CH_3CH_2NH_2 \longrightarrow$

(b) 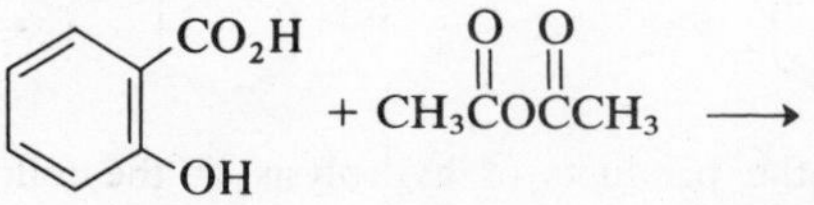 (o-hydroxybenzoic acid with CO_2H and OH) $+ CH_3\overset{O}{\overset{||}{C}}O\overset{O}{\overset{||}{C}}CH_3 \longrightarrow$

(c) $CH_3CH_2O{-}\bigcirc{-}NH_2 + CH_3\overset{O}{\overset{||}{C}}O\overset{O}{\overset{||}{C}}CH_3 \longrightarrow$ (phenacetin, a pain reliever)

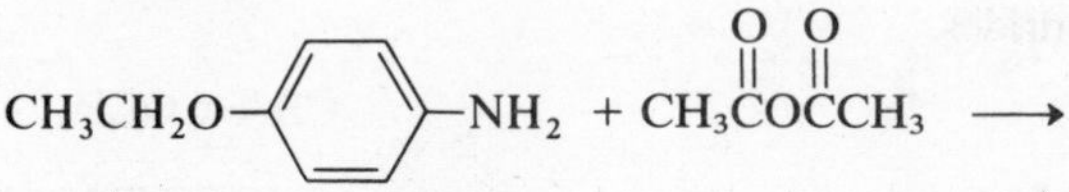

(d) $C_6H_5CO_2CH_3 + 2C_6H_5MgBr \longrightarrow$

(e) $C_6H_5CO_2CH_3 + LiAlH_4 \longrightarrow$

(f) $C_6H_5CO_2CH_3 + NH_3 \longrightarrow$

(g) 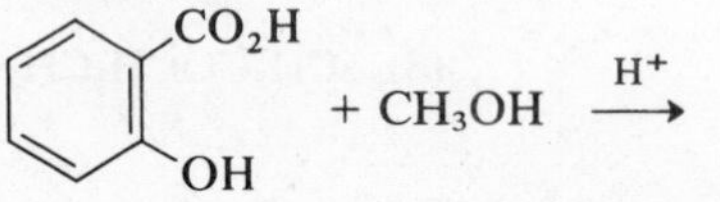 (o-hydroxybenzoic acid with CO_2H and OH) $+ CH_3OH \xrightarrow{H^+}$ (oil of wintergreen)

(h) $C_6H_5CO_2CH_2CH_3 + H_2O \xrightarrow{H_3O^+}$

(i) $CH_3\underset{CH_3}{\overset{|}{C}}HCO_2H + C_6H_5NHCH_3 \xrightarrow{\text{heat}}$

(j) $2CH_3\overset{O}{\overset{||}{C}}O\overset{O}{\overset{||}{C}}CH_3 + HOCH_2CH_2OH \longrightarrow$

(k) $\underset{CH_2OH}{\overset{CH_2OH}{\overset{|}{C}}}HOH + 3CH_3\overset{O}{\overset{||}{C}}Cl \longrightarrow$

(l) $O_2N{-}\bigcirc{-}CO_2H + SOCl_2 \longrightarrow$

(m) $\underset{\underset{CH_3}{|}}{\overset{\overset{CH_3}{|}}{CH_3-\overset{+}{N}-CH_2CH_2OH}} + CH_3-\overset{O}{\overset{||}{C}}-Cl \longrightarrow$

(n) $\overset{O}{\overset{||}{Cl-C-Cl}} + 2CH_3CH_2OH \longrightarrow$

(o) $\underset{\underset{H}{|}}{\underset{N}{\boxed{}}} + CH_3\overset{O\ O}{\overset{||\ ||}{COCCH_3}} \longrightarrow$

9.10 Draw structural formulas for the products formed by Claisen condensation of the following:

(a) $CH_3CH_2\overset{O}{\overset{||}{C}}OCH_2CH_3 \xrightarrow{\ NaOCH_2CH_3\ }$ (b) $C_6H_5CH_2\overset{O}{\overset{||}{C}}OCH_3 \xrightarrow{\ NaOCH_3\ }$

9.11 Draw structural formulas for the products of the following carbonyl condensation reactions.

(a) $\overset{O}{\underset{}{\boxed{}}} + C_6H_5\overset{O}{\overset{||}{C}}OCH_2CH_3 \xrightarrow{\ NaOCH_2CH_3\ }$

(b) $C_6H_5\overset{O}{\overset{||}{C}}CH_3 + CH_3CH_2O\overset{O}{\overset{||}{C}}OCH_2CH_3 \xrightarrow{\ NaOCH_2CH_3\ }$

9.12 Show how you might convert ethyl propanoate into each of the following:
(a) ethyl 2-methyl-3-oxopentanoate (b) 3-pentanone

9.13 Starting with any aldehyde, ketone, or ester, synthesize each of the following by using the aldol or Claisen condensation and any necessary subsequent steps.

(a) $\underset{\underset{CH_2-CH_3}{|}}{CH_3-CH_2-CH_2-\overset{\overset{OH}{|}}{CH}-CH-CH_2OH}$ (b) $\underset{\underset{CH_3}{|}}{H-\overset{O}{\overset{||}{C}}-CH-\overset{O}{\overset{||}{C}}-OCH_2CH_3}$

(c) $\boxed{}-\underset{\underset{OH}{|}}{CH}-CH_2-CH_2OH$ (d) $\boxed{}-CH=CH-\overset{O}{\overset{||}{C}}-CH_3$

(e) $\boxed{}-CH=CH-\overset{O}{\overset{||}{C}}-OH$

9.14 Starting with either diethyl malonate or ethyl acetoacetate and any other organic and inorganic reagents, show how you might synthesize the following:
(a) 3-methyl-2-hexanone
(b) cyclobutanecarboxylic acid
(c) methyl cyclopropyl ketone
(d) methyl 3-phenylpropanoate

9.15 Show reagents and conditions you would use for the following conversions. In addition to the given starting material, use any other organic or inorganic reagents as necessary.

253

(a) $CH_3CH_2CH_2CH_2CH_2OH \longrightarrow CH_3CH_2CH_2CH_2\overset{\overset{\displaystyle O}{\parallel}}{C}OCH_2CH_3$

(b) $CH_3CH_2CH_2CH_2CH_2OH \longrightarrow CH_3\overset{\overset{\displaystyle O}{\parallel}}{C}OCH_2CH_2CH_2CH_2CH_3$

(c) [benzene ring with CH_3] $\longrightarrow$ [benzene ring with $\overset{\overset{\displaystyle O}{\parallel}}{C}-NH_2$]

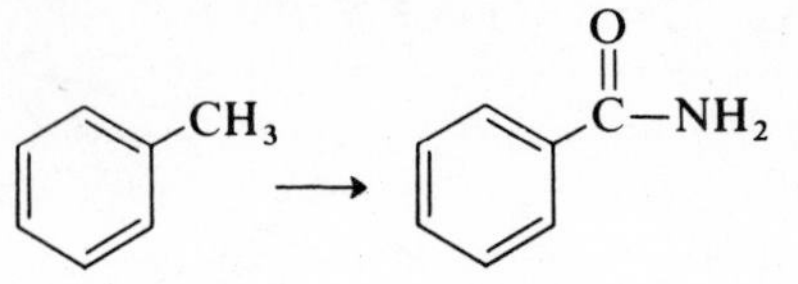

(d) [benzene ring with CH_2OH] $\longrightarrow$ [benzene ring with $CH_2-\overset{\overset{\displaystyle O}{\parallel}}{C}-Cl$]

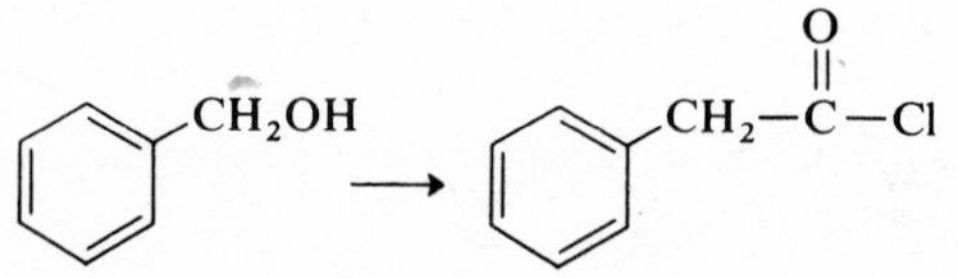

(e) [cyclohexene ring with CH_3] $\longrightarrow$ $CH_3-\overset{\overset{\displaystyle O}{\parallel}}{C}-CH_2-CH_2-CH_2-CH_2-\overset{\overset{\displaystyle O}{\parallel}}{C}-OCH_2CH_3$

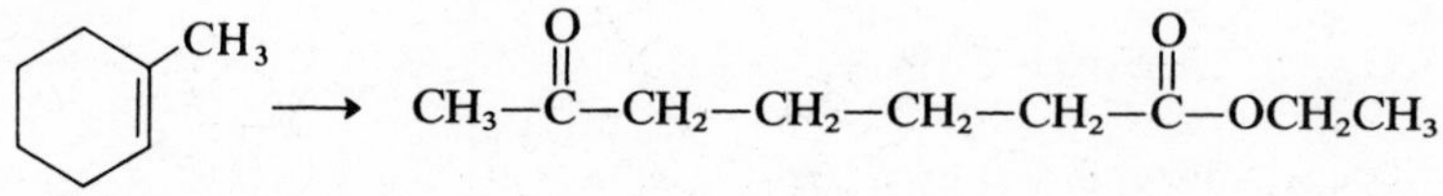

9.16 Suggest a combination of reagents that could be used to synthesize the following barbiturates: phenobarbital, pentobarbital, thiopental. (Structural formulas for each are given in Section 9.9.)

9.17 Compare acid-catalyzed ester formation with acid-catalyzed acetal formation, and list the similarities and differences between the two reactions.

9.18 The following compounds all contain a carbon-oxygen-carbon linkage. Compare the reactivity of each to water. (a) diethyl ether; (b) acetic anhydride; (c) ethyl acetate.

9.19 The following compounds are derivatives of carboxylic acids. Compare the reactivities of each to water. (a) ethyl acetate; (b) acetic anhydride; (c) acetamide.

9.20 The anions derived from ethyl acetoacetate and diethyl malonate are represented in the text as the following:

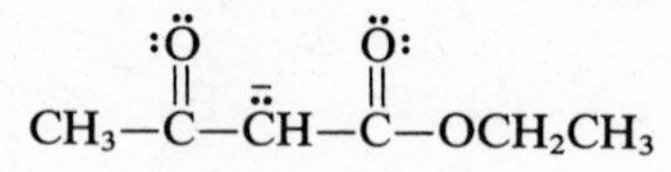

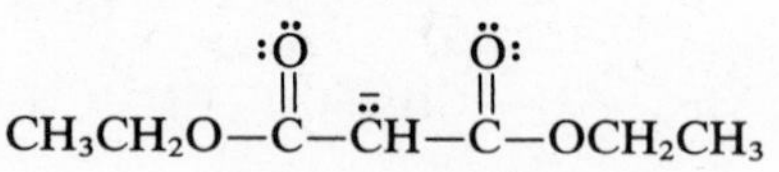

$CH_3-\overset{\overset{\displaystyle :\ddot{O}}{\parallel}}{C}-\overset{\displaystyle \bar{\ddot{C}}}{H}-\overset{\overset{\displaystyle \ddot{O}:}{\parallel}}{C}-OCH_2CH_3$ $CH_3CH_2O-\overset{\overset{\displaystyle :\ddot{O}}{\parallel}}{C}-\overset{\displaystyle \bar{\ddot{C}}}{H}-\overset{\overset{\displaystyle \ddot{O}:}{\parallel}}{C}-OCH_2CH_3$

the anion derived from   the anion derived from
ethyl acetoacetate   diethyl malonate

However, each anion is best represented as a resonance hybrid. Draw two additional contributing structures for each anion and show by the use of curved arrow how one contributing structure may be converted into the other two.

9.21 Explain why it is preferable to hydrolyze esters in aqueous base rather than in aqueous acid.

9.22 (a) Write an equation for the equilibrium established when acetic acid and *n*-propyl alcohol are refluxed in the presence of a few drops of concentrated sulfuric acid.

(b) Using the data on page 236, calculate the equilibrium constant for this reaction.

9.23 Propose a reasonable mechanism for the reaction of acetic anhydride with water; with ammonia to form acetamide; with methanol to form methyl acetate; with ethanethiol to form ethyl thioacetate.

9.24 If 15 grams of salicylic acid is reacted with an excess of acetic anhydride, how many grams of aspirin could be formed?

9.25 Hydrolysis of a neutral compound (A), $C_6H_{12}O_2$, yielded an acid (B) and an alcohol (C). Oxidation of the alcohol C by potassium dichromate in aqueous sulfuric acid gave an acid identical to B. What are the structural formulas of A, B, and C?

9.26 One biosynthetic route to the formation of aromatic compounds, including derivatives of benzene, is thought to involve the formation and cyclization of a polycarbonyl chain derived from acetate units. Examples are the biosynthesis of 6-methylsalicylic acid and orsellinic acid. 6-Methylsalicylic acid is formed as a metabolite of the mold *Penicillium urticae*. If the acetic acid supplied to the mold is labeled at the carboxyl group with carbon-14 (indicated in color below) the 6-methylsalicylic acid has the label pattern as shown.

Devise a reasonable scheme for the biosynthesis of each of these compounds starting from acetyl coenzyme A. (Hint: Use a combination of the aldol condensation, the Claisen condensation, dehydration, reduction and keto-enol tautomerism.)

9.27 The same polycarbonyl chain shown in brackets in problem 9.26 can also cyclize in an alternative manner to give phloroacetophenone.

phloroacetophenone

(a) Propose a sequence of reactions that might lead from the bracketed polycarbonyl compound to phloroacetophenone.
(b) If the polycarbonyl compound is labeled with carbon-14 as shown in the previous problem, predict the positions of the carbon-14 labels in phloroacetophenone.

9.28 Show how you might distinguish between the following pairs of compounds by a simple chemical test. In each case, tell what test you would perform, what you would expect to observe, and write an equation for each positive test.
(a) isopropyl formate and 2-methylpropanoic acid (isobutyric acid)
(b) butanoic acid and butanamide
(c) methyl benzoate and acetophenone (methyl phenyl ketone)
(d) methyl hexanoate and hexanal

(e) aspirin   and   phenacetin

9.29 List one major spectral characteristic that will enable you to distinguish between the following compounds.

(a) isopropyl formate and 2-methylpropanoic acid (IR, NMR)

(b) butanoic acid and butanamide (IR)

(c) methyl benzoate and acetophenone (NMR)

(d) methyl hexanoate and hexanal (IR, NMR)

(e) aspirin and phenacetin (IR, NMR)

(f) phenyl acetate and cyclohexyl acetate (IR, NMR, UV)

(g) methyl benzoate and benzamide (IR, NMR)

9.30 There are four isomeric esters of molecular formula $C_4H_8O_2$.

(a) The NMR spectrum of one of these esters shows signals at $\delta = 1.3$ (doublet), $\delta = 5.3$ (septet), and $\delta = 8.0$ (singlet). Draw the structural formula of this ester.

(b) Draw the structural formula of the other ester whose NMR spectrum will also show a singlet at $\delta = 8.0$.

(c) The NMR spectra of the remaining two esters will each show three signals; a singlet, a triplet, and a quartet. Show how these esters can be distinguished by the chemical shift of the singlet; by the chemical shift of the quartet.

IV

Mini-Essay

Nylon and Dacron

Nylon and Dacron

Shortly after World War I, a number of creative and far-sighted chemists recognized the need for developing basic knowledge of polymer chemistry. One of the most productive of these pioneers was Wallace M. Carothers of E. I. du Pont de Nemours & Co., Inc. In the early 1930s, Carothers and his associates began fundamental research into the reactions of aliphatic dibasic acids and glycols (for example, adipic acid and ethylene glycol) and obtained polyester products of high molecular weight, over 10,000. These products were tough, opaque solids that melted at moderate temperatures to give clear, viscous liquids which could be drawn into fibers. However, the melting points of these polyester fibers were too low for them to be of use as textile fibers and they were not investigated further (that is, not for another decade). Next, Carothers turned his attention to the reactions of dibasic acids and diamines, and in 1934 synthesized Nylon 66, the first purely synthetic fiber. Nylon 66 was so-named because its synthesis uses two different organic intermediates, a six-carbon dicarboxylic acid and a six-carbon diamine.

Equimolar amounts of adipic acid and hexamethylene diamine dissolved in aqueous alcohol react to form a 1:1 salt, commonly called nylon salt.

$$\underset{\substack{\text{adipic acid}}}{HOC(CH_2)_4COH} + \underset{\substack{\text{hexamethylene}\\\text{diamine}}}{H_2N(CH_2)_6NH_2} \longrightarrow \underset{\substack{\text{1:1 salt}\\\text{(nylon salt)}}}{^-OC(CH_2)_4CO^-H_3N^+(CH_2)_6NH_3^+}$$

The 1:1 salt (nylon salt) is then heated in an autoclave to about 250°C. Because this is a closed system containing both water and alcohol vapor, the internal pressure rises to about 250 pounds per square inch (about 15 atmospheres). As the heat and pressure build up, water molecules are lost as condensation takes place between $-CO_2^-$ and $-NH_3^+$ groups to form amide bonds.

$$n\text{ 1:1 salt (nylon salt)} \xrightarrow[\text{heat}]{-H_2O} \underset{\text{Nylon 66}}{\left[C(CH_2)_4C-NH(CH_2)_6NH \right]_n}$$

As condensation proceeds and more water is produced, steam is bled from the autoclave to maintain a constant pressure. The temperature is gradually raised to about 275°C, and as steam continues to be removed, the internal pressure falls to that of the atmosphere. After a total reaction time of 2 to 4 hours, polymer molecules of molecular weights 10,000 to 20,000 are formed.

259

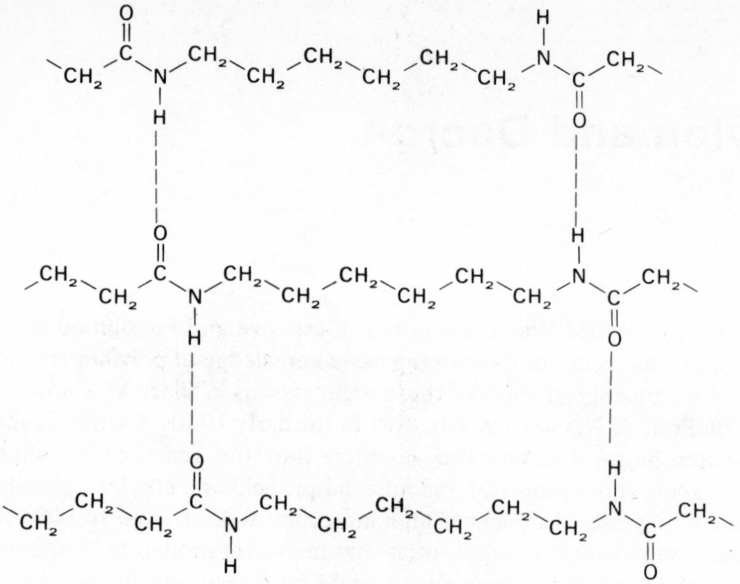

FIGURE 1 Structure of cold-drawn Nylon 66 fiber. Hydrogen bonds between adjacent chains hold molecules together in a crystalline structure.

As a first stage in fiber production, crude nylon, in the molten state at 260 to 270°C, is melt-spun into fibers and carefully cooled to room temperature. At this stage, the polymer is best described as unoriented crystalline. Crystallinity of polymers is not quite the same as crystallinity of such common substances as copper sulfate and glucose. These substances show obvious crystalline structure. The characteristic property of the crystalline state is a high degree of order—the atoms are arranged regularly in a three-dimensional lattice which is built up of a simple unit repeated over and over again. With a crystal of a low-molecular-weight substance like glucose, there is complete order, i.e., 100% crystallinity. Although polymers do not crystallize in the conventional sense, they often have regions where chains are precisely oriented with respect to one another. Such regions are called crystallites. Melt-spun Nylon 66 fibers have many crystallite regions, but they are randomly oriented and the fiber in this state is not suitable for textile use. However, the character of the fibers can be improved dramatically by drawing them at room temperature to about four times their original length. In this cold-drawing process, the fiber elongates and the crystalline regions are drawn together and oriented in the direction of the fiber axis. Hydrogen bonds between the oxygen atoms of one chain and the N—H groups of another chain hold the chains together (Figure 1). The effects of orientation of the polymer molecules on the physical properties can be profound; both tensile strength and stiffness increase with increasing orientation. Cold-drawing is an important step in the production of all synthetic fibers.

Of the two raw materials required for the synthesis of Nylon 66, only adipic acid was commercially available at the time du Pont's management decided to begin commercial production of the fiber. du Pont's chemical engineers' first task therefore was to develop an economical process for the large-scale production of hexamethylene diamine. This was accomplished by the transformation of phenol to adipic acid, followed by reaction with ammonia to form a salt, thermal decomposition of the salt to form a diamide, and finally, catalytic hydrogenation of the diamide to form hexamethylene diamine (HMDA).

$$phenol \xrightarrow{3H_2} cyclohexanol \xrightarrow{oxid} cyclohexanone \xrightarrow{oxid} HOC(CH_2)_4COH \xrightarrow{2NH_3}$$

adipic acid

$$NH_4^+ {}^-OC(CH_2)_4CO^- NH_4^+ \xrightarrow{heat} H_2NC(CH_2)_4CNH_2 \xrightarrow{4H_2} H_2N(CH_2)_6NH_2$$

diammonium adipate   adipamide   HMDA

The first commercial plant for the synthesis of hexamethylene diamine went into operation in 1939, using phenol as the starting material. Within a few years, however, the raw material base for the synthesis of HMDA was shifted to benzene which can be reduced to cyclohexane. In turn, this cycloalkane is oxidized to cyclohexanol, cyclohexanone, and finally, adipic acid. Recall that benzene is a primary petrochemical available by the catalytic reforming of naphtha. Following World War II, du Pont embarked on a major expansion of its Nylon 66 capacity, a program which demanded a major expansion in facilities for the production of HMDA. Should the company continue to make this key intermediate from benzene or might it be able to find a more economical raw material base and at the same time a more economical route for the synthesis of HMDA? Within a few years, two new commercial processes were developed, neither of which depended on benzene or cyclohexane as a starting material. The first of these used furfural, a heterocyclic aromatic aldehyde obtained by heating cellulose-containing materials like oat hulls, corncobs, or rice hulls with HCl. In a series of several steps, furfural is converted to 1,4-dichlorobutane and then to hexamethylene diamine.

$$wastes\ containing\ cellulose \xrightarrow[heat]{HCl} furfural \longrightarrow ClCH_2CH_2CH_2CH_2Cl \longrightarrow HMDA$$

furfural   1,4-dichlorobutane

Thus, this synthesis of hexamethylene diamine shifted at least part of the raw material base of nylon from benzene, a primary petrochemical, to furfural and agricultural byproducts.

The second major synthesis of hexamethylene diamine remained petroleum-based, but shifted from one primary petrochemical to another—from benzene to butadiene. Recall that butadiene is a component of the olefinic gasses produced in the thermal cracking of natural gas. Chlorination of butadiene under carefully controlled conditions forms 1,4-dichlorobutane, one of the same compounds produced in the furfural-based synthesis. By 1953, du Pont had closed down its facilities for making HMDA from adipic acid and was making this intermediate from both furfural and butadiene.

Nylon 66, has been the primary nylon fiber synthesized in the United States, Canada, the United Kingdom, and France. However, it is Nylon 6

261

that has become the primary nylon fiber synthesized in many other countries, particularly Germany, Italy, and Japan. The manufacture of Nylon 6 uses only one starting material, caprolactam, a six-carbon cyclic amide. During the synthesis of Nylon 6, a part of the caprolactam is hydrolyzed and then the partially hydrolyzed mixture is heated to 250°C to drive off water and bring about polymerization.

caprolactam   Nylon 6

The starting material for the synthesis of caprolactam is benzene. Catalytic hydrogenation yields cyclohexane which is in turn oxidized to cyclohexanone. Reaction of cyclohexanone with hydroxylamine, or alternatively, direct oxidation of cyclohexane with nitrous acid, yields cyclohexanone oxime. Acid-catalyzed rearrangement of this oxime yields caprolactam.

benzene   cyclohexanone
                                 oxime

Why is Nylon 66 the primary nylon fiber produced in the United States, Canada, and the United Kingdom, and France, while Nylon 6 is the primary nylon fiber in Germany, Italy, and Japan? The answer lies chiefly in the availability of different starting materials. In the United States, butadiene is readily available through the thermal cracking of natural gas, itself a vast natural resource. In Europe and Japan, natural gas is not as plentiful and, therefore, these countries are forced to depend on petroleum for not only the aromatics but also for the simple alkenes. Since they do not have as ready a source of butadiene and hence hexamethylene diamine, chemical companies in these countries find it more economical to synthesize Nylon 6 rather than Nylon 66.

By the 1940s, the relationships between molecular structure and bulk physical properties were well understood, and the polyester condensations were reexamined. Recall that Carothers and his associates had already concluded that the polyester fibers from aliphatic dibasic acids and ethylene glycol were not suitable for textile use because they were too low melting. Winfield and Dickson at the Calico Printers Association in England reasoned, quite correctly it turned out, that a greater resistance to rotation in the polymer backbone would confer a higher degree of crystallinity, stiffen the polymer chain, and thereby lead to a more acceptable polyester fiber. To create this stiffness in the polymer chain, they used terephthalic acid, an aromatic dicarboxylic acid.

$$HOC\text{—}\langle\text{—}\rangle\text{—}COH + HOCH_2CH_2OH \xrightarrow[\substack{-H_2O}]{\substack{condensation \\ polymerization}}$$

terephthalic
acid

ethylene
glycol

$$\left(\text{—}CH_2CH_2O\text{—}\overset{O}{\underset{\|}{C}}\text{—}\langle\text{—}\rangle\text{—}\overset{O}{\underset{\|}{C}}\text{—}O\right)_n$$

poly(ethylene terephthalate)
Dacron, Mylar, Terylene

Poly(ethylene terephthalate), formed from ethylene glycol and terephthalic acid, has proven tremendously successful and is marketed under the trade names Dacron, Mylar, and Terylene.

Just as with nylon, the crude polyester is first spun into fibers and then cold-drawn. The outstanding feature of Dacron is its stiffness (about four times that of Nylon 66), very high strength, and a remarkable resistance to creasing and wrinkling. Because the Dacron fibers are much harsher (due to their stiffness) to the touch than wool, they are usually blended with wool or cotton. Also the polyester is heat-settable, hence permanent creases may be introduced in the manufacturing of garments which then can be used repeatedly without ironing.

What are the sources of ethylene glycol and terephthalic acid, the two key organic intermediates for the synthesis of polyester fibers? Ethylene glycol is available by air oxidation of ethylene to ethylene oxide followed by hydrolysis (Section 5.5). However, in the early 1940s, terephthalic acid was not available commercially in either the quantities or at a cost that would permit production of polyester fibers at prices competitive with the natural fibers, cotton and wool, or with the semi-synthetic fiber, rayon (Section 11.9). Therefore, du Pont chemists were compelled to devise a synthesis for terephthalic acid. As a starting material, they chose p-xylene, available along with benzene from catalytic reforming of naphtha and other petroleum distillation fractions. Nitric acid oxidation of p-xylene yielded the desired terephthalic acid of sufficiently high purity and low cost to permit commercial production of polyester fibers beginning in 1949 to 1951.

$$CH_2\text{=}CH_2 \longrightarrow HO\text{—}CH_2\text{—}CH_2\text{—}OH\text{———}$$

ethylene

ethylene glycol

condensation
polymerization

→ Dacron

$$CH_3\text{—}\langle\text{—}\rangle\text{—}CH_3 \longrightarrow HO\text{—}\overset{O}{\underset{\|}{C}}\text{—}\langle\text{—}\rangle\text{—}\overset{O}{\underset{\|}{C}}\text{—}OH\text{———}$$

p-xylene

terephthalic
acid

Thus, the starting materials for the synthesis of Dacron polyester are ethylene and p-xylene, both primary petrochemicals.

In 1975, the United States production of nylon exceeded 2,200 billion pounds while that of polyester fiber exceeded 1,550 billion pounds. Production of the third major man-made fiber, the acrylic fibers, was 620 billion pounds.

263

REFERENCES

The American Chemical Society. *Chemistry In the Economy*. 1973.

Billmyer, F. W. *Textbook of Polymer Chemistry*. Interscience Publishers, New York, 1957.

Kaufman, M. *Giant Molecules*. Doubleday and Company, Garden City, New York, 1968.

Amines

10.1 Structure and Nomenclature

Amines are derivatives of ammonia in which one or more of the hydrogens are replaced by alkyl or aryl groups. Amines are classified as primary, secondary, or tertiary according to the number of hydrocarbon groups attached to the nitrogen atom. A nitrogen with four hydrocarbon groups attached is positively charged and known as a quaternary ammonium salt. The simpler aliphatic amines are named by specifying the alkyl groups attached to the nitrogen atom and adding the ending -amine

$$CH_3-NH_2$$
methylamine
(primary)

$$CH_3CH_2-\underset{\underset{H}{|}}{N}-CH_2CH_3$$
diethylamine
(secondary)

$$CH_3CH_2\underset{\underset{CH_3}{|}}{CH}-NH_2$$
sec-butylamine
(primary)

$$(CH_3)_3N$$
trimethylamine
(tertiary)

$$HOCH_2CH_2-NH_2$$
ethanolamine
(2-aminoethanol)
(primary)

$$CH_3-\underset{\underset{CH_3}{|}}{\overset{\overset{CH_3}{|}}{N^+}}-CH_3 \quad Cl^-$$
tetramethylammonium
chloride
(quaternary ammonium salt)

Aryl amines are named as derivatives of the parent member of the family, aniline.

aniline

2-nitroaniline
o-nitroaniline

2-methoxy-4-ethyl-
N-methylaniline

In the IUPAC system, the —NH$_2$ group may be treated as a substituent and named as an <u>amino group</u>.

2-amino-1-phenylpropane
amphetamine

p-aminobenzoic acid

$H_2NCH_2CH_2CH_2CO_2H$

4-aminobutanoic acid
γ-aminobutyric acid

If the amine nitrogen atom is incorporated into a ring, the resulting amine is called a <u>heterocyclic amine</u>. Piperidine and pyrrolidine, shown below, both undergo typical reactions of secondary aliphatic amines, and consequently, we can include them in our discussion of the chemistry and physical properties of aliphatic amines.

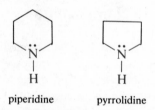

piperidine   pyrrolidine

We have already discussed the structure and nomenclature of several classes of heterocyclic aromatic amines in Chapter 6.

Structural formulas of a variety of naturally occurring amines of both plant and animal origin are shown in Figure 10.1. These are chosen to illustrate something of the structural diversity and range of physiological activity of amines, their value as drugs, and their importance in nutrition. <u>Coniine</u> (2-propylpiperidine) from the water hemlock is highly toxic. It can cause weakness, labored respiration, paralysis, and eventually death. The levorotatory isomer of coniine is the toxic substance in the "poison hemlock" used in the execution of Socrates. <u>Nicotine</u> is one of the chief heterocyclic amine bases of the tobacco plant. In small doses, nicotine is a stimulant. However, in larger doses it causes depression, nausea, and vomiting. In still larger doses, nicotine acts as a poison. Solutions of nicotine are often used as insecticides. Note that nicotinic acid, an oxidation product of nicotine, is one of the water-soluble vitamins required in man for proper nutrition.

<u>Papaverine</u> is one of the many analgesic substances obtained from extracts of the opium poppy. Both <u>morphine</u> and its methylated derivative,

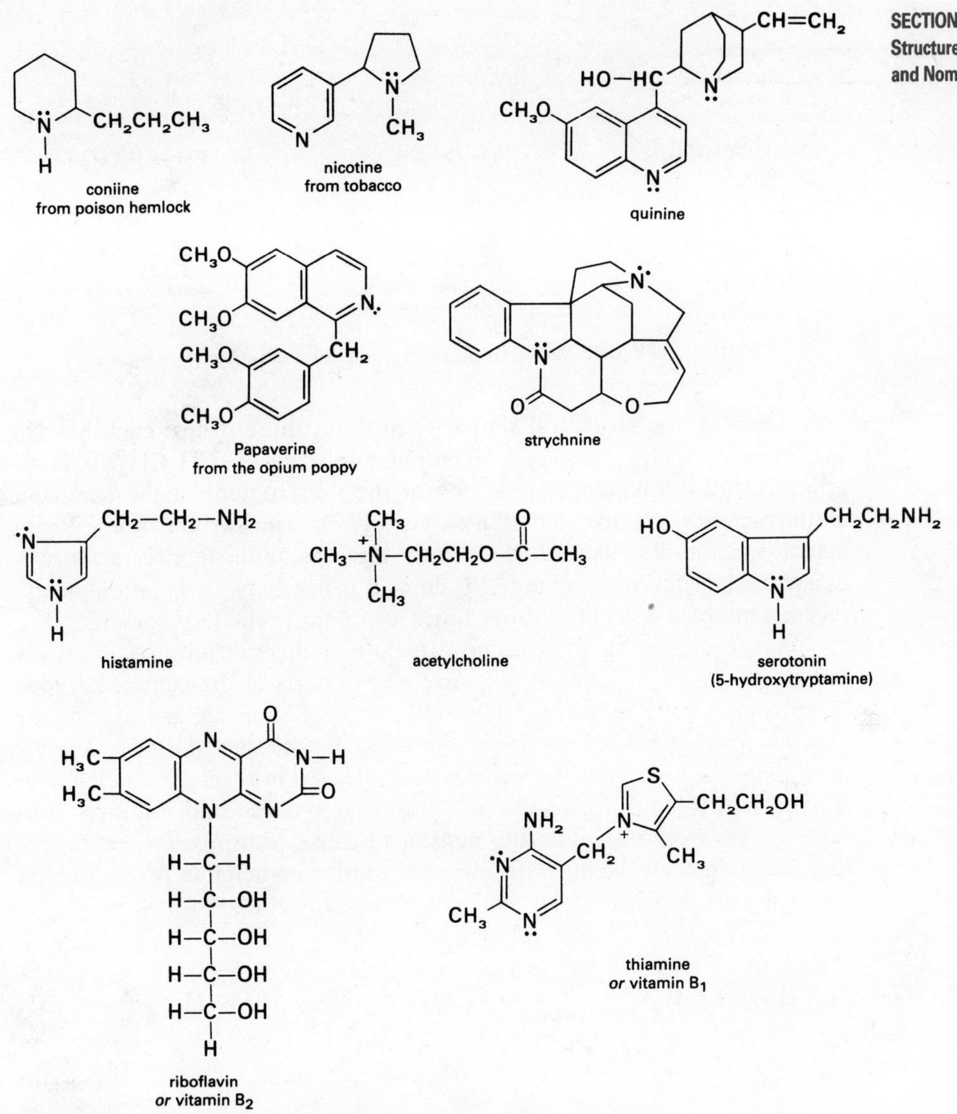

FIGURE 10.1 Several naturally occurring amines of plant and animal origin.

codeine, are also isolated from the opium poppy. For more information on the structure and physiological activity of morphine and the search for safer and nonaddictive analgesics, see the mini-essay "Alkaloids". Quinine, isolated from the bark of the cinchona tree in South America has been long used as a medicine for the treatment of malaria.

Histamine, formed by the decarboxylation of the amino acid histidine (Table 13.1), is a toxic substance present in all tissues of the body combined in some manner with proteins. Extensive production of histamine occurs during hypersensitive allergic reactions and the symptoms of this release are unfortunately familiar to most of us, in particular those who suffer from hay fever. The search for antihistamines—drugs that will inhibit the effects of histamine—has led to the synthesis of several drugs whose trade names are well known. Structural formulas for three of the more widely used antihistaminic drugs are shown in Figure 10.2.

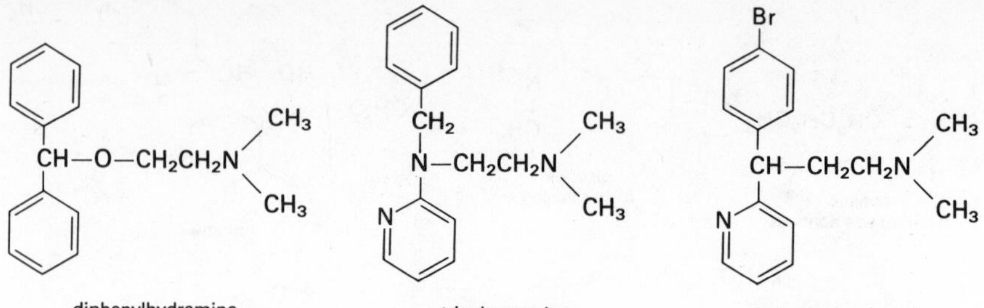

diphenylhydramine (Benadryl) | tripelennamine (Pyribenzamine) | dexbrompheniramine (Disomer)

FIGURE 10.2 Three synthetic antihistamines.

Observe the structural similarity in these three drugs: each has (1) two aromatic rings and (2) a dimethylaminoethyl, $-CH_2CH_2N(CH_3)_2$, group. Dexbrompheniramine is one of the most potent of the available antihistaminics. Notice that it has one asymmetric carbon atom. Pharmacological studies have shown clearly that the antihistaminic activity is due almost exclusively to the (+)-isomer. It is nearly twice as potent as the racemic mixture and about thirty times as potent as the (−)-isomer.

Serotonin and acetylcholine are both neurotransmitter substances important in human physiology; serotonin in parts of the central nervous system mediating affective behavior, acetylcholine in certain motor neurons responsible for causing contraction of voluntary muscles. Acetylcholine is stored in synaptic vesicles and released in response to electrical activity in the neuron; it diffuses across the synapse and interacts with receptor sites on a neighboring neuron to cause transmission of a nerve impulse. After interacting with receptor sites, acetylcholine is deactivated through hydrolysis catalyzed by the enzyme acetylcholinesterase.

$$\underset{\text{acetylcholine}}{CH_3-\overset{\overset{CH_3}{|}}{\underset{\underset{CH_3}{|}}{N^+}}-CH_2CH_2O-\overset{\overset{O}{\|}}{C}CH_3} \xrightarrow[\text{acetylcholinesterase}]{H_2O} \underset{\text{choline}}{CH_3-\overset{\overset{CH_3}{|}}{\underset{\underset{CH_3}{|}}{N^+}}-CH_2CH_2O-H} + \underset{\text{acetate}}{{}^-O-\overset{\overset{O}{\|}}{C}CH_3}$$

Inhibition of this enzyme leads to increased concentrations of acetylcholine in the synapse and increased stimulation of the particular neurons. In response to this overload, the neurons cease to fire, leading quickly to loss of response in voluntary muscles.

Among the natural acetylcholinesterase inhibitors is a substance contained in a plant extract called curare, a preparation long used by the Amazon Indians on arrow tips to kill or paralyze. In small doses, the extract can relax muscles without causing paralysis or death. A synthetic drug, decamethonium, is considerably less toxic than curare extract and is used medicinally as a muscle relaxant.

$$CH_3-\overset{\overset{CH_3}{|}}{\underset{\underset{CH_3}{|}}{N^+}}-CH_2CH_2CH_2CH_2CH_2CH_2CH_2CH_2CH_2CH_2-\overset{\overset{CH_3}{|}}{\underset{\underset{CH_3}{|}}{N^+}}-CH_3$$

hexamethyldecamethylenediammonium ion
decamethonium

There are several other classes of synthetic compounds that affect acetylcholine-mediated nerve transmission. Among the most widely known of these are the nerve gases and related insecticides. Diisopropyl fluorophosphate (DFP) and Tabun are potent inhibitors of acetyl-cholinesterase and a few milligrams of either can kill a man in a few minutes through paralysis and respiratory failure.

diisopropyl fluorophosphate (DFP)   Tabun

Several of the water-soluble vitamins contain cyclic amines. Riboflavin (Figure 10.1) contains a fused three-ring system known as flavin. One of the nitrogen atoms of this ring system is substituted with a five carbon chain derived from the sugar ribose, hence the name riboflavin. Thiamine or vitamin B_1 (Figure 10.1) contains a substituted pyrimidine ring as well as a five-membered ring containing one atom each of nitrogen and sulfur. You have already seen the structural formula of pyridoxal phosphate in Section 9.2. This substance, derived from pyridoxine or vitamin B_6 contains a substituted pyridine ring.

Even this brief introduction to amines and amine derivatives leads to the conclusion that these substances deserve attention for their biological activity. We have seen a few examples already. You can read about such familiar amines as atropine, cocaine, Novocaine, morphine, codeine and Demerol in the mini-essay "Alkaloids." There you can also read about several psychoactive amines such as mescaline, amphetamine, and LSD. The mini-essay, "Inside the Molecules of the Mind," discusses serotonin, norepinephrine and dopamine, all of which are transmitter substances in the central nervous system. In the same mini-essay you can read about the structure and mode of action of some of the major tranquilizers and antidepressants. In subsequent chapters, we will see many more amines that play vital roles in biological chemistry.

10.2   Physical Properties

Amines are polar compounds and, except for tertiary amines, they show evidence of hydrogen bonding in the pure liquid. Amines have higher boiling points than nonpolar compounds of comparable molecular weight, but lower boiling points than alcohols and carboxylic acids. (See Table 10.1.) All classes of amines are capable of forming hydrogen bonds with water.

10.3   Basicity of Amines

Amines, though weak bases compared to sodium hydroxide and other metal hydroxides, are strong bases compared to alcohols and water. All

TABLE 10.1 Physical properties of amines.

Name	bp (°C)	Solubility (g/100 g H$_2$O)	pK$_b$	pK$_a$
ammonia	−33	90	4.74	9.26
methylamine	−7	very	3.34	10.66
ethylamine	17	very	3.20	10.80
cyclohexylamine	134	slightly	3.34	10.66
benzylamine	185	very	4.67	9.33
dimethylamine	7	very	3.27	10.73
diethylamine	55	very	3.51	10.49
triethylamine	89	1.5	2.99	11.01
aniline	184	3.7	9.37	4.63
pyridine	116	very	8.75	5.25
imidazole	257	very	7.05	6.95

amines, whether soluble or insoluble in water, will react quantitatively with mineral acids to form salts. For example, <u>cyclohexylamine</u>, which is only slightly soluble in water, dissolves readily in dilute hydrochloric acid as the cyclohexylammonium salt.

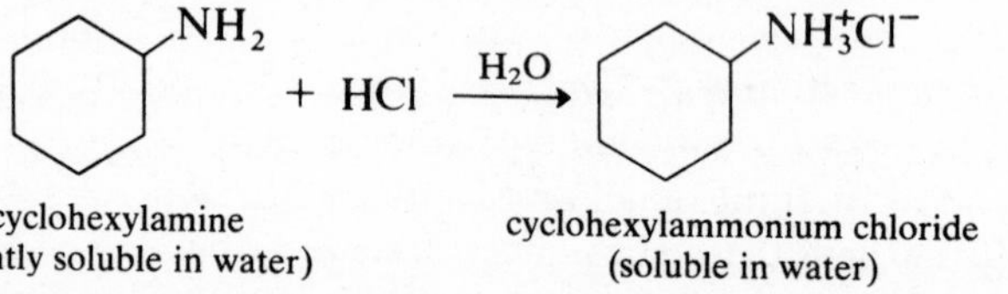

cyclohexylamine
(slightly soluble in water)

cyclohexylammonium chloride
(soluble in water)

The basicity of amines and the solubility of amine salts can be used to distinguish between amines and non-basic, water-insoluble compounds, and also to separate them. Following is a flowchart for the separation of aniline from methyl benzoate.

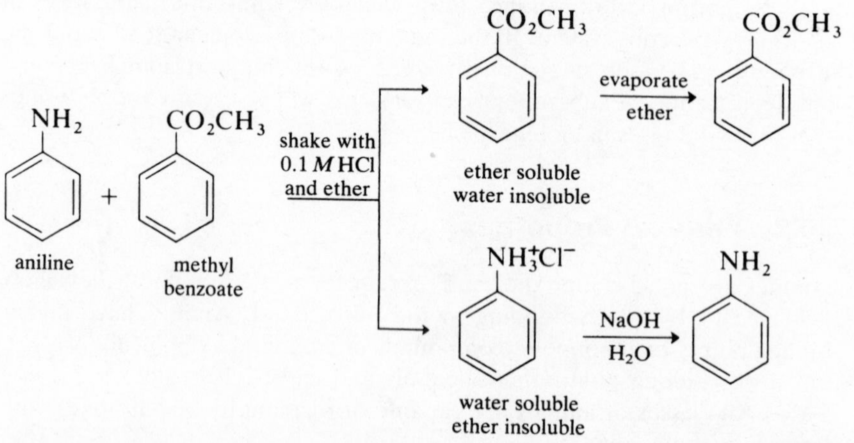

aniline

methyl benzoate

shake with 0.1 M HCl and ether

ether soluble
water insoluble

evaporate ether

water soluble
ether insoluble

NaOH
H$_2$O

Crystalline salts of amines can generally be obtained by reacting one equivalent of amine with an acid and evaporating the solution to dryness.

Amines react with water to give basic solutions. This basicity of amines is due to the presence of the unshared pair of electrons on nitrogen.

$$CH_3NH_2 + H_2O \rightleftharpoons CH_3NH_3^+ + OH^-$$

The equilibrium constant, K_b, for this reaction is given by the expression

$$K_b = \frac{[CH_3NH_3^+][OH^-]}{[CH_3NH_2]}$$

Values of pK_b for some primary, secondary, tertiary, heterocyclic, and aromatic amines are given in Table 10.1.

All aliphatic amines, whether primary, secondary, tertiary, or saturated heterocyclic amines, have about the same base strengths and do not differ appreciably in basicity from ammonia. The K_b for ethylamine is only about 25 times larger than that of ammonia.

Aromatic amines such as aniline are significantly less basic than aliphatic amines. The base dissociation constant of cyclohexylamine is larger than that of aniline by a factor of 10^6.

$$C_6H_{11}NH_2 + H_2O \rightleftharpoons C_6H_{11}NH_3^+ + OH^- \qquad K_b = 5 \times 10^{-4}$$
cyclohexyl-
amine

$$C_6H_5NH_2 + H_2O \rightleftharpoons C_6H_5NH_3^+ + OH^- \qquad K_b = 4.2 \times 10^{-10}$$
aniline

We can account for the markedly decreased basicity of aromatic amines compared to aliphatic amines by using the resonance theory in much the same manner as we did to account for the increased acidity of phenols compared to aliphatic alcohols. Recall that when comparing two similar equilibria, the position of equilibrium will be shifted toward the side favored by resonance. Both the free base and the anilinium ion are stabilized by resonance. For each, two Kekulé structures can be drawn. There are three additional contributing structures that can be drawn for the free base, and although they do involve the separation of unlike charge, they nevertheless make a contribution to the hybrid. No such structures involving the separation of unlike charge can be drawn for the anilinium ion because all electron pairs on nitrogen are used in forming sigma bonds.

aniline
(two Kekulé structures plus three with separation of unlike charge)

anilinium ion
(two Kekulé structures only)

Because of the greater resonance stabilization of aniline compared to the anilinium ion, the position of equilibrium is shifted to the left compared to that of cyclohexylamine; consequently, aniline is a weaker base than cyclohexylamine.

To extend our discussion of the basicity of the —NH$_2$ group one step further, we can also compare the basicities of cylohexylamine, aniline, and acetamide, all compounds containing the —NH$_2$ group.

$$\ddot{N}H_2$$ (cyclohexylamine) $\qquad$ $$\ddot{N}H_2$$ (aniline) $\qquad$ $$CH_3\overset{\overset{\displaystyle O}{\|}}{C}-\ddot{N}H_2$$

cyclohexylamine
$K_b = 5 \times 10^{-4}$

aniline
$K_b = 4.1 \times 10^{-10}$

acetamide
$K_b = 10^{-16}$ (approx.)

Although amines are bases (and aliphatic amines are stronger bases than aromatic amines), amides are neutral molecules. The basicity of $-NH_2$ containing molecules depends on the availability of the electron pair on nitrogen for bonding with a proton. In the case of aniline compared to cyclohexylamine, the electron pair is less available for bonding with a proton because of resonance interaction with the benzene ring. Much the same argument applies to amides. The amide group is best represented as a hybrid of two contributing structures.

$$R-C\overset{\displaystyle \ddot{O}:}{\underset{\underset{\displaystyle H}{|}}{\underset{\displaystyle \ddot{N}-H}{}}} \quad \longleftrightarrow \quad R-C\overset{\displaystyle \ddot{O}:^-}{\underset{\underset{\displaystyle H}{|}}{\underset{\displaystyle \overset{+}{N}-H}{}}}$$

While we cannot predict in advance just how important this resonance stabilization might be, the fact that amides are neutral molecules indicates that resonance stabilization is very significant.

One final note about the basicity of amines. Until recently it was common practice to list only K_b or pK_b values for amines. Now, however, it is more and more common to also list K_a and pK_a values for amines. In fact, it is only K_a and pK_a values that are used in discussing the acid-base properties of the amine group of an amino acid (Chapter 13). For this reason, Table 10.1 lists both pK_b and pK_a values for a variety of common amines.

To illustrate the difference between K_b, pK_b and K_a, pK_a for an amine, consider methylamine.

$$CH_3NH_2 + H_2O \rightleftharpoons CH_3NH_3^+ + OH^- \qquad K_b = 4.4 \times 10^{-4} \qquad pK_b = 3.34$$

$$CH_3NH_3^+ \rightleftharpoons CH_3NH_2 + H^+ \qquad K_a = 2.7 \times 10^{-11} \qquad pK_a = 10.66$$

Thus, pK_b measures directly the strength of CH_3NH_2 as a base. pK_a measures directly the strength of $CH_3NH_3^+$ as an acid. For perspective, you might compare the pK_a values for acetic acid and for methylamine.

$$CH_3CO_2H \rightleftharpoons CH_3CO_2^- + H^+ \qquad K_a = 1.8 \times 10^{-5} \qquad pK_a = 4.74$$

$$CH_3NH_3^+ \rightleftharpoons CH_3NH_2 + H^+ \qquad K_a = 2.7 \times 10^{-11} \qquad pK_a = 10.66$$

By using K_a and pK_a values for carboxylic acids and amines, we can compare them directly, for in each case we are looking at the dissociation of an acid to form a base and a proton. From a comparison of these K_a and pK_a values, it is obvious that acetic acid is a much stronger acid than methylammonium ion.

10.4 Preparation of Amines

One of the characteristic reactions of ammonia and amines is participation in nucleophilic displacement reactions, as illustrated by the synthesis of methylamine from methyl chloride and ammonia.

$$NH_3 + CH_3Cl \longrightarrow CH_3NH_3^+Cl^-$$

The reaction is a <u>nucleophilic displacement</u> of halide ion by the unshared pair of electrons on the nitrogen and belongs to the general class of S_N2 reactions. Note that the initial product of the reaction is a substituted ammonium salt. This salt is in equilibrium with other protonated species through transfer of a proton.

$$CH_3NH_3^+ + NH_3 \rightleftharpoons CH_3NH_2 + NH_4^+$$

Methylamine may undergo further alkylation to yield dimethylamine, trimethylamine, or a tetramethylammonium salt. The alkylation proceeds in stages and it is possible to maximize the yield of the primary, secondary, tertiary, or quaternary amine through control of the concentrations of the reactants and other reaction conditions.

Intramolecular alkylation to produce a pyrrolidine ring skeleton is illustrated by the following reaction, which is the terminal step in one laboratory synthesis of the alkaloid nicotine.

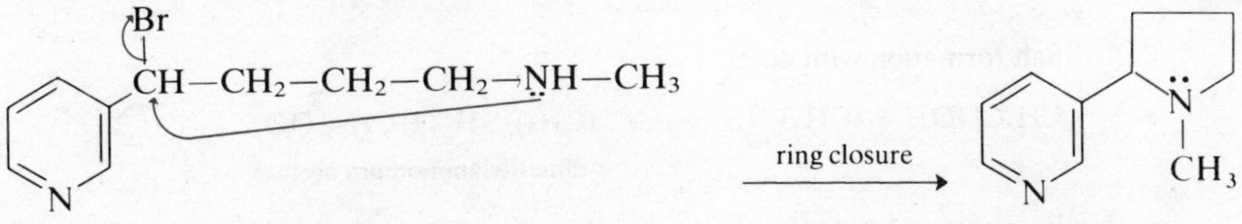

($\pm$)-nicotine

Aromatic amines are almost always made from reduction of aromatic nitro compounds. Commonly used reducing agents are iron and steam, zinc and hydrochloric acid, or hydrogen gas in the presence of a heavy metal catalyst such as platinum, palladium, or nickel.

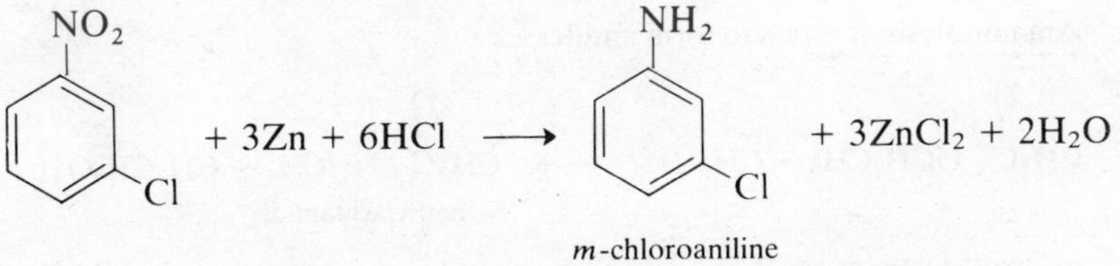

m-chloroaniline

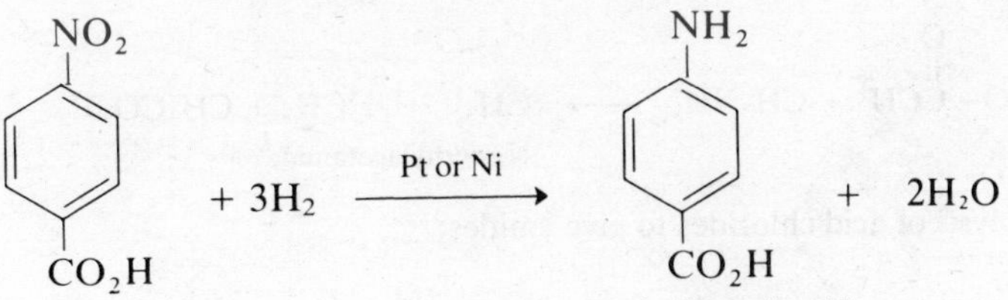

p-aminobenzoic acid

Since nitro groups can be introduced easily into an aromatic ring, nitration followed by reduction is one of the best methods for preparing aromatic amines. An especially valuable method for preparing pure primary, secondary, or tertiary aliphatic amines is the reduction of amides with lithium aluminum hydride.

$$CH_3-\underset{\underset{CH_3}{|}}{\overset{\overset{H_3C}{|}}{C}}-\overset{\overset{O}{||}}{C}-NHCH_3 \quad \xrightarrow[\text{ether}]{LiAlH_4} \quad CH_3-\underset{\underset{CH_3}{|}}{\overset{\overset{CH_3}{|}}{C}}-CH_2-NHCH_3$$

The starting amides are available from ammonolysis of esters, anhydrides, and acid halides (Chapter 9). This synthesis is particularly useful for the preparation of secondary and tertiary amines having different alkyl or aryl substituents.

10.5 Reactions of Amines

In our earlier discussions of the chemistry of the functional derivatives of carboxylic acids we have already encountered many of the more important reactions of ammonia. Amines, which resemble ammonia in having an unshared pair of electrons on the nitrogen atom, behave like ammonia in many ways: they form salts with acids, they can catalyze reactions where a basic catalyst is needed, and (except for tertiary amines) they can react with an ester or an acid anhydride to form amides.

1. Salt formation with acids:

$$CH_3COOH + (CH_3)_2NH \longrightarrow (CH_3)_2NH_2^+ + CH_3COO^-$$
dimethylammonium acetate

2. Pyrolysis of ammonium salts of carboxylic acids to give amides:

$$CH_3COO^- + (CH_3)_2NH_2^+ \xrightarrow[200°C]{heat} CH_3\overset{\overset{O}{||}}{C}-N\overset{\diagup CH_3}{\diagdown CH_3} + H_2O$$
N,N-dimethylacetamide

3. Ammonolysis of esters to form amides:

$$CH_3\overset{\overset{O}{||}}{C}-OCH_2CH_3 + CH_3NH_2 \longrightarrow CH_3\overset{\overset{O}{||}}{C}-NHCH_3 + CH_3CH_2OH$$
N-methylacetamide

4. Ammonolysis of anhydrides of carboxylic acids to give amides:

$$CH_3\overset{\overset{O}{||}}{C}-O-\overset{\overset{O}{||}}{C}CH_3 + CH_3NH_2 \longrightarrow CH_3\overset{\overset{O}{||}}{C}-NHCH_3 + CH_3COOH$$
N-methylacetamide

5. Ammonolysis of acid chlorides to give amides:

$$CH_3-\overset{\overset{O}{||}}{C}-Cl + H_2N-\bigcirc \longrightarrow CH_3\overset{\overset{O}{||}}{C}-NH-\bigcirc + HCl$$
N-phenylacetamide
acetanilide

Amines also react with derivatives of sulfuric and phosphoric acids to form amide derivatives. Following are the structural formulas of benzenesulfonamide and *p*-aminobenzenesulfonamide (more commonly known as sulfanilamide).

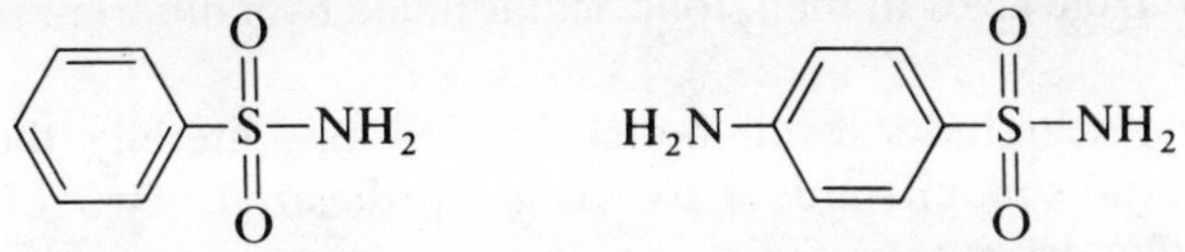

benzenesulfonamide

p-aminobenzenesulfonamide
sulfanilamide

The discovery of the medicinal use of sulfanilamide and its derivatives was a milestone in the history of chemotherapy for it represents one of the first rational investigations of synthetic organic molecules as potential drugs to fight infection. Sulfanilamide was first prepared in 1908 in Germany during an investigation of azo dyes (Section 10.6), but it was not until 1932 that its possible therapeutic value was realized. In that year the azo dye, Protonsil, was prepared and patented in Germany. In research over the next two years the German scientist G. Domagk observed that mice with streptococcal septicemia (an infection of the blood) could be treated with Protonsil. He also observed the remarkable effectiveness of Protonsil in curing experimental streptococcal and staphylococcal infections in other laboratory animals. Furthermore, Domagk discovered that Protonsil is rapidly reduced in the cell to sulfanilamide, and that it is sulfanilamide and not Protonsil which is the actual antibiotic. His discoveries were honored in 1939 by the Nobel Prize in Medicine.

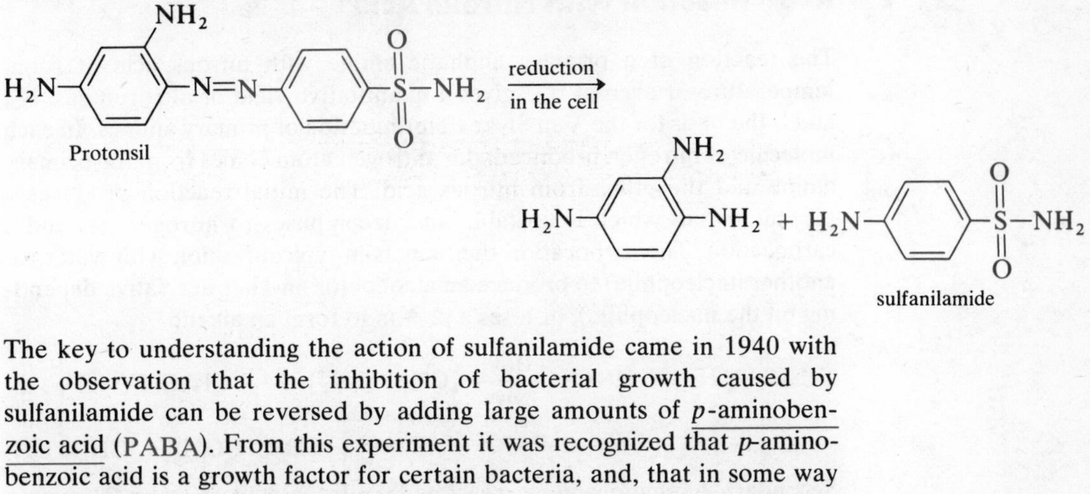

Protonsil

sulfanilamide

The key to understanding the action of sulfanilamide came in 1940 with the observation that the inhibition of bacterial growth caused by sulfanilamide can be reversed by adding large amounts of *p*-aminobenzoic acid (PABA). From this experiment it was recognized that *p*-aminobenzoic acid is a growth factor for certain bacteria, and, that in some way not then understood, sulfanilamide interferes with the bacteria's ability to use PABA.

There are obvious structural similarities between *p*-aminobenzoic acid and sulfanilamide.

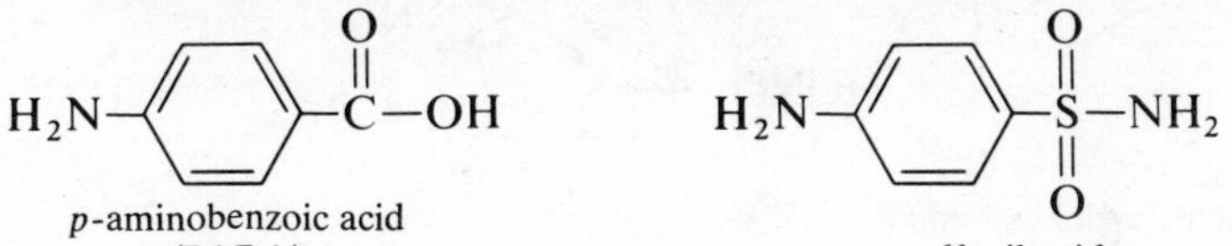

p-aminobenzoic acid
(PABA)

sulfanilamide

275

It now appears that sulfanilamide drugs inhibit one or more enzyme-catalyzed steps in the synthesis of folic acid from *p*-aminobenzoic acid. The ability of sulfanilamide to combat infections in man probably depends on the fact that man also requires folic acid but does not make it from *p*-aminobenzoic acid. In man, folic acid is made by a different biochemical pathway.

In the search for even better sulfa drugs, literally thousands of derivatives of sulfanilamide have been synthesized. Two of the more effective sulfa drugs are

sulfathiazole   sulfadiazine

Sulfa drugs were found to be effective in the treatment of tuberculosis, pneumonia, and diphtheria and helped usher in a new era in public health in the United States in the 1930s. During World War II, they were routinely sprinkled on wounds to prevent infection. These drugs were among the first of the new "wonder-drugs." As a footnote in history, the use of sulfa drugs to fight bacterial infection has been largely supplanted by an even newer wonder-drug, the penicillins. For the story of the discovery, development, and mode of action of this truly remarkable antibiotic, read the mini-essay "Penicillins."

10.6 Reaction with Nitrous Acid

The reaction of a primary aliphatic amine with nitrous acid at room temperature or even at 0°C gives a quantitative yield of nitrogen gas, N_2, and is the basis for the Van Slyke determination of primary amines. In each molecule of nitrogen produced, one nitrogen atom comes from the primary amine and the other from nitrous acid. The initial reaction produces a diazonium salt which is unstable and decomposes to nitrogen gas and a carbocation. The carbocation then reacts in typical fashion with water (or another nucleophile) to produce an alcohol (or another derivative depending on the nucleophile), or loses a proton to form an alkene.

$$CH_3CH_2NH_2 + HNO_2 \xrightarrow[\text{water}]{\text{HCl}} [CH_3CH_2N_2^+] \longrightarrow CH_3CH_2^+ + N_2$$

diazonium
ion $CH_3CH_2OH + CH_2{=}CH_2$

Secondary aliphatic amines react with nitrous acid to form N-nitroso compounds which are insoluble in aqueous acid and, therefore, separate as an oily layer. Tertiary aliphatic amines do not react with nitrous acid.

N-methylcyclohexylamine   N-nitroso-
                                       N-methylcyclohexylamine

In the case of aliphatic amines this reaction is of little value except for quantitative or qualitative analysis. Primary, secondary, and tertiary aliphatic amines are soluble in the aqueous media of this test. Evolution of nitrogen is a qualitative test for a primary amine. If the nitrogen gas is collected and measured it can be used as a quantitative test for a primary amine. Secondary amines do not liberate nitrogen but instead are converted into N-nitroso compounds which separate as an oily layer. Tertiary amines neither liberate nitrogen gas nor form N-nitroso compounds; they remain dissolved in the aqueous acid.

Primary aromatic amines react with cold nitrous acid and are converted to diazonium salts in the same way aliphatic amines are. The aromatic diazonium salts are stable at 0°C, but they too decompose with the evolution of nitrogen gas when their aqueous solutions are warmed to room temperature. In water, phenol is the organic product of the decomposition.

If the diazonium salt is warmed in the presence of cuprous chloride, cuprous bromide, or potassium iodide, an aryl halide is formed.

Thus, aromatic amines and the aromatic nitro compounds from which they can be readily synthesized provide a very convenient synthetic route to a large number of aromatic derivatives.

The diazonium ion can also attack aromatic rings in a typical electrophilic aromatic substitution reaction.

p-hydroxyazobenzene

Such reactions are often referred to as diazo coupling reactions. Notice that in this reaction the nitrogen atoms are retained. Because the diazonium ion is not a powerful electrophile, this reaction is limited in practice to coupling diazonium ions with phenols, aromatic amines, and a few other aromatic compounds.

The products of diazonium ion coupling reactions are called azo compounds. The simple azo compounds are yellow or orange; with increasing complexity and substitution on either of the aromatic rings an almost unlimited range of colors and shades can be produced. Azo compounds have found extensive use as dyes and indicators. Following is shown the structural formula of the commercial dye, congo red.

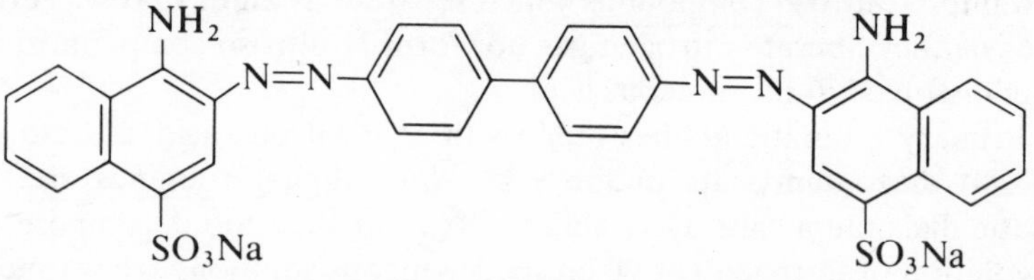

congo red

Azo compounds containing weakly basic amino groups or weakly acidic phenolic hydroxyl groups are useful as acid-base indicators. One such indicator, methyl orange, is formed by the reaction of diazotized sulfanilic acid with N,N-dimethylaniline.

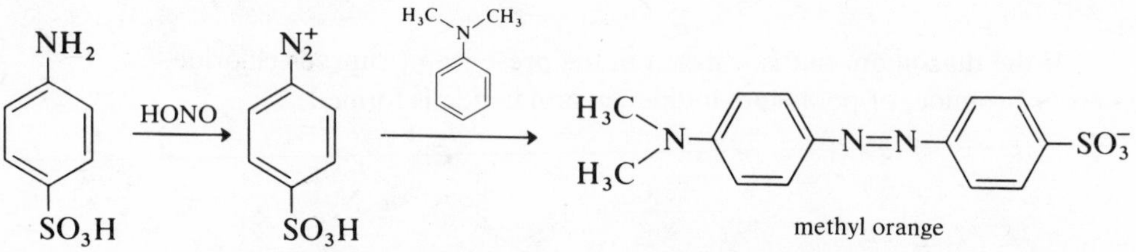

methyl orange

At pH 3.1 methyl orange is red; at pH 4.4 it is yellow-orange. Therefore this azo dye is widely used as an indicator in acid-base titrations where the end point of the titration is in the pH range 3.1 to 4.4.

10.7 Reaction with Ninhydrin

The detection of very small quantities of amino acids is facilitated by reaction with the reagent, ninhydrin. The reaction is very complex. Amino acids in general react to produce a purple-colored anion. The intensity of the color is directly proportional to the concentration of the anion, and hence may be used as a quantitative measure of the concentration of the amino acid in an unknown sample.

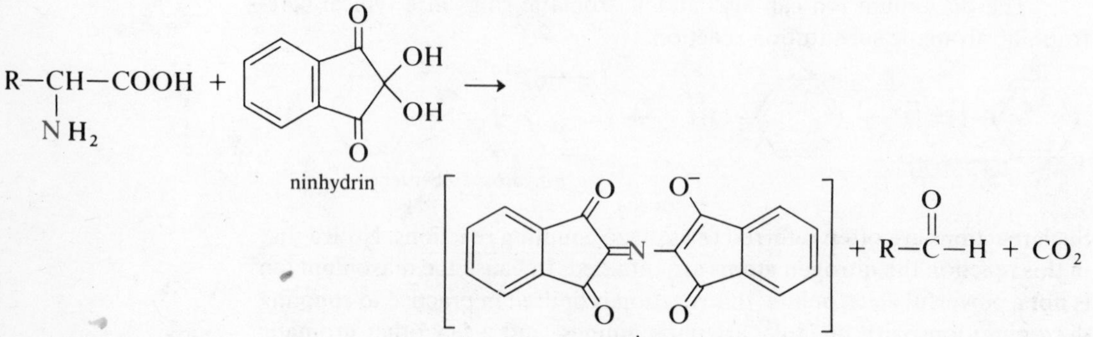

ninhydrin

anion
(purple colored)

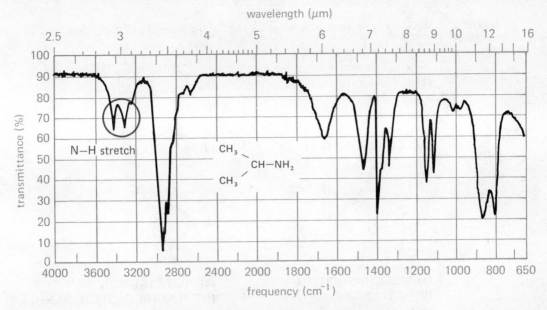

FIGURE 10.3 An infrared spectrum of isopropylamine.

10.8 Spectroscopic Properties

All primary and secondary amines show characteristic absorption between 3200 and 3500 cm^{-1} in the infrared region of the spectrum associated with stretching of the N—H bond. For primary amines, the N—H stretching frequency appears as a split peak at about 3500 cm^{-1}. Compare for example, the infrared spectrum of isopropyl amine (Figure 10.3). For secondary amines, the N—H stretching frequency appears as a single peak at about 3500 cm^{-1}. Tertiary amines lack an N—H bond and therefore show no absorption in this region.

Saturated amines, whether they be primary, secondary, or tertiary, show no absorption in the ultraviolet-visible region.

Hydrogens bound to nitrogen give NMR signals in the range $\delta = 1.0$ to 3.0. These signals are often broad and very difficult to detect. As is also the case with protons on O—H groups (Section 5.15), the signals of N—H protons of primary and secondary amines are not split by protons on adjacent carbon atoms and, therefore, generally appear as singlets.

PROBLEMS

10.1 Write structural formulas for the following compounds.

(a) diethylamine	**(b)** aniline
(c) cyclohexylamine	**(d)** pyrrole
(e) pyridine	**(f)** tetraethylammonium iodide
(g) 2-aminoethanol (ethanolamine)	**(h)** 2-aminopropanoic acid (alanine)
(i) tetramethylammonium hydroxide	**(j)** trimethylammonium benzoate
(k) *p*-methoxyaniline	**(l)** N,N-dimethylacetamide
(m) N-methylaniline	

10.2 Give an acceptable name for each of the following compounds.

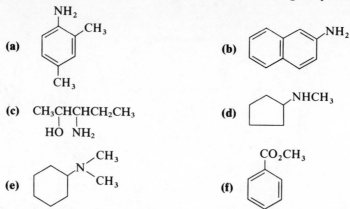

(a)

(b)

(c) CH₃CHCHCH₂CH₃
        | |
       HO   NH₂

(d)

(e)

(f)

(g) CH₃CH₂CH₂CH₂NH₂

(h) (CH₃CH₂)₂NCH₃

(i) (CH₃)₄N⁺OH⁻

(j) H₂NCH₂CH₂CH₂CH₂CH₂NH₂

10.3 Draw structural formulas for the eight isomeric amines of molecular formula $C_4H_{11}N$. Name each. Label each as primary, secondary, or tertiary.

10.4 Using the values of K_b in Table 10.1, determine if methylamine is a stronger or weaker base than triethylamine; than aniline.

10.5 Arrange the following compounds in order of increasing basicity.

(a)

(i)   (ii)   (iii)

(b)

(i)   (ii)

10.6 How would you account for the fact that pyrrole is considerably less basic than pyridine?

+ H₂O ⇌ + OH⁻ $K_b = 1.7 \times 10^{-9}$

+ H₂O ⇌ + OH⁻ $K_b = 2.5 \times 10^{-14}$

10.7 For the following compounds describe a procedure you would use to separate and to isolate each in a pure state.

CH₃CH₂CH₂CH₂CH₂CH₂OH

10.8 We have used the resonance theory to explain the fact that phenol is *more* acidic than ethanol by a factor of 10^6. We have also used the resonance theory to explain the fact that aniline is *less* basic than cyclohexylamine by a factor of 10^6. Compare the reasoning involved and explain how the substitution of a benzene ring can have such different impact on the two systems.

10.9 Consider the compounds of butylamine, $CH_3CH_2CH_2CH_2NH_2$, and 1-butanol. One of these compounds has a boiling point of 117°C, the other a boiling point of 78°C. Which of the two would you predict to have the boiling point of 117°C? Explain.

10.10 Would you predict aniline to be most soluble in water, 0.1 M NaOH, or 0.1 M HCl? Explain.

10.11 Draw structural formulas for the major products of the following reactions.

(a) $CH_3CHNH_2 + HCl \longrightarrow$
$\quad\;\;\; |$
$\quad\;\; CH_3$

(b) $(CH_3)_3N + H_2SO_4 \longrightarrow$

(c) 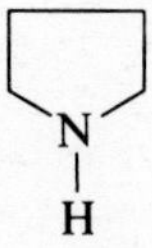 [pyrrolidine with N—H] $+ CH_3I \longrightarrow$

(d) 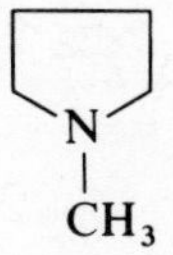 [pyrrolidine with N—CH$_3$] $+ CH_3I \longrightarrow$

(e) 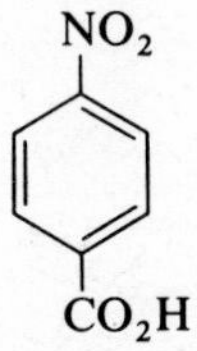 [benzene ring with NO$_2$ and CO$_2$H] $+ 3H_2 \xrightarrow{Pt}$

(f) $CH_3CHCH_2-\overset{\displaystyle H}{\underset{}{N}}-\overset{\displaystyle O}{\overset{\displaystyle ||}{C}}CH_3 + LiAlH_4 \longrightarrow$
$\quad\;\;\; |$
$\quad\;\; CH_3$

(g) $CH_3CH_2-NH-CH_2CH_3 + CH_3\overset{\displaystyle O}{\overset{\displaystyle ||}{C}}OH \longrightarrow$

(h) 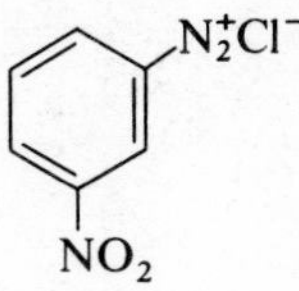 [benzene ring with $N_2^+Cl^-$ and NO$_2$] $+ Cu_2Br_2 \longrightarrow$

(i) 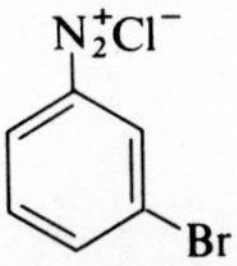 [benzene ring with $N_2^+Cl^-$ and Br] $+ H_2O \longrightarrow$

(j) 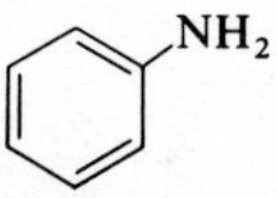 [benzene ring with NH$_2$] $+ CH_3\overset{\displaystyle O}{\overset{\displaystyle ||}{C}}Cl \longrightarrow$

10.12 Show reagents and conditions you would use to convert aniline to the following compounds.

(a) 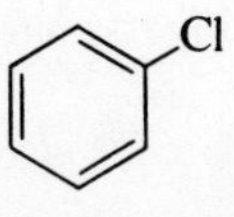 [benzene ring with Cl]
(b) 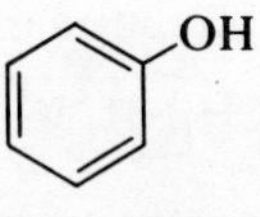 [benzene ring with OH]
(c) 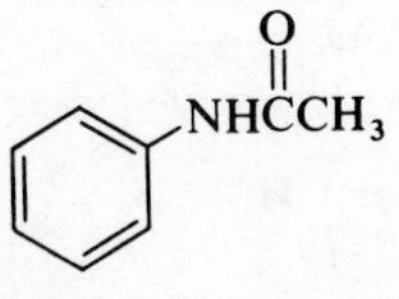 [benzene ring with $NH\overset{\displaystyle O}{\overset{\displaystyle ||}{C}}CH_3$]
(d) 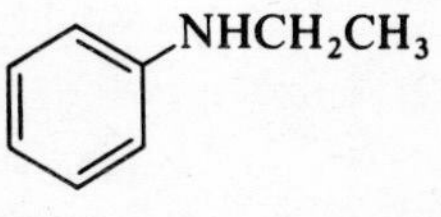 [benzene ring with $NHCH_2CH_3$]

10.13 Show reagents and conditions you would use to convert acetone to the following compounds.

(a) $\begin{array}{c} CH_3 \\ \;\;\;\;\; \diagdown \\ \;\;\;\;\;\;\;\;\; CH-OH \\ \;\;\;\; \diagup \\ CH_3 \end{array}$

(b) $\begin{array}{c} CH_3 \\ \;\;\;\;\; \diagdown \\ \;\;\;\;\;\;\;\;\; CH-NH_2 \\ \;\;\;\; \diagup \\ CH_3 \end{array}$

(c) $\begin{array}{c} CH_3 \\ \;\;\;\;\; \diagdown \\ \;\;\;\;\;\;\;\;\; CH-CO_2H \\ \;\;\;\; \diagup \\ CH_3 \end{array}$

(d) $\begin{array}{c} CH_3 \;\;\;\;\;\; O \\ \;\;\;\;\; \diagdown \;\;\;\;\;\; || \\ \;\;\;\;\;\;\;\;\; CH-CNH_2 \\ \;\;\;\; \diagup \\ CH_3 \end{array}$

10.14 Write equations to show how the following amides can be prepared by the reaction of (1) an ester and an amine, (2) an acid anhydride and an amine, and (3) an acid chloride and an amine.

(a)
$$CH_3CHCH_2CH_2\overset{\overset{\displaystyle O}{\|}}{C}-NH_2$$
$$|$$
$$CH_3$$

(b)
$$C_6H_5CH_2\overset{\overset{\displaystyle O}{\|}}{C}-NH-CH_3$$

(c) 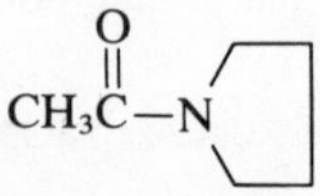
$$CH_3\overset{\overset{\displaystyle O}{\|}}{C}-N\bigg\langle$$

10.15 Write structural formulas formed when each of the amides in the previous problem is reduced with $LiAlH_4$.

10.16 Write structural formulas for an amide, which on reduction with $LiAlH_4$, will give the following amines.

(a) $CH_3CH_2CH_2CH_2NH_2$

(b) $CH_3CHCH_2NHCH_2CH_3$
$$|$$
$$CH_3$$

(c) 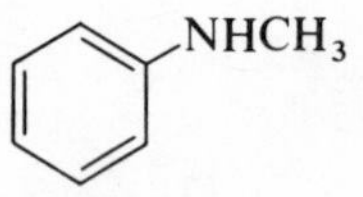
NHCH$_3$ (attached to benzene ring)

(d) 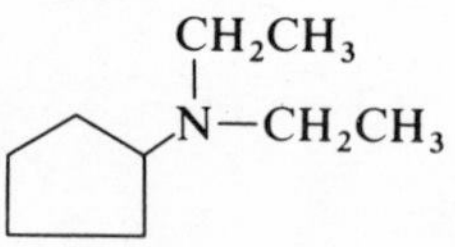
$$\text{(cyclopentane ring)}-N<\begin{array}{l}CH_2CH_3\\CH_2CH_3\end{array}$$

10.17 Show reagents and conditions you would use to carry out the following conversions.

(a) 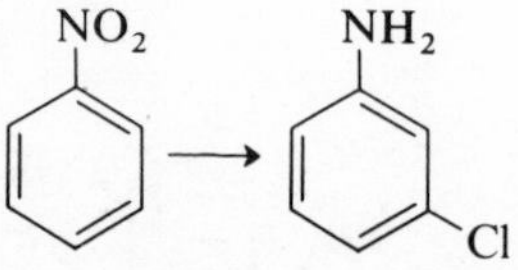 NO$_2$ (benzene) → NH$_2$ (benzene with Cl)

(b) 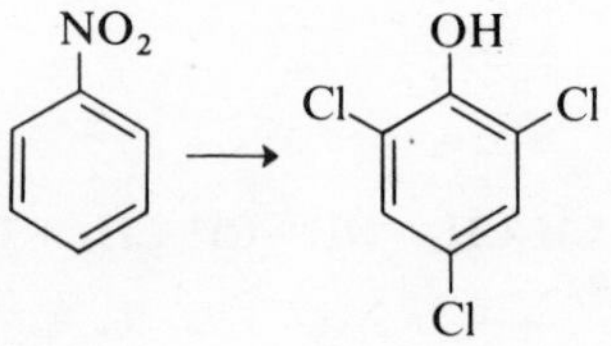 NO$_2$ (benzene) → OH (benzene with Cl, Cl, Cl)

(c) 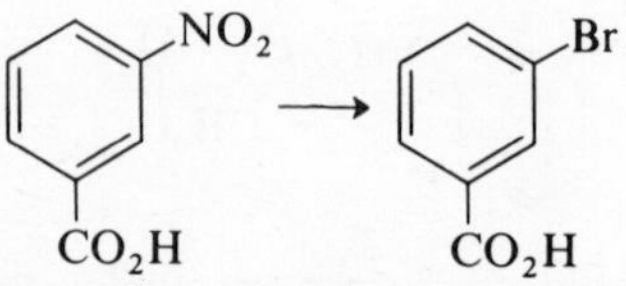 NO$_2$, CO$_2$H (benzene) → Br, CO$_2$H (benzene)

(d) 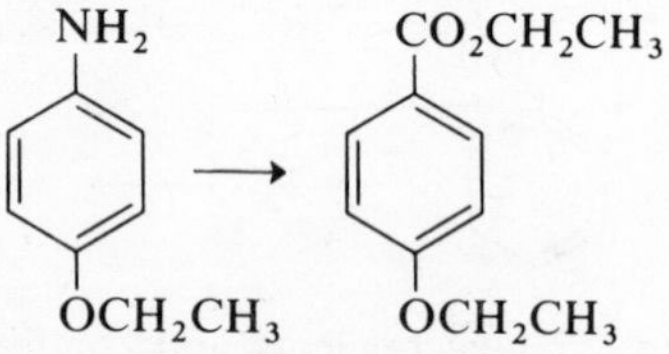 NH$_2$, OCH$_2$CH$_3$ (benzene) → CO$_2$CH$_2$CH$_3$, OCH$_2$CH$_3$ (benzene)

10.18 Describe a simple chemical test by which you could distinguish between the following pairs of compounds. In each case, state what you would do, what you would expect to observe, and write a balanced equation for all positive results.

(a) 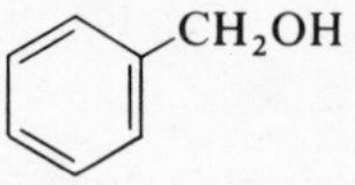 (benzene)CH$_2$OH   and   (benzene)CH$_2$NH$_2$

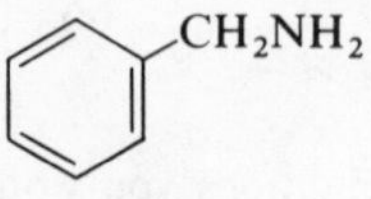

(b) 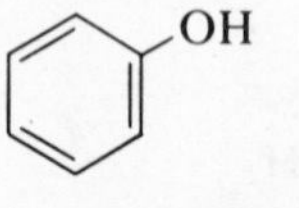 (benzene)OH   and   (benzene)NH$_2$

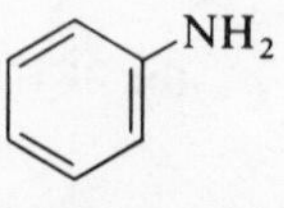

(c)   cyclohexanol and cyclohexylamine
(d)   ammonium butanoate and butanamide
(e)   trimethylacetic acid and 2,2-dimethylpropylamine
(f)   propylamine and methylethylamine
(g)   N-methylpiperidine and cyclohexylamine

282

(h) methylpropylamine and dimethylethylamine

(i) N-methylaniline and *p*-toluidine

10.19 List one major spectral characteristic that will enable you to distinguish between the following compounds.

(a) 1-aminohexane (*n*-hexylamine) and triethylamine (IR)

(b) aniline and benzamide (IR)

(c) aniline and cyclohexylamine (UV)

(d) methylpropylamine and dimethylethylamine (IR)

(e) *tert*-butylamine and dimethylethylamine (IR, NMR)

10.20 In Problem 10.3 you drew structural formulas for the eight isomeric amines of molecular formula $C_4H_{11}N$. Which of these isomeric amines has the following NMR spectra?

(a) a singlet (9H) at $\delta = 1.1$ and a singlet (2H) at $\delta = 1.3$

(b) a singlet (1H) at $\delta = 0.8$, a triplet (6H) at $\delta = 1.1$, and a quartet (4H) at $\delta = 2.6$

V

Mini-Essays

Alkaloids

Inside the Molecules of the Mind

Alkaloids

The term alkaloid was proposed early in the 19th century for a group of nitrogen-containing organic substances that could be extracted from bark, roots, leaves, berries, and fruits of plants. Although almost nothing was known at that time about the structural formulas of these substances, it was known that they all were amines and that they would react with acids to form salts. They were "alkali-like"; hence, the name alkaloid. Most of these alkaloids possessed another striking characteristic—they showed significant pharmacological activity when administered to animals. Some are extremely poisonous, yet others have important medical usages; some stimulate the central nervous system, yet others cause paralysis. In this mini-essay we will look at only a few of the more than 2500 alkaloids that have been identified so far from plants. In so doing we shall try to show something of the range of structural diversity and physiological activity that characterize alkaloids and to indicate the important role of these natural products in chemotherapy and psychopharmacology.

Atropine and cocaine are representatives of a group of alkaloids known as tropane alkaloids, so classified because of the unique ring structure common to members of this group. The tropane ring consists of a six-membered piperidine ring (atoms 1 to 6) with a two-carbon bridge fused between carbons 2 and 6.

atropine   cocaine

Atropine is a deadly poison. In dilute solutions (0.5 to 1.0%) it causes dilation of the pupil of the eye and is used in ophthalmic examinations. Cocaine is a stimulant to the central nervous system. In small doses it decreases fatigue, gives a general feeling of calm and well-being, and increases mental activity. Prolonged use of cocaine, however, produces deep depression and addiction. Cocaine was for a time used as a local anesthetic, but when its potential to cause addiction was recognized, efforts were made to develop other local anesthetics. In attempting to do this, the chemist (or biochemist, or pharmacologist) asks the questions: What is the relationship between the structure of cocaine and its physiological effect? And, is it possible to separate the anesthetic effects from the addiction?

If these questions could be answered, then it should be possible to prepare synthetic alkaloid-like drugs that incorporate only those structural elements essential for anesthetic function, and at the same time eliminate the undesirable effects. In practice, the process involves a good bit of guesswork (sometimes highly educated) about what parts of the molecule are essential for each function along with attempts to synthesize a variety of new chemicals, each incorporating one or more of the essential (?) structural features. In the case of cocaine, synthetic efforts and pharmacological testing pointed to the presence of a benzoate ester of an aminoalcohol as the feature essential for the anesthetic action of cocaine. Among the many cocaine substitutes resulting from this research effort are the widely used local anesthetics procaine and lidocaine.

procaine
(Novocaine)

lidocaine
(Xylocaine)

The commonly administered form of each is the hydrochloride salt. Procaine hydrochloride is marketed under the trade name Novocaine, lidocaine hydrochloride under the trade name Xylocaine.

Of all the alkaloids, probably the most widely known and completely studied are the so-called morphine alkaloids. Opium, obtained from the opium poppy (*Papaver somniferum*), contains about 10% morphine and 0.5% codeine.

	R	R'
morphine	H	H
codeine	CH_3	H
heroin	$\overset{O}{\overset{\|}{C}}CH_3$	$\overset{O}{\overset{\|}{C}}CH_3$

Morphine was the first alkaloid isolated in a pure state. That was in 1805. Yet is was not until 1925, fully 120 years later, that the correct structural formula of morphine was deduced. This structural formula was fully confirmed by independent laboratory synthesis of morphine in 1952. Morphine contains four fused rings, one of them aromatic, with a fifth nitrogen-containing ring fused across this skeleton. Codeine is a methylated morphine. Both the phenolic and alcoholic —OH groups of morphine are particularly important because many semisynthetic derivatives can be made by relatively simple modifications of these two functional groups. The most well known of these semisynthetic morphines is heroin or diacetyl morphine, made by treating morphine with acetic anhydride.

For the relief of severe pain of virtually every kind, morphine and its synthetic derivatives remain the most potent drugs known. In 1680, Syndenham wrote, "Among the remedies which it has pleased Almighty God to give man to relieve his sufferings, none is so universal and so efficacious as opium." Even today, morphine remains the standard against which newer analgesics are measured. The stereochemical specificity of morphine is dramatically illustrated by the observation that (+)-morphine, available by laboratory synthesis, has been found to possess none of the pharmacological properties of the naturally occurring (−)-morphine.

Morphine is indeed one of the most potent analgesics known to man. Yet it has its side effects; it can depress respiration, produce hallucinations, and lead to addiction. Here, chemists have asked the same type of question they asked about cocaine. Is it possible, using the structural features of morphine as a guide, to design a drug that possesses the desirable and powerful analgesic effects of morphine and yet is free from its undesirable side effects. Through research, chemists have developed a number of useful synthetic analgesics related to morphine, the oldest and perhaps best known of which is meperidine (Demerol). See Figure 1. Meperidine was at first thought to be free of many of the morphine-like undesirable side effects. However, it is now clearly recognized that meperidine is definitely addictive.

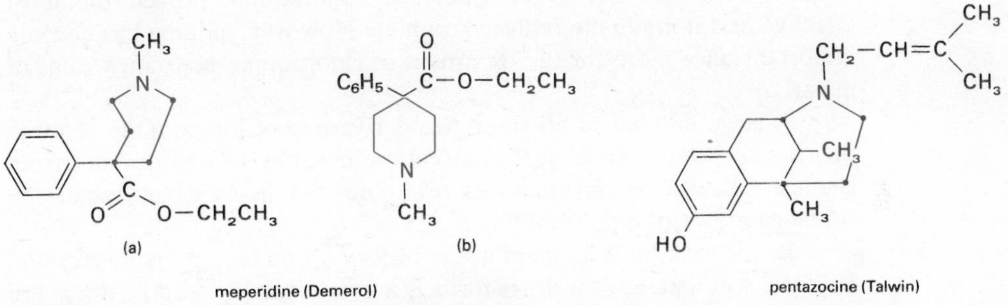

(a) meperidine (Demerol)   (b)   pentazocine (Talwin)

FIGURE 1 Meperidine (Demerol) and pentazocine (Talwin). Meperidine is drawn in (a) to emphasize its structural similarity to morphine and in (b) to emphasize the piperidine ring.

Among the newest of the morphine substitutes is pentazocine (Figure 1). Notice that it too has a striking structural similarity to morphine. Pentazocine is a highly effective analgesic for the relief of moderate to severe pain. When administered orally, it rarely produces respiratory depression and patients rarely experience withdrawal symptoms when drug treatment is stopped.

Of the alkaloids containing the quinoline or reduced quinoline ring system, the best known is undoubtedly quinine.

For years, quinine, administered as the sulfate salt, was the traditional drug for the treatment of malaria. However, uncertainties about the

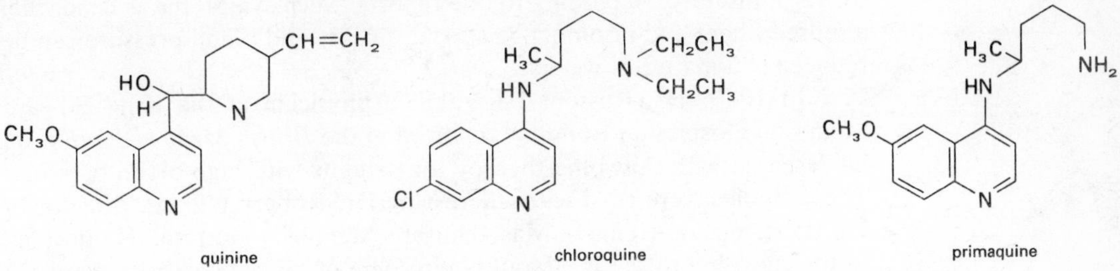

quinine   chloroquine   primaquine

FIGURE 2 Quinine, an alkaloid used for the treatment of malaria; and two synthetic antimalarials, chloroquine and primaquine.

289

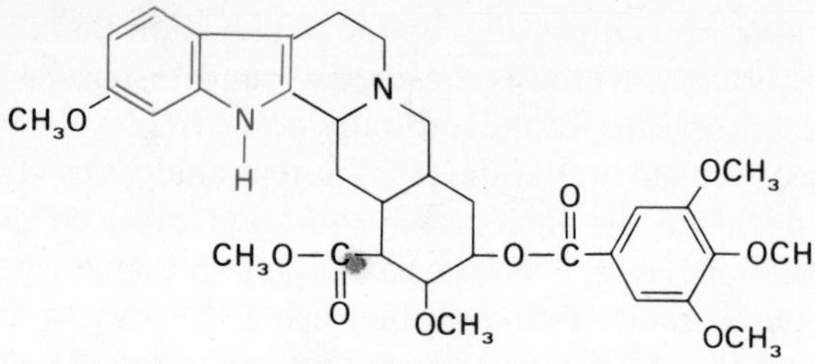

FIGURE 3 Reserpine, also commonly classed as a rauwolfia alkaloid indicating its natural occurrence. The indole ring is shown in color.

continued sources of supply of this natural product and desires for even more effective drugs generated an intensive research effort to find synthetic antimalarials. Among the most effective of those synthesized were chloroquine and primaquine, each of which bear obvious structural similarities to quinine. Of the synthetic antimalarials, chloroquine proved the most effective and it gradually replaced quinine. However, quinine has become important once more for the treatment of chloroquine-resistant strains of malaria.

A large number of alkaloids are derivatives of indole. One of these that has received a great deal of attention in the past three decades from chemists as well as physicians is reserpine, the first recognized of the so-called major tranquilizers (Figure 3).

Let us look for a moment at the history of the discovery of reserpine and its development as a drug, for it is a good example of how drugs are discovered. Reserpine is one of the many alkaloids present in the roots of the climbing shrub *Rauwolfia serpentina* and its effects are by no means new to man. For at least 3000 years, the people of India have used preparations of this root to treat a variety of afflictions: as an antidote for snake bites, as treatment for certain forms of insanity, and as a means of reducing fevers. Rauwolfia is called *sarpagandha* in Sanskrit, referring to its use as an antidote for snakebite; it is called *chandra* in Hindi meaning "moon" and referring to its calming effect on certain forms of insanity; and it is called "snakeroot" or "serpentwood" in English referring to the snakelike appearance of its roots.

The modern story of rauwolfia began in the late 1920s when Indian scientists and physicians began to study it and other botanical preparations used by native practitioners. By 1931, chemists had isolated a crystalline powder from dry rauwolfia root and physicians reported that use of this root-powder extract brought relief in cases of acute insomnia accompanied by fits of insanity. According to one report, "Such symptoms as headache, a sense of heat and insomnia disappear quickly and blood pressure can be reduced in a matter of weeks. . . ."

In 1949, Dr. Rustom Jal Vakil, a physician at the King Edward Memorial Hospital in Bombay, reported in the *British Medical Journal* on his research with rauwolfia therapy for patients with high blood pressure. These studies were read with interest by Dr. Robert Wilkins, director of the Hypertension Clinic at Massachusetts Memorial Hospital. He decided to try rauwolfia to see if it would help some of his patients who were not responding to other medications. Wilkins obtained a supply of whole-root extract tablets from India and began experimental tests of the drug. In

1952 he reported that rauwolfia is a slow-acting drug, and confirmed the reports from India of its mildly hypotensive (blood pressure lowering) effect. Further he reported that rauwolfia has a type of sedative action quite different from that of any other drug known at the time. Unlike the barbiturates and other standard sedatives, rauwolfia does not produce grogginess, stupor, or lack of coordination. Patients on rauwolfia therapy appeared to be relaxed, quiet, and tranquil. Reports such as this generated interest among psychiatrists and soon rauwolfia was recognized as an entirely new class of drug for the treatment of mental illness. It was the first of the so-called "tranquilizers." (Read the essay, "Inside the Molecules of the Mind.")

At the same time this clinical research was being conducted, chemists had undertaken the task of isolating the active substance(s) from rauwolfia. They succeeded in 1952. Because the active component of snakeroot was first isolated from *Rauwolfia serpentina*, the new drug was named reserpine.

Thus, within the span of a little more than two decades, rauwolfia had been advanced from a folk remedy to a widely used and highly effective drug for the treatment of high blood pressure and certain forms of mental illness.

Along with reserpine and reserpine-like alkaloids, another group of organic nitrogenous bases (Figure 4), either containing the indole nucleus or containing structural features suggestive of the indole nucleus, have been the subject of intense research. All of these drugs characteristically produce behavioral aberrations (hallucinations, delusions, disturbances in thinking, and changes in mood) and are classed together as psychotomimetics.

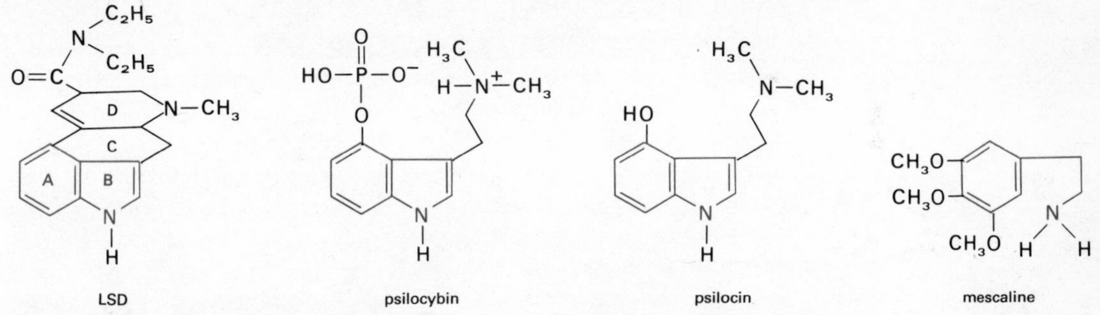

LSD   psilocybin   psilocin   mescaline

FIGURE 4 Structural similarities among several psychotomimetics. The indole ring appears in LSD, psilocybin, and psilocin. Mescaline is drawn so as to suggest the indole ring.

By far the most well known of the psychotomimetic drugs is lysergic acid diethylamide, LSD. The principal source of lysergic acid is ergot, a fungus that grows parasitically on rye. The rye grains attacked by the fungus develop into dark brown horn-shaped pegs projecting from the ripening ears. Lysergic acid and other alkaloids isolated from ergot are known collectively as ergot alkaloids.

Psilocybin is the hallucinogenic principal of *Psilocybe aztecorum*, the narcotic mushroom of the Aztecs.

Mescaline is obtained from the cactus known as peyote or mescal (*Lophophora williamsii*). The drug was named for the Mescalero Apaches of the Great Plains who developed a religious peyote rite.

In this mini-essay, we have considered only a few of the alkaloids that abound in nature. But these few should convince you of the structural diversity and wide range of physiological activities in animals of these plant substances. Furthermore, these few examples should illustrate for you how chemists, provided with clues from nature, have been able to synthesize drugs that are far more suitable for specific functions than anything so far discovered in nature.

Inside the Molecules
of the Mind

What is it that leads some people to erupt in rage over a small and seemingly insignificant incident while others remain cool and controlled in an otherwise infuriating situation? How is it that some cope with grief such as the death of a loved one while others slip into depression?

In recent years, scientists have amassed a wealth of experimental evidence showing that our emotions have deepseated and exceedingly complex biochemical roots that can be explained—if only partially and imperfectly at the moment—in terms of brain biochemistry. In this essay, we will explore a small bit of this current research.

How can we find out what goes on inside the working brain? How can we begin to unravel the mysteries of brain biochemistry and behavior? Thirty years ago this seemed a difficult if not impossible task. Then in the 1950s, the almost simultaneous introduction of two new and powerful drugs virtually revolutionized the treatment of emotional disorders and gave scientists a means to probe the very questions we have just asked.

The first of these drugs was reserpine, a substance isolated from the roots of the climbing shrub, *Rauwolfia serpentina*, and used for over 30 centuries by the people of India for the treatment of a variety of afflictions. (For the story of the development of reserpine from a folk remedy to an effective drug for the treatment of hypertension in just two decades, read the mini-essay, "Alkaloids.")

What was remarkable about reserpine was that it could suppress anxiety and modify disturbed emotional behavior in doses that were not profoundly hypnotic. Thus, reserpine was recognized as an entirely new class of drug for the treatment of mental illness and became one of the first of the so-called major tranquilizers.

reserpine

The second drug introduced in the 1950s was chlorpromazine (Thorazine), a drug whose great value in the treatment of mental illnesses was discovered in an interesting and quite unexpected manner. Pharmacologists had developed a host of drugs to block the actions of histamine, a substance within the body which appears to play an important role in many forms of allergy. Henri-Marie Laborit, a French anesthesiologist, experimented with one of these synthetic antihistamines, promethazine (Phenergan), in an attempt to find a means of preventing or reducing the incidence of shock following surgery. For this purpose, the promethazine was not especially effective. However, Laborit observed a particularly striking side effect. Patients given promethazine showed a sedation quite different from that produced by barbiturates. Laborit reported these observations and suggested this property might be helpful to psychiatrists in treating disturbed patients. Promethazine was the immediate forerunner of chlorpromazine, a drug which soon was found to be more effective than any previous treatment for the relief of some of the major symptoms of schizophrenia, especially the bizarre behavior and thinking, delusions, and hallucinations.

Chlorpromazine is but one of a class of tranquilizing drugs known as phenothiazines. They all have in common the three ring system of chlorpromazine with atoms of sulfur and nitrogen (as indicated by the letters thia and az) in the middle ring. They differ only in the nature of the side chain attached to the ring nitrogen atom and substitution of either hydrogen or another group in place of the chlorine atom.

promethazine
(Phenergan)
an antihistamine

chlorpromazine
(Thorazine)
a tranquilizer

The discovery of the tranquilizing effects of reserpine and chlorpromazine was followed within a few years by the discovery of another drug, iproniazid (Marsilid), which also had an effect on mood. Iproniazid was found to be very effective in the treatment of tuberculosis. However, in some patients, it caused excitement. Therefore, iproniazid was replaced by other drugs that were equally effective in treating tuberculosis, but without such side effects. Yet, what was an unwanted side effect in the treatment of tuberculosis was recognized as a desired effect in another medical situation.

iproniazid
(Marsilid)
an antidepressant

imipramine
(Tofranil)
a tricyclic antidepressant

amitriptyline
(Elavil)
a tricyclic antidepressant

Soon iproniazid became the basis for an effective treatment of depression. Thus, iproniazid became one of the first of the so-called antidepressants. Within a short time, an even more effective group of antidepressants was developed, the tricyclic antidepressants such as imipramine and amitriptyline.

The interest aroused by the discovery of drugs which act selectively to alter moods and behavior focused new attention on possible biochemical foundations for disturbances in behavior, and as is often the case in science, one field of research seemed to fertilize another. While the use of tranquilizers and antidepressants was being studied, neuroscientists had made important discoveries about the physical and chemical processes of nerve function. One major new concept was that of the synapse, a highly specialized junction between one nerve cell and another and through which information is carried. Even more important was the demonstration that chemical substances called neurotransmitters carry the messages across the synapse. The concept of a chemical neurotransmitter and its importance for medicine and psychiatry was recognized with the award of the 1970 Nobel prize in medicine and physiology to Julius Axelrod, Bernard Katz, and Ulf von Euler for their contribution to the understanding of acetylcholine and the catecholamines as neurotransmitters.

Out of these findings in the neurosciences, medicine, and psychiatry grew the hypothesis that if there are indeed biochemical disturbances underlying mental disturbances, then drugs which bring relief from these illnesses might do so by altering the chemical environment of the synapses. In examining how this might happen, we will focus our attention on one neurotransmitter, norepinephrine, and its possible role in mediating emotional states. Certainly, this is not the only neurotransmitter involved in the central nervous system. It may not even be the most important. Furthermore, much of the evidence for the involvement of norepinephrine in emotional states by no means excludes the operation of other transmitter substances such as dopamine and serotonin. Therefore, we will describe norepinephrine not as the one biochemical substrate for emotion, but rather as a model of how such a substance might operate in the central nervous system.

Norepinephrine, dopamine, and serotonin are transmitter substances commonly referred to as monoamines because each contains a primary amine functional group. Norepinephrine and dopamine are also referred to as catecholamines because of their structural similarity to catechol (*o*-hydroxyphenol). The main source of norepinephrine within nerve cells is biosynthesis from the amino acid tyrosine (Table 12.1). Once synthesized within the nerve cell, norepinephrine is stored in granules called synaptic or storage vesicles. These storage vesicles are found in highest

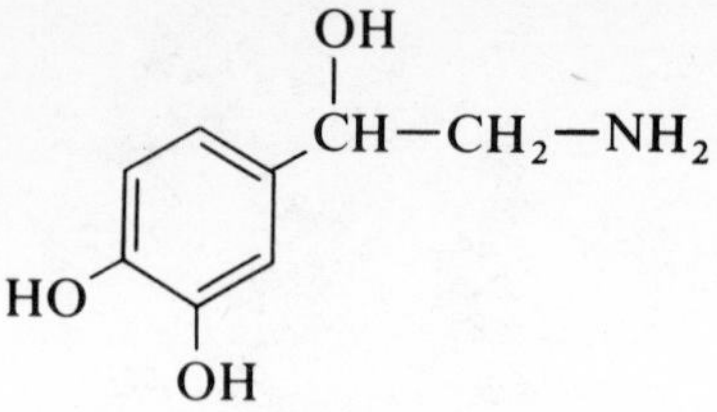

norepinephrine
(noradrenalin)

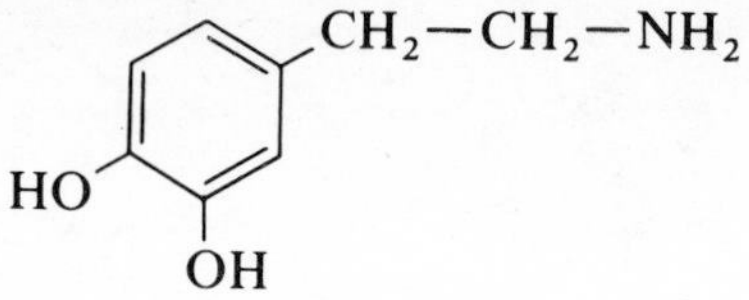

dopamine

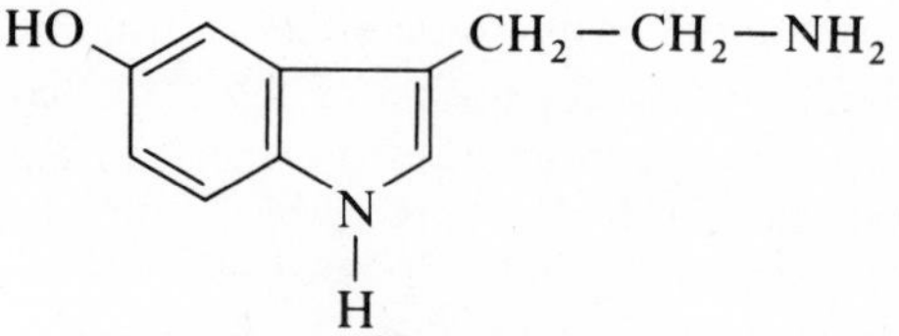

serotonin
(5-hydroxytryptamine)

concentration at the presynaptic nerve endings. Norepinephrine is effectively protected from destruction while it is bound in these vesicles. However, on release from storage vesicles into the intracellular space, either spontaneously or by the action of certain drugs, intracellular norepinephrine is susceptible to oxidation catalyzed by the enzyme monoamine oxidase (MAO). The resulting oxidation product, 3,4-dihydroxymandelic acid, leaves via the blood and is excreted (Figure 1).

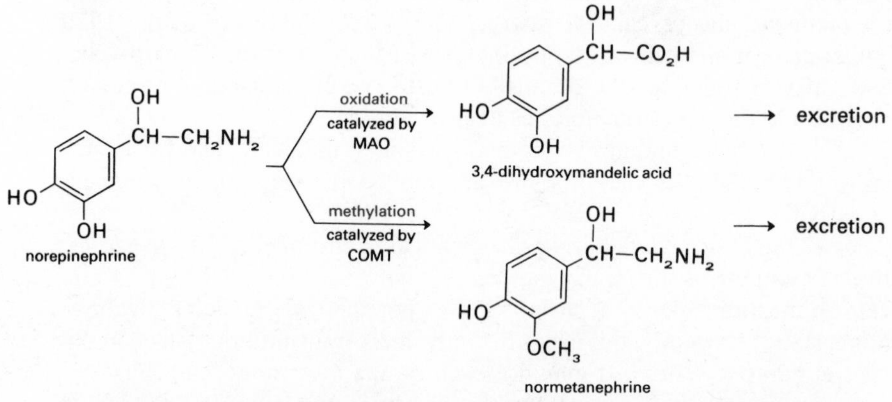

FIGURE 1 Pathways for the degradation and excretion of norepinephrine.

The role of norepinephrine in the central nervous system is that of a chemical mediator of communication. It is released from storage vesicles into the synapse in response to nerve stimulation, diffuses across the synapse, interacts with a specific receptor site on the postsynaptic membrane, and by some means only poorly understood, influences the excitability of the adjacent nerve cell. After interaction with the receptor site, norepinephrine is inactivated in either of two ways. First, a considerable percentage of it is taken back into nerve endings and stored for reuse later. Second, it is methylated in a reaction catalyzed by the enzyme catecholamine-O-methyl transferase (COMT). The methylated norepinephrine leaves via the blood stream and is excreted (Figure 1). The

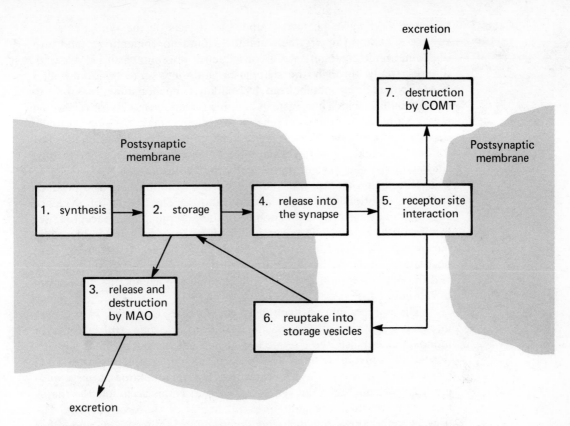

FIGURE 2 The metabolism of norepinephrine and its action as a transmitter substance.

synthesis, storage, role of norepinephrine as a transmitter substance, and its destruction and excretion are illustrated in Figure 2.

Obviously the metabolism of norepinephrine and its action as a transmitter substance represent a multiplicity of finely tuned biochemical processes. Any factors, be they psychological, physiological, or pharmacological, that disturb this dynamic balance could alter brain biochemistry. And as far as we are concerned in this essay, they could also alter our moods and emotional states.

A current hypothesis is that norepinephrine is the neurotransmitter in those parts of the brain that are intimately connected with mood and, furthermore, that all of the drugs that have a significant effect on mood have one or another effect on norepinephrine. The major tranquilizer drugs act either by depleting the supply of available norepinephrine or by reducing the effectiveness of its interaction at the specific receptor sites. Conversely, the major antidepressants act either by increasing the amount of norepinephrine released into the synapse or by enhancing its accumulation at appropriate receptor sites.

First, let us consider the tranquilizers. Recall that reserpine produces a marked tranquilization of overexcited states in man. In some individuals it even produces severe depressions of mood that are regarded by many as indistinguishable from naturally occurring depression. The effects of reserpine have been studied in animals and man and suggest that the drug, by some mechanism not as yet understood, acts on the synaptic vesicle storage sites and impairs their ability to bind norepinephrine (and also dopamine and serotonin). As a consequence norepinephrine is released, diffuses freely through the cytoplasm and onto mitochondrial-bound MAO, where it is metabolized by oxidative deamination to 3,4-dihydroxymandelic acid. Thus, instead of being stored in vesicles for release on nerve stimulation, the norepinephrine supply may become virtually depleted. Even though there is continuing synthesis, the depletion may last for days or weeks, probably because reserpine attaches itself to the storage sites and impairs uptake of newly synthesized transmitter.

Recall that chlorpromazine and other phenothiazines also produce sedation, emotional quieting, and relaxation. The weight of evidence suggests that their principal action is on the norepinephrine receptor sites in susceptible neurons. They combine with receptor sites and render them incapable of responding to norepinephrine. The result is similar to a decreased supply of the transmitter substance. Although both reserpine and the phenothiazines are powerful tranquilizing agents, the mechanism of action of each is quite different.

The norepinephrine hypothesis was further reinforced when it was discovered that the antidepressant, iproniazid, is a powerful inhibitor of monoamine oxidase, the enzyme responsible for the intracellular destruction of norepinephrine and other brain monoamines. Inhibition of MAO permits an increase in the concentration of norepinephrine at nerve endings and presumably at the synapse as well. It is this accumulation that is thought to produce the antidepressant action of the drug in man. The tricyclic antidepressants including imipramine, amitriptyline and other closely related compounds are the drugs most widely used today for the treatment of depression because they are safer to use than MAO inhibitors and are more acceptable to patients. The mechanism by which these drugs counteract depression in humans is not known. However, it appears that they promote accumulation of norepinephrine at receptor sites by inhibiting the reuptake of intercellular norepinephrine. Thus, they artificially increase the concentration of norepinephrine at the receptor sites and thereby potentiate its action.

We should also mention one additional class of drugs which has been useful over the years in treating a variety of psychological and behavioral disorders, namely, the amphetamines. The structural formula of amphetamine and two of its derivatives are

$$C_6H_5-CH_2-\overset{\overset{\textstyle CH_3}{|}}{CH}-NH_2$$
amphetamine

$$C_6H_5-CH_2-\overset{\overset{\textstyle CH_3}{|}}{CH}-NH-CH_3$$
metamphetamine

methylphenidate
(Ritalin)

It now seems clear that many of the physiological and behavioral effects of amphetamine and related compounds are the result of the drug's actions on brain and peripheral catecholamine-containing nerve cells.

First, amphetamine appears to block the reuptake of norepinephrine (and possibly dopamine as well) into storage vesicles for later reuse. Second, it stimulates the release of norepinephrine from storage sites. Both of these actions result in an increased amount of transmitter substance and increased synaptic activity.

MINI-ESSAY
Inside
the Molecules
of the Mind

The proposed relations of these psychoactive drugs to biogenic amine activity are summarized in Table 1.

TABLE 1 Proposed relationships between psychoactive drugs and biogenic amine activity.

Tranquilizers: drugs which inhibit synaptic activity
    a. inhibit vesicle storage: reserpine
    b. block receptor sites: chlorpromazine and related phenothiazines

Antidepressants: drugs which enhance synaptic activity
    a. inhibit monoamine oxidase: iproniazid, amphetamine
    b. inhibit amine reuptake: tricyclic antidepressants such as imipramine and amitriptyline
    c. amine releasers: amphetamine

In summary, a significant body of pharmacological evidence is at least consistent with the hypothesis that the effects of the major tranquilizers and antidepressants are related to their effects on norepinephrine and that norepinephrine plays an important role in mental and behavioral states. But by no means does this evidence prove the hypothesis. It is important to bear in mind that no drug produces just one effect. Each of the drugs we have presented has the ascribed effect (for example, reserpine depletes norepinephrine stores) and a host of side effects as well. The problem in the use of drugs as probes in the study of behavior is to separate and identify the various effects of the drug. Finally, we must be wary of oversimplification in a subject as complex and multifaceted as emotional behavior. It is unlikely that norepinephrine, or any other single amine for that matter, is entirely responsible for a specific emotional state. Rather it is more likely that the important factors are the interaction of certain amines at particular sites in the central nervous system together with a host of environmental and psychological determinants of emotion. The unravelling of these factors promises man a wealth of information and insight into the relationships between brain and mind.

REFERENCES

Axelrod, J. "Neurotransmitters," *Scientific American*, vol. 230 (1974), pp. 58–71.

Cooper, J., Bloom, F., and Roth, R. *The Biochemical Basis of Neuropharmacology*, 2nd ed. Oxford University Press, New York, 1974.

Katz, Bernard. *Nerve, Muscle, and Synapse*. McGraw-Hill Book Company, New York, 1966.

McGeer, Patrick L. "The Chemistry of Mind," *American Scientist*, vol. 59, 221 (1971).

Quarton, G. C., Melnechuk, T., and Schmitt, T. O., eds, *The Neurosciences—A Study Program*. The Rockefeller University Press, New York, 1967.

Schildkraut, J. J., and Kety, S. S. "Biogenic Amines and Emotion," *Science*, vol. 156, 21 (1967).

Schildkraut, J. J. "Neuropharmacology of the Affective Disorders," *Annual Review of Pharmacology*, vol. 13, 427 (1973).

Carbohydrates

11.1 Introduction

Carbohydrates are polyhydroxyaldehydes or polyhydroxyketones or substances which on hydrolysis give polyhydroxyaldehydes and/or polyhydroxyketones.

These substances are among the most abundant constituents of the plant and animal world and serve many vital functions: as a storehouse of chemical energy (glucose, starch, glycogen); as supportive structural components in plants (cellulose); as essential components in the mechanisms of genetic control of development and growth of living cells (ribose and deoxyribose).

The name carbohydrate is an old one, and was derived from the early observations that many members of this class have the empirical formula $C_n(H_2O)_m$ and were termed hydrates of carbon. For example, grape sugar (glucose) has the molecular formula $C_6H_{12}O_6$. Cane sugar (sucrose) has the formula $C_{12}H_{22}O_{11}$. Starch and cellulose are very large molecules of variable molecular weight having the empirical formula $(C_6H_{10}O_5)_x$, where x may be as large as several hundred thousand. It soon became clear that not all compounds having the properties of carbohydrates have this general formula. For example, now we know of carbohydrates that contain nitrogen in addition to the elements of carbon, hydrogen and oxygen. Although the term *carbohydrate* is not fully descriptive, it is firmly rooted in the chemical nomenclature and has persisted as the name of this class of compounds.

Carbohydrates are often referred to as saccharides because of the sweet taste of the simpler members of the family, the sugars (Latin, *saccharum*, "sugar").

The chemistry of the carbohydrates is essentially the chemistry of two functional groups, the hydroxyl group and the carbonyl group. Furthermore, we encounter derivatives of these two functional groups in the form of hemiacetals, acetals, esters, and so on. However, the fact that there are only two functional groups in these substances belies the complexity and challenge of carbohydrate chemistry. All but the most simple carbohydrates contain multiple chiral centers, and it has been a challenge to the chemist to develop techniques to study the stereochemistry of these substances. In addition, all carbohydrates contain multiple hydroxyl groups and accordingly multiple reactive sites. It has been a challenge to develop techniques for reaction at selected functional groups within the molecule.

11.2 Monosaccharides

Monosaccharides, or the simple sugars, are carbohydrates having the general formula $C_nH_{2n}O_n$, where n varies from three to eight. They are grouped together according to the number of carbons they contain.

$C_3H_6O_3$	triose	$C_6H_{12}O_6$	hexose
$C_4H_8O_4$	tetrose	$C_7H_{14}O_7$	heptose
$C_5H_{10}O_5$	pentose	$C_8H_{16}O_8$	octose

While examples of heptoses and octoses do occur in nature, they are far less abundant than the smaller monosaccharides.

Glyceraldehyde (I) and dihydroxyacetone (II) are the smallest molecules termed carbohydrates and are the only possible trioses.

$$
\begin{array}{cc}
\text{CHO} & \text{CH}_2\text{OH} \\
| & | \\
\text{H---C---OH} & \text{C=O} \\
| & | \\
\text{CH}_2\text{OH} & \text{CH}_2\text{OH} \\
\text{glyceraldehyde} & \text{dihydroxyacetone} \\
\text{I} & \text{II}
\end{array}
$$

Glyceraldehyde has one chiral carbon atom and, therefore, can exist as a pair of enantiomers, Ia and Ib:

$$
\begin{array}{cc}
\text{CHO} & \text{CHO} \\
| & | \\
\text{H---C---OH} & \text{HO---C---H} \\
| & | \\
\text{CH}_2\text{OH} & \text{CH}_2\text{OH} \\
\text{Ia} & \text{Ib}
\end{array}
$$

Because the representations of three-dimensional structures are often cumbersome, a convention for picturing the molecules in the plane of the paper has been adopted. It is the so-called Fischer projection method. According to this convention, the carbon chain is written vertically on the page with carbon 1 toward the top of the page. Horizontal groups (those written to the left and right of any chiral carbon atom) are understood to project above the plane of the paper; vertical groups (those written above and below any chiral carbon atom) are understood to project below the plane of the paper. Applying the Fischer projection rules to the two three-dimensional representations of glyceraldehyde gives structures Ia' and Ib'.

$$
\begin{array}{ccc}
\text{CHO} & \qquad & \text{CHO} \\
| & & | \\
\text{H---C---OH} & & \text{HO---C---H} \\
| & & | \\
\text{CH}_2\text{OH} & & \text{CH}_2\text{OH} \\
\text{Ia'} & & \text{Ib'}
\end{array}
$$

Structures Ia and Ib are mirror images and show identical reactivities except in the presence of reagents which themselves are asymmetric. One of the enantiomers has a specific rotation of $+20.9°$ while the other has a specific rotation of $-20.9°$. The question is, which of the two structures corresponds to the enantiomer having the specific rotation of $+20.9°$? Is Ia dextrorotatory and Ib levorotatory, or is the situation reversed? The answer is that Ia is the structure of the enantiomer that is dextrorotatory. This assignment was made quite arbitrarily by the German chemist Emil Fischer in 1891. (Of course, he had a $50:50$ chance of being correct in this assignment.) In 1952 his choice was proved to be the correct one by a Dutch scientist using a special technique of X-ray analysis. By convention, structure Ia is called D -glyceraldehyde where the D refers to the configuration at the chiral carbon. Structure Ib is designated as L -glyceraldehyde. The configuration of D-glyceraldehyde serves as a reference point for the assignment of configuration to other carbohydrates.

The Fischer projection formulas were the first, and until recently the only, scheme for designating absolute configuration. While this convention has proved extremely useful in carbohydrate and also amino acid chemistry, it has not been widely used in other areas of organic chemistry. A newer and much more general set of rules for specifying absolute configuration about a chiral carbon atom was proposed by Cahn, Ingold, and Prelog (Section 4.9). According to these rules, D-glyceraldehyde becomes R-glyceraldehyde and L-glyceraldehyde becomes S-glyceraldehyde. While the newer R,S convention is more general in its applications, the older D,L convention has persisted and most certainly will continue to be used in carbohydrate chemistry; for this reason we shall use it in this chapter and throughout the remainder of the text whenever we discuss carbohydrate chemistry.

D-Glyceraldehyde is the parent member of the aldose (ald ehyde + -ose) family of monosaccharides. Aldoses contain one or more hydroxymethylene (—CHOH) groups between the aldehyde carbonyl and the terminal —CH$_2$OH group.

Dihydroxyacetone is the parent member of the ketose (ket one + -ose) family of monosaccharides. General formulas for D-aldoses and D-ketoses are

$$
\begin{array}{ccc}
& & \text{CH}_2\text{OH} \\
& & | \\
\text{CHO} & & \text{C=O} \\
| & & | \\
\text{(CHOH)}_n & & \text{(CHOH)}_n \\
| & & | \\
\text{H---C---OH} & & \text{H---C---OH} \\
| & & | \\
\text{CH}_2\text{OH} & & \text{CH}_2\text{OH} \\
\text{D-aldose} & & \text{D-ketose}
\end{array}
$$

Tables 11.1 and 11.2 show the names and structural formulas for all trioses, tetroses, pentoses, and hexoses of the D-series.

TABLE 11.1 The isomeric D-ketopentoses and D-ketohexoses derived from di-hydroxyacetone and D-erythrulose.

```
              CH₂OH
              │
              CO
              │
              CH₂OH
           dihydroxyacetone

              CH₂OH
              │
              CO
              │
         H—C—OH
              │
              CH₂OH
           D-erythrulose

       CH₂OH              CH₂OH
       │                  │
       CO                 CO
       │                  │
  H—C—OH             HO—C—H
       │                  │
  H—C—OH             H—C—OH
       │                  │
       CH₂OH              CH₂OH
     D-ribulose          D-xylulose


  CH₂OH        CH₂OH         CH₂OH        CH₂OH
  │            │             │            │
  CO           CO            CO           CO
  │            │             │            │
 H—C—OH      HO—C—H        H—C—OH       HO—C—H
  │            │             │            │
 H—C—OH      H—C—OH        HO—C—H       HO—C—H
  │            │             │            │
 H—C—OH      H—C—OH        H—C—OH       H—C—OH
  │            │             │            │
  CH₂OH        CH₂OH         CH₂OH        CH₂OH
 D-psicose    D-fructose     D-sorbose    D-tagatose
```

D-Ribose and 2-deoxy-D-ribose are the most important pentoses.

```
     CHO              CHO
     │                │
  H—C—OH           H—C—H
     │                │
  H—C—OH           H—C—OH
     │                │
  H—C—OH           H—C—OH
     │                │
     CH₂OH            CH₂OH
   D-ribose        2-deoxy-D-ribose
```

These pentoses are present as intermediates in metabolic pathways and are important building blocks in nucleic acids, D-ribose in ribonucleic acid (RNA) and 2-deoxy-D-ribose in deoxyribonucleic acid (DNA).

Hexoses are very common in nature and play major physiological roles. Glucose is by far the most common hexose monosaccharide. It is known as dextrose because it is optically active and dextrorotatory. It is also referred to as grape sugar, blood sugar, and corn sugar—names that

TABLE 11.2 The isomeric D-aldotetroses, D-aldopentoses, and D-aldohexoses derived from D-glyceraldehyde.

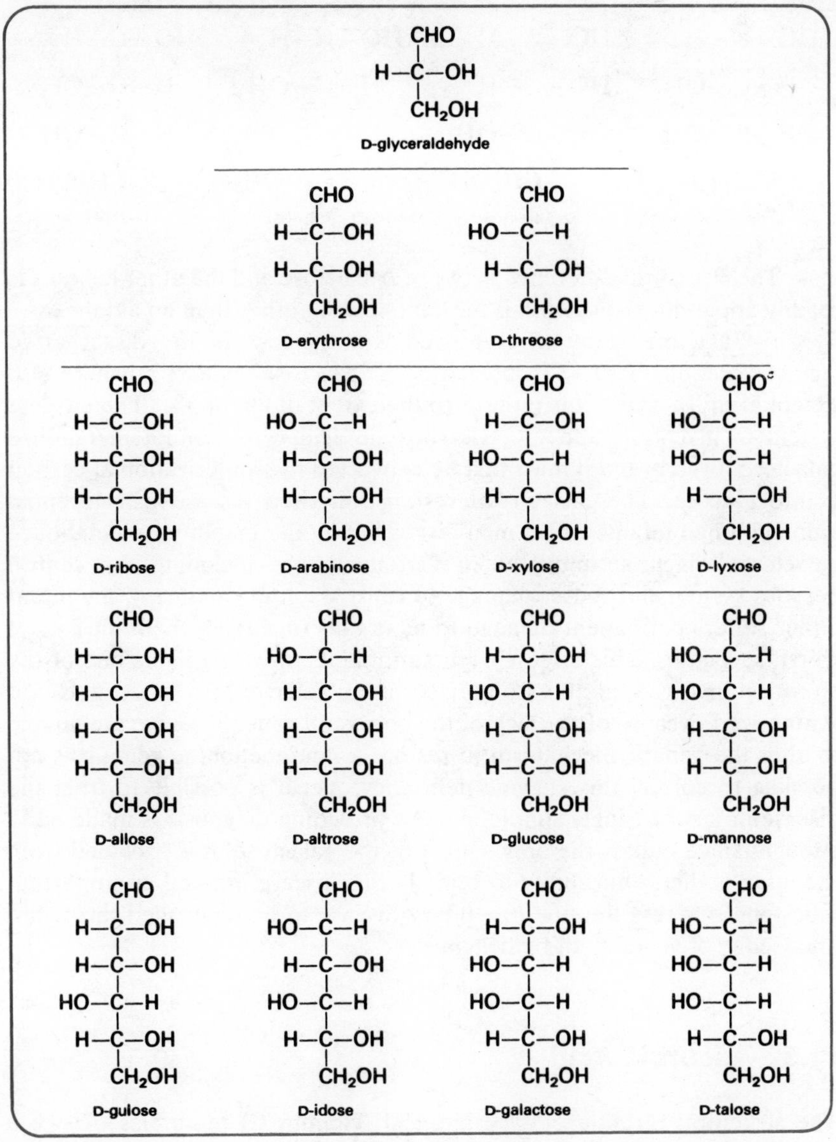

SECTION 11.2
Monosaccharides

CHO
|
H—C—OH
|
CH₂OH

D-glyceraldehyde

CHO	CHO
H—C—OH	HO—C—H
H—C—OH	H—C—OH
CH₂OH	CH₂OH
D-erythrose	D-threose

CHO	CHO	CHO	CHO
H—C—OH	HO—C—H	H—C—OH	HO—C—H
H—C—OH	H—C—OH	HO—C—H	HO—C—H
H—C—OH	H—C—OH	H—C—OH	H—C—OH
CH₂OH	CH₂OH	CH₂OH	CH₂OH
D-ribose	D-arabinose	D-xylose	D-lyxose

CHO	CHO	CHO	CHO
H—C—OH	HO—C—H	H—C—OH	HO—C—H
H—C—OH	H—C—OH	HO—C—H	HO—C—H
H—C—OH	H—C—OH	H—C—OH	H—C—OH
H—C—OH	H—C—OH	H—C—OH	H—C—OH
CH₂OH	CH₂OH	CH₂OH	CH₂OH
D-allose	D-altrose	D-glucose	D-mannose

CHO	CHO	CHO	CHO
H—C—OH	HO—C—H	H—C—OH	HO—C—H
H—C—OH	H—C—OH	HO—C—H	HO—C—H
HO—C—H	HO—C—H	HO—C—H	HO—C—H
H—C—OH	H—C—OH	H—C—OH	H—C—OH
CH₂OH	CH₂OH	CH₂OH	CH₂OH
D-gulose	D-idose	D-galactose	D-talose

clearly indicate its source in an uncombined state. Glucose is the carbo-hydrate used by cells; mannose, galactose and fructose must first be con-verted into glucose for utilization as an energy source. Glucose is stored in the liver as the polysaccharide glycogen (Section 11.9). The blood normally contains 65 to 110 mg (0.06 to 0.1%) of glucose per 100 ml, but in diabetics the level may be much higher.

Inspection of the structural formula of D-glucose reveals that the molecule contains four chiral carbons. D-Glucose and its mirror image L-glucose represent two of the sixteen ($2^4 = 16$) possible stereoisomers of this structure. Of these sixteen, only D-glucose, D-galactose, and D-mannose are widely distributed in nature. The other thirteen isomers have been prepared in the laboratory and characterized.

```
      CHO              CHO              CHO             CH₂OH
       |                |                |                |
  H—C—OH           H—C—OH          HO—C—H             C=O
       |                |                |                |
  HO—C—H           HO—C—H          HO—C—H          HO—C—H
       |                |                |                |
  H—C—OH           HO—C—H           H—C—OH           H—C—OH
       |                |                |                |
  H—C—OH            H—C—OH           H—C—OH           H—C—OH
       |                |                |                |
    CH₂OH            CH₂OH            CH₂OH            CH₂OH
   D-glucose        D-galactose      D-mannose        D-fructose
```

The structural difference between D-fructose and the other hexoses is readily apparent. D-Fructose is a 2-ketohexose rather than an aldohexose.

D-Galactose is found combined with glucose in the disaccharide lactose (Section 11.8). It is interesting that lactose appears nowhere else except in milk, where it is present to the extent of about 5%. The enzyme system in glycolysis and the tricarboxylic acid cycle cannot metabolize galactose directly and it must first be converted by isomerization at carbon 4 into glucose. There is an interesting inherited disease galactosemia among human infants which manifests itself by the inability to metabolize galactose, which accumulates in various tissues including the central nervous system and causes damage to cells. Without treatment, the infant either suffers permanent damage to its organs (including the brain), or at worst, succumbs. This disease occurs in infants in whose livers one of the enzymes involved in the isomerization of galactose to glucose is not synthesized because of the lack of the necessary gene. Since we are unable to alter the genetic mechanism to restore a nonfunctioning gene, it is not possible to correct this enzyme deficiency. Yet it is possible to treat the disease in a surprisingly simple manner providing diagnosis is made early enough; since milk is the only source of the galactose, it is excluded from the infant's diet. Our ability to remedy this disease marked an important milestone, for it is the first hereditary disease to be controlled through a knowledge of genetics and biochemistry.

11.3 Ascorbic Acid

The structural formula of ascorbic acid (vitamin C) resembles that of a monosaccharide, and in fact this vitamin is synthesized biochemically from the D-glucose. Enzyme-catalyzed oxidation of D-glucose by NAD^+ gives D-glucuronic acid. This four-electron oxidation requires two moles of NAD^+ per mole of D-glucose. A selective enzyme-catalyzed reduction of the —CHO group of D-glucuronic acid gives L-gulonic acid. The carbon atom 1 of what was D-glucose is now —CH_2OH, and carbon atom 6 of D-glucose is now —CO_2H. According to the Fischer convention, the molecule should be renumbered so that the —CO_2H group is carbon atom 1. Therefore, the Fischer projection formula is turned 180° in the plane of the paper so that the —CO_2H group is uppermost and appears as carbon 1. This molecule, L-gulonic acid, in turn forms a cyclic ester (a lactone), and is oxidized by molecular oxygen in an enzyme-catalyzed reaction to L-ascorbic acid. Man and other primates, and guinea pigs, lack the enzyme system necessary to convert L-gulonic acid to L-ascorbic acid and, therefore, are unable to synthesize ascorbic acid. For humans, ascorbic acid is a necessary dietary supplement.

The biochemical function of ascorbic acid is still not completely understood. It probably functions as a biological oxidation–reduction reagent. Whatever the biological roles of ascorbic acid may be, its most clearly established function is in the maintenance of <u>collagen</u> (Section 13.13), an intercellular material of bone, dentine, cartilage, and connective tissue.

Severe deficiency of ascorbic acid produces <u>scurvy</u>. A most vivid description of this vitamin deficiency disease was given by Jacques Cartier in 1536 when it afflicted his men during the exploration of the Saint Lawrence River.

> Some did lose all their strength, and could not stand on their feet.... Others also had all their skin spotted with spots of blood of a purple color; then did it ascend up to their ankles, knees, thighs, shoulders, arms, and necks. Their mouths became stinking, their gums so rotten that all of the flesh did fall off, even to the roots of the teeth, which did also almost fall out.

This is scurvy! Green or fresh vegetables and ripe fruits are the best remedies and the best means of preventing the disease. (These foods are of course sources of vitamin C.) This was recognized by the English physician James Lind, who in 1753 urged the inclusion of lemon or lime juice in the diet of British sailors, hence the nickname limeys.

307

There is good evidence that severe ascorbic acid deficiency as seen in scurvy is the result of impaired synthesis of collagen, the protein material of bone and connective tissue. Collagen synthesized in the absence of ascorbic acid cannot properly form fibers, thereby resulting in such typical symptoms as fragility of blood vessels and hemorrhaging, loosening of the teeth, poor wound healing, and so on. We will discuss the structure and function of collagen more fully in Chapter 13.

11.4   Amino Sugars

Amino sugars are formed through the substitution of an hydroxyl group by an amino group on a monosaccharide. There are three naturally occurring amino sugars, D-glucosamine, D-mannosamine, and D-galactosamine. Of these, D-glucosamine is the most common.

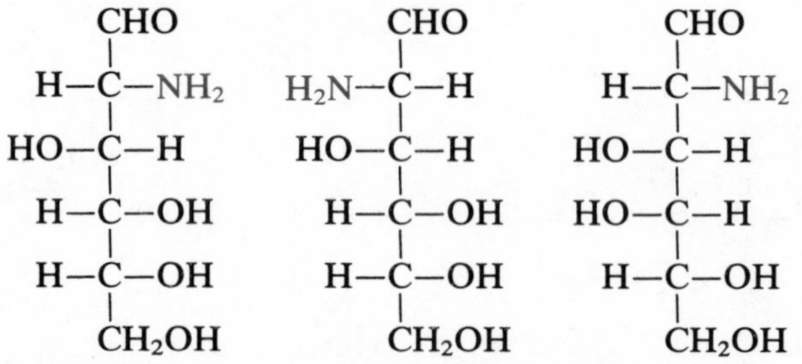

$$
\begin{array}{ccc}
\text{CHO} & \text{CHO} & \text{CHO} \\
\text{H}-\text{C}-\text{NH}_2 & \text{H}_2\text{N}-\text{C}-\text{H} & \text{H}-\text{C}-\text{NH}_2 \\
\text{HO}-\text{C}-\text{H} & \text{HO}-\text{C}-\text{H} & \text{HO}-\text{C}-\text{H} \\
\text{H}-\text{C}-\text{OH} & \text{H}-\text{C}-\text{OH} & \text{HO}-\text{C}-\text{H} \\
\text{H}-\text{C}-\text{OH} & \text{H}-\text{C}-\text{OH} & \text{H}-\text{C}-\text{OH} \\
\text{CH}_2\text{OH} & \text{CH}_2\text{OH} & \text{CH}_2\text{OH} \\
\text{D-glucosamine} & \text{D-mannosamine} & \text{D-galactosamine}
\end{array}
$$

In D-glucosamine, and in D-mannosamine and D-galactosamine as well, the —OH group on carbon 2 is replaced by an —NH$_2$ group. In most cases where these monosaccharides occur in nature, the —NH$_2$ groups are acetylated. D-Amino sugars are essential parts of many polysaccharides including chitin, the hard, shell-like covering of lobsters, crab, shrimp and other crustaceans. Furthermore they are an integral part of the cell wall structure of certain bacteria and of heparin, a naturally occurring blood anticoagulant.

Since D-glucose occupies a central position in carbohydrate metabolism and since it is the most common and most thoroughly investigated carbohydrate, we shall examine its structure and properties in some detail. However, much of this discussion applies to other sugars as well.

11.5   Mutarotation and the Cyclic Structure of Glucose

Although glucose shows many of the typical reactions of aldehydes, it fails to give a positive Schiff's test (a qualitative color reaction test characteristic of aldehydes). This failure led Tollens in 1883 to propose a cyclic hemiacetal structure for glucose. This formula was not generally accepted due to the (then) well-known instability of hemiacetals.

In 1893, Fischer attempted to convert glucose into a dimethyl acetal (Section 7.8) by treatment with methanol, in accordance with the known reaction:

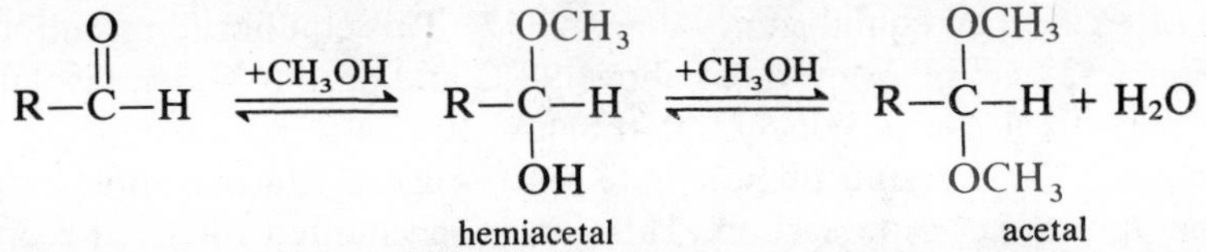

$$\underset{\substack{O \\ \| \\ R-C-H}}{} \xrightleftharpoons{+CH_3OH} \underset{\substack{OCH_3 \\ | \\ R-C-H \\ | \\ OH \\ \text{hemiacetal}}}{} \xrightleftharpoons{+CH_3OH} \underset{\substack{OCH_3 \\ | \\ R-C-H + H_2O \\ | \\ OCH_3 \\ \text{acetal}}}{}$$

Surprisingly, the product obtained had the formula $C_7H_{14}O_6$ rather than $C_8H_{18}O_7$ expected for the dimethyl acetal and contained only one methyl group. It was called a methyl glucoside. This substance showed no aldehydic properties, and Fischer correctly regarded it as a true acetal, a cyclic acetal structure involving one of the hydroxyl groups from the sugar itself and one molecule of methanol. The proposed partial structures shown below indicate the presence of one new chiral carbon called the anomeric carbon and suggested that glucose should yield two isomeric glucosides. The second methyl glucoside was isolated the following year by van Eckenstein. The designation α-methyl D-glucoside was applied to the isomer having the higher (more positive) specific rotation. The other was designated β-methyl D-glucoside. For the moment we shall represent these two methyl glucosides in partial formulas. In each, the stereochemistry at carbon 1, the new chiral center, is represented according to the Fischer convention.

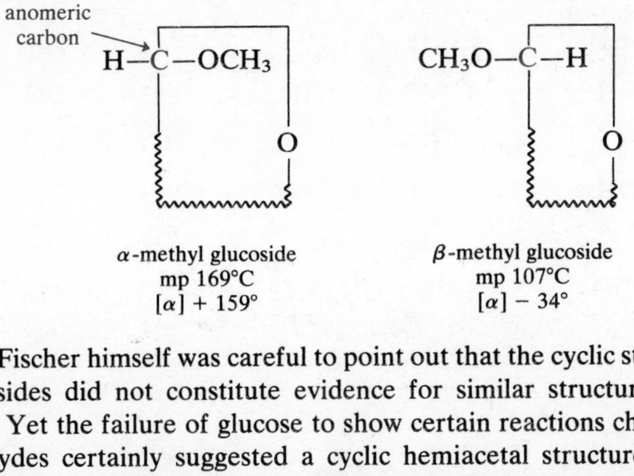

anomeric carbon

$$H-C-OCH_3 \qquad\qquad CH_3O-C-H$$

$$O \qquad\qquad\qquad\qquad O$$

α-methyl glucoside
mp 169°C
$[\alpha] + 159°$

β-methyl glucoside
mp 107°C
$[\alpha] - 34°$

Fischer himself was careful to point out that the cyclic structure of the glucosides did not constitute evidence for similar structures of glucose itself. Yet the failure of glucose to show certain reactions characteristic of aldehydes certainly suggested a cyclic hemiacetal structure for glucose. Other monosaccharides also react with alcohols in the presence of a trace of mineral acid to form anomeric α- and β-cyclic acetal structures. The term glycoside is a general name for this type of substance. Particular glycosides are named by dropping the suffix -e from the name of the parent saccharide and adding -ide, for example, glucoside, riboside, galactoside.

The next indication of a cyclic structure of the free sugar came from observations of another phenomenon, namely, mutarotation, and from the isolation of two isomeric forms of D-glucose. One form (α), best isolated by crystallization of glucose from alcohol-water below 30°C, has a specific rotation of +112°. The other (β), obtained when an aqueous solution of glucose is evaporated at temperatures above 98°C, has a specific rotation of +19°.

309

A freshly prepared solution of α-glucose shows an initial rotation of +112°, which gradually decreases to a constant value of +52°. This change, first observed in 1846, is known as mutarotation. A solution of β-glucose also undergoes mutarotation, the specific rotation changing from an initial value of +19° to an equilibrium value of +52°. This equilibrium solution of rotation +52° can be evaporated above 98°C and pure β-glucose isolated. Alternatively, it can be concentrated below 30°C and pure α-glucose can be isolated. These results demonstrate that α- and β-glucose differ only at carbon 1, the aldehyde carbon. This can be accounted for by the cyclic hemiacetal structure. The molecular event responsible for the observed change in the optical rotation is attributed to the interconversion of the two forms in solution with the free aldehyde as an intermediate.

It was demonstrated in 1901 that the configuration at carbon 1 of α-methyl glucoside is the same as that of α-glucose, and that of β-methyl glucoside corresponds to β-glucose. Fischer had observed that the hydrolysis of α-methyl glucoside is catalyzed by the enzyme maltase, but not emulsin, and that the hydrolysis of β-methyl glucoside is catalyzed by emulsin but not maltase. It was subsequently shown that α-methyl glucoside is hydrolyzed to α-glucose.

The next question is to decide the size of the ring, in other words, which oxygen of the carbon chain is bonded to the carbon 1 in the cyclic hemiacetal and in the cyclic acetal. Conclusive evidence that the ring is actually six-membered was presented in 1926. Thus, we may complete the formulas of α-D-glucose, β-D-glucose and the corresponding methyl·

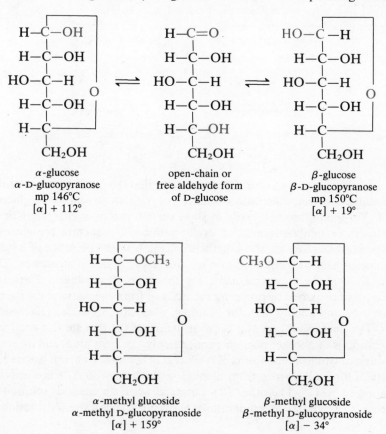

α-glucose
α-D-glucopyranose
mp 146°C
[α] + 112°

open-chain or
free aldehyde form
of D-glucose

β-glucose
β-D-glucopyranose
mp 150°C
[α] + 19°

α-methyl glucoside
α-methyl D-glucopyranoside
[α] + 159°

β-methyl glucoside
β-methyl D-glucopyranoside
[α] − 34°

glucosides as shown. Also shown is the open-chain or free aldehyde form of
D-glucose. This form is intermediate in the interconversion between
α-glucose and β-glucose. This interconversion is the molecular change
underlying mutarotation.

Both α- and β-glucose have the six-membered ring structure of
pyran, and in the systematic nomenclature, are designated as gluco-
pyranoses, for example, α-D-glucopyranose. The glucosides are properly
designated as pyranosides, for example, α-methyl D-glucopyranoside. A
sugar having a five-membered cyclic structure is designated as a furanose
(in the case of a hemiacetal) or a furanoside (in the case of an acetal). For
example, ribose as it is found combined in RNA is a five-membered ring
and is properly called a β-D-ribofuranoside. (See Figure 14.3.)

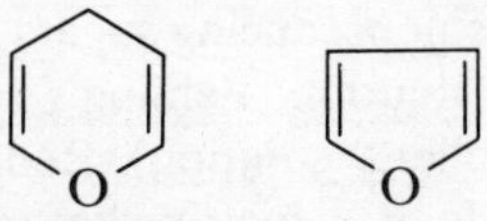

pyran   furan

11.6 Stereorepresentations of Sugars

There are three common ways for representing the stereochemistry of
glucose and other carbohydrates. We have already encountered the Fischer
projection representation. While this type of representation does show
clearly the configuration at each carbon it is grossly inaccurate in its
representation of bond angles and the geometry of the molecule. In the
Haworth projection, the cyclic forms of carbohydrates are represented as
regular hexagons or pentagons, as shown below for the examples of
glucose, ribose and fructose.

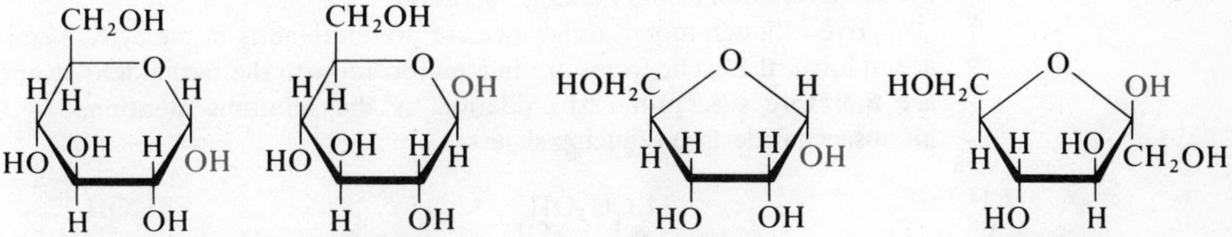

α-D-glucose   β-D-glucose   α-D-ribose   β-D-fructose
α-D-glucopyranose   β-D-glucopyranose   α-D-ribofuranose   β-D-fructofuranose

Although useful, the Haworth projection formulas are not accurate in
that they show the six-membered ring as being planar. The ring is more
accurately represented as a strain-free chair conformation, in which the

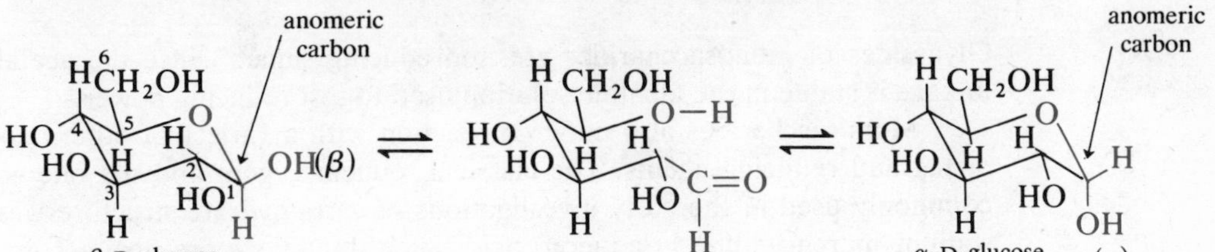

β-D-glucose   α-D-glucose
β-D-glucopyranose   α-D-glucopyranose

substituents on carbons 2, 3, 4, and 5 of the ring all lie in equatorial positions. The structural formulas of α-glucose and β-glucose and the phenomenon of mutarotation are shown in this type of representation.

11.7 Properties of Monosaccharides

The monosaccharides are colorless, crystalline solids and sweet to the taste. Because of the possibilities for hydrogen bonding they are very soluble in water, but only slightly soluble in ethanol, and quite insoluble in non-hydroxylic solvents such as ether, chloroform, and benzene.

Monosaccharides (and carbohydrates, in general) are conveniently classified as reducing or as nonreducing sugars according to their behavior toward Cu^{2+} (Benedict's solution, Fehling's solution) or toward Ag^+ in ammonium hydroxide (Tollens' solution). Reducing sugars contain a free or potentially free aldehyde or α-hydroxyketone. As we saw in Chapter 7, an aldehyde is very susceptible to oxidation to the corresponding acid, even by such weak oxidizing agents as silver ion. In Tollens' test, the deposition of a silver mirror indicates the presence of an aldehyde group. A second reagent, Fehling's solution, is prepared by mixing a solution of copper sulfate with an alkaline solution of a salt of tartaric acid. The resulting deep blue solution contains copper ion complexed with tartrate. When this solution reacts with a sugar containing an aldehyde or an α-hydroxyketone, the cupric ion is reduced to cuprous ion, which precipitates as brick red cuprous oxide.

$$RCHO + Cu^{2+} \longrightarrow RCO_2^- + Cu_2O$$

brick red
precipitate

Benedict's solution contains cupric ion complexed with citrate and serves the same function as the Fehling's solution.

Even though monosaccharides are predominantly in the cyclic hemiacetal form, the cyclic forms are in equilibrium with the free aldehyde and are therefore susceptible to oxidation by the solutions mentioned. All monosaccharides are reducing sugars.

hemiacetal form   open-chain form   carboxylic acid

Glycosides of monosaccharides are nonreducing sugars, since the acetal linkage is stable in the alkaline solution used to test reducing power.

Monosaccharides also undergo reaction with a variety of other oxidizing and reducing agents. The classical reducing agent and the one so commonly used in the early investigations of carbohydrate structure was sodium–mercury alloy in aqueous acid. Such reductions are now accomplished using sodium borohydride.

$$\begin{matrix} \text{CHO} \\ | \\ \text{(CHOH)}_4 \\ | \\ \text{CH}_2\text{OH} \end{matrix} \quad \xrightarrow[\text{Na/Hg at pH 4}]{\text{NaBH}_4 \text{ or}} \quad \begin{matrix} \text{CH}_2\text{OH} \\ | \\ \text{(CHOH)}_4 \\ | \\ \text{CH}_2\text{OH} \end{matrix}$$

D-glucose — D-glucitol sorbitol

These reduction products are known as <u>alditols</u>. For example, D-glucose forms D-glucitol (also known as sorbitol), D-threose forms D-threitol.

Oxidation of aldoses by Fehling's, Benedict's, or Tollens' solutions forms the corresponding carboxylic acids. This selective oxidation of the aldehyde group can also be accomplished using a halogen (most commonly bromine or iodine) in base. The monocarboxylic acid products are known as <u>aldonic acids</u>. For example, D-glucose forms D-gluconic acid, D-threose yields D-threonic acid.

$$\begin{matrix} \text{CHO} \\ | \\ \text{(CHOH)}_4 \\ | \\ \text{CH}_2\text{OH} \end{matrix} \quad \xrightarrow[\text{or Br}_2 \text{ in base}]{\text{Ag}^+ \text{ or Cu}^{2+}} \quad \begin{matrix} \text{CO}_2\text{H} \\ | \\ \text{(CHOH)}_4 \\ | \\ \text{CH}_2\text{OH} \end{matrix}$$

D-glucose — D-gluconic acid

More vigorous oxidation with warm nitric acid leads to oxidation of the —CHO group and also the terminal —CH$_2$OH group. These dicarboxylic acids are known as <u>aldaric acids</u>. For example, nitric acid oxidation of either D-glucose or D-gluconic acid yields D-glucaric acid.

$$\begin{matrix} \text{CHO} \\ | \\ \text{(CHOH)}_4 \\ | \\ \text{CH}_2\text{OH} \end{matrix} \quad \xrightarrow[\text{heat}]{\text{HNO}_3} \quad \begin{matrix} \text{CO}_2\text{H} \\ | \\ \text{(CHOH)}_4 \\ | \\ \text{CO}_2\text{H} \end{matrix}$$

D-glucose — D-glucaric acid

11.8 Disaccharides

Most carbohydrates in nature contain more than one hexose unit. Sucrose (cane sugar), lactose (milk sugar), maltose, and cellobiose each consist of two hexose units and are called disaccharides. In a <u>disaccharide</u> the two monosaccharides are linked together by acetal formation (Section 7.8). Condensation of the hydroxyl function of the hemiacetal group of one monosaccharide with an hydroxyl group of another monosaccharide forms the bond (called a <u>glycosidic bond</u>) joining the two saccharide units. Since glycosides are acetals they may be hydrolyzed to the constituent monosaccharides by dilute mineral acid or by the appropriate enzymes such as maltase and emulsin. We shall look at the structures of four disaccharides: maltose, cellobiose, lactose, and sucrose.

<u>Maltose</u> derives its name from the fact that it occurs mainly in malt liquors, the juice expressed from sprouted barley and other cereal grains. In yeast the enzyme amylase or maltase hydrolyzes the starch in seeds to maltose in about 18% yield. Further hydrolysis of maltose by the enzyme maltase yields only glucose. Maltose is formed from two molecules of

glucose joined together by an α-glycosidic bond between carbon 1 of one glucose and the carbon 4 of the second glucose unit. In maltose, glucose is joined to glucose by an α-1,4-glycosidic bond.

α-D-glucopyranose unit β-D-glucopyranose unit

maltose (from degradation of starch)

Cellobiose, a reducing sugar, is one of the major fragments isolated after extensive hydrolysis of cellulose (cotton fiber for example). Further hydrolysis affords two molecules of glucose. Cellobiose differs from maltose only in that the two D-glucose units are joined by a β-1,4-glycosidic linkage.

β-D-glucopyranose unit β-D-glucopyranose unit

cellobiose (from controlled hydrolysis of cellulose)

Another disaccharide of special interest is lactose. It is the major sugar of milk and is present to the extent of 5 to 8% in human milk and 4 to 6% in cow's milk. Hydrolysis of lactose affords equal amounts of D-glucose and D-galactose. In lactose, galactose is joined to glucose by a β-1,4-glycosidic bond.

β-D-galactopyranose unit β-D-glucopyranose unit

β-lactose (from milk of mammals)

Sucrose is the most abundant disaccharide. It is common table sugar and is readily obtained from the juice of cane sugar or sugar beet. Hydrolysis of sucrose affords one molecule of D-glucose and one of D-fructose. Sucrose is not a reducing sugar. Since both glucose and fructose are reducing sugars, both monosaccharides must be present as the glycosides, that is, the anomeric hemiacetal carbons of both glucose and fructose must be involved in the formation of the glycosidic bond in sucrose.

$$\overset{6}{C}H_2OH$$

α-1,2-glycoside
bond

α-D-glucopyranose   β-D-fructofuranose
unit   unit

sucrose (cane sugar)

Sucrose is a glycoside in which the carbon 1 of glucose is bonded by an α-glycosidic bond to the carbon 2 of fructose, that is, in sucrose, glucose is joined to fructose by an α-1,2-glycosidic bond. Glucose is in the pyranoside form while fructose is in the furanoside form.

11.9 Polysaccharides

Polysaccharides are grouped together into two general classes: those that are insoluble and form the skeletal structure of plants, and those that constitute the reserve sources of simple sugars and are liberated as required by certain enzymes in the organism. Both types are high-molecular-weight polymers.

Starch is the reserve carbohydrate for plants. It is found in all plant seeds and tubers, and is the form in which glucose is stored for use by the plant in its metabolic processes. Starch can be separated into two fractions, amylose and amylopectin. Although amylose and amylopectin are both polymers of glucose, and yield maltose on partial hydrolysis and only glucose on complete hydrolysis, they differ in several respects. Amylose has a molecular weight range of 10,000 to 50,000 (60 to 300 glucose units) and amylopectin has a molecular weight range of 50,000 to 1,000,000 (300 to 6000 glucose units). X-ray diffraction studies of amylose show a continuous, unbranched chain of glucose units with the carbon 1 of one glucose linked by an α-glycosidic bond to the carbon 4 of the adjacent glucose.

Amylopectin has a highly branched structure. It contains the same type of repetitive sequence of α-1,4-glycosidic bonds as does amylose but the chain lengths vary only from about 24 to 30 units. In addition, there is considerable branching from this linear network. At branch points a new chain is started by an α-1,6-glycosidic linkage between the carbon 1 of one glucose unit and a carbon 6 hydroxyl of another glucose unit. (See p. 316.)

Plants and animals synthesize these large molecules at the expense of considerable energy. Why? If a plant seed requires a large amount of food as stored carbohydrate, why is it not sufficient to pack into it a certain amount of glucose? The reason is rooted in a very elementary principle of physical chemistry. Osmotic pressure is proportional to the molar concentration, not the molecular weight of a solute. If we assume that a thousand glucose monomers are assembled into a starch macromolecule, then we can predict that a solution of starch containing 1 gram of starch per 10 ml of

315

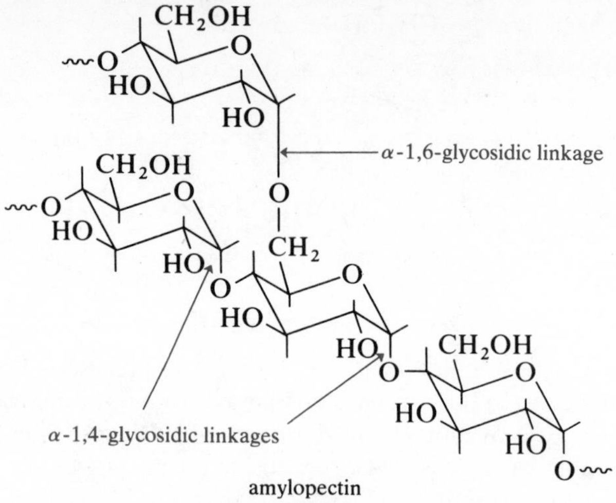

α-1,6-glycosidic linkage

α-1,4-glycosidic linkages

amylopectin

solution will have one one-thousandth the osmotic pressure of 1 gram of glucose in the same volume of solution. This feat of packaging is of tremendous advantage because it reduces the strain on various membranes enclosing such macromolecules.

Glycogen serves as the reserve carbohydrate for animals. Like amylopectin, glycogen is a nonlinear polymer of glucose units joined by α-1,4- and α-1,6-glycoside bonds, but it has a lower molecular weight and a more highly branched structure. The chief source of glycogen is ingested starch or glycogen, which is then hydrolyzed by amylases in the intestinal tract to glucose. Since the amount of glucose absorbed into the blood stream during digestion is greater than that required for immediate use by the body, most of the excess glucose is converted into glycogen (a process called glycogenesis) and stored in the liver and muscles. The total amount of glycogen in the body of a well-nourished adult is about 350 grams divided nearly equally between the liver and muscle.

Insulin (Figure 13.7), the hormone produced in the pancreas, stimulates the oxidation of glucose by the glycolytic and tricarboxylic acid pathways and also promotes the synthesis of glycogen. Injection of insulin into a normal animal results in a decrease in concentration of blood sugar usually accompanied by a small increase in the glycogen concentrations of the liver and muscle. Conversely, a deficiency of insulin, as in diabetes mellitus, causes a gradual depletion of glycogen stores in the liver and an increase in blood sugar.

Cellulose is the most widely distributed skeletal polysaccharide. It constitutes almost half of the cell wall material of wood. Cotton is almost pure cellulose. Controlled hydrolysis of cellulose affords cellobiose as the only disaccharide. Vigorous hydrolysis affords only D-glucose. This and other evidence demonstrates that cellulose is a linear polymer of glucose units joined together by β-1,4-glycosidic linkages. The molecular weight of cellulose is approximately 400,000, corresponding to 2800 glucose units. X-ray analysis indicates that cellulose fiber consists of bundles of parallel oriented chains in which the pyranose rings lie alternately in different orientations as shown in the following diagram. The chains are held together by hydrogen bonding between the hydroxyls of the adjacent chains. This type of arrangement of parallel chains into bundles and the resultant hydrogen bonding gives cellulose fibers a high mechanical strength and a chemical stability.

cellobiose unit

cellulose chain

Man and other carnivorous animals are unable to utilize cellulose as a foodstuff even though, like starch, it is a polymer of glucose. In the evolutionary scheme of things man either lost or never carried the genetic information required to make the enzyme system capable of catalyzing the hydrolysis of the β-glycosidic linkage. Our digestive systems contain only the α-amylases and hence we must use starch and glycogen as our sources of glucose. On the other hand, many bacteria and microorganisms do possess the enzyme systems capable of digesting cellulose. Termites are fortunate to have such bacteria in their intestine and can use wood as their principal food. Ruminants (cud-chewing animals) can also digest grasses and wood because of the presence of micro-organisms within the specially constructed alimentary system.

Cellulose, whether from cotton, wood pulp, or other natural source, is such a readily available and relatively inexpensive raw material that besides processing it into cotton and linen fibers for textiles and into paper goods, it has been possible to modify it chemically to produce certain widely used semi-synthetic materials. The most important of these modified celluloses are cellulose nitrate and cellulose acetate.

Nitration of cellulose by a mixture of nitric and sulfuric acids yields a polynitrate ester called <u>cellulose nitrate</u> or nitrocellulose. The extent of nitration varies with the conditions of the process used. Following is a partial structural formula of fully nitrated cellulose.

a nitrate
ester

cellulose nitrate
(partial formula)

Note that there are three nitrate esters per glucose unit. Guncotton approaches this degree of nitration. Less completely nitrated celluloses are used in the manufacture of the common moldable plastic <u>celluloid</u>. In this process, partially nitrated cellulose is mixed with alcohol and camphor and then molded or rolled into sheets. The molded or rolled article is then heated at a slightly elevated temperature to evaporate the alcohol. The finished material is a tough, hard plastic. One of the first uses for this plastic was as a substitute for ivory billiard balls. Celluloid deserves special mention in the chemistry and technology of plastics for it was the first (and until about 1920 the only) plastic to be manufactured on an industrial scale. Its major disadvantage is that it is highly flammable.

Fibers made from regenerated and chemically modified cellulose were the first of the man-made fibers to become and remain commercially important. Because of its insolubility in any of the convenient and inexpensive solvents, cellulose must be chemically modified and made soluble before it can be processed into a fiber. If the modifying groups are subsequently removed in the processing, the resulting fiber is called regenerated cellulose. If the modifying groups appear in the resulting fiber, this is indicated in the name of the fiber, as for example in the name acetate rayon.

In one industrial process, cellulose is reacted with carbon disulfide to form an alkali-soluble xanthate ester. A solution of cellulose xanthate is then extruded into dilute sulfuric acid which brings about hydrolysis of the xanthate ester and precipitation of the free or regenerated cellulose. Extruding the regenerated cellulose as a filament produces "viscose rayon" threads; extruding it as sheets produces cellophane.

In another industrial process, cellulose is acetylated with acetic anhydride. The extent of acetylation varies with the conditions. Acetylated cellulose is dissolved in a suitable solvent, precipitated, and then drawn into fibers. This material is known as acetate rayon. Cellulose acetate, acetylated to the extent of about 80%, became commercial in Europe about 1920 and in the United States a few years later. Cellulose triacetate, which has about 97% of the hydroxyls acetylated, became commercial in the United States in 1954. Acetate fibers rank fourth in production in the United States, surpassed only by polyester, nylon, and rayon fibers.

PROBLEMS

11.1 Define carbohydrate in terms of the functional groups present. Literally the term *carbohydrate* is derived from "hydrates of carbon." Show by reference to molecular formulas the origin of this term.

11.2 Draw perspective formulas for D-glyceraldehyde and L-glyceraldehyde. Explain the meaning of the designation D- and L- as used to specify the stereo-chemistry of glyceraldehyde.

11.3 List the rules for drawing the so-called Fischer projection formulas. Draw Fischer projection formulas for D- and L-glyceraldehyde.

11.4 Draw the four stereoisomers of 2, 3, 4-trihydroxybutanal. Label them I, II, III, and IV. Which are enantiomers? diastereomers? (To check your answer, refer to Section 4.6.)

11.5 What is the difference in structure between D-ribose and 2-deoxy-D-ribose? Draw structural formulas for the open-chain form of each.

11.6 A careful study of D-ribose in aqueous solution shows that it exists as an equilibrium mixture of both furanoses and pyranoses. Draw suitable stereo-representations of the four possible cyclic hemiacetals

(a) α-D-ribofuranose   **(b)** β-D-ribofuranose
(c) α-D-ribopyranose   **(d)** β-D-ribopyranose

11.7 One of the important techniques for establishing relative configurations among the isomeric straight-chain aldoses has been to convert both terminal carbon atoms into the same functional group. This can be done either by selective oxidation or reduction. As a specific example, nitric acid oxidation of D-erythrose leads to the formation of meso-tartaric acid. Oxidation of D-threose under similar conditions leads to the formation of optically active D-tartaric acid.

$$\text{D-threose} \xrightarrow[\text{oxidation}]{\text{HNO}_3} \text{D-tartaric acid}$$

$$\text{D-erythrose} \xrightarrow[\text{oxidation}]{\text{HNO}_3} \text{meso-tartaric acid}$$

Using this information, show which of the structural formulas (a) or (b) is D-erythrose; which of the structural formulas (a) or (b) is D-threose. Check your answer by referring to Table 11.2.

```
        CHO              CHO
         |                |
   H—C—OH          HO—C—H
         |                |
   H—C—OH          H—C—OH
         |                |
      CH₂OH            CH₂OH

       (a)              (b)
```

11.8 As you can see from Table 11.2, there are four isomeric D-aldopentoses. Suppose that each is reduced with sodium borohydride. Which of the four will yield optically inactive D-alditols? Which will yield optically active D-alditols?

11.9 Draw a Fischer projection formula for D-glucose. State the total number of possible stereoisomers of this structural formula. Of this number of stereoisomers, only D-glucose, D-galactose, and D-mannose occur in nature. Draw Fischer projection formulas for D-galactose and D-mannose.

11.10 There are three common conventions for representing the stereochemistry of carbohydrates. They are (1) the Fischer projection, (2) the Haworth projection, and (3) the chair/boat projection formulas. Draw α-D-glucose according to the rules of each of these conventions. Do the same for β-D-glucose.

11.11 Using a set of molecular models, build the following:
(a) D-glucose in the open chain form.
(b) Using this molecular model, show the reaction of the —OH on carbon 5 with the aldehyde of carbon 1 to form a cyclic hemiacetal.
(c) Show that either α-D-glucose or β-D-glucose can be formed depending on the direction from which the —OH group interacts with the aldehyde group.

11.12 Explain the conventions α- and β- as used to designate the stereochemistry of the cyclic forms of carbohydrates.

11.13 Draw structural formulas for the open-chain and cyclic forms of D-fructose.

11.14 Explain the phenomenon of mutarotation with reference to carbohydrates. By what means is it detected?

11.15 A freshly prepared solution of α-D-glucose has a specific rotation of +112°; a similarly prepared solution of β-D-glucose has a specific rotation of +19°. On mutarotation, the specific rotation of either solution changes to an equilibrium value of +52°. Calculate the percent of β-D-glucose in the equilibrium mixture.

11.16 Treatment of D-glucose in dilute aqueous alkali at room temperature yields an equilibrium mixture of D-glucose, D-mannose, and D-fructose. Account for this transformation.

11.17 Ketones cannot be oxidized by mild oxidizing agents. However, both dihydroxyacetone and fructose give a positive Benedict's test and are therefore classed as reducing sugars. Using structural formulas show how these monosaccharides react to give a positive test.

11.18   Fischer attempted to convert D-glucose into its dimethylacetal according to the following reaction:

(a) $C_6H_{12}O_6 + 2CH_3OH \xrightarrow{H^+} C_8H_{18}O_7 \qquad + H_2O$
         D-glucose   D-glucose dimethylacetal

However, the reaction that takes place is instead

(b) $C_6H_{12}O_6 + CH_3OH \xrightarrow{H^+} C_7H_{14}O_6 \qquad + H_2O$
         D-glucose   methyl D-glucoside

Draw a structural formula for the expected product $C_8H_{18}O_7$ and the observed product $C_7H_{14}O_6$. Reaction (b) gives two isomeric methyl glucosides designated as α-methyl D-glucoside and β-methyl D-glucoside. Draw Fischer, Haworth, and chair projection formulas for each of these and label them accordingly.

11.19   Draw Haworth and chair projection formulas for the disaccharides sucrose, lactose, cellobiose and maltose. Name the constituent monosaccharides and label as α or β the stereochemistry of the glycosidic linkages. State which of these disaccharides are reducing sugars, which will undergo mutarotation. State one important natural source of each of these sugars.

11.20   Draw Haworth and chair projection formulas to show how D-glucose is linked in the polysaccharides cellulose and starch. State the major function of each of these polysaccharides in the biological world.

11.21   What is the difference in meaning between the terms *glucoside* and *glycoside*?

11.22   Propose a likely structure for each of the following polysaccharides:
(a)   Alginic acid, isolated from seaweed, is used as a thickening agent in ice cream and other foods. Alginic acid is known to be a polymer of D-mannuronic acid units in the pyranoside form and the units are thought to be joined together by β-1,4-glycoside bonds.
(b)   Pectic acid is the main constituent of pectin responsible for the formation of jellies from fruits and berries. Pectic acid is known to be a polymer of D-galacturonic acid units in the pyranoside form and the units are thought to be joined together by α-1,4-glycoside bonds.

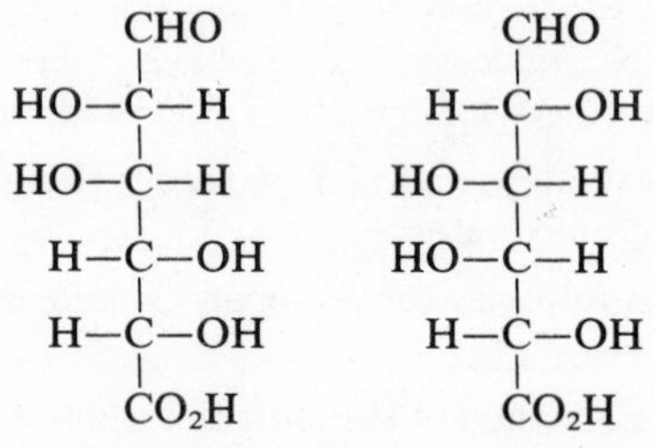

$$
\begin{array}{cc}
\text{CHO} & \text{CHO} \\
\text{HO--C--H} & \text{H--C--OH} \\
\text{HO--C--H} & \text{HO--C--H} \\
\text{H--C--OH} & \text{HO--C--H} \\
\text{H--C--OH} & \text{H--C--OH} \\
\text{CO}_2\text{H} & \text{CO}_2\text{H} \\
\text{D-mannuronic acid} & \text{D-galacturonic acid}
\end{array}
$$

11.23   Trehalose is found in young mushrooms and is the chief carbohydrate in the blood of certain insects. Trehalose is a disaccharide consisting of two glucose units joined by an α-1,1-glycoside bond.

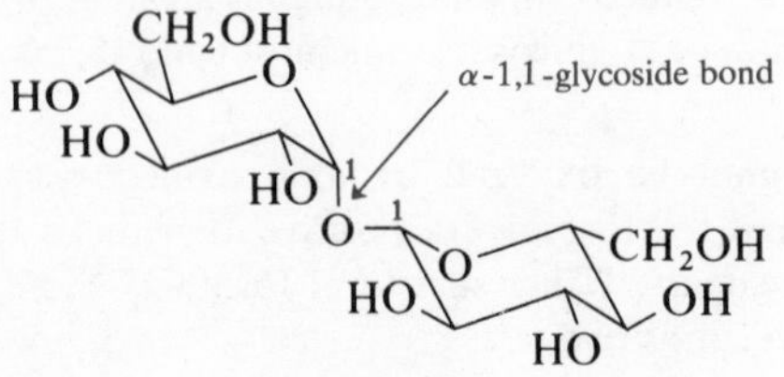

On the basis of its structural formula would you expect trehalose to
(a) be a reducing sugar?
(b) undergo mutarotation?

11.24 Raffinose is the most abundant trisaccharide in nature.

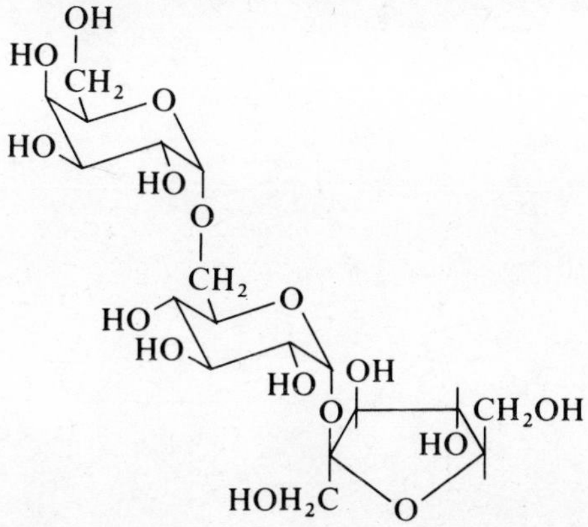

(a) Name each of the three monosaccharide units in raffinose.
(b) There are two glycoside bonds in raffinose. Describe each as you have already done for other disaccharides (that is, an α-1,2-glycoside bond in sucrose; a β-1,4-glycoside bond in cellobiose).

11.25 Draw Haworth and chair representations for the α- and β-pyranose forms of D-glucosamine, D-mannosamine, and D-galactosamine.

11.26 Shown below is a Fischer projection formula for N-acetyl-D-glucosamine.

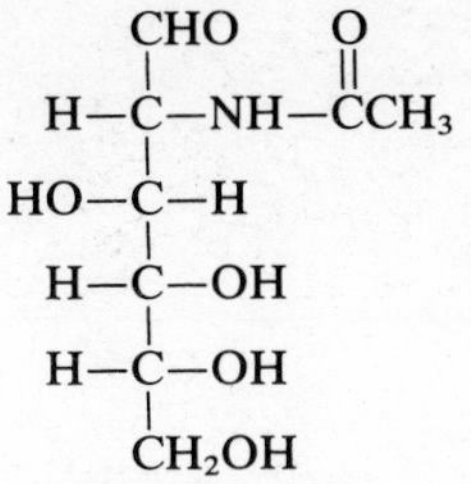

N-acetyl-D-glucosamine

(a) Draw Haworth and chair representations for the β-pyranose form of this sugar.
(b) Draw Haworth and chair representations for the disaccharide formed by joining two β-pyranose units of N-acetyl-D-glucosamine together by a β-1,4-glycoside bond. (If you have done this correctly, you have drawn a structural formula for the repeating dimer of chitin, the polysaccharide component of chitin.)

Mini-Essay

**Clinical Chemistry:
the Search for
Specificity**

Clinical Chemistry—
The Search for
Specificity

In this essay, let us look at what is probably the analytical procedure most often performed in the clinical chemistry laboratory, namely, the determination of glucose in blood, urine, and other biological fluids. Interest in this problem has been stimulated by the high incidence of the disease, diabetes mellitus. It is estimated that there are approximately 2 million recognized diabetics in the United States, and probably another 2 million more who are undiagnosed. Therefore, the overall incidence of this disease in the general population is nearly 2%. Diabetes mellitus is characterized by insufficient blood levels of the polypeptide hormone insulin (Section 13.8). Deficiency of insulin results in the inability of glucose to enter muscle and liver cells, and, therefore, increased levels of blood glucose (hyperglucosemia); impaired metabolism of fats and proteins; ketosis; and possibly diabetic coma. Thus it is critical for the early diagnosis and effective management of this disease to have a rapid and reliable procedure for the determination of blood glucose. Over the past 50 years or more, a great many such tests have been developed. We will discuss only three of these, each chosen to illustrate something of the problems involved in developing suitable clinical laboratory tests and how these problems can be solved. Furthermore, these tests will illustrate the use of both chemical and enzymatic techniques in the modern clinical chemistry laboratory.

The first of the glucose tests we will discuss is based on the fact that glucose is a reducing sugar, and specifically that it will reduce ferricyanide ion (yellow in hot alkaline solution) to ferrocyanide ion (colorless in hot alkaline solution).

$$\text{glucose} + 2\text{Fe(CN)}_6^{3-} \longrightarrow \text{gluconic acid} + 2\text{Fe(CN)}_6^{4-}$$

$$\underset{\substack{\text{ferricyanide} \\ \text{ion}}}{} \qquad\qquad\qquad \underset{\substack{\text{ferrocyanide} \\ \text{ion}}}{}$$

The decrease in absorbance of the test solution, measured at 420 nm, is proportional to the concentration of glucose in the sample. A modification of this method, introduced in 1928 and for many years the standard procedure for the determination of blood glucose, involved carrying out the reaction in the presence of excess ferric ion. Under these conditions, the ferrocyanide ion formed on oxidation of glucose reacts further with Fe^{3+} to form ferric ferrocyanide, or as it is more commonly known, Prussian blue. In this procedure, the absorbance is directly proportional to the concentration of glucose in the test sample.

325

$$3Fe(CN)_6^{4-} + 4Fe^{3+} \longrightarrow Fe_4[Fe(CN)_6]_3$$

ferric ferrocyanide
(Prussian blue)

Yet, while these and other oxidative methods can be used to measure glucose concentration, they suffer from the distinct disadvantage that they also react with a whole series of other reducing substances normally found in blood. In other words, the ferricyanide and other oxidative methods as well give false positive results. The term saccharoid was introduced by Benedict in 1931 for these nonglucose, nonfermentable substances. The known saccharoids include ascorbic acid, uric acid, certain amino acids, and phenols. In addition, other aldoses will of course also give false positive results. For these reasons, the oxidative procedures first developed often gave values as much as 30% or more higher than the so-called "true glucose" value.

How to make the determination of blood glucose more specific? Quite naturally, a considerable effort was devoted, with varying degrees of success, to eliminating or at least minimizing the interference of saccharoid substances. However, a more satisfactory approach in the search for specificity lay in attacking the problem in a completely different way, namely, by taking advantage of a chemical reactivity of glucose other than its property as a reducing sugar. One of the most successful and widely used of these nonoxidative methods involves reaction of glucose with o-toluidine to form a blue-green Schiff base. The absorbance of this Schiff base can be measured at 625 nm and is directly proportional to glucose concentration.

glucose   o-toluidine   a Schiff base (blue-green)

The procedure calls for mixing the test sample with o-toluidine, heating in a boiling water bath for 8 minutes, and then measuring the absorbance. The o-toluidine method can be applied directly to serum, plasma, cerebrospinal fluid, and urine and to samples as small as 20 microliters (20×10^{-6} liter). In addition, it does not give false positive results with other reducing substances because the procedure itself does not involve oxidation. However, galactose and mannose, and to a lesser extent lactose and xylose, are potential sources of false positive results for they also react with o-toluidine to give colored Schiff bases. This is generally not a problem though, for these mono- and disaccharides are normally present in serum and plasma in only very low concentrations.

The search for even greater specificity in glucose determinations has led in recent years to the introduction of enzyme-based assay procedures. What is needed is an enzyme that will catalyze some reaction of glucose but, at the same time, will not catalyze a comparable reaction of any other substance that may also be present in biological fluids. The enzyme glucose oxidase does just this. It catalyzes the oxidation of β-D-glucose to D-gluconic acid.

$$\beta\text{-D-glucose} + O_2 \xrightarrow{\text{glucose oxidase}} \text{D-gluconic acid} + H_2O_2$$

Note that this enzyme is highly specific for β-D-glucose and recall from Section 11.5 that glucose in solution exists as 36% of the α-form and 64% β-form. Therefore, complete oxidation of glucose requires mutarotation and interconversion of the α- and β-forms. Because glucose oxidase is so highly specific for β-D-glucose, measurement of blood glucose can be made even in the presence of other mono- or disaccharides that might also be present in plasma, serum, or urine. Direct determination can be made quantitative by coupling this oxidation with a reaction in which the H_2O_2 generated is used to oxidize another substance whose concentration then can be determined spectrophotometrically. In one procedure, H_2O_2 is reacted with iodide ion to form molecular iodine, I_2.

$$2I^- + H_2O_2 + 2H^+ \longrightarrow I_2 + 2H_2O$$

The absorbance at 420 nm is used to calculate iodine concentration and then glucose concentration. In another procedure, H_2O_2 is used to oxidize the aromatic amine _o_-toluidine to a colored product. The enzyme peroxidase is used to catalyze this second oxidation. The concentration of the colored oxidation product is determined spectrophotometrically.

$$o\text{-toluidine} + H_2O_2 \xrightarrow{\text{peroxidase}} \text{oxidized } o\text{-toluidine} + H_2O$$
$$\text{(colored)}$$

Recall that the first of the assay methods we looked at in this mini-essay was also an oxidative procedure. However, it gave positive errors in the presence of "saccharoid" substances. In a sense, the search for specificity has now turned full circle for at the present time the most highly specific and accurate assay, and the one which is said to give "true" glucose values, is also an oxidative method. Unlike the earlier ferricyanide method, this newer oxidative method is highly specific because it is catalyzed by a highly specific enzyme, β-D-glucose oxidase.

A number of commercially available test kits employ the glucose oxidase reaction for qualitative determination of glucose in the urine. One of these, Clinistix (Ames Co., Elkhart, Ind.), consists of a filter paper strip impregnated with glucose oxidase, peroxidase, and _o_-toluidine. The test end of the paper is dipped in urine, removed, and examined after 10 seconds. A blue color will develop if the concentration of glucose in the urine exceeds about 100 mg/ml.

REFERENCES

Henry, R. J., Cannon, D. O., and Winkelman, J. W., editors. _Clinical Chemistry, Principles and Technics,_ 2nd ed. Harper & Row, 1974.
Tietz, N., editor. _Fundamentals of Clinical Chemistry._ W. B. Saunders, 1976.

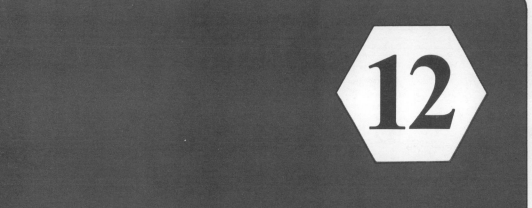

Lipids

12.1 Introduction

The term lipid designates a rather heterogeneous class of naturally occurring organic substances. Unlike the various classes of organic compounds we have studied so far (for example, alcohols, carboxylic acids, ketones, amines, carbohydrates), lipids as a class are grouped together not by the presence of a distinguishing functional group or structural feature, but rather on the basis of common solubility properties. Lipids are all insoluble in water and highly soluble in one or more organic solvents including ether, chloroform, benzene, and acetone. In fact, these four solvents are often referred to as "lipid-solvents" or "fat-solvents."

Lipids are widely distributed in the biological world and play a wide variety of roles in both plant and animal tissue. In the human body, lipids function as storage forms of energy, metabolic fuels, structural components of cell membranes, emulsifying agents, vitamins, and regulators of metabolism. This heterogeneous class of compounds is generally divided into four major groups, based primarily on certain common structural features.

1. Simple lipids: esters of fatty acids with various alcohols. These include the neutral fats, oils, and waxes.
2. Compound lipids: esters of fatty acids with alcohols plus other groups. These include the phospholipids, glycolipids, and lipoproteins.
3. Derived lipids: substances derived from hydrolysis of compounds in groups 1 and 2 that still retain the general physical characteristics of

lipids. These include the saturated and unsaturated fatty acids, as well as alcohols and steroids.

4. <u>Miscellaneous lipids</u>: These include the fat-soluble vitamins E and K as well as certain terpenes such as the carotenes.

In this chapter we shall discuss the structure and biological function of representative members of each group.

12.2 Fats and Oils

We are all familiar with fats and oils, for we encounter them every day in such things as milk, cream, butter, lard, oleomargarine, the liquid vegetable oils such as corn, cottonseed, and soybean oils, and many other foods. The common neutral fats and oils are <u>triesters of glycerol</u> and are the most abundant naturally occurring lipids. The only distinction between a fat and an oil is one of melting point; <u>fats</u> are by definition solids at room temperature while <u>oils</u> are liquid at room temperature.

Since neutral fats and oils are triesters of glycerol, they are more correctly called <u>triglycerides</u> or <u>triacylglycerols</u>. There are many different types of triglycerides depending on the identity and the position of the three fatty acid components. Those containing only one kind of fatty acid are called simple triglycerides. Examples are <u>glyceryl tripalmitate</u> and <u>glyceryl tristearate</u>, commonly known as <u>tripalmitin</u> and <u>tristearin</u>. Simple triglycerides rarely occur in nature.

$$
\begin{array}{ll}
\underset{\displaystyle CH_2-O-\overset{\displaystyle O}{\overset{\|}{C}}-(CH_2)_{14}CH_3}{} & \underset{\displaystyle CH_2-O-\overset{\displaystyle O}{\overset{\|}{C}}-(CH_2)_{16}CH_3}{} \\[2ex]
CH-O-\overset{\displaystyle O}{\overset{\|}{C}}-(CH_2)_{14}CH_3 & CH-O-\overset{\displaystyle O}{\overset{\|}{C}}-(CH_2)_{16}CH_3 \\[2ex]
CH_2-O-\overset{\displaystyle O}{\overset{\|}{C}}-(CH_2)_{14}CH_3 & CH_2-O-\overset{\displaystyle O}{\overset{\|}{C}}-(CH_2)_{16}CH_3
\end{array}
$$

glyceryl tripalmitate   glyceryl tristearate
tripalmitin   tristearin

Those containing two or more fatty acid components are called mixed triglycerides. In general, a fat or oil does not consist of one pure glyceride, but rather is a mixture (often complex) of glycerides. The composition of a fat is usually expressed in terms of the acids obtained from it by hydrolysis. Some fats and oils contain mainly one or two fatty acids: <u>olive oil</u> is 83% oleic acid, 6% palmitic acid, 4% stearic acid, and 7% linoleic acid; <u>palm oil</u> is 43% palmitic acid, 43% oleic acid, and 10% linoleic acid. <u>Butter fat</u> consists of a complex mixture of esters of at least 14 different acids; it differs from most other fats in having appreciable amounts of lower-molecular-weight fatty acids (Table 12.1).

The melting points of fats are determined by their fatty acid components. In general, the melting point increases with the number of carbons in the hydrocarbon chain and with the degree of saturation of the acid components. Triglycerides rich in low-melting fatty acids such as oleic,

TABLE 12.1 Fatty acid composition of butter.

Carbon Atoms	Fatty Acid	%	Carbon Atoms	Fatty Acid	%
4	butanoic	3.0	18	stearic	9.2
6	hexanoic	1.4	20	arachidic	1.3
8	octanoic	1.5	12	lauroleic	0.4
10	decanoic	2.7	14	myristoleic	1.6
12	lauric	0.7	16	palmitoleic	4.0
14	myristic	12.1	18	oleic	29.6
16	palmitic	25.3	18	linoleic	3.6

linoleic, or linolenic acid (Table 8.4) generally are liquid at room temperature and accordingly are called oils. Those rich in saturated fatty acids such as palmitic or stearic acid generally are semisolid or solid at room temperature and accordingly are called fats.

For a variety of reasons, in part convenience and in part dietary preference, a major industry has developed for the conversion of oils to fats. The process involves catalytic hydrogenation of some or all of the double bonds of the fatty acid constituents of a triglyceride and is called "hardening." If all of the double bonds are saturated, the product is hard and brittle. In practice, the degree of hardening is carefully controlled to produce a fat of the desired consistency. The resulting fats are sold for kitchen use (for example, Crisco, Spry). Oleomargarine and other butter substitutes are prepared by hydrogenation of cottonseed, soybean, corn, or peanut oils. The resulting product is often churned with milk and artificially colored to give it a flavor and consistency resembling that of butter.

On exposure to air, most triglycerides develop an unpleasant odor and flavor and are said to become rancid. In part this is the result of slight hydrolysis of the fat and the production of lower-molecular-weight fatty acids. For example, the odor of rancid butter is due to the presence of butanoic acid formed by the hydrolysis of butterfat. These same low-molecular-weight, volatile substances can be formed by oxidation and cleavage of double bonds in the unsaturated fatty acid side chains. The rate of oxidation and the production of rancidity vary with the individual fat, largely because of the presence of certain naturally occurring lipid-soluble substances which inhibit this oxidation. These substances are known as antioxidants. One of the most common lipid antioxidants is vitamin E.

12.3 Waxes

Waxes are related to the triglycerides in both structure and properties. They are esters of fatty acids and monohydroxylic alcohols. The leaf waxes are found in the leaves and stems of plants and serve to prevent or retard loss of moisture by evaporation. They typically consist of esters of fatty acids and alcohols each having from 16 to 34 carbon atoms. Carnauba wax, which coats the leaves of the carnauba palm native to Brazil, is largely myricyl cerotate, $C_{25}H_{51}CO_2C_{30}H_{61}$. Beeswax, secreted from the wax glands of the bee, is largely myricyl palmitate, $C_{15}H_{31}CO_2C_{30}H_{61}$. Waxes are generally harder, more brittle, and less greasy to the touch than fats and find application in polishes, cosmetics, ointments, and other pharmaceutical preparations.

12.4 Phospholipids

These lipids, known as phospholipids, phosphoglycerides, or phosphatides, are the second most abundant kind of naturally occurring lipid. They are found almost exclusively in plant and animal cell membranes, which typically consist of about 40 to 50% phospholipid and 50 to 60% protein.

The most abundant phospholipids contain glycerol and fatty acids, as do the simple fats. In addition, they also contain phosphoric acid and a low-molecular-weight alcohol. The most common of these low-molecular-weight alcohols are choline, ethanolamine, serine, and inositol.

$$HOCH_2CH_2\overset{+}{\underset{CH_3}{\overset{CH_3}{N}}}CH_3 \qquad HOCH_2CH_2NH_3^+ \qquad HOCH_2\underset{NH_3^+}{CH}CO_2^-$$

choline   ethanolamine   serine   inositol

The most abundant phosphoglycerides in higher plants and animals are phosphatidyl choline, more commonly known as lecithin, and phosphatidyl ethanolamine or cephalin.

$$R_2C-O-CH \quad \begin{matrix} CH_2-O-CR_1 \\ | \\ | \\ CH_2-O-P-O-CH_2CH_2\overset{+}{N}CH_3 \\ | \\ O^- \end{matrix}$$

a phosphatidyl choline
a lecithin

$$R_2C-O-CH \quad \begin{matrix} CH_2-O-CR_1 \\ | \\ | \\ CH_2-O-P-O-CH_2CH_2NH_3^+ \\ | \\ O^- \end{matrix}$$

a phosphatidyl ethanolamine
a cephalin

The lecithins and cephalins are a family of compounds that differ only in the nature of the fatty acid components. The fatty acids most common in these membrane phosphoglycerides are palmitic and stearic acids (both fully saturated) and oleic acid (one double bond in the hydrocarbon chain).

12.5 Cell Membranes

We know that the membranes of plant and animal cells are typically composed of 40 to 50% phospholipid and 50 to 60% protein. Yet there are wide variations in phospholipid–protein content even between different types of cells within the same organism. For example, myelin, the membrane that surrounds specific types of nerve cells and serves as an insulator, contains only about 18% protein. Membranes that are active in transporting specific molecules into and out of cells contain about 50%

protein. Membranes actively involved in the transformation of energy, such as those of mitochondria and chloroplasts, contain up to 75% protein.

We also know that the cell membrane is an important feature of cell structure and that it serves a number of functions essential to the life of the cell. First, it is a mechanical support which separates the contents of the cell from the external environment. Second, the cell membrane provides structural support for certain proteins which are responsible for transporting ions and polar molecules across cell membranes and for the multitude of processes by which a cell communicates with its external environment. Some membrane proteins serve as "pumps" and "gates" to transport certain molecules across membranes but exclude others. Others act as "receptor sites" by which certain molecules on the outside of the cell communicate messages to the inside of the cell. For example, the poly-peptide hormone, insulin, regulates the cellular uptake of glucose in certain target cells yet does not cross the cell membrane. Instead, insulin reacts with specific receptor proteins on the outer surface of the membrane and the receptor proteins in turn communicate the insulin-borne message to the inside of the cell. Still other membrane-bound proteins catalyze specific intracellular processes. Obviously, the cell membrane is more than an impervious, mechanical barrier separating the cell contents from the external environment. It is a highly specialized structure that performs a multitude of tasks with great precision and accuracy.

The question of the detailed molecular structure of cell membranes is one of the most challenging problems in biochemistry today. Despite intensive research, many aspects of cell membrane structure and activity still are not understood. Let us begin the discussion of a possible model for cell membrane structure by first considering the shapes of phospholipid molecules and the possible arrangement of phospholipids in aqueous solution.

Phospholipids are elongated, almost rod-like molecules with the two nonpolar fatty acid hydrocarbon chains lying essentially parallel to one another and with the polar choline, serine, ethanolamine or inositol and phosphate groups pointing in the other direction (Figure 12.1).

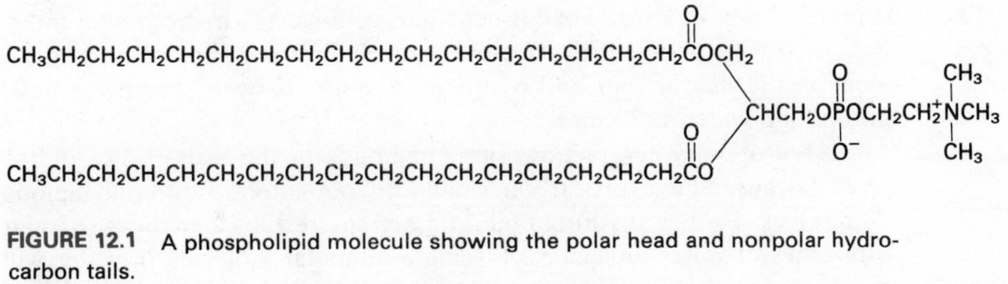

FIGURE 12.1 A phospholipid molecule showing the polar head and nonpolar hydrocarbon tails.

Next let us consider what happens when phospholipid molecules are placed in an aqueous medium. Recall that when soap molecules are placed in water, they form micelles (Section 8.10) in which the polar head groups interact with water molecules and the nonpolar hydrocarbon tails cluster within the micelle and are removed from contact with water. One possible arrangement for phospholipids also is micelle formation (Figure 12.2).

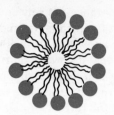

FIGURE 12.2 Micelle
formation of phospholipids
in aqueous medium.

Another arrangement which also satisfies the requirement that polar groups interact with water and nonpolar groups avoid water is the so-called lipid bilayer or bimolecular sheet. A schematic diagram of a lipid bilayer is shown in Figure 12.3.

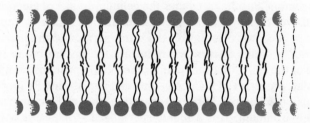

FIGURE 12.3 Lipid bilayer or bimolecular sheet formed by phospholipid molecules in aqueous medium.

The favored structure for most phospholipids in aqueous solution is the lipid bilayer rather than the micelle because micelles can only grow to a limited size before holes begin to appear in the outer polar surface. Lipid bilayers can grow to almost infinite size and provide a boundary surface for the cell, whatever its size.

It is important to realize that the self-assembly of phospholipid molecules into the lipid bilayer is a spontaneous process and the information necessary to assemble individual molecules into bilayers is inherent in the very structures of the phospholipid molecules themselves. Specifically, this information is contained in the particular combination of a polar head and nonpolar tails in each molecule. This spontaneous self-assembly is driven by two types of noncovalent interactions: (1) hydrophobic interactions when nonpolar chains unite to exclude water molecules, and (2) electrostatic interactions and hydrogen bonding between the polar head groups and water molecules.

Hydrophobic interactions are not bonds in the usual sense of the word because the hydrocarbon chains do not interact or bond among themselves. Rather, hydrophobic interactions result from the particular properties of water molecules. Placing a nonpolar molecule in water will cause the surrounding water molecules to become more ordered and restricted in their motions, that is, the entropy of the solution is decreased. Removing the nonpolar molecule from water will leave the water molecules more free to move about, that is, the entropy of the solution is increased. Thus, clustering together of nonpolar molecules or groups occurs not because the nonpolar groups interact with themselves but rather because the surrounding water molecules can return to a more random and disordered state.

As we might expect from these structural characteristics, lipid bilayers are highly impermeable to ions and most polar molecules, for it would take a great deal of energy (work) to transport an ion or a polar molecule through the nonpolar interior of the bilayer. One exception is water, which readily passes in and out of the lipid bilayer. Glucose passes through lipid bilayers 10,000 times more slowly than water, and sodium ion 1,000,000,000 times more slowly than water.

The most satisfactory current model for the arrangement of proteins and phospholipids in plant and animal cell membranes is the so-called fluid-mosaic model. According to this model, the membrane phospholipids form a lipid bilayer which is highly impermeable to virtually all polar molecules and ions. The membrane proteins are imbedded in this bilayer. Some proteins are exposed to the aqueous environment on the outer surface of the membrane, others provide channels that penetrate from the outer to the inner surface of the membrane, while still others are imbedded within the lipid bilayer. Four possible protein arrangements are shown schematically in Figure 12.4.

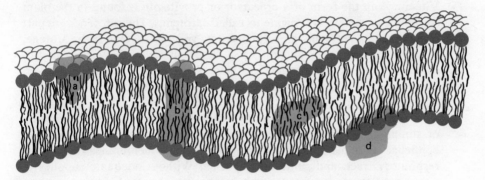

FIGURE 12.4 Fluid-mosaic model of a cell membrane showing the lipid bilayer and membrane proteins oriented (a) on the outer surface, (b) penetrating the entire thickness of the membrane, (c) embedded within the membrane, and (d) on the inner surface of the membrane.

The fluid-mosaic model is consistent with the evidence provided by chemical analysis and electron microscope pictures of cell membranes. However, this model does not explain just how membrane proteins act as pumps and gates for the transport of ions and molecules across the membrane, or how they act as receptors for hormone-borne messages and communications between one cell and another. Nor does it explain how enzymes bound on membrane surfaces catalyze the reactions they do. All of these questions are very active areas of research today.

12.6 Fat-Soluble Vitamins

Vitamins are divided into two broad classes on the basis of solubility: those that are water-soluble, and those that are fat-soluble. The water-soluble vitamins include ascorbic acid, nicotinamide, pantothenic acid, thiamine, riboflavin, pyridoxine, lipoic acid, biotin, the folic acid group, and vitamin B_{12}. In most cases, the water-soluble vitamins serve as components of

335

coenzymes and the role of these coenzymes is well understood. The fat-soluble vitamins include vitamins A, D, E, and K. At the present time, the molecular basis of their function is only poorly understood.

Vitamin A, or retinol, is a primary alcohol of molecular formula $C_{20}H_{30}O$. Notice that its structural formula can be divided into four isoprene units and, therefore, it can be classed as a terpene.

$$H_3C \quad CH_3 \qquad CH_3 \qquad\qquad CH_3$$

$$CH \qquad C \qquad CH \qquad C \qquad CH_2OH$$

$$CH \qquad CH \qquad CH \qquad CH$$

$$CH_3$$

vitamin A
retinol

Vitamin A alcohol occurs only in the animal world, where the best sources are cod-liver oil and other fish-liver oils, animal liver, and dairy products. Vitamin A in the form of a precursor or provitamin is found in the plant world in a group of plant pigments called carotenes. The carotenes are part of the pigments of many green and yellow vegetables. One such carotene, β-carotene, is shown in Section 3.5 in the discussion of the terpene hydrocarbons. The carotenes themselves have no vitamin-A activity. However, after ingestion, they are cleaved in the intestinal mucosa at the central double bond into vitamin A.

A deficiency of vitamin A or vitamin-A precursors leads to a slowing or stopping of growth. Probably the major action of this vitamin is on epithelial cells, particularly those of the mucous membranes of the eye, respiratory tract, and genitourinary tract. Without adequate supplies of vitamin A, these mucous membranes become hard and dry, a process known as keratinization. One of the first and most obvious effects of vitamin-A deficiency is on the eye. The cells of the tear glands become keratinized and stop secreting tears, and the external surface of the eye becomes dry, dull, and often scaly. Without tears to remove bacteria, the eye is much more susceptible to serious infection. If this condition is not treated in time, blindness results. The mucous membranes of the respiratory, digestive, and urinary tracts also become keratinized in vitamin-A deficiency and susceptible to infection.

A less serious condition but one frequently seen in humans whose diets contain insufficient vitamin A is night blindness. This is the inability to see in dim light or to adapt to a decrease in light intensity.

Vitamin D is a secondary alcohol of molecular formula $C_{28}H_{44}O$. Its primary effect is on calcium metabolism. Vitamin D increases the absorption of calcium ions (Ca^{2+}) from the intestinal tract and is necessary for normal calcification of bone tissue. Rickets is a vitamin-D deficiency disease in which the growing parts of bones are affected, producing bowlegs, knock-knees, and enlarged joints. Because of its relationship to rickets, vitamin D is often called the antirachitic vitamin.

There are at least ten different compounds with antirachitic activity. These are designated D_1, D_2, and so on. The two of greatest importance are D_2 or ergocalciferol of vegetable origin, and D_3 or cholecalciferol of animal origin.

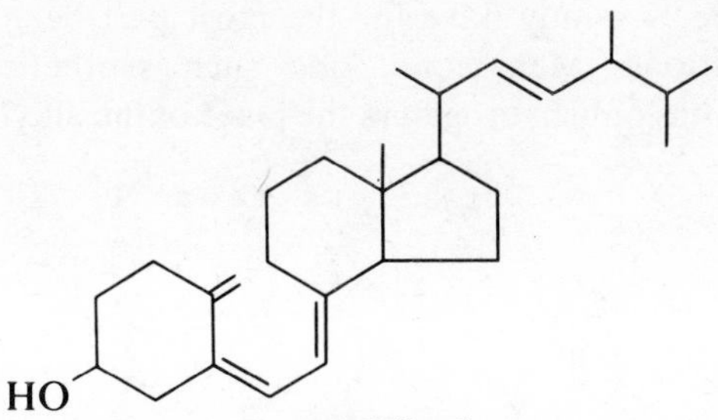

ergocalciferol
D_2

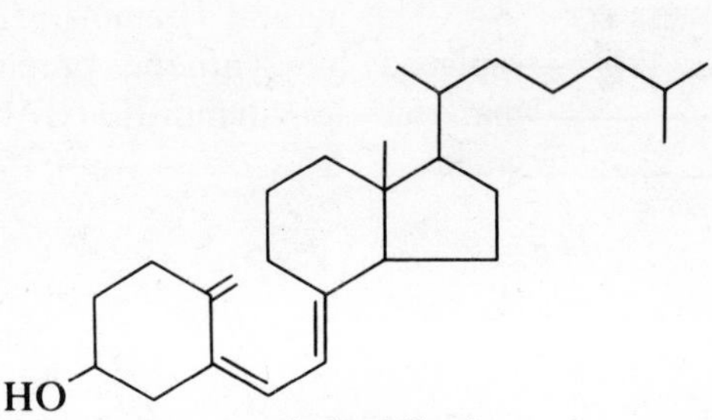

cholecalciferol
D_3

D_3 is the form of the vitamin found in fish-liver oils, the richest natural source. Vitamin D_3 is also produced in the skin by irradiation of 7-dehydrocholesterol. It is for this reason that exposure to sunlight greatly increases the vitamin-D content of the body.

Vitamin E is actually a group of about seven compounds of similar structure. Of these, α-tocopherol has the greatest potency. Notice that part of the structural formula of α-tocopherol can be divided into four isoprene units.

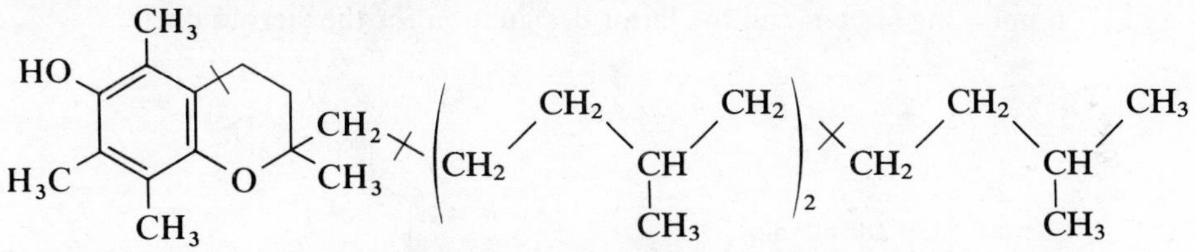

vitamin E (α-tocopherol)

Vitamin E was first recognized in 1922 as a dietary substance essential for normal reproduction in rats. From this observation came its name (Greek, *tocopherol*, promoter of childbirth). Vitamin E occurs in fish oil, in other oils such as cottonseed and peanut oil, and in green leafy vegetables. The richest source of vitamin E is wheat germ oil. In the body, vitamin E functions as an antioxidant in that it inhibits the oxidation of unsaturated lipids by molecular oxygen. In addition it is necessary for the proper development and functioning of membranes in red blood cells, muscle cells, and other cells.

Vitamin K was discovered in 1935 as a result of a study of newly hatched chicks that had a fatal hemorrhagic disease. This condition could be prevented and cured by the administration of a substance found in hog liver and in alfalfa. It was later discovered that the delayed clotting time of the blood was caused by a deficiency of prothrombin. It is now known that vitamin K is essential for the synthesis of prothrombin in the liver. The natural vitamin has a long, branched alkyl chain of usually 20 or 30 carbon atoms.

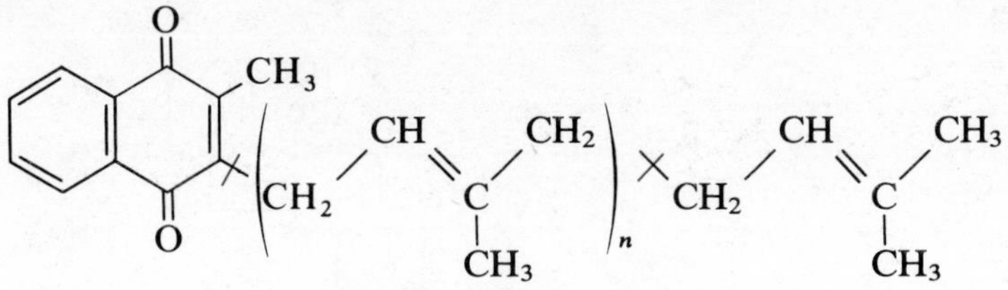

vitamin K_2 (n may be 5, 6, or 8)

The natural vitamins of the K family have for the most part been replaced by synthetic preparations. Menadione, one such synthetic material with vitamin-K activity, has only hydrogen in the place of the alkyl chain.

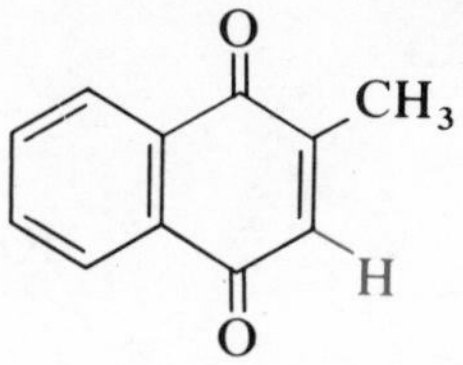

menadione

12.7 Steroids

All steroids contain four fused carbon rings: three rings of six carbon atoms and one ring of five carbon atoms. These 17 carbon atoms make up the structural unit known as the steroid nucleus. Figure 12.5 shows both the numbering system and the letter designation for the steroid nucleus.

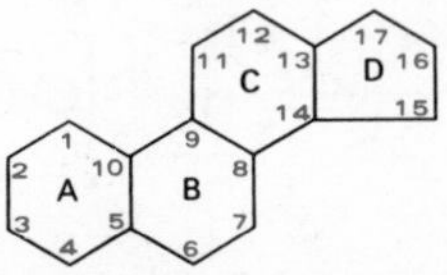

FIGURE 12.5 The steroid nucleus.

The steroid nucleus is found in a number of extremely important biomolecules. For our discussion, we will divide these into four groups: cholesterol, the adrenocorticoid hormones, the sex hormones, and bile acids.

12.8 Cholesterol

Cholesterol is a white, optically active solid found in varying amounts in practically all living organisms except bacteria. In animal cells it serves (1) as an essential component of membrane structures, and (2) as the precursor of bile acids, steroid hormones, and vitamin D. In man, the central and peripheral nervous systems have a very high cholesterol content (about 10% of dry brain weight). Human plasma contains an average of 50 mg of free cholesterol per 100 ml and about 170 mg of cholesterol esterified with fatty acids. Gallstones are almost pure cholesterol.

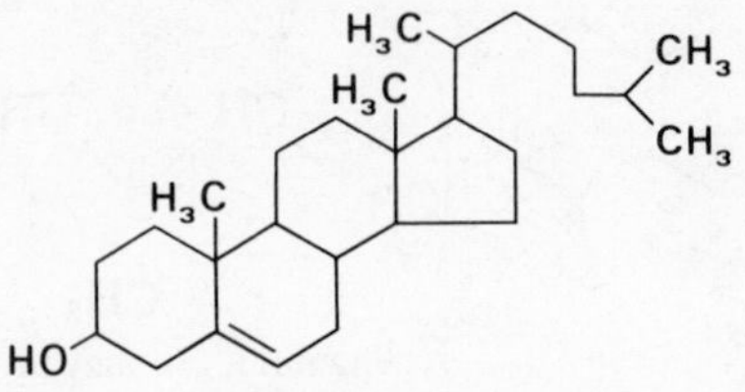

FIGURE 12.6 Cholesterol.

Most organisms, including man, have the ability to synthesize cholesterol in a complex series of reactions beginning with acetate molecules in the form of acetyl coenzyme A.

$$\text{acetate units in the form of acetyl CoA} \xrightarrow[\text{enzyme-catalyzed reactions}]{\text{complex series of}} \text{cholesterol}$$

In fact, biosynthesis of cholesterol provides the major portion of man's and other animals' need for this steroid.

Since it is relatively easy to measure the concentration of cholesterol in serum, a great deal of information has been collected in attempts to correlate serum levels with various diseases. One of these diseases, arteriosclerosis or "hardening of the arteries," is among the most common diseases of aging. With increasing age, humans normally develop decreased capacity to metabolize fat, and therefore, cholesterol concentration in the tissues increases. When arteriosclerosis is accompanied by the build-up of cholesterol and other lipids on the inner surfaces of arteries, the condition is known as atherosclerosis and results in a decrease in the diameter of the channels through which blood must flow. This decreased diameter together with increased turbulence leads to a greater probability of clot formation within the channel. If the channel is blocked by a clot, cells may be deprived of oxygen and die. The death of tissue in this way is called infarction.

It is well known that those suffering from atherosclerosis usually have high blood cholesterol levels. However, this is not always the case. Not all patients with atherosclerosis have high blood cholesterol levels, and not all persons with high cholesterol levels have atherosclerosis. Consequently, there are different theories on the origin and treatment of this disease. Blood cholesterol levels do respond directly to the amount and type of fat in the diet. People on low-fat diets have lower cholesterol levels than those on higher fat diets and the incidence of atherosclerosis in these people is lower. In addition, the amount of carbohydrate and the relative proportion of unsaturated to saturated fatty acids in the diet significantly affect blood cholesterol levels in man.

12.9 General Characteristics of Hormones

Communication is one of the most important problems in any multicellular organism. For example, in the human body it is absolutely essential that the various tissues communicate with each other so that each can perform its particular role with maximum efficiency. For this communication there are two prime channels. One is the central nervous system. The other is by chemical means and involves the synthesis and release of substances called hormones. Actually, the two systems, the neurologic and the hormonal, are closely interrelated; working together they regulate our metabolism, growth, and development.

Hormones are produced by special glands or cells such as the adrenals, ovaries, testes, pituitary, and the thyroid. They are generally classified into three groups:

1. Steroids. These all have the four-ring steroid nucleus and are derived from cholesterol.
2. Derivatives of amino acids. These include thyroxine and epinephrine which are derived from the aromatic amino acid tyrosine.
3. Polypeptides and proteins (Section 13.8).

In this chapter we will examine the structural formulas of two classes of steroid hormones: the adrenocorticoid hormones and the sex hormones. In the next chapter we will study hormones related to amino acids and to polypeptide and protein hormones.

12.10 Adrenocorticoid Hormones

The cortex of the adrenal gland is stimulated by the pituitary hormone ACTH to synthesize several hormones that affect (1) water and electrolyte balance and (2) carbohydrate and protein metabolism. Those that control mineral balance are called mineralocorticoids; those that control glucose and carbohydrate balance are called glucocorticoid hormones. Both groups of hormones are derived from cholesterol and have the four-ring nucleus common to all steroids.

Aldosterone is the most effective of the mineralocorticoid hormones secreted by the adrenal cortex. This hormone acts on kidney tubules to stimulate the resorption of sodium ions and thus regulates water and electrolyte metabolism. An adult on a diet of normal sodium content secretes about 0.1 to 0.2 mg per day of aldosterone.

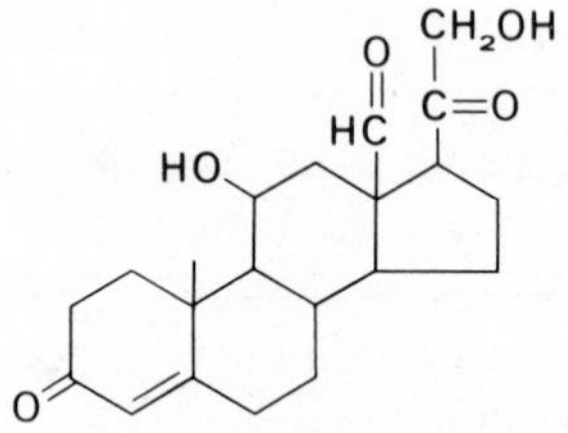

FIGURE 12.7
Mineralocorticoid
hormone, aldosterone.

aldosterone

Cortisol is the principal glucocorticoid hormone of the adrenal cortex which secretes about 25 mg per day. Cortisol effects the metabolism of carbohydrates, proteins, and fats; it effects water and electrolyte balance; and it effects inflammatory processes within the body. In the presence of cortisol, the synthesis of protein in muscle tissue is depressed, protein degradation is increased, and there is an increase in the supply of free amino acids in both muscle cells and blood plasma. The liver, in turn, is stimulated to use the carbon skeletons of certain of the amino acids for the synthesis of glucose and glycogen. (The biosynthesis of glucose from noncarbohydrate substances, such as amino acids, is called gluco-neogenesis.) Thus, cortisol and other glucocorticoid hormones act to increase the supply of glucose and liver glycogen at the expense of body protein. Cortisol also has some mineralocorticoid action, that is, it promotes resorption of sodium ions by the tubules of the kidney and water retention. However, it is far less potent as a mineralocorticoid than is aldosterone.

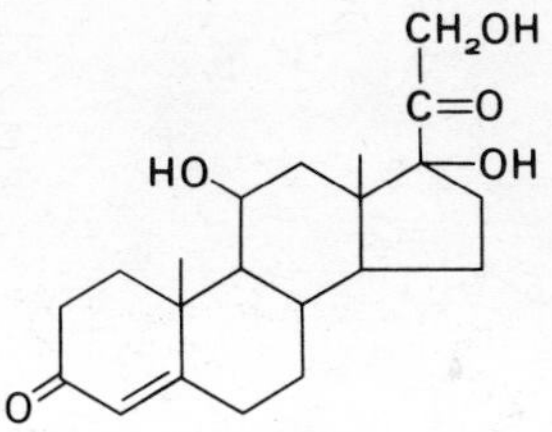

FIGURE 12.8 Glucocorticoid hormones.

Cortisol and its oxidation product, cortisone (Figure 12.8), are probably best known for their use in clinical medicine as remarkably effective anti-inflammatory agents. They are used in the treatment of a host of inflammatory diseases including acute attacks of rheumatoid arthritis and bronchial asthma, and inflammations of the eye, colon, and other organs. Laboratory research has produced a series of semisynthetic steroid hormones (for example, prednisolone Figure 12.8) which are even more potent than cortisone in treating inflammatory diseases. Many of these semisynthetic hormones have an additional advantage over cortisone in that they do not at the same time stimulate sodium retention and fluid accumulation.

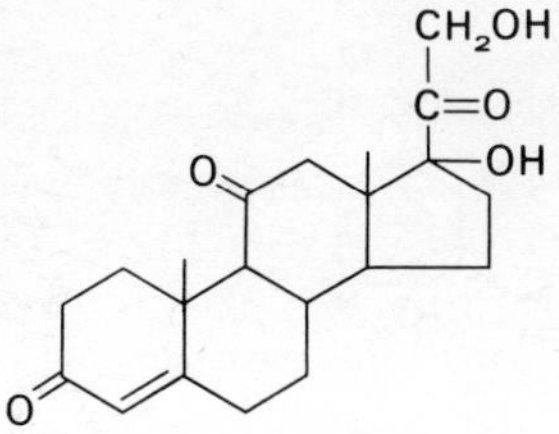

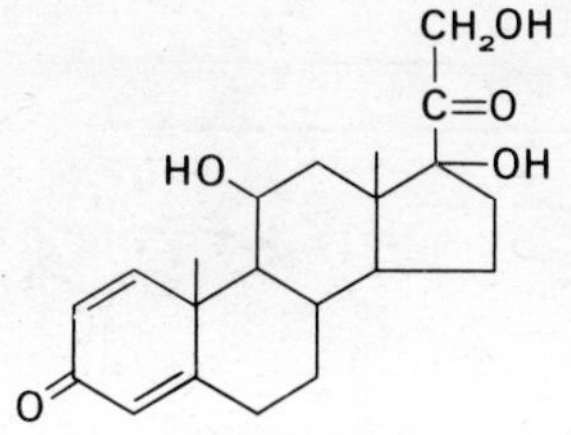

12.11 Sex Hormones

The testes in the male and ovaries in the female, besides producing spermatozoa or ova, also produce steroid hormones which control secondary sex characteristics, the reproductive cycle, and the growth and development of accessory reproductive organs.

Of the male sex hormones or <u>androgens</u>, <u>testosterone</u> is the most important. It is produced in the testes from cholesterol. The chief function of testosterone is to promote normal growth of the male reproductive organs and development of the characteristic deep voice, pattern of facial and body hair, and male type of musculature.

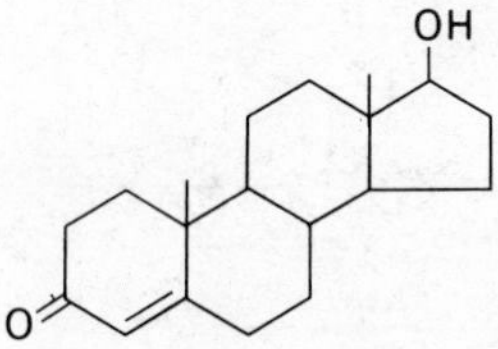

FIGURE 12.9 Testosterone.

In the female there are two types of sex hormones of particular importance, progesterone and a group of hormones known as <u>estrogens</u>. Changing rates of secretion of these hormones cause the periodic change in the ovaries and uterus known as the <u>menstrual cycle</u>. Immediately following menstrual flow, increased estrogen secretion causes growth of the lining of the uterus and ripening of the ovum. <u>Estradiol</u> is one of the most important estrogens, which are also responsible for development of the female secondary sex characteristics.

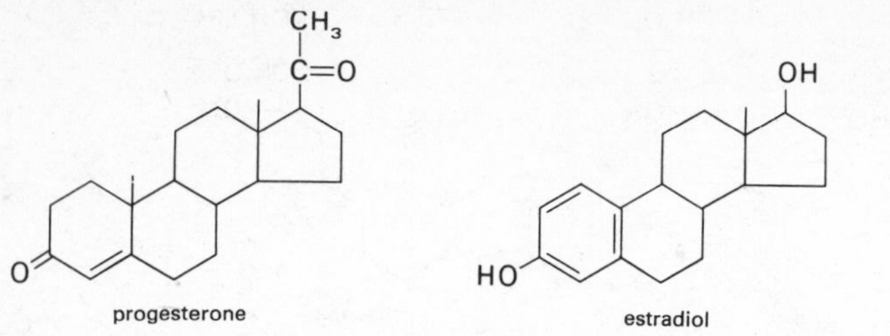

FIGURE 12.10 Progesterone and estradiol, two female sex hormones.

Progesterone is synthesized by the oxidation of cholesterol. Its secretion just prior to ovulation prevents other ova from ripening and also prepares the uterus for implantation and maintenance of a fertilized egg. If conception does not occur, progesterone production decreases and menstruation occurs. If fertilization and implantation do occur, production of progesterone continues and helps to maintain the pregnancy. One of the consequences of continued progesterone production is prevention of ovulation during pregnancy.

Once the role of progesterone in inhibiting ovulation was understood, its potential as a possible contraceptive drug was realized. Unfortunately, progesterone itself is relatively ineffective when taken orally and injection often produces local irritation. As a result of massive research programs, a large number of synthetic steroids that could be administered orally became available in the early 1960s. When taken regularly, these drugs prevent ovulation yet allow most women a normal menstrual cycle. Some of the most effective contain a progesterone-like analog such as ethynodiol diacetate (Figure 12.11) combined with a smaller amount of an estrogen-like material. The small amount of estrogen prevents irregular menstrual flow ("breakthrough bleeding") during prolonged use of contraceptive pills.

FIGURE 12.11 Ethynodiol diacetate a progesterone analog widely used in oral contraceptive preparations.

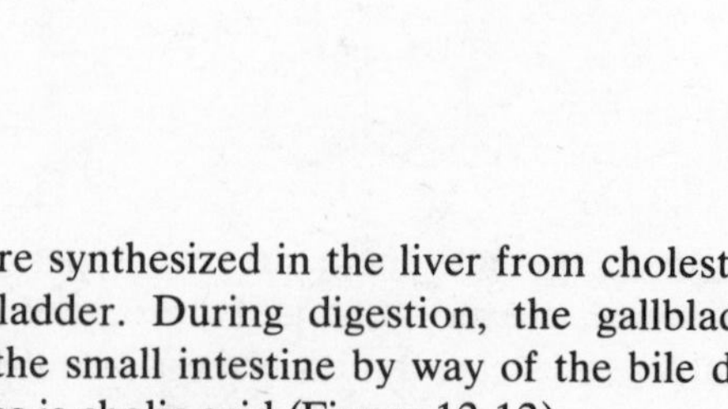

12.12 Bile Acids

These important compounds are synthesized in the liver from cholesterol and then stored in the gallbladder. During digestion, the gallbladder contracts and supplies bile to the small intestine by way of the bile duct. The primary bile acid in humans is cholic acid (Figure 12.12).

Bile acids have several important functions. First, they are products of the breakdown of cholesterol and thus are a major pathway for the elimination of cholesterol from the body via the feces. Second, because they are able to emulsify fats in the intestine, bile acids aid in the digestion

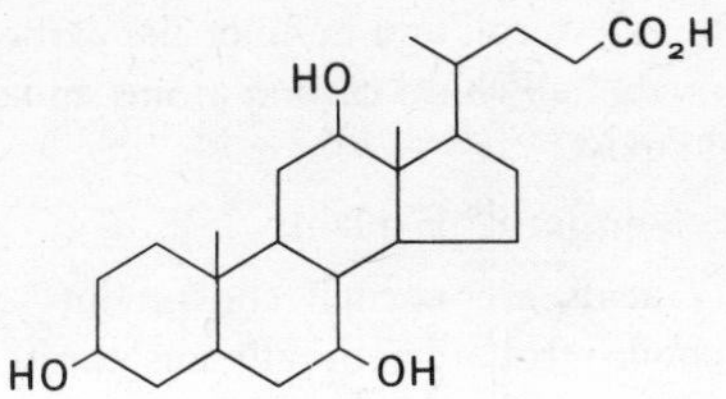

FIGURE 12.12 Cholic acid, an important constituent of human bile.

and absorption of dietary fats. Third, they can dissolve cholesterol by the formation of cholesterol–bile salt micelles or cholesterol–lecithin–bile salt micelles. In this way cholesterol, whether it is from the diet, synthesized in the liver, or removed from circulation by the liver, can be solubilized.

PROBLEMS

12.1 List six major functions of lipids in the human body. Name and draw a structural formula for a lipid representing each function.

12.2 How many isomers (including stereoisomers) are possible for a triglyceride containing one molecule each of palmitic, stearic, and oleic acid?

12.3 What is meant by the term "hardening" as applied to fats and oils?

12.4 Saponification (Section 8.10) is the alkaline hydrolysis of naturally occurring fats and oils. A saponification number is the number of mg of potassium hydroxide required to saponify 1 g of a fat or oil. Calculate the saponification number of tristearin (p. 330), molecular weight 890.

12.5 The saponification number of butter is about 230; that of oleomargarine is about 195.
(a) Calculate the average molecular weight of butter; of oleomargarine.
(b) Show that these molecular weight values are in agreement with the fact that butter differs from oleomargarine in having appreciable amounts of lower-molecular-weight fatty acids (Table 12.1).

12.6 Calculate the percentage of unsaturated fatty acids in butter fat. Compare this with the percentage in olive oil.

12.7 Draw a structural formula for a phosphatidyl serine; a phosphatidyl inositol.

12.8 Draw structural formulas for the products of complete hydrolysis of a lecithin; a cephalin.

12.9 Two of the major noncovalent forces directing the organization of biomolecules in aqueous solution are the tendencies to (1) arrange polar groups so that they can interact with water by hydrogen bonding, and (2) arrange nonpolar molecules or groups so that they are shielded from water. Show how these forces direct micelle formation by soap molecules and lipid bilayer formation by phospholipids.

12.10 Describe the major features of the fluid-mosaic model of cell membrane structure.

12.11 Describe three major functions of the protein components of cell membranes.

12.12 Examine the structural formula of vitamin A and state the number of *cis-trans* isomers possible for this molecule.

12.13 Describe the symptoms of severe vitamin-A deficiency.

12.14 Examine the structural formulas of vitamins A, D_2, and D_3, E, and K_2. Based on their structural formulas, would you expect them to be more soluble in water or in olive oil? Would you expect them to be soluble in blood plasma?

12.15 Draw the structural formula of cholesterol, and number the carbon atoms in accord with the IUPAC convention. Label all chiral carbon atoms and state the total number of stereoisomers that could exist.

12.16 Are humans able to synthesize cholesterol? Explain.

12.17 Esters of cholesterol and fatty acids are normal constituents of blood plasma. The fatty acids esterified with cholesterol are generally unsaturated. Draw the structural formula for cholesteryl oleate.

12.18 Cholesterol is an important component of the lipid fraction of cell membranes. However, its precise function in membranes is unknown. How do you think a cholesterol molecule might be oriented in a cell membrane according to the fluid-mosaic model?

12.19 Name three general groups of hormones.

12.20 Name the six functional groups in cortisol; in aldosterone.

12.21 Examine the structural formulas of testosterone, a male sex hormone, and progesterone, a female sex hormone. What are the similarities in structure between the two? the differences?

12.22 Why are progesterone and the estrogens called "sex hormones"?

12.23 Describe how a combination of progesterone and estrogen analogs functions as an oral contraceptive.

12.24 Examine the structural formula of cholic acid and account for the fact that this and other bile acids are able to emulsify fats and oils.

Amino Acids and Proteins

13.1 Introduction

<u>Proteins</u>, as much as any other class of compounds, are inseparable from life itself. These remarkable molecules function in a variety of separate and distinct ways. Certain <u>fibrous proteins</u> serve as major elements of the structural support system for living organisms. For example, the keratins are the structural proteins of hair, skin, nails, and claws. Collagen is the structural protein of the connective tissue of flesh, tendons, and muscle. Myosin and actin are major components of the myofilaments of skeletal muscle. Other proteins, the <u>enzymes</u>, catalyze the myriad cellular reactions vital to maintenance and growth of an organism. Certain large globular proteins known as <u>antibodies</u> function as one of the major defense mechanisms of the body. <u>Polypeptide hormones</u> like insulin, glucagon, and oxytocin play essential roles in the regulation of physiological processes.

In this chapter we will examine a few of the more important properties of proteins and explore the question of what it is that gives these molecules such a wide range of properties. To do this we shall have to look at what proteins are and how they are put together.

13.2 Amino Acids

Amino acids are substances that contain both a <u>carboxyl group</u> and an <u>amino group</u>. While many types of amino acids are known in nature, it is the α-amino acids that are most significant in the biological world for they are the fundamental units from which proteins are constructed. The general formula of an α-amino acid is shown in Figure 13.1.

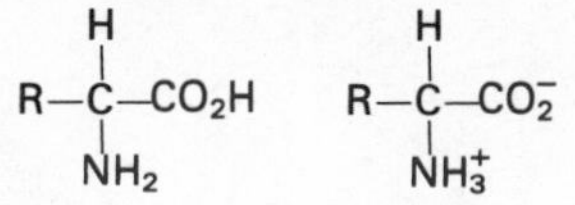

$$R-\underset{\underset{NH_2}{|}}{\overset{\overset{H}{|}}{C}}-CO_2H \qquad R-\underset{\underset{NH_3^+}{|}}{\overset{\overset{H}{|}}{C}}-CO_2^-$$

(a) (b)

FIGURE 13.1 General formula for α-amino acids.

Although Figure 13.1(a) is a common way of writing structural formulas for amino acids, it is not an accurate representation for it shows an un-ionized acidic group ($-CO_2H$) and an unprotonated basic group ($-NH_2$) within the same molecule. These acidic and basic groups will of course react to form an internal salt or zwitterion, Figure 13.1(b). Note that the zwitterion form of an amino acid has no net charge; it contains one positive and one negative charge. In the remainder of this text we will use the zwitterion formulas for amino acids. Furthermore, when discussing amino acids and proteins important in biological systems, we will show all amino and carboxyl groups as they are ionized at pH 7.4, the physiological pH.

It is readily apparent from Figure 13.1(b) that if the R— group is something other than hydrogen, the amino acids of protein origin contain a chiral carbon atom adjacent to the carboxylic acid. For example, the amino acid serine, where R is $-CH_2OH$, may exist in two forms, one the mirror image of the other. It has been established that all the amino acids that occur naturally in proteins have the same configuration about the chiral α-carbon. With D-glyceraldehyde as a standard, the naturally occurring amino acids have the opposite, or L-configuration. This relationship is illustrated for L-serine.

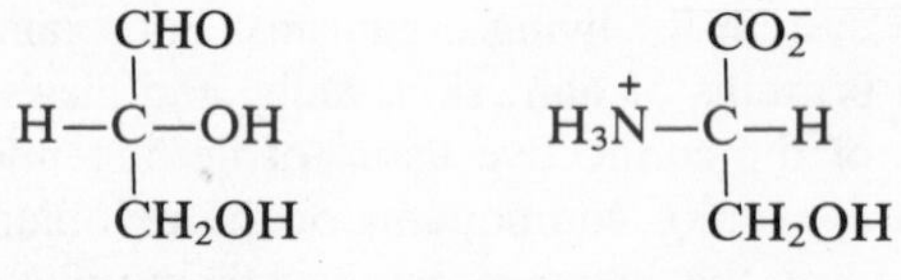

$$\underset{\text{D-glyceraldehyde}}{H-\underset{\underset{CH_2OH}{|}}{\overset{\overset{CHO}{|}}{C}}-OH} \qquad \underset{\text{L-serine}}{H_3\overset{+}{N}-\underset{\underset{CH_2OH}{|}}{\overset{\overset{CO_2^-}{|}}{C}}-H}$$

While D-amino acids are not found in proteins and are not a part of the metabolism of higher organisms, several are important in the structure and metabolism of lower forms of life. As an example, both D-alanine and D-glutamic acid are structural components of the cell walls of certain bacteria. For a discussion of the cell wall structure of these bacteria, see the mini-essay "The Penicillins."

Table 13.1 shows the names, structural formulas, and standard three-letter abbreviations for each of the 20 common amino acids found in proteins. In this table the amino acids are grouped into three categories according to the nature of their side chains. The nonpolar category includes glycine, alanine, valine, leucine, isoleucine, and proline with aliphatic hydrocarbon chains; phenylalanine with an aromatic side chain; and methionine. The polar but neutral category includes serine and threonine with hydroxyl groups; asparagine and glutamine with amide groups;

TABLE 13.1 The 20 common amino acids of protein origin, grouped by categories.

SECTION 13.2
Amino Acids

Nonpolar Side Chains

$$H-\overset{\overset{\displaystyle NH_3^+}{|}}{CH}-CO_2^-$$

glycine (gly)

$$CH_3-\overset{\overset{\displaystyle NH_3^+}{|}}{CH}-CO_2^-$$

L-alanine (ala)

$$\overset{\displaystyle H_3C}{\underset{\displaystyle H_3C}{>}}CH-\overset{\overset{\displaystyle NH_3^+}{|}}{CH}-CO_2^-$$

L-valine (val)

phenyl—$CH_2-\overset{\overset{\displaystyle NH_3^+}{|}}{CH}-CO_2^-$

L-phenylalanine (phe)

$$\overset{\displaystyle H_3C}{\underset{\displaystyle H_3C}{>}}CH-CH_2-\overset{\overset{\displaystyle NH_3^+}{|}}{CH}-CO_2^-$$

L-leucine (leu)

$$\overset{\displaystyle H_3C}{\underset{\displaystyle CH_3-CH_2}{>}}CH-\overset{\overset{\displaystyle NH_3^+}{|}}{CH}-CO_2^-$$

L-isoleucine (ile)

$$\begin{array}{c} H_2C-CH_2 \\ H_2C \qquad CH-CO_2^- \\ \overset{+}{N} \\ H \quad H \end{array}$$

L-proline (pro)

$$CH_3-S-CH_2-CH_2-\overset{\overset{\displaystyle NH_3^+}{|}}{CH}-CO_2^-$$

L-methionine (met)

Polar but Uncharged Side Chains

$$HO-CH_2-\overset{\overset{\displaystyle NH_3^+}{|}}{CH}-CO_2^-$$

L-serine (ser)

$$CH_3-\overset{\overset{\displaystyle OH}{|}}{CH}-\overset{\overset{\displaystyle NH_3^+}{|}}{CH}-CO_2$$

L-threonine (thr)

$$H_2N-\overset{\overset{\displaystyle O}{||}}{C}-CH_2-\overset{\overset{\displaystyle NH_3^+}{|}}{CH}-CO_2$$

L-asparagine (asn)

$$H_2N-\overset{\overset{\displaystyle O}{||}}{C}-CH_2-CH_2-\overset{\overset{\displaystyle NH_3^+}{|}}{CH}-CO_2^-$$

L-glutamine (gln)

$$HS-CH_2-\overset{\overset{\displaystyle NH_3^+}{|}}{CH}-CO_2^-$$

L-cysteine (cys)

$$HO-\text{(ring)}-CH_2-\overset{\overset{\displaystyle NH_3^+}{|}}{CH}-CO_2^-$$

L-tyrosine (tyr)

$$\text{(indole)}-CH_2-\overset{\overset{\displaystyle NH_3^+}{|}}{CH}-CO_2^-$$

L-tryptophan (trp)

Polar Charged Side Chains

$$^-O-\overset{\overset{\displaystyle O}{||}}{C}-CH_2-\overset{\overset{\displaystyle NH_3^+}{|}}{CH}-CO_2^-$$

L-aspartic acid (asp)

$$^-O-\overset{\overset{\displaystyle O}{||}}{C}-CH_2-CH_2-\overset{\overset{\displaystyle NH_3^+}{|}}{CH}-CO_2^-$$

L-glutamic acid (glu)

$$\text{(imidazole)}-CH_2-\overset{\overset{\displaystyle NH_3^+}{|}}{CH}-CO_2^-$$

L-histidine (his)

$$H_3\overset{+}{N}-CH_2-CH_2-CH_2-CH_2-\overset{\overset{\displaystyle NH_3^+}{|}}{CH}-CO_2^-$$

L-lysine (lys)

$$H_2N-\overset{\overset{\displaystyle NH_2^+}{||}}{C}-NH-CH_2-CH_2-CH_2-\overset{\overset{\displaystyle NH_3^+}{|}}{CH}-CO_2^-$$

L-arginine (arg)

tyrosine with a phenolic side chain; tryptophan with a heterocyclic aromatic amine side chain; and cysteine with a sulfhydryl group. In the charged polar category are aspartic acid and glutamic acid with acidic side chains; and histidine, lysine, and arginine with basic side chains. Note that at 7.4, the pH of cellular fluids and blood plasma, aspartic acid and glutamic acid bear net negative charges because of the ionization of the side-chain carboxyl groups; lysine and arginine bear net positive charges because of the protonation of the side-chain amino groups.

A few proteins contain special amino acids. For example, both L-hydroxyproline and L-5-hydroxylysine are important components of collagen but are found in very few other proteins. These and all other special amino acids are formed after the protein is constructed by modification of one of the 20 common amino acids already incorporated into the protein.

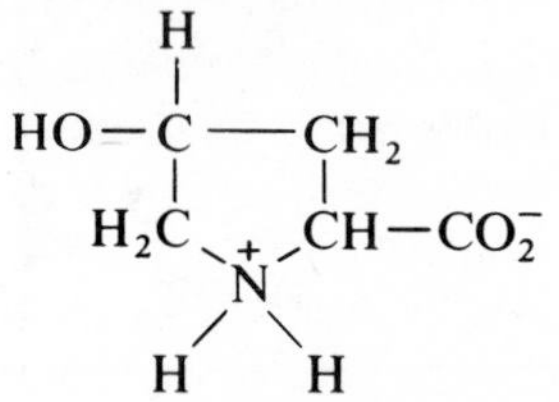

L-hydroxyproline

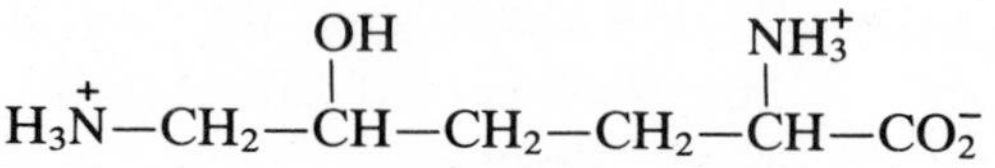

L-5-hydroxylysine

In addition to those amino acids listed in Table 13.1, there are a number of important nonprotein-derived amino acids, many of which are either metabolic intermediates or parts of nonprotein biomolecules. Several of these are shown in Table 13.2. Ornithine and citrulline each are part of the metabolic pathway that converts excess ammonia to urea. Gamma-aminobutyric acid (GABA) is present in the free state in brain tissue. Its function in the brain is largely unknown.

TABLE 13.2 Several nonprotein-derived amino acids.

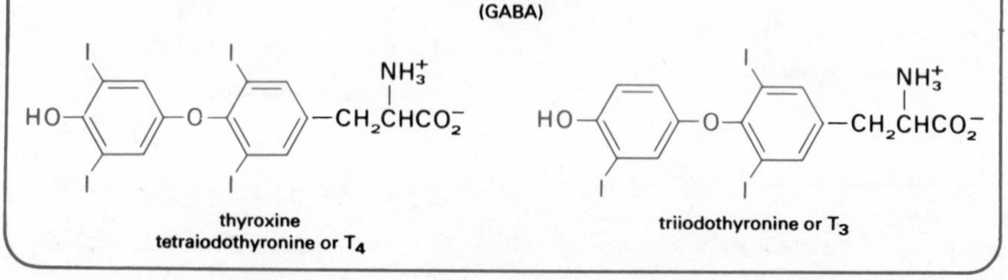

Thyroxine, one of several hormones derived from the amino acid tyrosine, was first isolated from thyroid tissue in 1914. In 1952, triiodothyronine, a compound identical to thyroxine except that it contains only three atoms of iodine, was also discovered in the thyroid. Tri-iodothyronine is even more potent than thyroxine. Although the exact mechanism of action of these thyroid hormones is not known, they are essential for the proper regulation of cellular metabolism. The levorotatory isomer of each is significantly more active than the dextrorotatory isomer.

13.3   Essential Amino Acids

Of the 20 amino acids required by the body for the production of proteins, adequate amounts of about 10 can be synthesized by enzyme-catalyzed reactions starting from carbohydrate or lipid fragments and a source of nitrogen. For the remaining amino acids, either there are no biochemical pathways available for their synthesis, or the available pathways do not provide adequate amounts for proper nutrition. Accordingly, these amino acids must be supplied in the diet, and are therefore called "essential" amino acids. In reality, all amino acids are essential for normal tissue growth and development. However, the term "essential" is reserved for those that must be supplied in the diet.

During the late 1950s, C. W. Rose and his coworkers at the University of Illinois determined which amino acids are essential for fully grown young men by first feeding them a well-balanced mixture of the known amino acids in pure form (instead of in proteins) in an otherwise adequate diet. Next, different amino acids were left out of the diet, one at a time, and the effect on nitrogen balance was observed. Rose determined that for fully grown young men, eight amino acids are essential. The estimated minimum daily requirements of these are given in Table 13.3.

TABLE 13.3   Estimated daily requirements of the essential amino acids.

Essential Amino Acids	Minimum Daily Requirement		
	infants (mg/kg)	women (grams)	men (grams)
isoleucine	126	0.45	0.70
leucine	150	0.62	1.10
lysine	103	0.50	0.80
methionine	45	0.55	1.01
phenylalanine	90	1.12	1.40
threonine	87	0.30	0.50
tryptophan	22	0.16	0.25
valine	105	0.65	0.80

Note that since phenylalanine is the precursor of tyrosine, the figures given for phenylalanine assume the presence of adequate tyrosine. Similarly, the figures for methionine assume the presence of adequate cysteine. Anyone who consumes about 40 to 55 grams of protein daily in the form of meat, fish, cheese, or eggs certainly satisfies his need for these essential amino acids.

In later studies, histidine was shown to be essential for growth in infants, and there is some evidence that it may be needed by adults as well. Arginine can be synthesized by adults, but apparently the rate of internal synthesis is not fast enough to meet the needs of the body during periods of rapid growth and protein synthesis. Therefore, depending on the age and state of health, either eight, nine or ten amino acids may be essential for humans.

The relative usefulness of a dietary protein depends on how well its amino acid pattern matches that required for the formation of tissue protein in humans. For proper tissue maintenance and growth, all amino acids, both essential and nonessential, must be present at the same time. In this sense, tissue growth is an all-or-nothing process; if even one amino acid is missing, no protein at all is made.

Plant proteins generally vary more from the amino acid pattern we require than do animal proteins. Fortunately, however, not all plant proteins are deficient in the same amino acids. For example, beans are low in the sulfur-containing amino acids cysteine and methionine, yet they are high in lysine. Wheat has just the opposite pattern. By eating wheat and beans together, it is possible to increase by 33% the usable protein you would get by eating each of these foods separately.

The provision of a diet adequate in protein and the essential amino acids is a grave problem in the world today, especially in areas of Asia, Africa, and Latin America. The overriding dimension of this problem is poverty and the inability to select foods of adequate protein and caloric content. The best overall source of calories is the cereal grains, which not only provide calories but proteins as well. When these are supplemented by animal protein or the right selection of plant protein, the diet is adequate for even the most vulnerable. However, as income decreases, there is less animal protein in the diet and even the cereal grains are often replaced by cheaper sources of calories such as sugar or tubers, which have either very little or no protein. The poorest 25% of the people in the world consume diets with caloric and protein content that falls below, often dangerously below, the calculated minimum daily requirements.

Those most apt to show the symptoms of too little food, too little protein, or both are young children in the years immediately following weaning. There is failure to grow properly and a wasting of tissue. This sickness is called marasmus, a name derived from a Greek word meaning "to waste away." The muscles become atrophied and the face develops a wizened "old man" look. Another disease, kwashiorkor, leads to tragically high death rates among children. As long as a child is breast fed it is healthy. At weaning (often forced when a second child is born), the first child's diet suddenly is switched to starch and inadequate sources of protein. Such children develop bloated bellies and patchy, discolored skin, and are often doomed to short lives.

One attack on the problem of quantity and quality of protein is the breeding of new varieties of cereal grains with higher protein content, better protein quality, or both. Alternatively, it is possible to supplement present cereals or their derived products with the deficient amino acids—principally lysine for wheat, lysine or lysine plus threonine for rice, and lysine plus tryptophan for corn. New methods of chemical synthesis and fermentation now provide a cheap source of these amino acids, thus making the economics of food fortification entirely practical. In another attack on the problem of protein malnutrition, several special high-protein, low-cost infant foods have been developed by teams of nutritionists. These take advantage of locally available foods that have supplementary amino acid compositions and conform as closely as possible to cultural food preferences. Clearly, advances in chemistry and food technology have provided the means to eradicate most hunger and malnutrition. What remains is for political and social systems to put this knowledge into practice.

13.4 Titration of Amino Acids

Glycine and all other monoamino monocarboxylic amino acids contain —COOH and —NH_3^+ groups that can ionize in aqueous solution, and each group will have a specific ionization or dissociation constant. The value of these dissociation constants can be determined by titration. In order to illustrate how this might be done, imagine that a solution contains 1 mole of glycine and that enough strong acid is added so that each ionizable group on glycine is fully protonated. Next this solution is titrated with aqueous sodium hydroxide; the volume of base added and the pH of the resulting solution are recorded and then plotted as shown in Figure 13.2.

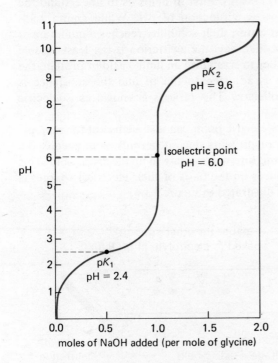

FIGURE 13.2 Titration of glycine with sodium hydroxide.

The most acidic group and the one to react first with added sodium hydroxide is —COOH. At pH 2.4, enough sodium hydroxide has been added to half-neutralize the —COOH group.

The pH at which this half-neutralization occurs is equal to the pK_a of the acid; accordingly, the pK_1 of glycine (the pK_a of the —COOH group) is

351

2.4. When 1 full mole of sodium hydroxide has been added, glycine is in the zwitterion form and has no net charge. The pH at which a molecule has no net charge is called its isoelectric point, and is denoted by pI. Accordingly, the isoelectric point of glycine is 6.0. Addition of more sodium hydroxide converts the $-NH_3^+$ group to $-NH_2$. Half-neutralization of $-NH_3^+$ occurs at pH 9.6 and accordingly the pK_2 of glycine (the pK_a of the $-NH_3^+$ group) is 9.6. This acid–base behavior of the glycine molecule is shown in Figure 13.3.

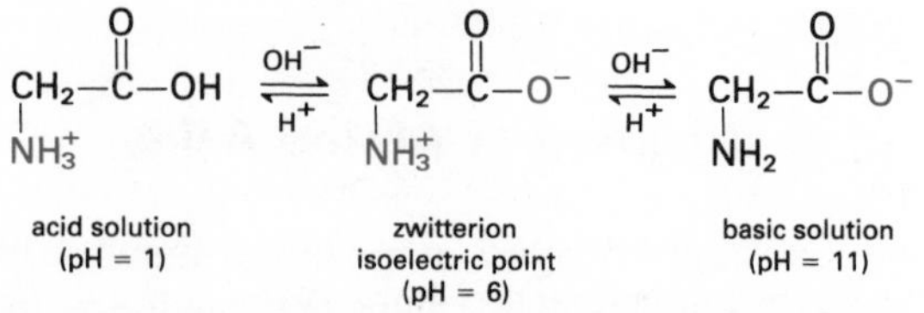

$$
\underset{NH_3^+}{CH_2}-\overset{O}{\overset{\|}{C}}-OH
\quad \underset{H^+}{\overset{OH^-}{\rightleftharpoons}} \quad
\underset{NH_3^+}{CH_2}-\overset{O}{\overset{\|}{C}}-O^-
\quad \underset{H^+}{\overset{OH^-}{\rightleftharpoons}} \quad
\underset{NH_2}{CH_2}-\overset{O}{\overset{\|}{C}}-O^-
$$

FIGURE 13.3 Acid-base behavior of glycine.

acid solution (pH = 1)   zwitterion isoelectric point (pH = 6)   basic solution (pH = 11)

The monoamino monocarboxylic acids have isoelectric points in the pH range 5.5 to 6.5 (6.0 for glycine). The basic amino acids have isoelectric points at higher pH values (for example, 9.74 for lysine), and the acidic amino acids have isoelectric points at lower pH values (for example, 3.22 for glutamic acid).

An understanding of this acid–base behavior of amino acids and proteins is important for two reasons. First, it helps us to understand the solubility of these molecules as a function of pH. While amino acids generally are quite soluble in water, their solubility reaches a minimum at the isoelectric point. Put another way, the zwitterion is the least soluble form of an amino acid. In order to crystallize an amino acid or protein, the pH of an aqueous solution is adjusted to the pI and the substance is precipitated, filtered, and collected. This process is known as isoelectric precipitation.

Second, knowing the isoelectric point can also be useful for separating mixtures of amino acids or proteins since it permits us to predict the way each component of the mixture will migrate in an electrical field. This process of separating substances on the basis of their electrical charges is called electrophoresis and is illustrated in Figure 13.4.

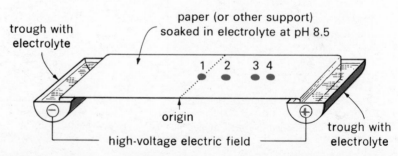

FIGURE 13.4 Paper electrophoresis of a protein mixture carried out at pH 8.5. Component 1 does not migrate from the origin, and therefore, must have an isoelectric point of 8.5. The other three components of the mixture move toward the positive electrode, and therefore, have net negative charges at pH 8.5. Component 4 has the greatest negative charge density. From David S. Page, *Principles of Biological Chemistry* (Willard Grant Press, Boston, 1976).

In paper electrophoresis, a paper strip saturated with an aqueous buffer of predetermined pH serves as a bridge between two electrode vessels. A sample of protein or amino acid is applied as a spot. When an electrical potential is applied, the protein or amino acid molecules migrate toward the electrode carrying the opposite charge. Molecules having a greater density of either positive or negative charge will move more rapidly to the opposite electrode. Any molecule already at its isoelectric point will remain stationary at the origin. After separation is complete, the strip is dried; to fix the separated components on the paper and to render them visible, the paper is generally sprayed with a dye. For example, ninhydrin can be used to make amino acids visible as purple spots.

Electrophoretic separations also can be carried out using starch, agar, certain plastics, and cellulose acetate as solid supports. This technique is extremely important in biochemical research and also is an invaluable tool in the clinical chemistry laboratory. For a discussion of the electrophoretic screening of blood samples for sickle-cell anemia, read the mini-essay "Abnormal Human Hemoglobins."

13.5 Reactions of Amino Acids

All amino acids contain two functional groups in common, a primary amine (except for proline where the amine is secondary) and a carboxylate group, and all show the reactions typical of these functional groups.

Amino acids react with alcohols to form esters. This is conveniently accomplished by adding the amino acid to a solution of alcohol containing dissolved hydrogen chloride gas.

$$R-\underset{\underset{NH_3^+}{|}}{CH}-\overset{\overset{O}{\|}}{C}-OH + CH_3CH_2OH \xrightarrow{HCl} R-\underset{\underset{NH_3^+}{|}}{CH}-\overset{\overset{O}{\|}}{C}-OCH_2CH_3 + H_2O$$

This type of esterification must be carried out under conditions acidic enough to first convert the carboxylate anion to the free acid. The ester is then isolated as the hydrochloride salt.

Amino acids can be acylated by reaction with an acid anhydride, for example, acetic anhydride.

$$R-\underset{\underset{NH_3^+}{|}}{CH}-\overset{\overset{O}{\|}}{C}-O^- + CH_3\overset{\overset{O}{\|}}{C}-O-\overset{\overset{O}{\|}}{C}CH_3 \xrightarrow{Na_2CO_3} R-\underset{\underset{HN-COCH_3}{|}}{CH}-\overset{\overset{O}{\|}}{C}-O^-$$

For acylation the $-NH_3^+$ is not a nucleophile and must first be converted by a weak base into the free $-NH_2$ group which then reacts readily with acetic anhydride. Sodium carbonate can be used for this purpose.

α-Amino acids also react with ninhydrin (Section 10.7) to yield a purple colored anion with an absorption maximum at 570 nm in the visible region of the spectrum. Only proline, in which the α-amino group is secondary, fails to undergo this reaction.

13.6 The Peptide Bond

We owe our earliest formulation of the manner in which amino acids are bonded in proteins to the great German chemist Emil Fischer. In 1902, he postulated that proteins are long sequences of amino acids joined in linear fashion by amide linkages between the carboxyl group of one amino acid and the α-amine of another. Such linkages between amino acids are called peptide bonds. Figure 13.5 illustrates the peptide bond formed between glycine and alanine in the peptide glycylalanine.

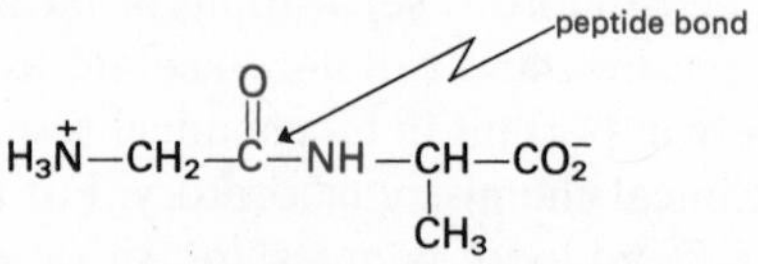

$$\overset{+}{H_3N}-CH_2-\overset{\overset{O}{\|}}{C}-NH-CH-CO_2^-$$

peptide bond

glycylalanine

FIGURE 13.5 The peptide bond in glycylalanine.

   A peptide such as glycylalanine contains two amino acids and is called a dipeptide. Those containing larger numbers of amino acids are called tripeptides, tetrapeptides, pentapeptides, and so on. Peptides containing 10 or more amino acids generally are called polypeptides.

   Proteins are biological macromolecules, generally of molecular weight 5000 or greater, that consist of one or more polypeptide chains. In addition, many proteins also contain nonamino acid groups as integral parts of their structure. For example, myoglobin (Figure 13.16) has within its structure a porphyrin ring which is essential for its biological activity.

   By convention, polypeptides generally are written beginning with the free $-NH_3^+$ group on the left and proceeding to the right toward the free terminal $-CO_2^-$ group. Alternatively, the standard abbreviations of the amino acids may be used, each connected by an arrow. The tail of the arrow indicates the amino acid contributing the carboxyl group to the peptide bond and the head of the arrow indicates the amino acid contributing the amino group.

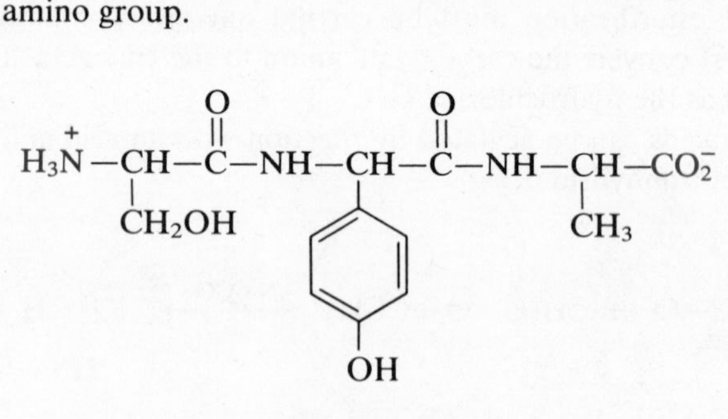

$$\overset{+}{H_3N}-CH-\overset{\overset{O}{\|}}{C}-NH-CH-\overset{\overset{O}{\|}}{C}-NH-CH-CO_2^-$$

ser → tyr → ala

   Given the fact that proteins are linear sequences of amino acids joined by peptide bonds, the next questions we might ask are: What is the sequence of amino acids along the polypeptide chain? And what is the detailed three-dimensional arrangement of the polypeptide chain within the protein molecule?

13.7 Amino Acid Sequence

To appreciate the problem of deciphering the sequence of amino acids along a polypeptide chain, we need only imagine the incredibly large number of different chemical words (proteins) that can be constructed with a 20-letter alphabet, where words can range from under 10 letters to well into hundreds of letters. With only three amino acids there are six entirely different tripeptides possible. If we choose glycine, alanine, and serine, the six possible tripeptides are:

$$gly \rightarrow ala \rightarrow ser \qquad ala \rightarrow ser \rightarrow gly$$

$$gly \rightarrow ser \rightarrow ala \qquad ser \rightarrow gly \rightarrow ala$$

$$ala \rightarrow gly \rightarrow ser \qquad ser \rightarrow ala \rightarrow gly$$

For a polypeptide containing one each of the 20 different amino acids, the number of possible molecules runs to $20 \times 19 \times 18 \times \cdots \times 2 \times 1$ or about 2×10^{18}. With larger polypeptides and proteins, the number of possible arrangements becomes truly astronomical! Until recently the possibility of decoding the exact amino acid sequence for any protein seemed staggeringly complicated. Of the various chemical methods developed for determining the amino acid sequence of a polypeptide chain, the most widely used was introduced in 1950 by Pehr Edman of the University of Lund, Sweden. In this procedure, a polypeptide chain is reacted with phenylisothiocyanate, C_6H_5—N=C=S, to derivatize the terminal —NH_2 group. Treatment of the derivatized polypeptide chain with dilute acid catalyzes selective cleavage and removal of the N-terminal amino acid as a substituted phenylthiohydantoin (Figure 13.6).

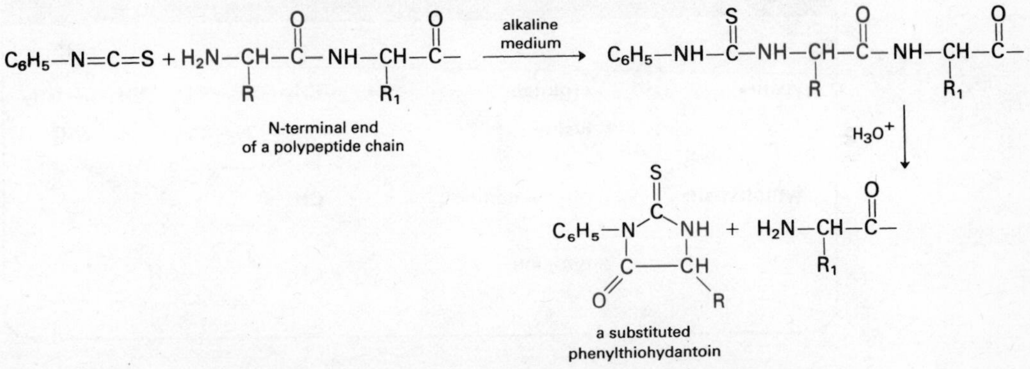

FIGURE 13.6 The Edman degradation. Reaction of a polypeptide chain with phenyl-isothiocyanate followed by hydrolysis in dilute acid selectively removes the N-terminal amino acid as a substituted phenylthiohydantoin.

The special value of the Edman degradation is that it can be used to cleave the N-terminal amino acid from the polypeptide chain without at the same time affecting any other bonds in the chain. In this sense, it is a non-destructive degradation. Furthermore, it can be repeated on the now-shortened polypeptide chain and the next amino acid in the sequence cleaved and identified. In principle, the Edman degradation can be repeated over and over again until the complete sequence of amino acids in a polypeptide chain is determined. In practice, it is now possible to sequence the first 40 or so amino acids in a polypeptide chain by this method.

If it is not possible to sequence all amino acids in a polypeptide chain by repeated Edman degradations, the chain is subjected instead to partial hydrolysis to yield a series of smaller fragments. Such partial hydrolysis is most often carried out using certain enzymes which catalyze the hydrolysis of only specific types of peptide bonds.

$$
\underset{\substack{\text{selective}\\\text{cleavage}}}{\text{NH}-\underset{\text{R}}{\text{CH}}-\overset{\overset{\text{O}}{\|}}{\text{C}}-\text{NH}-\underset{\text{R}_1}{\text{CH}}-\overset{\overset{\text{O}}{\|}}{\text{C}}-}\; + \; \text{H}_2\text{O} \;\xrightarrow{\text{enzyme}}\; -\text{NH}-\underset{\text{R}}{\text{CH}}-\overset{\overset{\text{O}}{\|}}{\text{C}}-\text{OH} \; + \; \text{H}_2\text{N}-\underset{\text{R}_1}{\text{CH}}-\overset{\overset{\text{O}}{\|}}{\text{C}}-
$$

The two enzymes most commonly used for this type of selective hydrolysis are trypsin and chymotrypsin. Trypsin catalyzes the hydrolysis of peptide bonds in which the carboxyl group is contributed by either arginine or lysine; chymotrypsin catalyzes the hydrolysis of peptide bonds where the carboxyl group is contributed by either phenylalanine or tyrosine (Table 13.4).

TABLE 13.4 Specific cleavage of peptide bonds catalyzed by trypsin and chymotrypsin.

Enzyme	Catalyzes the hydrolysis of peptide bonds formed by the carboxyl group of	Side chain (R—group) of the amino acid undergoing selective cleavage
trypsin	arginine	$-\text{CH}_2-\text{CH}_2-\text{CH}_2-\text{NH}-\overset{\overset{\text{NH}_2^+}{\|}}{\text{C}}-\text{NH}_2$
	lysine	$-\text{CH}_2-\text{CH}_2-\text{CH}_2-\text{CH}_2-\text{NH}_3^+$
chymotrypsin	phenylalanine	$-\text{CH}_2-\bigcirc$
	tyrosine	$-\text{CH}_2-\bigcirc-\text{OH}$

When enough smaller fragments have been obtained and sequenced, it is then possible to match overlapping amino acid patterns and thereby deduce the complete sequence of a polypeptide chain.

13.8 Polypeptide and Protein Hormones

In this section we shall look at the structure and function of four polypeptide and protein hormones: insulin, vasopressin, oxytocin, and ACTH.

These are by no means all of the important human polypeptide and protein hormones. However, they will illustrate something of the structure and function of this class.

Insulin contains a total of 51 amino acid units in two polypeptide chains (Figure 13.7). Note that the A chain consists of 21 amino acids with glycine as the N-terminal amino acid. The B chain consists of 30 amino acids with phenylalanine at the N-terminus. The two chains are cross-linked by two disulfide bonds. The sequence of amino acids in insulin reveals no particular pattern. It appears to be just one of an almost infinite variety of possible sequences. Yet somewhere in the sequence lies the key to its physiological potency in the regulation of blood glucose levels. Where in the sequence this potency lies we do not know.

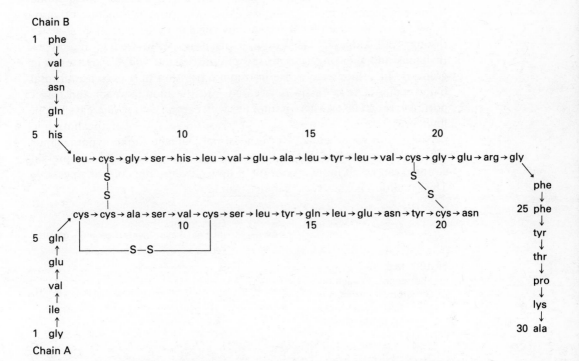

FIGURE 13.7 The amino acid sequence of bovine insulin and the position of the —S—S— cross-linkages. Chain A is composed of 21 amino acids, has glycine (gly) at the $-NH_3^+$ terminus and asparagine (asn) at the $-CO_2^-$ terminus. Chain B is composed of 30 amino acids, has phenylalanine (phe) at the $-NH_3^+$ terminus and alanine (ala) at the $-CO_2^-$ terminus.

The structures of human, beef, pig, sheep, horse, and sperm whale insulin have been determined, and they reveal some interesting species characteristics. Chain B is the same in each of these insulins. The only differences are in the amino acids at positions 8, 9, and 10 in the A chain (Table 13.5). Apparently these amino acids do not have unique roles and substitution is possible without radically altering the physiological behavior of insulin. This is fortunate since diabetics who become allergic to one type of insulin may begin using insulin from another source and thereby avoid allergic reactions.

357

TABLE 13.5 Amino acid variations in insulin Chain A.

Animal	Amino acid residues of Chain A		
	8	9	10
man	threonine	serine	valine
beef	alanine	serine	valine
pig	threonine	serine	isoleucine
sheep	alanine	glycine	valine
horse	threonine	glycine	isoleucine
sperm whale	threonine	serine	isoleucine

Both oxytocin and vasopressin (Figure 13.8) are nonapeptide hormones secreted by the pituitary gland. The primary function of vaso-pressin is to increase blood pressure by regulating the excretion of water through the kidneys. The hormone itself is synthesized in the hypo-thalamus and then migrates to the pituitary gland where it is stored in granules. In response to a rise in osmotic pressure or a decrease in total blood volume, vasopressin is released into the blood stream and trans-ported to the kidneys where it interacts with certain cells to decrease loss of fluid in the urine. Conversely, a decrease in secretion of the hormone permits more water to be lost in the urine. Damage to the hypothalamus (the site of vasopressin synthesis) or to the pituitary (the site of vasopressin storage) can result in the excretion of up to several liters of urine per day. This condition is known as diabetes insipidus.

FIGURE 13.8 Two peptide hormones. In each the carboxyl terminus is glycinamide, indicated by glyNH$_2$.

$$cys \rightarrow tyr \rightarrow ile \rightarrow gln \rightarrow asn \rightarrow cys \rightarrow pro \rightarrow leu \rightarrow glyNH_2$$
$$\underset{\text{S—S}}{\rule{0pt}{0pt}}$$

bovine oxytocin

$$cys \rightarrow tyr \rightarrow phe \rightarrow gln \rightarrow asn \rightarrow cys \rightarrow pro \rightarrow arg \rightarrow glyNH_2$$
$$\underset{\text{S—S}}{\rule{0pt}{0pt}}$$

bovine vasopressin

Oxytocin acts on the smooth muscles of the uterus to enhance contraction and is often administered at childbirth to induce delivery. Oxytocin also acts on the smooth muscles of lactating mammary glands to stimulate milk ejection.

Notice that the structures of vasopressin and oxytocin are similar. In fact, they differ only in the amino acids at positions 3 and 8. Because of this similarity, their physiological actions overlap. For example, vasopressin stimulates contraction of the pregnant uterus, but not nearly so strongly as oxytocin. Vasopressin also stimulates milk ejection, but again not nearly so strongly as oxytocin. By comparing the physiological functions of these two polypeptide hormones, we can see that only slight changes in structure can lead to major changes in function.

Adrenocorticotropic hormone (ACTH) is one of the most important hormones synthesized by the anterior lobe of the pituitary gland. It consists of a single polypeptide chain of 39 amino acids with no disulfide cross-links and has a molecular weight of 4,500 (Figure 13.9).

ser—tyr—ser—met—glu—his—phe—arg—trp—gly—lys—pro—val—gly—lys—lys—arg—arg—pro—val—

21   24 25   30   39
—lys—val—tyr—pro—asp—ala—gly—glu—asp—gln—ser—ala—glu—ala—phe—pro—leu—glu—phe

FIGURE 13.9   The amino acid sequence of bovine ACTH.

The major function of ACTH is to stimulate the biosynthesis of steroid hormones in the cortex of the adrenal gland. This hormone has been prepared in pure form from the pituitary glands of sheep, pigs, and cattle. It has been possible to cleave ACTH at specific sites and to study the biological activity of various fragments. These studies have revealed that the full biological activity of the hormone is present in the amino acid sequence 1 to 24. In other words, the amino acid sequence 25 to 39 is not necessary for the hormonal action of ACTH. The pituitary gland stores about 300×10^{-6} grams of ACTH and secretes it at the rate of 10×10^{-6} grams per day. The level of ACTH in circulating blood plasma is 0.03×10^{-9} grams per ml.

13.9   Conformations of Protein Chains

In the late 1930s, Linus Pauling and his collaborators set out to determine by means of X-ray crystallographic studies the geometry of the peptide bond and to gain a clearer understanding of the patterns of protein chain folding. Pauling began with the study of crystalline amino acids and simple di- and tripeptides. One of his first discoveries was that the peptide bond itself is planar. As shown in Figure 13.10, the four atoms of the peptide bond and the two alpha carbons joined to the peptide bond all lie in the same plane.

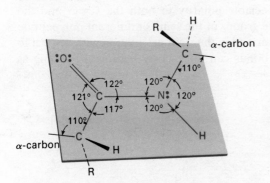

FIGURE 13.10   Planarity of the peptide bond. Bond angles about the carbonyl carbon and the amide nitrogen are all approximately 120°.

Had we been asked in Chapter 1 to describe the geometry of the peptide bond, we would have predicted bond angles of approximately 120° about the carbonyl carbon and 109.5° about the amide nitrogen. This prediction is in good agreement with the observed bond angles of 120° for the carbonyl carbon; however, the geometry of the amide nitrogen is unexpected. To account for this observed geometry, Pauling proposed that the peptide bond is more accurately represented as a <u>resonance hybrid</u> of two important contributing structures.

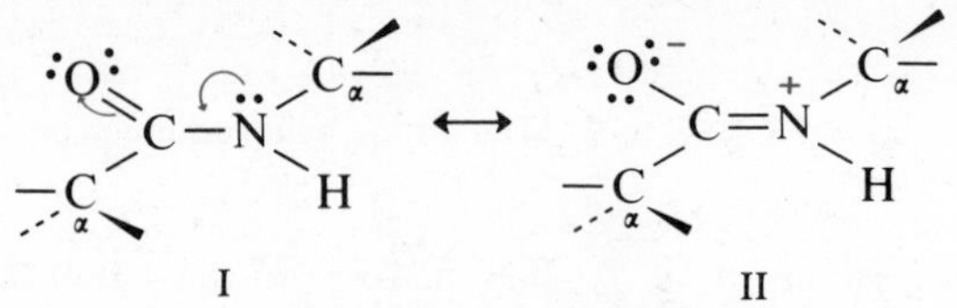

I   II

The hybrid, of course, is neither of these, but in it the carbon–nitrogen bond has at least partial double-bond character. Accordingly, in the hybrid the six-atom group is planar.

There are two possible planar arrangements of the atoms of the peptide bond. In one configuration, the two α-carbons are at diagonally opposite corners of the amide plane. This is called a *trans* configuration.

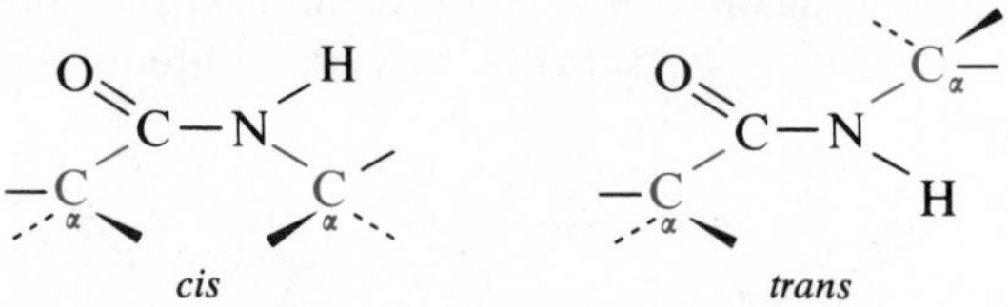

cis   *trans*

In the other, a *cis* configuration, the two α-carbons are *cis* to each other in the amide plane. The *cis* configuration appears slightly less favorable, probably because the bulkier α-carbons are in closer proximity. In the *trans* configuration, the bulky α-carbons are maximally separated from each other. The *cis* configuration is found in only a few polypeptides.

Pauling reasoned that since amide bonds were planar in the model compounds he chose for study, then there is every reason to believe that they are also planar in natural proteins, and this would be one major factor in determining the folding patterns of polypeptide chains. He further recognized the importance of hydrogen bonding in stabilizing folding patterns. Of particular importance is the hydrogen bonding between the amide groups. There is appreciable polarity to both the C=O and the N—H bonds. When two amide groups of the same or different polypeptide chains lie close enough together and in the right orientation with respect to each other, the two peptide bonds can interact by hydrogen bonding as shown in Figure 13.11.

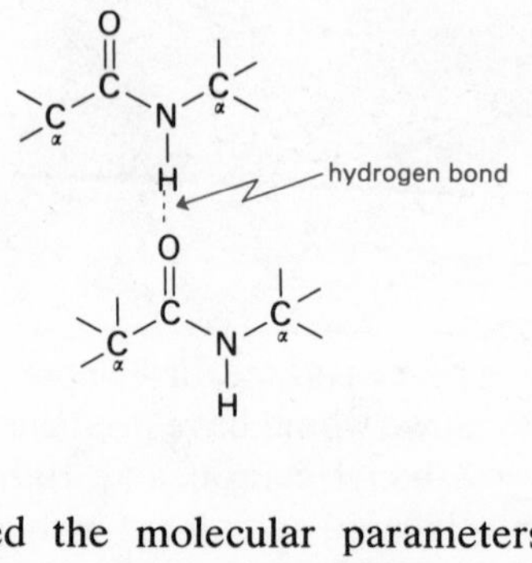

FIGURE 13.11   Hydrogen bonding between amide groups.

Once Pauling had determined the molecular parameters of the peptide bond, he then set out by very careful model building to determine which arrangements of polypeptide chains would be particularly stable. He assumed that the folding patterns of greatest stability in natural proteins would have (1) all amide bonds *trans*-planar, and (2) a maximum of hydrogen-bonded interaction between amide groups. On the basis of this model building, Pauling proposed that two folding patterns should be of particular stability: the <u>α-helix</u> and the <u>antiparallel β-pleated sheet</u>.

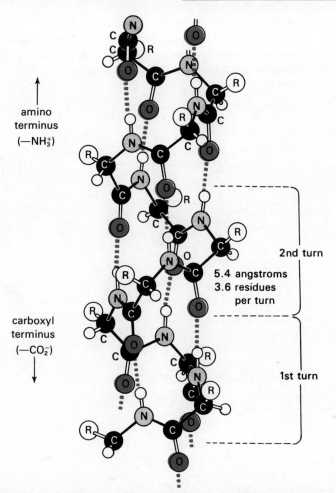

amino
terminus
$(—NH_3^+)$

carboxyl
terminus
$(—CO_2^-)$

2nd turn

5.4 angstroms
3.6 residues
per turn

1st turn

FIGURE 13.12 Ball-and-stick model of the α-helix, showing intrachain hydrogen bonding. There are 3.6 amino acid residues per turn (a distance of 5.4 angstroms along the axis of the helix). Redrawn from *Introduction to Molecular Biology*, by G. H. Haggis, D. Michie, A. R. Muir, K. B. Roberts, and P. M. B. Walker (Longmans Green and Company, Harlow, England, 1964).

In the helix pattern, shown in Figure 13.12, the polypeptide chain is coiled in a spiral, as if wound around an imaginary cylinder. As you study this diagram, note four major features.

1. The planar character of each peptide bond is maintained.
2. There are 3.6 amino acid residues per turn of the helix.
3. Each peptide bond N—H group points roughly upward parallel to the axis of the helix, and each peptide bond C=O group points roughly downward.
4. The R— groups point outward from the cylinder in such a way that the interaction between side chains is minimized.

361

From an examination of molecular models, Pauling also concluded that for L-amino acids, the right-handed helix is more stable than the left-handed helix. Right-handed means that if you turn the helix clockwise, it twists away from you. In this sense, a right-handed helix is analogous to the right-hand threads of a common wood or machine screw.

Almost immediately after Pauling reported the results of his model building and proposed the α-helix structure, other workers reexamined available X-ray crystallographic data and proved the presence of the α-helix in hair keratin. It soon became obvious that the α-helix is one of the fundamental patterns of folding in polypeptide chains.

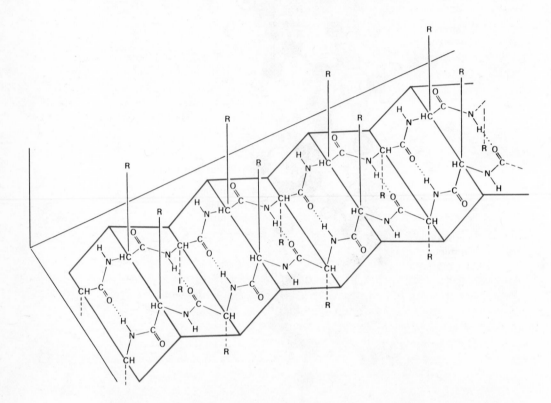

FIGURE 13.13 β-Pleated sheet conformation with two polypeptide chains running antiparallel. Hydrogen bonding between chains is indicated by dotted lines. From P. Karlson, *Introduction to Modern Biochemistry*, 3rd ed. (Academic Press, New York, 1968).

The antiparallel β-pleated sheet (Figure 13.13), proposed in 1951, is built of extended polypeptide chains with neighboring chains running in opposite (antiparallel) directions and with a maximum of interchain hydrogen bonding. To date, extensive regions of β-pleated sheet structure have been found only in silk fibroin. However, many proteins do have one or more short regions of β-sheet structure.

Note that the pleated sheet conformation is stabilized by hydrogen bonding between —NH groups of one chain and C=O groups of an adjacent chain. By comparison, the α-helix is stabilized by hydrogen bonding between —NH and C=O groups within the same polypeptide chain.

13.10 Three-Dimensional Shape of Proteins

As it occurs in the native state, each type of protein has a characteristic three-dimensional shape. This shape is generally described in terms of four levels of molecular organization or architecture. The term primary structure refers to the sequence of amino acids in a polypeptide chain and also to the location of any disulfide bonds. In this sense then, primary structure is a complete description of all covalent bonding in a polypeptide or protein. As an example, a description of the primary structure of the insulin molecule is simply a statement of the amino acid sequence in the A and B chains and the location of the three disulfide bonds as given in Figure 13.7.

Secondary structure is a description of the conformation of the various sections of α-helix, β-pleated sheet, collagen-like triple helix, and other types of periodic structure, usually in localized regions of the molecule.

Tertiary structure refers to the overall folding of the polypeptide chain and the relative arrangement of the secondary structure elements. Actually there is no sharp dividing line between what is considered secondary and tertiary structure. Between the extremes though, secondary structure refers to the spatial arrangement of amino acids close to one another on the polypeptide chain, while tertiary structure refers to the three-dimensional arrangement of all atoms, including not only those close together on the polypeptide chain but also those quite far apart on the linear sequence.

Quaternary structure is a term applied only to those proteins that contain two or more polypeptide chains and refers to the way in which these chains are packed together. For example, a description of the quaternary structure of normal hemoglobin (Hb A) would state that it is a tetramer composed of two α-chains and two β-chains, and would describe how these four polypeptide chains are arranged together to form the tetramer. The major factor stabilizing quaternary structure is hydrophobic interaction. We shall discuss quaternary structure in more detail in Section 13.15.

Proteins are classified in two broad categories depending on their conformation. The fibrous proteins are stringy, physically tough, and generally insoluble in water and most solvents. As we shall see presently, they are composed of elongated, rod-like chains joined together by several types of cross-linkages to form stable, insoluble structures. There are three major classes of fibrous proteins: the keratins of skin, wool, claws, horn, scales, and feathers; the silks; and the collagens of tendons and hides.

Globular proteins, the second broad category, are generally spherical and, as their name implies, globular in shape. They tend to be soluble in water and aqueous solutions. Nearly all enzymes, antibodies, hormones, and transport proteins are globular. In the following sections we shall discuss three types of fibrous proteins (silk, hair, and collagen) and one globular protein (myoglobin). For each type of protein we will be concerned with the relationships between macroscopic properties and molecular structure, and the particular factors that stabilize the secondary, tertiary, and quaternary structures of these and other proteins.

13.11   Silks—The Pleated Sheet Structure

In terms of bulk physical properties, <u>silk</u> is characterized by great strength and flexibility, and also by a resistance to stretching. At the level of molecular architecture, silk is composed of long polypeptide chains that are stretched or extended along the length of the fiber axis. The distribution of amino acids in silk is unique in that most of the polypeptide chain is built from just three amino acids: glycine (45%), alanine (30%), and serine (12%). In addition, there are also small amounts of most of the other amino acids. Chemical studies of the amino acid sequence of silk fiber have revealed that the hexapeptide unit

$$-(gly-ser-gly-ala-gly-ala)-$$

repeats for long distances in the polypeptide chain. Notice that every other amino acid in this hexapeptide unit is glycine.

The polypeptide chains of silk are arranged in the extended β-pleated sheet. In this arrangement, the amino acid side chains alternate, first above the plane of the sheet, the next below the plane, the next above the plane, and so on. In one type of silk, every other amino acid is glycine and all of the R— groups on one side of the plane are hydrogen atoms. Since alanine and serine make up most of the rest of the amino acids of silk, most of the amino acid side chains on the other side of the sheet are methyl or hydroxymethyl groups.

In silk fiber, the sheets are stacked with glycine facing glycine (hydrogen against hydrogen) and with the alanines and serines also facing each other. The bulk properties of silk are a consequence of this molecular structure and arrangement. The silk fiber is very strong because tension is borne by the covalent bonds of the polypeptide chains themselves. Silk is resistant to stretching because the chain is already extended as far as it can go. Since there are no strong forces of attraction between the side chains of the amino acid units, the sheets themselves are quite flexible and slip past each other easily.

13.12   Wool—The α-Helix Structure

The second great class of fibrous proteins is illustrated by <u>hair</u> and <u>wool</u>. This type of fiber is very flexible and in contrast to silk is quite extensible under tension and elastic, so that when tension is released the fiber snaps back to its original condition. At the molecular level, the basic structural unit of hair is the polypeptide chain wound in an α-helix conformation. Furthermore, there are several levels of structural organization built from the simple α-helix. First, it appears that three α-helices are arranged together to form a larger interwoven coil called a <u>protofibril</u>. In the protofibril, each α-helix is itself coiled slightly in much the manner that three strands of fiber are twisted together in rope or cable (Figure 13.14). These protofibrils are then arranged in bundles to form an 11-stranded cable called a <u>microfibril</u>. These in turn are imbedded in a larger matrix

FIGURE 13.14 The supracoiling of three α-helices in hair and wool to form a protofibril.

that ultimately forms the hair fiber. The α-helices themselves are cross-linked by disulfide bonds between cysteine side chains.

Regarding the macroscopic properties of the fiber itself, we have already noted that wool can be stretched and will spring back on release of tension. What happens in the stretching process is an elongation of the hydrogen bonds along turns of the α-helix. The major force causing the stretched wool fiber to return to its original length is the reformation of the hydrogen bonds in the α-helices. The α-keratins of horns and claws have essentially the same structure but with a much higher content of cysteine and a greater degree of disulfide bridge cross-linking between the helices. These additional disulfide bonds greatly increase the resistance to stretch and produce the hard keratins of horn and claw.

13.13 Collagen—The Triple Helix

The third major class of fibrous proteins contains the collagens. Collagens are constituents of skin, bone, teeth, blood vessels, tendons, cartilage, and connective tissue. In fact, they are the most abundant of all proteins in higher vertebrates, making up almost 30% of the total body protein mass in humans. The distinctive physical property of collagen is that it forms long, insoluble fibers of very high tensile strength. Table 13.6 lists the collagen content of several tissues. Note that bone, Achilles' tendon, skin, and the cornea of the eye are largely collagen.

Tissue	Collagen (% dry weight)
bone, mineral free	88
Achilles' tendon	86
skin	72
cornea	68
cartilage	46–63
ligament	17
aorta	12–24

TABLE 13.6 Collagen content of some tissues.

Because of its abundance and wide distribution in vertebrates and because it is associated with a variety of diseases and the problems of aging, more is known about collagen than probably any other fibrous protein. The collagen molecule is very large and it has a distinctive amino acid composition. One-third of the amino acids in collagen are glycine and another 20% are proline and hydroxyproline. Tyrosine is present in very low amounts and the essential amino acid tryptophan is absent entirely. Cysteine is also absent, so there are no disulfide cross-links in collagen. When collagen fibers are boiled in water, they are converted into soluble gelatins. Gelatin itself has no biological food value because it lacks the essential amino acid tryptophan.

Collagen chains contain an unusually high percentage of proline and hydroxyproline, amino acids in which the α-amino groups are incorporated into a five-membered ring. This ring formation places a constraint on the rotation about the α-carbon and as a consequence, proline and hydroxyproline molecules cannot be rotated properly in a polypeptide chain to fit into an α-helix. Furthermore, because of the relatively high concentrations of amino acids with bulky side chains, these proteins of collagen cannot pack together in the form of β-pleated sheets. However, they can fold into another type of conformation that is particularly stable and unique to collagen. In this conformation, three protein strands wrap around each other to form a left-handed superhelix which looks much like a three-stranded rope.

FIGURE 13.15   The collagen triple helix.

This three-stranded conformation is called the collagen triple helix and the unit itself is called tropocollagen. Collagen fibers are formed when many tropocollagen molecules line up side-by-side in a regular pattern and are then cross-linked by the formation of new covalent bonds. Recall from Section 11.3 that one of the effects of severe ascorbic acid deficiency is impaired synthesis of collagen. Without adequate supplies of vitamin C, cross-linking of tropocollagen strands is inhibited with the result that they do not unite to form stable, physically tough fibers.

The extent and type of cross-linking vary with age and physiological conditions. For example, the collagen of rat Achilles' tendon is highly cross-linked while that of the more flexible tendon of rat tail is much less highly cross-linked. Further, it is not clear when, if ever, the process of cross-linking is completed. Some believe it is a process that continues throughout life, producing increasingly stiffer skin, blood vessels, and other tissues which then contribute to the medical problems of aging and the aged.

13.14   Myoglobin—A Globular Protein

Myoglobin is the first of the globular proteins for which a three-dimensional structure was determined. This feat represented a milestone in the study of the molecular architecture of proteins, and for this pioneering research, J. C. Kendrew shared in the Nobel Prize in Chemistry in 1963. Myoglobin is a relatively small globular protein (molecular weight 16,700) which consists of a single polypeptide chain of 153 amino acids. The complete amino acid sequence (the primary structure) of the chain is known. Myoglobin also contains an iron-porphyrin unit (Figure 13.16). It is found in cells of skeletal muscle and is particularly abundant in diving mammals such as seals, whales, and porpoises. Myoglobin and its structural relative <u>hemoglobin</u> are the oxygen transport and storage molecules of vertebrates. Hemoglobin binds molecular oxygen in the lungs and transports it to myoglobin in the muscle cells. Myoglobin stores the molecular oxygen until it is required for metabolic oxidation.

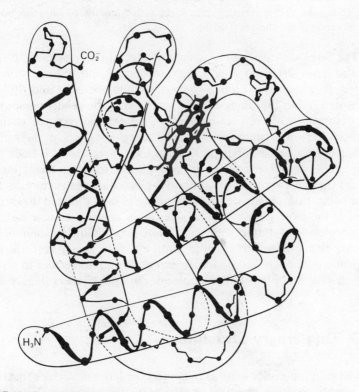

FIGURE 13.16 The three-dimensional structure of myoglobin as deduced from X-ray analysis.   The heme group is shown in color;   the —NH$_3^+$ terminus at the lower left, the —CO$_2^-$ terminus at the upper left. Reproduced from R. E. Dickerson, in H. Neurath, Ed., *The Proteins*, Vol. II (Academic Press, New York, 1964).

The three-dimensional structure of myoglobin was deduced from X-ray analysis. The first analysis, completed in 1957, revealed that the single peptide chain of myoglobin is bent and folded into a very compact shape with little empty space in the interior of the molecule. A more detailed analysis revealed the exact location of the atoms of the peptide backbone and also of the R— groups. The important structural features of myoglobin are:

1. The backbone consists of eight relatively straight sections of α-helix, each separated by a bend in the polypeptide chain. The longest α-helix section has 23 amino acid residues, the shortest has 7. Some 75% of the amino acid residues are found in these eight regions of α-helix.
2. The peptide bonds are planar with each peptide N—H and C=O lying *trans* to each other.

367

3. The inside of the molecule consists almost entirely of nonpolar side chains such as those of leucine, valine, methionine, and phenylalanine. Thus, hydrophobic interactions appear to be a major factor in determining the three-dimensional shape of myoglobin, that is, its tertiary structure.

4. The polar side chains of glutamate, aspartate, glutamine, asparagine, lysine, and arginine are on the outside of the molecule and in contact with the aqueous environment. The only two polar side chains that point toward the interior of the molecule are those of two histidines. These side chains can be seen in Figure 13.16 as five-membered rings that point inward toward the heme group.

The three-dimensional structures of several other globular proteins have also been determined and their secondary and tertiary structures analyzed. It is clear that globular proteins contain α-helix and β-pleated sheet structure but that there is wide variation in the relative amounts of each. Lysozyme with 129 amino acids in a single polypeptide chain has only about 25% of its amino acids in α-helix regions. Cytochrome with 104 amino acids in a single polypeptide chain has no regions of α-helix structure but does contain several regions of β-pleated sheet conformation. Yet whatever the proportions of α-helix and β-pleated sheet structure, virtually all nonpolar side chains of globular proteins are directed toward the interior of the molecule while polar side chains are on the outer surface of the molecule and in contact with water. Note that this arrangement of polar and nonpolar groups in globular proteins very much resembles the arrangements of polar and nonpolar groups of soap molecules in micelles (Figure 8.4) and of phospholipid molecules in lipid bilayers (Figure 12.3).

13.15 Quaternary Structure

Many protein molecules exist as oligomers, or large molecules formed by the assembly of either identical or closely related subunits. This arrangement of protein monomers into an aggregation is known as quaternary structure (Section 13.10). Most proteins of molecular weight greater than 50,000 consist of two or more noncovalently linked chains. Hemoglobin is a good example of such a protein. It exists as a tetramer of four separate protein monomers, two so-called alpha chains of 141 amino acids each and two so-called beta chains of 146 amino acids each. The molecular weights, number of monomeric subunits, and biological functions of several proteins with quaternary structure are shown in Table 13.7.

While there is some stabilization of these quaternary structures by ionic interactions and hydrogen bonding, the major factor stabilizing the aggregation of protein subunits is hydrophobic interaction. Even when the separate monomers fold into a compact three-dimensional shape so as to expose polar side chains to the aqueous environment and at the same time shield the nonpolar side chains from water, there are still one or more hydrophobic "patches" on the surface and in contact with water. These patches can be shielded from water if two or more monomers assemble so that their hydrophobic patches are in contact. Thus hydrophobic interactions, which are so important in directing the self-assembly of lipid

TABLE 13.7   Quaternary structure of selected proteins.

Protein	Mol Wt	Number of Subunits	Subunit Mol Wt	Biological Function
insulin	11,466	2	5,733	a hormone regulating glucose metabolism
hemoglobin	64,500	4	16,400	oxygen transport in blood plasma
alcohol dehydrogenase	80,000	4	20,000	an enzyme of alcoholic fermentation
lactic dehydrogenase	134,000	4	33,500	an enzyme of anaerobic glycolysis
aldolase	150,000	4	37,500	an enzyme of anaerobic glycolysis
fumarase	194,000	4	48,500	an enzyme of the tricarboxylic acid cycle
tobacco mosaic virus	40,000,000	2,130	17,500	plant virus coat

monomers into lipid bilayers, also direct the self-assembly of protein monomers. In addition, this information that directs self-assembly is inherent in the primary structure of the protein monomers themselves.

13.16   Denaturation

Globular proteins as found in living organisms usually are soluble substances with definite three-dimensional shapes and definite biological activity. Globular proteins are also remarkably sensitive to changes in their environment. Even relatively mild changes in temperature or solvent composition for only short periods of time will cause them to lose some or all of their biological activity, that is, to denature.

Denaturation is a physical change, the most observable result of which is loss of biological activity. At the molecular level, it is the result of unfolding or disruption of secondary, tertiary, and quaternary structure of the biologically active conformation of the native protein. With the exception of cleavage of disulfide bonds, denaturation stems from changes in conformation through disruption of noncovalent interactions, for example, hydrogen bonds, salt bridges, and hydrophobic interactions. Common denaturing agents include:

1. Heat. Most globular proteins denature when heated above 50 to 60°C. For example, boiling or frying an egg causes the egg white protein to denature and form an insoluble mass.
2. Large changes in pH. Adding concentrated acid or alkali to a protein in aqueous solution causes changes in the charged character of the ionizable side chains and interferes with ionic or salt interactions. For example, in certain clinical chemistry tests it is necessary to first remove any protein material. This is done by adding trichloroacetic acid (a strong organic acid) to denature and precipitate any protein present.

369

3. Detergents. Treatment of protein in aqueous solution with sodium dodecylsulfate (SDS), a common detergent, causes the native conformation to unfold and exposes the nonpolar protein side chains. These side chains are then stabilized by hydrophobic interaction with the long hydrocarbon chain of the detergent.

4. Organic solvents such as alcohols, acetone, or ether. These solvents can participate in hydrogen bonding and can disrupt the hydrogen bonding in the native protein.

5. Mechanical treatment. Most globular proteins denature in aqueous solution if they are stirred or shaken vigorously. An example is the whipping of egg whites to make a meringue.

6. Urea and guanidine hydrochloride.

$$\underset{\text{urea}}{\overset{\displaystyle \overset{O}{\underset{\|}{}}}{H_2N-C-NH_2}} \qquad \underset{\substack{\text{guanidine} \\ \text{hydrochloride}}}{\overset{\displaystyle \overset{NH_2^+ \quad Cl^-}{\underset{\|}{}}}{H_2N-C-NH_2}}$$

These reagents cause disruption of protein hydrogen bonding and hydrophobic interactions.

Denaturation may be partial or it may be complete. It may also be reversible or irreversible. The hormone insulin can be denatured with $8M$ urea and the three disulfide bonds cleaved. If the urea is then removed and the disulfide bonds reformed, the resulting molecule has less than 1% of its former biological activity. In this case, the denaturation has been both complete and irreversible. As another example, treating the enzyme aldolase, a tetramer (Table 13.7), with $4M$ urea separates and fully denatures the four subunits. If the urea is then removed from the solution, the protein recovers about 70% of its biological activity. In this example, the subunits not only refold with the proper secondary and tertiary conformations but also assemble with the proper quaternary structure; the denaturation has been complete but reversible.

13.17 Enzymes

One of the unique characteristics of the living cell is its ability to carry out complex reactions rapidly and with remarkable specificity. The principal agents for these transformations are a group of protein biocatalysts called enzymes. Some of these enzymes function as catalysts by themselves, while others require the presence of a metal ion (for example, Mg^{2+}, Zn^{2+}) or some other nonprotein molecule (for example, NAD^+).

It is now clear that all enzymes are proteins, that the particular properties of any given enzyme are determined by the sequence of its amino acid residues, and that the catalyst action is carried out within a discrete region of the enzyme known as the active site.

Basically, enzymes function as catalysts in much the same way as the common inorganic or organic laboratory catalysts. A catalyst, whatever the kind, combines with a reactant to "activate" it. In the case of enzyme catalysis, the reactant or reactants are generally referred to as substrates. This activated complex or enzyme-substrate complex as it is more commonly known, then undergoes a chemical change to form product(s) and regenerate the enzyme.

$$E \ + \ S \ \longrightarrow \ E-S \ \longrightarrow \ E \ + \ P$$

enzyme substrate   enzyme-substrate   enzyme product
                            complex

As catalysts, enzymes are far superior to their nonbiological laboratory counterparts in three major ways: (1) enzymes have enormous catalytic power, (2) they are able to discriminate between very closely related molecules, and (3) the activity of many enzymes is regulated. Let us look at each of these unique characteristics in more detail.

First, enzymes have enormous power to increase the rate of chemical reactions. In fact, most of the reactions that occur readily in living cells would occur too slowly to support life in the absence of these biocatalysts. Consider, for example, the enzyme carbonic anhydrase, which catalyzes the reaction of carbon dioxide and water to produce carbonic acid. At physiological pH (7.1 to 7.4), carbonic acid ionizes to form H^+ and bicarbonate ion.

$$CO_2 + H_2O \ \underset{\text{anhydrase}}{\overset{\text{carbonic}}{\rightleftharpoons}} \ H_2CO_3$$

Carbonic anhydrase increases the rate of hydration of carbon dioxide almost 10^7 times compared to the uncatalyzed reaction. Red blood cells are especially rich in this enzyme and for this reason they are able to promote the rapid interconversion of carbon dioxide and bicarbonate.

Second among their unique properties is that enzymes are highly specific in the reactions they catalyze. A given enzyme will generally catalyze only one reaction or a single type of reaction. In other words, competing reactions and by-products such as we find under laboratory conditions are not observed in enzyme-catalyzed reactions. We have already discussed in Section 4.11 how an enzyme might catalyze a reaction of (+)-glyceraldehyde but not of its enantiomer (−)-glyceraldehyde, and how an enzyme might catalyze the reduction of pyruvate to form (+)-lactate but not (−)-lactate. These are but two examples of specificity. In more general terms, there are ranges of enzyme specificity. Recall from Section 13.7 that both trypsin and chymotrypsin catalyze the hydrolysis of peptide bonds but each is highly selective for those involving only certain amino acids.

Third among their unique properties is the fact that the activities of many enzymes can be regulated. Mechanisms exist in living cells to both increase and decrease the activities of many enzymes and most often these mechanisms involve the interaction of enzymes with certain small molecules within the living cell or tissue.

The potential of enzymes for enormous catalytic power, for great specificity, and for regulation have important consequences for the living cell. Any living cell contains literally thousands of different molecules and there is an almost infinite number of chemical reactions that are thermodynamically possible in this mix. Yet the cell, by virtue of its enzymes, can not only select which chemical reactions will take place but, by regulating the activities of key enzymes, it can also control the rates of these reactions and how much of any given product is formed. In this regard, enzymes are truly remarkable catalysts!

PROBLEMS

13.1 Define and give an example of each of the following:

(a) α-amino acid   (b) tripeptide   (c) peptide bond
(d) fibrous protein   (e) polypeptide   (f) globular protein
(g) nonprotein-derived   (h) zwitterion   (i) isoelectric point
     amino acid

13.2 Amino acids of protein origin are designated as L-amino acids. Explain the meaning of the designation L- as it is used to indicate the stereochemistry of amino acids. What is the structural relationship between L-serine and D-glyceraldehyde? L-phenylalanine? D-glucose? D-serine?

13.3 Draw structural formulas for two amino acids that contain more than one chiral carbon atom.

13.4 Name the essential amino acids for man. Why are they termed "essential"? Compare meat, fish, and the cereal grains in terms of their ability to supply the essential amino acids.

13.5 Consider the amino acids alanine, lysine, and aspartic acid. Draw the structural formulas for the form of each that you would expect to predominate at 7.4, the physiological pH; at pH 1.0; at pH 11.0.

13.6 Glutamic acid is an acidic amino acid with three ionizable groups. The pK_a values for each are shown below.

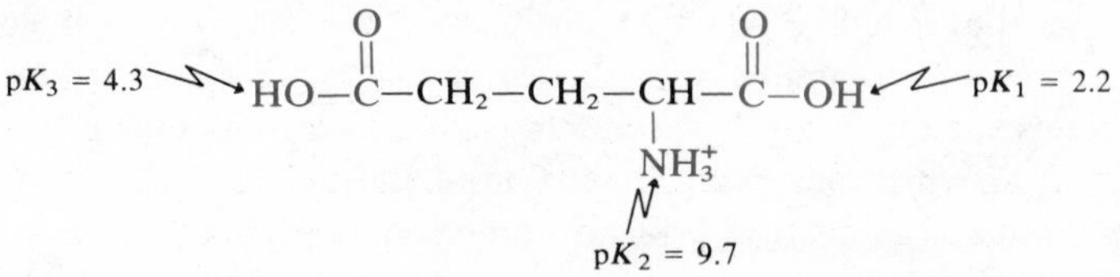

$$pK_3 = 4.3 \searrow \quad HO - \overset{\overset{\displaystyle O}{\|}}{C} - CH_2 - CH_2 - CH - \overset{\overset{\displaystyle O}{\|}}{C} - OH \swarrow pK_1 = 2.2$$
$$\underset{\underset{\displaystyle pK_2 = 9.7}{\overset{\displaystyle N}{\nwarrow}}}{\overset{\displaystyle |}{NH_3^+}}$$

(a) Which of these ionizable groups is the strongest acid; the weakest acid?
(b) Estimate the net charge on glutamic acid at pH 1.0; at pH 7.4; at pH 11.0.

13.7 Would you expect an aqueous solution of lysine to be acidic, basic, or neutral? Explain your reasoning. (In thinking about this problem, first consider the effect on pH of the carboxyl and the α-amino groups together in the zwitterion form and then the effect of the terminal amino group.)

13.8 Draw structural formulas for glutamic acid and glutamine. One of these amino acids has an isoelectric point of 5.7, the other has an isoelectric point of 3.2. Which amino acid has which isoelectric point? Explain your reasoning.

13.9 Examine the amino acid composition of bovine insulin (Figure 13.7) and list the total number and kind of acidic and basic groups in the molecule. Would you predict the isoelectric point of insulin to be nearer that of the acidic amino acids (pH 2.0 to 3.0), the neutral amino acids (pH 5.5 to 6.5), or the basic amino acids (pH 9.5 to 11.0)? Do the same for bovine ACTH.

13.10 Would you predict bovine oxytocin or bovine vasopressin to have the higher isoelectric point?

13.11 Write the structural formula for the tripeptide, glycylserylaspartic acid. Write the structural formula for an isomeric tripeptide.

13.12 How many tetrapeptides can be constructed from the 20 amino acids (a) if each of the amino acids is used only once in the tetrapeptide; (b) if each amino acid can be used up to four times in the tetrapeptide?

13.13 The following are amino acid sequences for several tripeptides. Indicate which will be cleaved by trypsin, which by chymotrypsin.

(a) ala-asp-lys   **(b)** glu-arg-ser   **(c)** phe-met-trp
(d) lys-ala-asp   **(e)** arg-tyr-gly   **(f)** asp-lys-phe
(g) ser-phe-asp   **(h)** lys-cys-tyr   **(i)** gly-glu-asp

13.14 Following is the primary structure of an undecapeptide.

gly-glu-arg-gly-phe-phe-tyr-thr-pro-lys-ala

Write amino acid sequences for all fragments formed when this polypeptide is digested with:

(a) trypsin   **(b)** chymotrypsin   **(c)** both trypsin and chymotrypsin

13.15 Deduce the amino acid sequence of a pentapeptide from the following experimental results.
    amino acid composition: arg, glu, his, phe, trp (listed in alphabetical order)
    Edman degradation: glu
    digestion with chymotrypsin:
        fragment A: glu, his, phe
        fragment B: arg, trp
    digestion with trypsin:
        fragment C: arg, glu, his, phe
        fragment D: trp

13.16 Deduce the primary structure of a hexapeptide from the following experimental results:
    amino acid composition: lys, pro, pro, tyr, val, val (alphabetical order)
    Edman degradation: pro
    digestion with trypsin:
        fragment E: pro, lys, val
        fragment F: pro, tyr, val
    digestion with chymotrypsin:
        fragment G: lys, pro, tyr, val, val
        fragment H: pro

13.17 Deduce the amino acid sequence of an octapeptide from the following experimental results.
    amino acid composition: ala, arg, leu, lys, met, phe, ser, tyr
    Edman degradation: ala
    digestion with trypsin:
        fragment I: ala, arg
        fragment J: leu, met, tyr
        fragment K: lys, phe, ser
    digestion with chymotrypsin:
        fragment L: ala, arg, phe, ser
        fragment M: leu
        fragment N: lys, met, tyr

13.18 Using structural formulas, show how the theory of resonance accounts for the fact that the peptide bond is planar.

13.19 In constructing models of arrangements of polypeptide chains that would be particularly stable, Pauling assumed that for maximum stability (1) all amide bonds would be *trans* and coplanar, and (2) there would be a maximum of hydrogen bonding between amide groups. Examine the β-pleated sheet (Figure 13.13) and

373

the α-helix (Figure 13.12) and convince yourself that in each of these, and amide bonds are planar and that each carbonyl is hydrogen bonded to an amide hydrogen.

13.20 Examine the structure of the α-helix. Are the amino acid side chains arranged all inside the helix, all outside the helix, or randomly oriented?

13.21 Characterize the amino acid composition of silk. Silk fiber is very strong and quite resistant to stretching. How are these properties accounted for in terms of the molecular structure?

13.22 If a significant number of the glycines in silk were replaced by amino acids having much larger side chains, what effect would you expect this substitution to have on the stacking of the pleated sheets?

13.23 To what aspects of protein structure do each of the following terms refer: primary structure; secondary structure; tertiary structure; quaternary structure?

13.24 What is the function of collagen? Describe (a) the macroscopic physical properties and (b) the molecular structure of collagen.

13.25 Examine the structure of myoglobin (Figure 13.16) and identify (a) the amino terminal end of the polypeptide chain, (b) the carboxyl terminal end of the polypeptide chain, and (c) the eight relatively straight sections of α-helix.

13.26 Consider a typical globular protein in aqueous medium at pH 7.4. Which of the following amino acids would you expect to find on the outside and in contact with water; which on the inside?

(a)	glutamic acid	**(b)**	glutamine	**(c)**	arginine
(d)	serine	**(e)**	valine	**(f)**	phenylalanine
(g)	lysine	**(h)**	isoleucine	**(i)**	threonine

13.27 Assume that the same globular protein of Problem 13.26 is imbedded in a lipid bilayer, as in Figure 12.4. Which amino acids would you expect to find now on the outside in contact with the interior of the lipid bilayer; which on the inside?

13.28 What is meant by the term denaturation?

13.29 Account for the fact that the solubility of globular proteins is a function of pH, and that solubility is a minimum when the pH of the solution equals the isoelectric point (pI) of the protein.

13.30 Insulin and ACTH are both globular proteins, soluble in aqueous media. Calculate the percentage of polar amino acids (both neutral and charged) in each of these proteins. Also calculate the percentage of polar amino acids in the repeating hexapeptide of silk (Section 13.11). What generalization might you make about the relative percentages of polar groups in globular proteins compared to fibrous proteins?

13.31 Would you expect to be able to separate bovine and human insulin by paper electrophoresis; sperm whale and human insulin?

13.32 Myoglobin and hemoglobin are globular proteins. Myoglobin consists of a single polypeptide chain of 153 amino acids. Hemoglobin is composed of four polypeptide chains, two of molecular weight 141 and two of molecular weight 146. The three-dimensional structure of myoglobin and hemoglobin polypeptide chains is very similar. Yet, myoglobin exists as a monomer in aqueous solution while the four polypeptide chains of hemoglobin self-assemble to form a tetramer. Which polypeptide chains, those of myoglobin or hemoglobin, would you predict to have a higher percentage of nonpolar amino acids?

The Penicillins

The phenomenon of the inhibition of growth of one sort of microorganism by the metabolic products of another was observed very early in the history of microbiology. Several specific substances were isolated and shown to stop the growth of a number of species of bacteria, and in some cases even to kill them. These substances were properly regarded as antibiotics, at least in the literal sense of the word, for they did inhibit bacterial growth. However, these substances usually did not distinguish between microorganisms and mammalian tissue, and as a consequence were often highly toxic to animals. The term antibiotic, as it is understood today, implies that the substance not only shows toxicity to microorganisms but also that the substance is more toxic to microorganisms than to mammalian tissues. Of course, this distinction is sharper in some classes of antibiotics than in others.

It is common knowledge that the most successful of the antibiotics are the penicillins. These truly remarkable drugs are almost completely innocuous to all living materials except for certain groups of bacteria. As a result of the tremendous research that has been done on the structure, the chemistry, the large-scale commercial manufacture, and the mechanism of action of the penicillins, it is safe to say that we have a clearer understanding of the penicillins than of any other antibiotic.

The discovery of penicillin was purely fortuitous, but its development and therapeutic application represent the results of a well-planned and carefully executed program that brought about one of the major advances in medical science. In 1929, Sir Alexander Fleming published his now famous observations that colonies of staphylococci lysed on a plate which had become contaminated with the mold *Penicillium notatum*. Some unknown substance from the mold was bactericidal not only for staphylococci but for many other common disease-causing organisms as well. Because the mold belonged to the genus *Penicillium*, Fleming named the antibacterial substance penicillin. His efforts to extract the penicillin failed and very little was made of this discovery for nearly a decade. In 1939, the outbreak of hostilities stimulated an intensive search for new chemotherapeutic agents, and in Great Britain the potential of Fleming's penicillin was reinvestigated. Within a few months, crude preparations were available and many of the chemical, physical, and antibacterial properties of penicillin were determined. Preliminary results were so impressive that production was undertaken. Because large-scale manufacture in Great Britain was impossible due to the exigencies of the war, a vast cooperative research program was undertaken in the United States. By 1943, penicillin was authorized for use by the medical service of the U.S. Armed Forces, and by 1949, the antibiotic was available for widespread use in almost

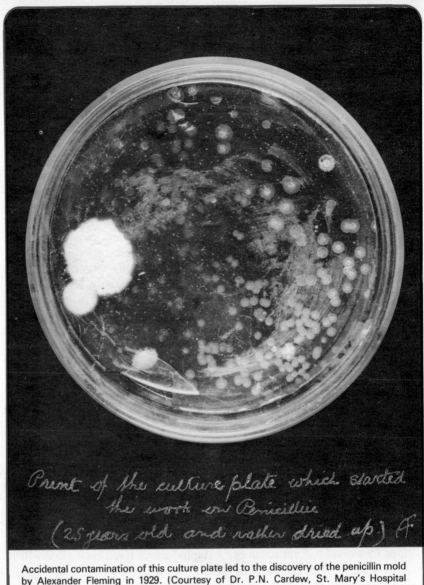

Print of the culture plate which started the work on Penicillin (25 years old and rather dried up) F.

Accidental contamination of this culture plate led to the discovery of the penicillin mold by Alexander Fleming in 1929. (Courtesy of Dr. P.N. Cardew, St. Mary's Hospital Medical School, University of London)

unlimited quantity. Thus, penicillin progressed in the span of two decades from a chance observation in a research laboratory to a therapeutic agent of enormous importance.

Preliminary investigations of the structure and chemistry of penicillin presented a confusing picture because of discrepancies in the analytical and degradative results obtained in different laboratories. These discrepancies were resolved once it was discovered that *P. notatum* produces different kinds of penicillin depending on the nature of the medium in which the mold was grown. Initially, six different penicillins were recognized and all proved ultimately to be acyl derivatives of 6-aminopenicillanic acid. Of the six, penicillin G (benzyl penicillin) became the most widely used and the standard against which others were compared.

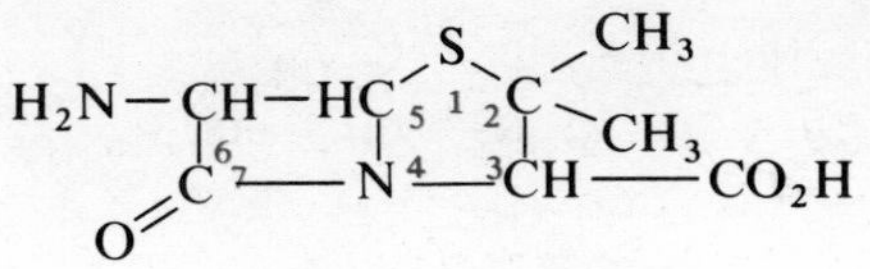

6-aminopenicillanic acid

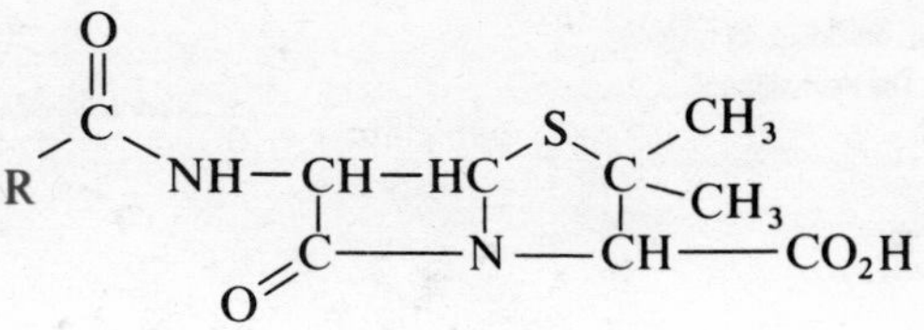

penicillin G

$R = C_6H_5-CH_2-$; benzylpenicillin

The basic structure of penicillin consists of a five-membered ring containing a nitrogen and a sulfur (a thiazolidine ring) fused to a four-membered ring containing a cyclic amide (a β-lactam). The structural integrity of these two rings is essential for the biological activity of penicillin, and cleavage of either ring leads to products devoid of antibacterial activity.

Penicillin undergoes a variety of chemical reactions. We shall look at only two of these reactions, each of which has important consequences for therapeutic use of this antibiotic. Treatment of penicillin with strong mineral acid brings about hydrolysis to penicillamine and a penaldic acid (Figure 1). The rupture of the β-lactam ring involves hydrolysis of an amide bond, and the rupture of the thiazolidine ring can be formulated as the hydrolysis of the nitrogen–sulfur analog of an acetal. Penaldic acid, like other compounds with a carboxylic acid beta to a carbonyl group, undergoes ready decarboxylation, and gives CO_2 and a penicilloaldehyde. Penicillamine, penaldic acid, and penicilloaldehyde are devoid of antibacterial activity. Penicillin G is rapidly inactivated by this type of hydrolysis in the acid conditions of the stomach.

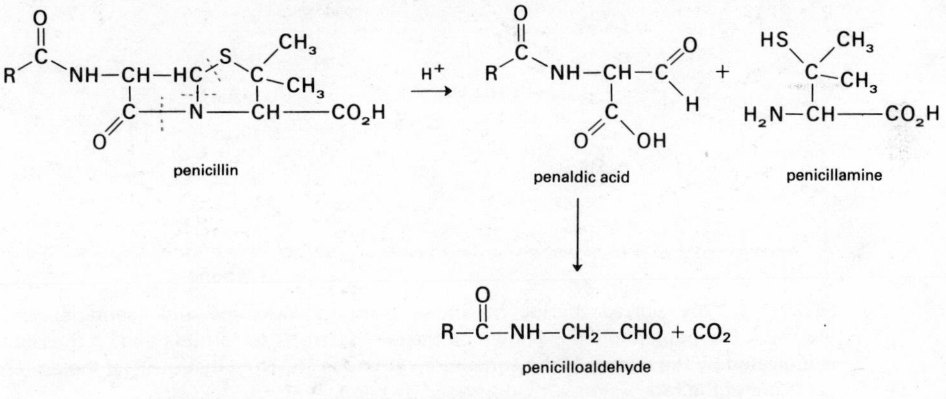

FIGURE 1 Hydrolysis of penicillin by strong acid.

A second important reaction is the selective hydrolysis of the β-lactam ring catalyzed by the enzyme penicillinase (Figure 2). Penicilloic acid, like other ring cleavage products, has no antibacterial activity. The main basis for natural bacterial resistance to penicillin is the ability of resistant strains to synthesize penicillinase.

All penicillins owe their antibacterial activity to a common mechanism which inhibits the biosynthesis of a vital part of the bacterial cell wall. Essentially all bacteria are surrounded by a layer of strengthening

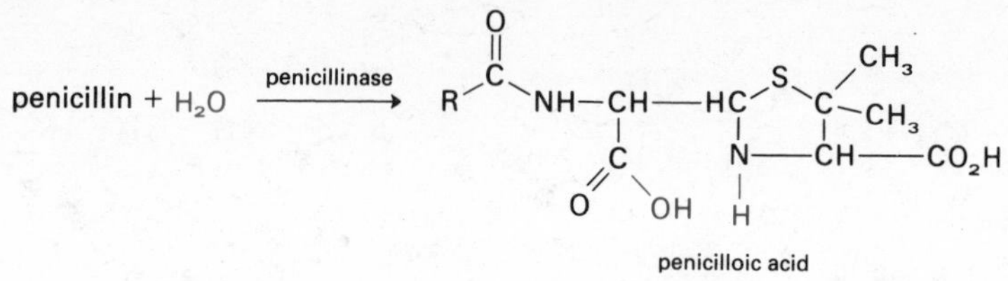

$$\text{penicillin} + H_2O \xrightarrow{\text{penicillinase}}$$

penicilloic acid

FIGURE 2   Hydrolysis of penicillin catalyzed by penicillinase.

material known as mucopeptide or glycopeptide. This mucopeptide is the principal mechanical support of the entire cell and the viability of the bacterial cell depends on the integrity of the cell wall. It is well known that the concentration of smaller-molecular-weight substances within the cell is often considerably greater than in the surrounding medium, giving rise to a high osmotic pressure within the cell. In certain bacterial cells, this osmotic pressure has been estimated to be as high as 10 to 20 atmospheres. The cell wall, in providing a mechanical support, prevents the disruption of the cell.

The simplest type of mucopeptide is made up of a backbone of long polysaccharide chains cross-linked by short peptide chains (Figure 3). The polysaccharide chain itself is made up of alternating units of N-acetyl-D-glucosamine and N-acetyl-D-muramic acid. Each of the N-acetyl-D-muramic acids has a carboxyl group and in the mucopeptide, each carboxyl group is linked by an amide bond to a pentapeptide beginning with L-alanine.

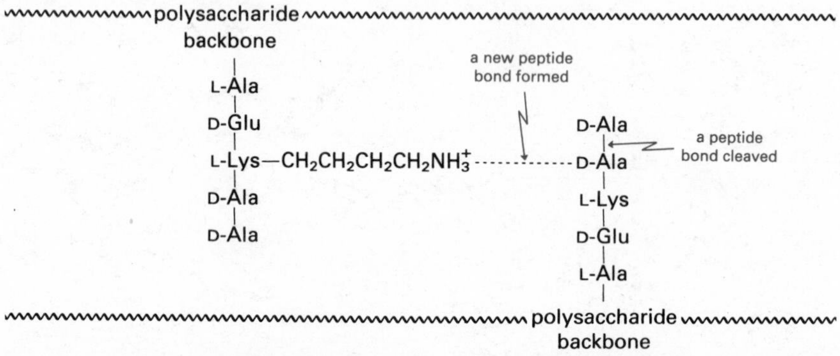

FIGURE 3   The polysaccharide backbone from mucopeptide with pentapeptides attached. The pentapeptide is shown on the left by structural formula and on the right is indicated by the names of the constituent amino acids. Note that three of the amino acids are of the rare D series.

The final reaction sequence in construction of the cell wall matrix is formation of amide bonds between carboxyl and amino groups from adjacent mucopeptide chains. In staphylococci, this is thought to occur first by hydrolysis of the terminal D-alanine, and then peptide bond formation between the remaining D-alanine of one chain and the amino group of lysine from another chain. Thus, each long polysaccharide chain is cross-linked to other polysaccharide chains by short lengths of polypeptide chain, and this matrix of each cell wall probably forms one enormous molecule—a "bag-shaped macromolecule."

It is the formation of the final cross-linked mucopeptide macromolecule that is inhibited by the penicillins, and a number of hypotheses about the selective action of the antibiotic have been suggested. One theory is that penicillin is structurally similar to D-alanyl-D-alanine, the terminal amino acid of the peptide units that must be cross-linked (Figure 4). Penicillin is thought to bind selectively to the active site of the enzyme complex that catalyzes peptide bond hydrolysis of the terminal D-alanine and peptide bond formation in cross-linking, thus making the enzyme complex unavailable for cell wall synthesis. Hence, penicillin's antibacterial activity at the molecular level appears to derive from selective enzyme inhibition.

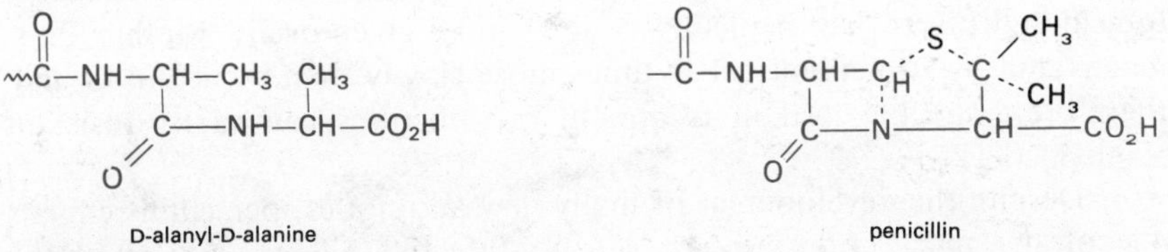

D-alanyl-D-alanine   penicillin

FIGURE 4   Penicillin and D-alanyl-D-alanine, drawn to suggest a structural similarity between the two.

The fact that this type of rigid cell wall construction is unique to bacteria and is not present in mammalian cells no doubt accounts for the lack of toxicity of penicillin to patients even when it is administered in massive doses. Penicillin is a remarkable drug that does its work with great efficiency and selectivity. However, severe allergic reaction in some patients is a serious problem. A significant proportion of the population (perhaps as high as 5 to 10% in some countries) has become hypersensitive to penicillin, and this has tended to limit use of the drug. The factor responsible for the allergic reaction is not penicillin itself, but rather certain degradation products of the drug, particularly 6-aminopenicillanic acid. This allergic sensitivity is not regarded as toxicity in the usual pharmacological sense and is not related to the drug's effects on bacterial cell walls.

The susceptibility of penicillin G to hydrolysis by acid and the emergence of penicillinase-producing strains have provided incentive for the discovery and clinical development of various other natural and semisynthetic penicillins. The nature of the R— group has a major effect on the pharmacological properties, and a large number of new semisynthetic penicillins have been prepared either by the addition of substituted acetic acids ($R—CH_2—COOH$) to the fermentation medium, or by the direct acylation of 6-aminopenicillanic acid. By these means, new penicillins have been prepared which (1) possess greater acid stability than penicillin G, (2) are more resistant to the action of penicillinase, and (3) possess a broader spectrum of antibacterial activity than penicillin G. At the present time, the three most widely prescribed penicillins are ampicillin, penicillin G, and penicillin VK. The side chains of each of these along with that of methicillin are shown in Table 1.

Penicillin G is most effective against gram-negative bacteria. The semisynthetic ampicillin is a broader range antibiotic that attacks both

TABLE 1 Side chains (R—groups) of penicillin G and three semisynthetic penicillins.

gram-negative and gram-positive organisms. Penicillin VK is more stable to acid hydrolysis than penicillin G and can be given orally. Methicillin is inactivated by penicillinase 100 times more slowly than penicillin G and, therefore, may be used in treatment of infections caused by resistant staphylococci.

Despite the development of many new antibiotics, penicillin remains one of the most important of the anti-infective agents, and since the possibilities of adding new acyl groups to the essential unit, 6-amino-penicillanic acid, are manifold, new penicillins may be developed into even more important chemotheraputic agents. It seems certain that the final chapters on the development and clinical applications of the penicillins have yet to be written.

REFERENCES

Fleming, A. "History and Development of Penicillin," *Penicillin: Its Practical Applications.* The Blakiston Company, Philadelphia, 1946.

Gale, E. F., et al. *The Molecular Basis of Antibiotic Action.* John Wiley & Sons, New York, 1972.

Hoover, J. R. E. and R. J. Stedman. *Medicinal Chemistry*, 3rd edition, A. Burger, ed., part 1, pp. 371–408. John Wiley & Sons (Interscience), New York, 1970.

Weinstein, L. "The Penicillins," *The Pharmacological Basis of Therapeutics*, L. S. Goodman and A. Gilman, eds. The Macmillan Company, New York, 1970.

Inborn Errors of Metabolism

... the anomalies of which I propose to treat ... may be classed together as inborn errors of metabolism. Some of them are certainly, and all of them are probably, present from birth .... They are characterized by wide departures from the normal of the species far more conspicuous than any ordinary individual variations, and one is tempted to regard them as metabolic sports, the chemical analogues of structural malfunctions .... It may well be that the intermediate products formed at the various stages (in metabolism) have only momentary existence as such, being subjected to further change almost as soon as they are formed; and that the course of metabolism along any particular path should be pictured as in continuous movement rather than as a series of distinct steps. If any one step in the process fails, the intermediate product in being at the point of arrest will escape further change, just as when the film of a biograph is brought to a standstill the moving figures are left foot in the air. (Sir Archibald Garrod, *Inborn Errors of Metabolism*, 1909.)

In this remarkably prescient passage, Sir Archibald Garrod anticipated many of the developments in the rapidly unfolding study of the metabolic basis of inherited diseases. Garrod studied four inborn errors of metabolism—albinism, alcaptonuria, cystinuria, and pentosuria—and postulated that each results from a congenital failure of the body to carry out a particular reaction in a normal metabolic sequence. His predictions have been fully corroborated, and the inborn errors of metabolism of which he spoke have been traced to a partial or total lack of activity of a particular enzyme. In this essay we shall consider two of the diseases studied by Garrod, albinism and alcaptonuria, and a third recognized more recently, phenylketonuria. These three involve closely related metabolic pathways and illustrate the important principles of metabolic disorders.

Figure 1 summarizes several important pathways in the metabolism of phenylalanine and tyrosine. Under normal physiological conditions, the bulk of phenylalanine is oxidized to tyrosine, a reaction catalyzed by phenylalanine hydroxylase. In a second and minor oxidative pathway, phenylalanine is converted to phenylpyruvate, which accumulates in the blood and is excreted in the urine. This conversion of an α-amino acid to an α-ketoacid is called transamination and requires pyridoxal phosphate and pyridoxamine phosphate, forms of vitamin B_6, as coenzymes.

Two metabolic pathways for tyrosine are shown. In the first, tyrosine undergoes transamination to p-hydroxyphenylpyruvate followed by oxidation to homogentisate. Oxidation of homogentisate, initiated by the enzyme homogentisate oxidase, leads ultimately to fumarate and acetoacetate which can be disposed of via the Krebs cycle. A second major

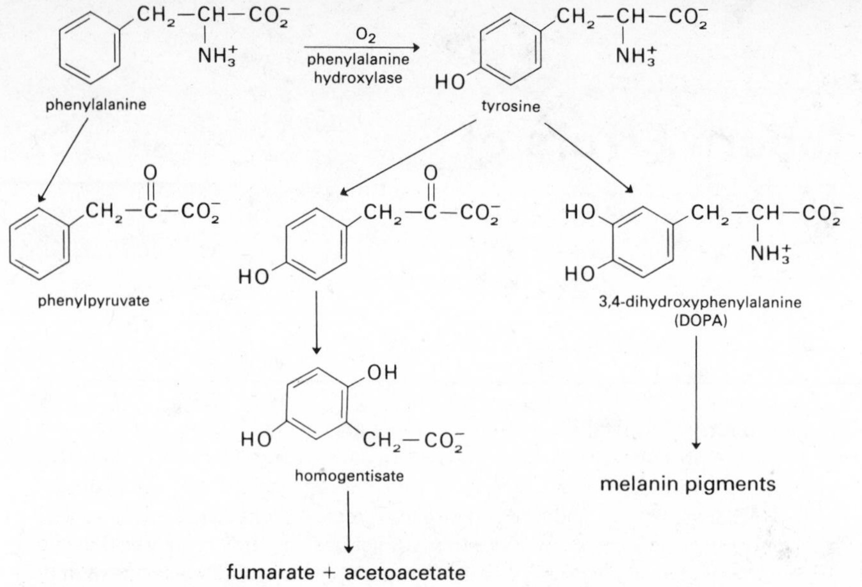

FIGURE 1 Important pathways in the metabolism of phenylalanine and tyrosine.

pathway of tyrosine metabolism is oxidation to 3,4-dihydroxyphenyl-alanine (DOPA). Next in a complex series of reactions initiated by tyrosinase, DOPA is converted into melanin, the polymeric pigment of hair, skin, and retina.

The characteristic feature of alcaptonuria is excretion in the urine of abnormally large amounts of homogentisate (from 3 to 5 grams/day but in some instances as high as 15 grams/day). This condition occurs about once in 200,000 births and is often discovered early in life from observation that an infant's diapers turn black on exposure to air. The darkening results from oxidation of homogentisate by atmospheric oxygen to give the pigment alcapton. Alcaptonuric patients experience no other symptoms early in life, but later there is a darkening of tendons and cartilage due to deposition of pigments which are often associated with the development of arthritis.

At the time Garrod took up the study of this disease, it was generally assumed that the large quantity of homogentisate excreted in the urine was the product of some type of infection in the intestine. Garrod, however, felt this hypothesis was inadequate to explain the symptoms. In 1901, he made a critical discovery. Healthy and apparently normal parents of his alcaptonuric patients often were blood relations. He realized immediately that this parental consanguinity was more than chance and wrote, "The facts here brought forward lend support to the view that alcaptonuria is what may be described as a 'freak' of metabolism. They can hardly be reconciled with the theory that it (alcaptonuria) results from an infection of the alimentary canal."

The time was right for this discovery. The work of Gregor Mendel had been rediscovered only a few years before and the challenge of contemporary biologists was to learn how Mendel's laws applied in the animal kingdom and to humans in particular. William Bateson, one of the pioneers of genetics, was quick to point out that the characteristic familial distribution of alcaptonuria discovered by Garrod was precisely what would be expected if the abnormality were determined by a rare recessive Mendelian factor. Garrod was the first to suggest that the basic defect in the disease is the inability to metabolize homogentisate. This was fully confirmed in 1958 with the demonstration that the liver and kidneys of alcaptonurics are deficient in the enzyme homogentisate oxidase.

Albinism was another of the inborn errors of metabolism considered by Garrod, who correctly concluded that the condition is due to lack of the enzyme necessary for the formation of melanin. Complete (universal) albinism is characterized by a total lack of melanin pigment in the hair, skin, and retina. In partial (generalized) albinism, the lack of melanin is restricted to particular areas of the hair, skin, or retina. The key enzyme involved in the formation of melanin from tyrosine is tyrosinase. It is the lack of this enzyme that results in albinism.

Of the three diseases discussed in this mini-essay, phenylketonuria (PKU) is by far the most crippling. It is estimated that this disease occurs once in every 10,000 to 20,000 births. PKU is characterized by an inherited defect in the ability of the body to metabolize phenylalanine. This defect has been traced to the absence or deficiency of phenylalanine hydroxylase, the enzyme which catalyzes the oxidation of phenylalanine to tyrosine. As a result, phenylalanine concentrations increase. A large portion of this increased supply of phenylalanine is either excreted as such in the urine, or converted in the liver to phenylpyruvate and then excreted. Some of the phenylalanine, however, accumulates in tissues where it is quite toxic, especially to developing brain tissue. As a result of this accumulation, there can be serious brain damage and progressive mental retardation. Without prompt and effective treatment, a PKU child may develop as an idiot with an I.Q. in the range 20 to 30. Fortunately, almost all of the clinical abnormalities of the disease, for example, excretion of phenylalanine in the urine, the attendant mental retardation, and so on, can be minimized and in many cases even eliminated by restricting the dietary intake of phenylalanine to a level absolutely essential for normal growth and development. In addition, the diet must be supplemented in tyrosine, which is an essential amino acid for phenylketonurics.

Because the effects of this potentially crippling disease can be minimized by early detection and prompt treatment including the use of a phenylalanine-free diet, it is critical to have a rapid and reliable method for PKU screening. The earliest screening procedure, and in fact the one used until quite recently, was discovered over four decades ago, even before PKU was identified clinically. Dr. Asbjörn Fölling of the University of Oslo School of Medicine discovered that the musty-smelling urine of two mentally retarded siblings turned blue-green rather than the expected red-brown when tested with 10% ferric chloride solution. Fölling speculated that the substance responsible for this unusual color reaction was somehow related to the children's mental retardation. Soon after he discovered that the substance was phenylpyruvate, formed in the kidney from phenylalanine. Thus, Fölling correctly identified one of the underlying biochemical characteristics of the disease. From the time of his discovery until the mid-1950s, this simple test—the color reaction of ferric chloride solution on a wet diaper—was the standard method of identifying PKU.

However, this test is unreliable because it is too insensitive and often fails to detect PKU early enough after birth.

Recently, two highly sensitive, rapid, and reliable tests for this disease have been developed. One test is microbiological, the other chemical. The microbiological procedure, known as the Guthrie test after its discoverer Dr. Robert Guthrie of the State University of New York at Buffalo, depends on the fact that the growth of a certain strain of the bacterium, *B. subtilis*, is inhibited by the substance β-thienylalanine. Guthrie discovered that the growth inhibition of β-thienylalanine can be counteracted by phenylalanine.

β-thienylalanine   L-phenylalanine

Notice the structural similarities between these two molecules. In the test procedure, concentrations of inhibitor are set so that if phenylalanine levels are normal (about 2 mg/100 ml) there will be no bacterial growth. Bacterial growth will occur, however, in the presence of elevated levels of phenylalanine. The extent of growth is proportional to the increase in phenylalanine concentration. The second method, and the one most often used for daily monitoring of phenylalanine levels of PKU patients, depends on the fact that phenylalanine, ninhydrin, and copper react to form a colored complex. The concentration of this colored complex and therefore of phenylalanine can be determined by fluorescence spectroscopy. In the analysis, light of wavelength 365 nm is shone on the sample. A part of this radiation is absorbed by the phenylalanine-ninhydrin-copper complex and reemitted at 515 nm. The intensity of the emission (that is, the fluorescence) at 515 nm is directly proportional to the concentration of the colored complex.

These three inborn errors of metabolism illustrate the typical consequences of a metabolic block. First, the normal products of a particular metabolic pathway are not formed, or are formed at a reduced rate, and it is the failure to form a product which is the significant feature of each disease. For example, failure to form melanin pigment is the characteristic feature of albinism. The alternative consequence of the enzyme deficiency is the accumulation of substrate of an enzyme, and it is this accumulation of substrate that gives rise to the pathological symptoms. In PKU, phenylalanine is not degraded in the normal manner and instead accumulates and is converted to phenylpyruvate. In many of these diseases the primary metabolic block may also lead to secondary biochemical changes, as in the mental retardation seen in PKU. The nature of these secondary changes is often obscure. For example, it is not at all clear how the absence of phenylalanine hydroxylase and the attendant increase in phenylalanine in the brain leads to the secondary biochemical changes that eventuate in mental retardation.

Beginning with the seminal observations of Garrod nearly 70 years ago, advances in the understanding of inborn errors of metabolism have been truly impressive. In the decades ahead, we can expect many more

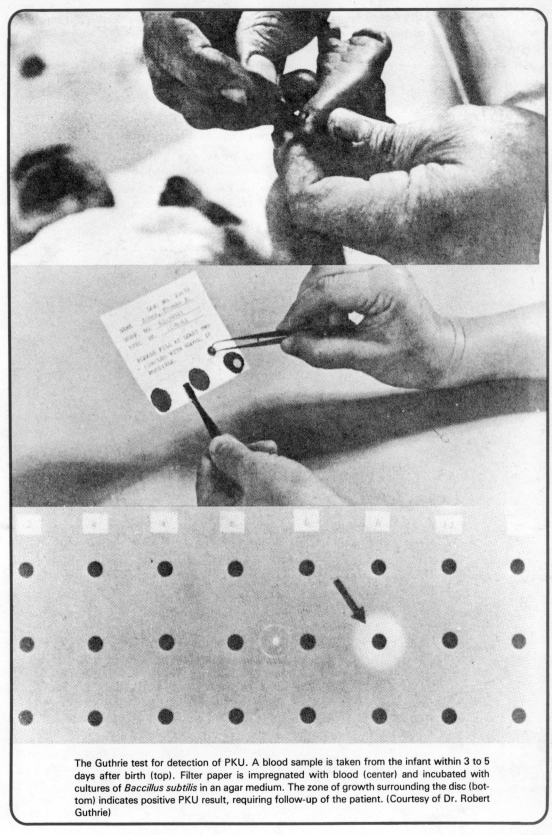

The Guthrie test for detection of PKU. A blood sample is taken from the infant within 3 to 5 days after birth (top). Filter paper is impregnated with blood (center) and incubated with cultures of *Baccillus subtilis* in an agar medium. The zone of growth surrounding the disc (bottom) indicates positive PKU result, requiring follow-up of the patient. (Courtesy of Dr. Robert Guthrie)

diseases to be understood in genetic and biochemical terms, perhaps even certain aspects of mental diseases.

REFERENCES

Garrod, A. E. *Inborn Errors of Metabolism*, 1909. Reprinted with a supplement by H. Harris (Oxford University Press, New York, 1963).

Henry, R. J., Cannon, C. C., and Winkelman, J. W., eds. *Clinical Chemistry: Principles and Technics*, 2nd ed. Harper and Row, New York, 1974.

Stanbury, J. B., Wyngaarden, A., and Fredrickson, D. S., eds. *The Metabolic Basis of Inherited Diseases*, 2nd ed. McGraw-Hill Book Company, New York, 1966.

Abnormal Human Hemoglobins

Hemoglobin is one of the most plentiful proteins in the body. In the bloodstream there are 5 billion red blood cells or erythrocytes, each packed with 280 million molecules of hemoglobin. Hemoglobin picks up molecular oxygen in the lungs and delivers it to all parts of the body for use in metabolic oxidation. The main component of normal adult hemoglobin (Hemoglobin A or Hb A) is the protein globin, molecular weight 64,500. Globin consists of four polypeptide chains, two identical alpha chains (141 amino residues) and two identical beta chains (146 residues). One heme group similar to that in myoglobin (Figure 13.16) is bound to each polypeptide chain. Hydrophobic interaction is the principal force holding the four polypeptide subunits in the tetrameric structure. M. F. Perutz elucidated the complete three-dimensional structure of hemoglobin and for this pioneering work he shared in the Nobel Prize in Chemistry in 1963.

The abnormal hemoglobins have attracted medical attention because of the anemias they produce. The most well known and widespread of these is the crippling and highly lethal disease known as sickle cell anemia. When combined with oxygen, the red blood cells of persons with this disease have the flat, concave, disk-like conformation of normal erythrocytes. However, when oxygen pressure is reduced, the red blood cells of these persons tend to distort into a sickle shape. This phenomenon is known as sickling. These distorted or sickled cells are easily destroyed, leading to anemia.

The clinical symptoms of sickle-cell anemia result from two stresses. First, the body eliminates sickled red cells in large numbers and its synthesizing capacity to produce new red cells is sorely taxed. There is overactivity in bone marrow and general development is poor. Second, the sickled cells tend to clump and interfere with blood circulation, causing a host of syndromes ranging from kidney damage to brain damage.

In most persons whose cells are capable of sickling, fewer than 1% of the red cells in venous circulation are normally sickled, and there are almost no harmful effects ascribable to the condition. These persons are said to have sickle-cell trait. However, about 2% of those persons whose cells are capable of sickling suffer from severe, chronic anemia resulting from excessive destruction of erythrocytes, a condition known as sickle-cell anemia. From 30 to 60% of the red blood cells in circulation in these persons are sickled. The sickling phenomenon stems from the presence of a single gene which, when inherited from one parent, produces the benign sickle-cell trait, but when inherited from both parents produces the crippling sickle-cell anemia.

Can this disease be treated? Careful medical care, including blood transfusion, certainly helps. But this is not a cure. Fortunately, there is now hope that new drugs will be found that can minimize sickling or possibly prevent it altogether.

The molecular basis for this abnormality was discovered by Linus Pauling. He isolated hemoglobin (Hemoglobin S or Hb S) from the blood

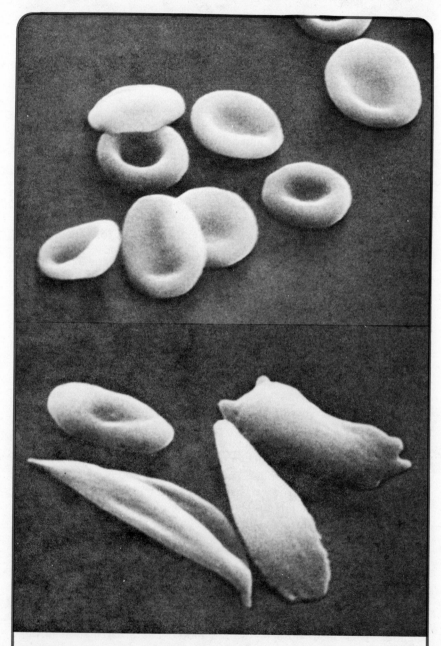

Normal red blood cells and cells that have sickled after discharge of oxygen. Magnification X 6750. (Courtesy of Dr. Marion I. Barnhart, Wayne State University School of Medicine)

of sickle-cell anemic patients, separated the colored heme from the globin, and compared each with corresponding fractions from the hemoglobin of normal persons. The heme from the two sources was identical. Next he examined the protein from each hemoglobin by paper electrophoresis (Section 13.4). Paper electrophoresis of normal and sickle-cell anemia hemoglobin at pH 6.9 indicated a significant difference between the two; normal hemoglobin moved as a negative ion while sickle-cell hemoglobin moved as a positive ion. The two types are quite distinct. The hemoglobin from persons with sickle-cell trait is a mixture: 60% normal hemoglobin (Hb A) and 40% sickle-cell hemoglobin (Hb S).

Pauling's discovery revealed that the cause of the sickling phenomenon lay in the protein. Later work showed that the abnormal hemoglobin functions perfectly satisfactorily in transporting molecular oxygen to the cells, but after it has discharged oxygen it changes shape; the deoxygenated hemoglobin molecules cluster together in stacks which distort the cell membrane. The erythrocyte sickles and becomes more rigid and more susceptible to rupture and enzymatic destruction. Pauling's studies also confirmed a direct relationship between the gene present and the characteristics of the hemoglobin found. Persons with sickle-cell trait, heterozygotes for the sickle-cell gene, synthesize both Hb A and Hb S. Those with sickle cell anemia, homozygotes for the gene, produce only Hb S. Thus, sickle cell anemia is a molecular disease of genetic origin.

Vernon Ingram of Cambridge, England, discovered that sickle-cell hemoglobin differs from normal hemoglobin in only a single amino acid residue of each beta chain. The alpha chains of the two are identical, but glutamic acid at position 6 of the normal beta chain is replaced by valine in Hb S. Apparently, the introduction of the hydrophobic amino acid, valine, makes possible new interactions between deoxygenated hemoglobin molecules which in turn allow them to stack into crystal-like structures. Thus, this single amino acid substitution makes the difference between healthy and sickling red blood cells. This was the first instance where a genetic disease could be traced to a single amino acid substitution in a specific protein.

The fact that there is so much natural selection pressure against this abnormal gene raises the question of why the sickle-cell trait is so common in populations in certain parts of the world. Assuming that the gene resulted from a single mutation, one must conclude, from the current wide distribution of the disease, that the deleterious gene mutation occurred long ago. Why then was it not erased from the pool of human genes long ago? The curious geographic distribution of the disease provides an important clue. The incidence of the disease is high (15% and over) among the populations of central Africa and southern India, areas where falciparum malaria is endemic. There is now considerable evidence that heterozygotes for the hemoglobin S gene have a greater resistance to malarial infection.

The malaria parasite lives in red blood cells for a part of its life cycle. Red blood cells containing hemoglobin S have a shortened lifetime and, consequently, the malaria parasite cannot mature. Thus, heterozygotes for the hemoglobin S gene have a greater resistance to malarial infection, but at the price of seeing one-fourth of their offspring afflicted with sickle-cell anemia.

The dramatic success in discovering the genetic and molecular basis for sickle-cell anemia spurred interest in searching for other abnormal hemoglobins. To date, several hundred have been isolated and the changes in primary structure have been determined. In the vast majority, there is but a single amino acid residue change in either the alpha or the beta chain, and each substitution is consistent with the change of a single nucleotide in one DNA codon (Section 14.11). Several abnormal hemoglobins are listed in Table 1.

TABLE 1 Abnormal human hemoglobins. Many of these names are derived from the location of their discovery.

| Hemoglobin Variant | Amino Acid Substitution | | |
	Position	From	To
alpha chain			
J-Paris	12	ala	asp
G-Philadelphia	68	asn	lys
M-Boston	58	his	tyr
Dakar	112	his	gln
beta chain			
S	6	glu	val
J-Trinidad	16	gly	asp
E	26	glu	lys
M-Hamburg	63	his	tyr

Although most of the abnormal hemoglobins differ from Hb A by only a single amino acid substitution, several have been discovered in which there are either insertions or deletions of amino acids. For example, in hemoglobin-Leiden, discovered in 1968, glutamic acid at position 6 in each beta chain is missing altogether. In hemoglobin-Gun Hill, discovered in 1967 in a 41-year-old man and one of his three daughters, there is a deletion of five amino acids in each beta chain. Thus, each beta chain is shortened to 141 amino acid residues.

$$
\begin{array}{lll}
 & 90 \qquad 92\ 93 \qquad\qquad 97\ 98 \\
\text{Hb A} & \text{glu-leu-his-cys-asp-lys-leu-his-val} \\
\text{Hb-Gun Hill} & \text{glu-leu-his-[- - - - - - - - - - - - - - - - -]-val}
\end{array}
$$

five amino
acids deleted

In hemoglobin-Grady, discovered in 1974 in a 25-year-old woman and her father, there is insertion of three amino acids in each alpha chain. Thus, each alpha chain is elongated to 144 amino acid residues.

$$
\begin{array}{lll}
 & 114 \qquad\qquad 118\ 119 \\
\text{Hb A} & \text{pro-ala-glu-phe-thr-pro-ala} \\
 & \qquad\qquad\qquad\ \ 118 \qquad\qquad\quad 119 \\
\text{Hb-Grady} & \text{pro-ala-glu-phe-thr-[glu-phe-thr]-pro-ala}
\end{array}
$$

three amino
acids inserted

Notice that in hemoglobin-Grady, the three amino acids inserted are a repeat of the amino acids at positions 116-117-118 in the normal alpha chain.

These abnormal hemoglobins are the clearest examples in man of the expression of mutant forms of a single gene, and there can be no doubt that many other kinds of physiological and morphological variations will be found to have similar structural bases. In particular, it is probable that some and perhaps most of the specific enzyme deficiencies underlying the inborn errors of metabolism (read the mini-essay, "Inborn Errors of Metabolism") will turn out to be due to specific amino acid substitutions, insertions, or deletions in critical enzymes and consequent loss of their activity. Conditions such as sickle-cell anemia, alcaptonuria, and phenylketonuria may be simply extreme examples of biochemical variation present everywhere in minor degree.

REFERENCES

Antonine, E. and M. Brunori. "Hemoglobin," *Annual Review of Biochemistry*, vol. 39, 977 (1970).

Beale, D. and H. Lehmann, "Abnormal Hemoglobins and the Genetic Code," *Nature*, vol. 207, 259 (1965).

Dayhoff, M. O. *Atlas of Protein Sequence and Structure*, vol. 5, National Biomedical Research Foundation, 1972.

Harkness, D. R. "Trends," *Biochemical Science*, vol. 1 (1976), pp. 73–76.

Huisman, T. H. J., et al., "Hemoglobin Grady: The First Example of a Variant with Elongated Chains Due to the Insertion of Residues," *Proceedings of the National Academy of Sciences*, vol. 71, 3270 (1974).

Ingram, V. "Gene Mutations in Human Hemoglobin. The Chemical Difference Between Normal and Sickle-Cell Hemoglobin," *Nature*, vol. 180, 326 (1957).

Morimoto, H., H. Lehmann, and M. F. Perutz. "Molecular Pathology of Human Hemoglobins," *Nature*, vol. 232, 408 (1971).

Pauling, L., H. A. Itano, S. J. Singer, and I. C. Wells. "Sickle-Cell Anemia, a Molecular Disease," *Science*, vol. 110, 543 (1949).

Perutz, M. F. "The Hemoglobin Molecule," *Scientific American*, November 1964.

Nucleic Acids

14.1 Introduction

Nucleic acids are a third great class of macromolecules or biopolymers which, like proteins and polysaccharides, are vital components of living materials. Of all the biopolymers, probably none are so important as the nucleic acids, for they are the genetic material itself. In this chapter we shall examine the structure of the mononucleotides, and the manner in which these small building blocks are bonded together to form giant nucleic acid molecules. Then we look at the three-dimensional structure of nucleic acids, and finally the manner in which genetic information is coded, used in the synthesis of proteins, and transmitted from cell to cell.

Controlled hydrolysis, catalyzed by aqueous acid, base, or specific enzymes, breaks nucleic acids into successively smaller fragments and ultimately into three characteristic components: (1) phosphoric acid, (2) a pentose, and (3) heterocyclic bases. Nucleic acids which on hydrolysis yield

nucleic acids (DNA, RNA)

H_2O | catalyst

mononucleotides

H_2O | catalyst

nucleosides + phosphoric acid

H_2O | catalyst

nitrogenous bases + D-ribose or 2-deoxy-D-ribose

FIGURE 14.1 Hydrolysis products of nucleic acids.

the pentose ribose are termed <u>ribonucleic acids</u> (RNAs); those yielding deoxyribose are termed <u>deoxyribonucleic acids</u> (DNAs).

Two classes of heterocyclic bases are found on hydrolysis of nucleic acids: those related to pyrimidine and those related to purine. The heterocyclic bases related to <u>pyrimidine</u> are cytosine (C), uracil (U), and thymine (T).

<u>Cytosine</u> has been found in all nucleic acids examined thus far. <u>Uracil</u> is present in all samples of RNA but is usually absent from DNA. Conversely, <u>thymine</u> is found in samples of DNA but is only rarely found in RNA. It often is possible to distinguish DNA from RNA by the presence or absence of thymine. In addition to these three major pyrimidines, several others (for example, 5-methylcytosine) occur in lesser amounts.

| pyrimidine | cytosine (C) | uracil (U) | thymine (T) |

The heterocyclic bases related to <u>purine</u> are adenine (A) and guanine (G).

| purine | adenine (A) | guanine (G) |

<u>Adenine</u> and <u>guanine</u> are universally distributed in all nucleic acids. Other purines, for example, N^6-methyladenine, occur in small amounts in certain specific nucleic acids. Note that the designation N^6 specifies that the methyl substituent is attached to the nitrogen atom on carbon-6 of the adenine ring.

14.2 Nucleosides

<u>Nucleosides</u> are a combination of a purine or pyrimidine base and a pentose, either D-ribose or 2-deoxy-D-ribose. Two nucleosides, <u>adenosine</u> and <u>2'-deoxycytidine</u>, are shown in Figure 14.2. Natural nucleosides are glycosides in which a nitrogen (N-9 of the purines or N-1 of the pyrimidines) is bonded by a glycoside linkage to the carbon 1' of ribose or deoxyribose. The pentose is always in the furanose form and the glycoside linkage is always beta (see Section 11.8). Like all glycosides, the

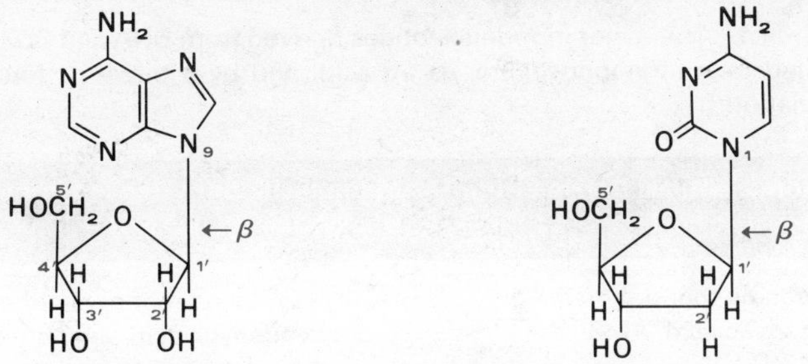

adenosine
(9-β-D-ribofuranosyladenine)

2'-deoxycytidine
(1-β-D-2'-deoxyribofuranosylcytosine)

FIGURE 14.2 Nucleosides: adenosine and 2'-deoxycytidine. Unprimed numbers are used for atoms of the purine and pyrimidine bases; primed numbers are used for atoms of the pentoses.

nucleosides are relatively stable to aqueous alkali but undergo hydrolysis in acid to form a base and a pentose.

14.3 Mononucleotides

The mononucleotides are phosphoric acid esters of nucleosides in which the phosphoric acid is esterified with one of the hydroxyls of the pentose. There are three hydroxyl groups in ribose and two free hydroxyl groups in deoxyribose and, in principle, the phosphoric ester can be formed with any one of these hydroxyls. Although all variations do exist in cells, the predominant ester is the 5'-phosphate. These compounds, illustrated below by the 5'-esters of adenosine, are named either as esters (adenosine monophosphate, abbreviated AMP) or as acids (adenylic acid). Note that the two protons on each phosphate ester are fully ionized.

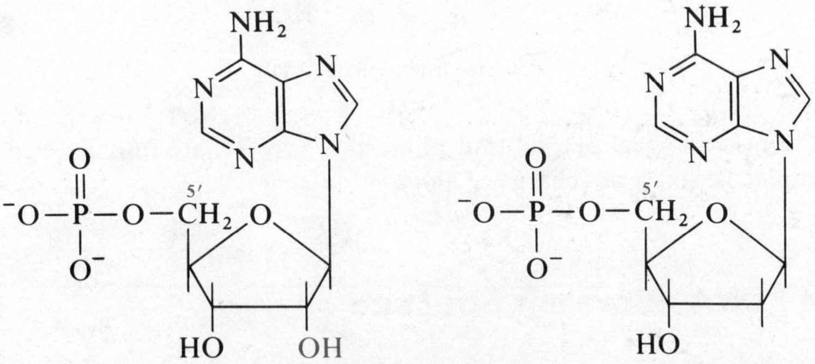

adenosine 5'-monophosphate

deoxyadenosine 5'-monophosphate

Table 14.1 lists the names of the major mononucleotides derived from ribonucleic and deoxyribonucleic acids.

All nucleoside monophosphates may be further phosphorylated to form nucleoside diphosphates and nucleoside triphosphates. In the case of the di- and triphosphates, the second and third phosphate groups are joined by anhydride bonds. Attachment of one molecule of phosphate to

TABLE 14.1 The major mononucleotides derived from DNA and RNA. Each is named as a monophosphate, as an acid, and by a three- or four-letter abbreviation.

Mononucleotides Derived from Ribonucleic acids	Mononucleotides Derived from Deoxyribonucleic acids
adenosine monophosphate adenylic acid, AMP	deoxyadenosine monophosphate deoxyadenylic acid, dAMP
guanosine monophosphate guanylic acid, GMP	deoxyguanosine monophosphate deoxyguanylic acid, dGMP
cytidine monophosphate cytidylic acid, CMP	deoxycytidine monophosphate deoxycytidylic acid, dCMP
uridine monophosphate uridylic acid, UMP	(uridine monophosphate not a major component of DNA)
(thymidine monophosphate not a major component of RNA)	deoxythymidine monophosphate deoxythymidylic acid, dTMP

adenosine by an anhydride bond forms adenosine diphosphate (ADP); attachment of two molecules of phosphate forms adenosine triphosphate (ATP), shown below.

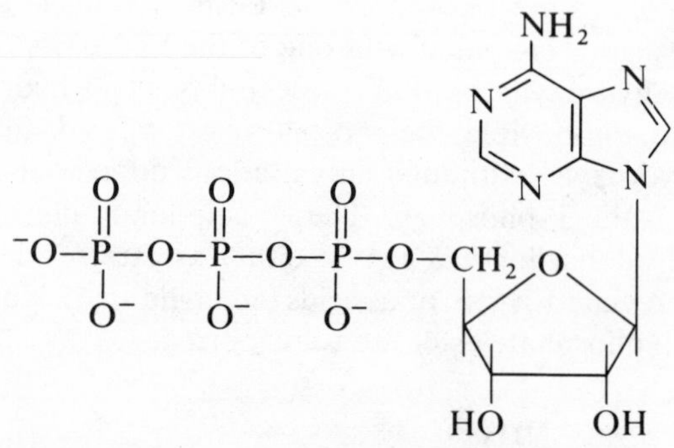

adenosine triphosphate (ATP)

At the physiological pH, all four phosphate protons are fully ionized and the molecule has a net charge of -4.

14.4 DNA—Covalent Structure

Deoxyribonucleic acid (DNA) consists of a backbone of alternating units of deoxyribose and phosphate in which the 3′-hydroxyl of one deoxyribose is joined to the 5′-hydroxyl of the next deoxyribose by a phosphodiester bond (Figure 14.3). This backbone is constant throughout the entire DNA molecule. A heterocyclic base, either adenine, guanine, thymine, or cytosine, is then attached to each deoxyribose by a β-glycoside bond.

Here we might compare the primary structures of DNA and a protein molecule and note that each consists of a backbone or constant portion and a variable portion. In DNA, the backbone consists of deoxyriboses linked

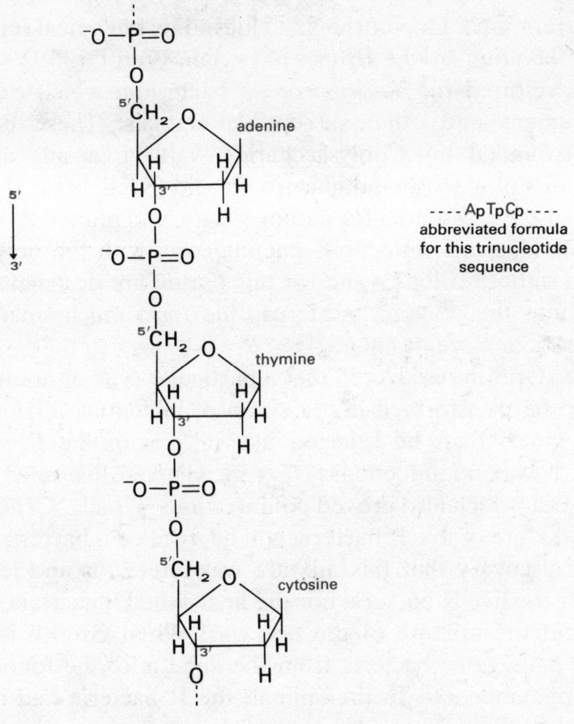

FIGURE 14.3 Deoxyribonucleic acid (DNA). Partial structural formula showing a tri-nucleotide sequence. In the abbreviated sequence shown (right), the bases of the trinucleotide are read from the 5′ end of the chain to the 3′ end as indicated by the arrow (far left).

by phosphodiester bonds; in a protein, the backbone consists of repeating peptide bonds. In DNA, the variable portion is the sequence of four heterocyclic bases along the backbone; in proteins, the variable portion is the sequence of amino acid side chains along the backbone.

Another characteristic of DNA molecules is that one end of the chain has the 5′-hydroxyl of deoxyribose free while the other end has the 3′-hydroxyl free. By convention, nucleic acid chains are always written beginning with the free 5′-hydroxyl group and proceeding toward the free 3′-hydroxyl group. Given this way of writing polynucleotide chains, it should be obvious that ATC and CTA are different trinucleotides, just as gly—ala—ser and ser—ala—gly are different tripeptides.

Naturally occurring DNA molecules are long and thread-like, and have very high molecular weights. For example, the chromosome of the bacteria *E. coli* is a single DNA molecule of molecular weight 2,300,000,000 (2.3 trillion). Its diameter is only 20×10^{-8} cm but its total length is 0.12 cm. By comparison, a molecule of collagen, one of the longest proteins, has a length of only 0.00003 cm.

14.5 DNA—Biological Function

In 1869, Miescher, a Swiss chemist, discovered nucleic acids in the nuclei of pus cells. However, it took many decades before the major function of

DNA was determined. One of the first clues to its biological function grew out of an observation by the British physician, Fred Griffith. Griffith had isolated and cultured the *Pneumococcus* bacterium which causes pneumonia in humans and other susceptible animals. These bacteria are normally surrounded by a polysaccharide wall or capsule made up of alternating units of glucose and glucuronic acid (Section 11.3). This polysaccharide capsule is essential for pathogenicity, and mutants without it are nonpathogenic. Normal, infectious pneumococci with the polysaccharide capsule form smooth colonies and for this reason are designated S. Those mutants without the polysaccharide capsule form much smaller, rough-looking colonies and are designated R.

In 1928, Griffith discovered that a particular type of noninfectious R mutant could be transformed into a strain of infectious S bacteria in the following manner. First he injected live mice with the R mutant and showed that it was noninfectious. Next he injected live mice with heat-killed S bacteria, which also proved noninfectious. Finally, Griffith injected mice with a mixture of live R bacteria and heat-killed S bacteria, and made the startling discovery that this mixture was infectious and lethal to the mice. Neither the live R bacteria nor the heat-killed S bacteria were lethal separately, but the mixture of the two was! When Griffith isolated and cultured the pathogenic bacteria from the dead mice, he found that they were live S pneumococci. In the animal, the R bacteria had been transformed into S bacteria.

Other biologists pursued the investigation of this transformation. They fractionated the components of S cells and tested the ability of each to transform R cells into S cells. In 1944, Oswald Avery, Colin MacLeod, and Maclyn McCarty reported that "a nucleic acid of the deoxyribose type is the fundamental unit of the transforming principle of the *Pneumococcus* Type III." Moreover, they found that DNA-induced transformation of R cells into S cells is a permanent, heritable characteristic. Thus, for the first time it was realized that genetic information is stored in DNA. This discovery has been confirmed over and over again by a wide range of biochemical investigations and, today, there is no doubt about the fact that genes are made of DNA.

14.6 The Three-Dimensional Structure of DNA

By 1950, there was general agreement that DNA molecules consist of chains of alternating units of deoxyribose and phosphate groups linked by phosphodiester bonds and with a heterocyclic base attached to each deoxyribose. Although the general structural formula of DNA was known, the precise sequence of bases along the chain was completely unknown for any particular DNA molecule. At one time it was thought that the heterocyclic bases occurred in equal ratios and that perhaps the four major bases might repeat in a regular pattern along the pentose-phosphate backbone of the molecule. However, more precise determinations of base composition revealed that the bases do not occur in equal ratios (Table 14.2).

TABLE 14.2 Comparison of base composition (in mole percent) of DNAs from several organisms.

Organism	A	G	C	T	A/T	G/C	purines pyrimidines
man	30.9	19.9	19.8	29.4	1.05	1.00	1.04
sheep	29.3	21.4	21.0	28.3	1.03	1.02	1.03
sea urchin	32.8	17.7	17.3	32.1	1.02	1.02	1.02
marine crab	47.3	2.7	2.7	47.3	1.00	1.00	1.00
yeast	31.3	18.7	18.7	32.9	0.95	1.09	1.00
E. coli	24.7	26.0	26.0	23.6	1.04	1.01	1.03

From consideration of data such as these, the following conclusions emerged:

1. The mole percent base composition of DNA in any organism is the same in all cells and is characteristic of the organism.
2. Base compositions vary from one organism to another. The DNAs of closely related species have similar base compositions, whereas the DNAs of widely differing species are likely to have widely different base compositions.
3. In nearly all DNAs, the mole percent of adenine equals that of thymine, and the mole percent of guanine equals that of cytosine.
4. The total number of purine residues (A + G) equals the total number of pyrimidine residues (C + T).

Additional information on the structure of DNA emerged from analysis of X-ray diffraction photographs of DNA fibers taken by Rosalind Franklin and Maurice Wilkins. It became clear that DNA molecules are long, fairly straight, and not more than a dozen atoms thick. Furthermore, despite the fact that the base composition of DNAs isolated from different organisms varied over a rather wide range, the DNA molecules themselves appeared to be remarkably uniform in thickness. Herein lay one of the major problems to be solved. How could the structure appear so regular in molecular dimensions when the ratios of bases differed so widely? There was also another problem to be solved. In what form is the genetic information stored in DNA molecules and how is it transmitted or replicated from one generation or cell to the next?

With this accumulated information the stage was set for the development of a hypothesis about DNA conformation. In 1953 F. H. C. Crick, a British physicist, and James D. Watson, an American biologist, postulated a precise model of the three-dimensional structure of the DNA molecule, a model in accord with the known dimensions of the structural units, the equivalences of certain base ratios, and the measured dimensions of intact molecules. This model not only accounted for many of the observed physical and chemical properties of DNA but also suggested a mechanism by which genetic information could be repeatedly and accurately replicated. Watson, Crick, and Wilkins shared the 1962 Nobel Prize in Physiology and Medicine for "their discoveries concerning the molecular structure of nucleic acids, and its significance for information transfer in living material."

The heart of their model is the postulate that a molecule of DNA consists of two polynucleotide chains both coiled in a right-handed manner about the same axis to form a double helix. To account for the observed base ratios and the constant thickness of the molecular strand, Watson and Crick postulated that the purine and pyrimidine bases project inward toward the axis of the helix and that the bases always pair in a very specific manner. Thymine is always hydrogen-bonded to adenine and cytosine is always hydrogen-bonded to guanine. The hydrogen bonding between base pairs is a major factor stabilizing the three-dimensional conformation of DNA (see Figure 14.4).

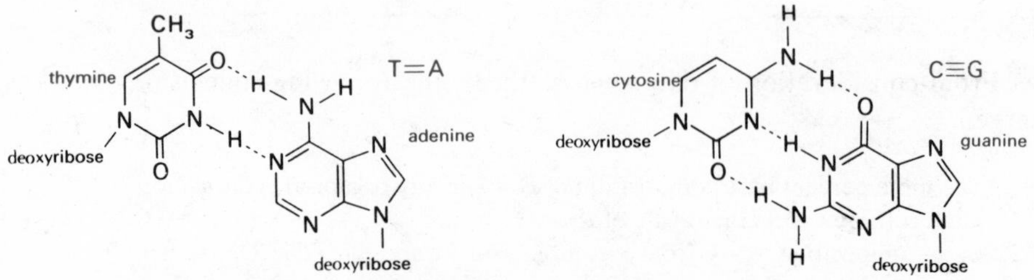

FIGURE 14.4 Hydrogen-bonded interaction between thymine and adenine; cytosine and guanine. The first couple is abbreviated as T=A (showing two hydrogen bonds) and the second couple as C≡G (showing three hydrogen bonds).

According to scale models, the dimensions of the thymine-adenine base pair are identical with those of the cytosine-guanine base pair, and the length of each couple is consistent with the thickness of the DNA strand. A significant fact arising from this model building is that no other base pairing is consistent with the observed regularities of DNA structure. A pair of pyrimidines would be too small to account for the observed thickness of the DNA molecule; a pair of purine bases would be too big to fit into the observed thickness.

The formation of such hydrogen-bonded base pairs with identical overall dimensions resolved the paradox presented by the X-ray data: the emergence of repetitive structure out of the disorder of random base composition. According to the model, the repeating units are not single bases of differing dimensions, but specific base pairs of identical dimensions.

To account for the periodicity observed from X-ray data, Watson and Crick postulated that the bases are closely stacked perpendicular to the axis of the helix at a distance of 3.4 angstroms between base pairs, and exactly ten base pairs are coiled in one complete turn of the helix. The distance (34 Å) between the repeat pattern in the crystal corresponds to one complete turn of the helix.

14.7 DNA Replication

At the time Watson and Crick postulated the model for the conformation of DNA, biologists were amassing evidence that DNA was in fact the hereditary material. Detailed studies of living cells revealed that during cell

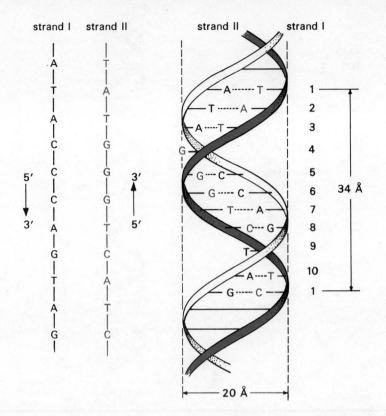

strand I   strand II   strand II   strand I

FIGURE 14.5 Abbreviated representation of the Watson–Crick double-helical model of DNA. On the left are shown two uncoiled complementary antiparallel polynucleotide strands. On the right the strands are twisted in a double helix of thickness 20 angstroms and a repeat distance of 34 angstroms along the axis of the double helix. There are 10 base pairs per complete turn of the helix.

division there is an exact copying or duplication of chromosomes, and that chromosomes contain DNA and some protein. The challenge posed to molecular biologists was this: How does the genetic material duplicate itself with such unerring fidelity?

One of the exciting things about the double helix model is that it immediately suggested how DNA might produce an exact copy of itself. The double helix consists of two parts, each the precise complement of the other. If the two strands separate and each serves as a template for the construction of its own new complement, then each new double strand is an exact replica of the original DNA. Because each new double-stranded DNA molecule contains one strand from the parent molecule and one newly synthesized strand, the process is said to be a semiconservative replication. This process is illustrated schematically in Figure 14.6.

This hypothesis of DNA replication was simple and at the same time revolutionary. Yet it was at this stage only a hypothesis. Could it be confirmed experimentally? Within a few years, two separate and ingenious lines of research led to the experimental confirmation of this bold hypothesis. Meselson and Stahl in 1957–1958 demonstrated conclusively that in living intact bacteria, *Escherichia coli* (*E. coli*), DNA is replicated in exactly the manner predicted by Watson and Crick.

403

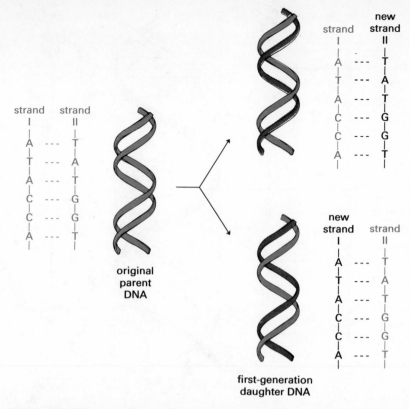

FIGURE 14.6 Schematic diagram of the replication mechanism by which DNA might duplicate itself. The double helix uncoils and each chain of the parent serves as a template for the synthesis of its complement. Each daughter DNA contains one strand from the original DNA and one newly synthesized strand.

The American biochemist Arthur Kornberg approached this problem in quite another manner by investigating the enzymatic mechanism of DNA replication. The Watson–Crick model suggested that DNA acts as a template, inducing an enzyme to synthesize new DNA. Kornberg reasoned that the starting materials for this enzyme-catalyzed DNA synthesis must be the four deoxyribonucleoside triphosphates. Accordingly he and his colleagues incubated dATP, dGTP, dTTP, and dCTP with crude extracts of *E. coli* and were able to demonstrate the synthesis of new DNA. The enzyme catalyzing this reaction is named DNA polymerase.

$$
\begin{matrix} n_1 \text{dATP} \\ n_2 \text{dGTP} \\ n_3 \text{dCTP} \\ n_4 \text{dTTP} \end{matrix} \xrightarrow[\text{DNA polymerase}]{\text{DNA (template)}} \text{DNA} \begin{cases} \text{dAMP} \\ \text{dGMP} \\ \text{dCMP} \\ \text{dTMP} \end{cases} + (n_1 + n_2 + n_3 + n_4)\text{pyrophosphate}
$$

The striking and characteristic property of DNA polymerase is that it requires the presence of some preexisting DNA and that the newly synthesized DNA has a base composition nearly identical to that of the added DNA. After painstaking and laborious purification procedures, Kornberg was able to isolate a small amount of crystalline DNA polymerase. More recently he has shown that DNA polymerase is capable of inducing the replication of an infectious bacterial virus and that the synthetic product has the infecting capacity of the original material.

Thus, the experiments of Meselson and Stahl on the synthesis of DNA in living, intact *E. coli* cells, and those of Kornberg on the enzymatic mechanism involving DNA polymerase have provided convincing evidence that DNA is replicated in accord with the prediction of the Watson–Crick model.

14.8 Ribonucleic Acid—RNA

Ribonucleic acids (RNAs) are similar to deoxyribonucleic acids in that they too consist of long, unbranched chains of nucleotides joined by phosphodiester bonds between the 3′-hydroxyl of one pentose and the 5′-hydroxyl of the next. Thus, their structure is much the same as that of DNA shown in Figure 14.3 except that (1) the pentose in RNA is ribose rather than deoxyribose, and (2) one of the four major bases normally is uracil rather than thymine. Like thymine, uracil can form a hydrogen-bonded base pair with adenine. However, it lacks the methyl group present in thymine.

Unlike DNA, RNA is distributed throughout the cell; it is present in the nucleus, the cytoplasm, and even in subcellular particles such as the mitochondria. Furthermore, cells contain three types of RNA: ribosomal RNA, transfer RNA, and messenger RNA. These three types differ in molecular weight and, as their descriptive names imply, they perform different functions within the cell.

Ribosomal RNA (rRNA) molecules have molecular weights of 0.5 to 1.0 million and comprise up to 85 to 90% of the total cellular ribonucleic acid. In the cytoplasm of the cell are located a large number of subcellular particles called ribosomes, which contain about 60% rRNA and 40% protein. The intact ribosomal particles are an essential component of the protein-synthesizing machinery of the cell.

Transfer RNA (tRNA) has the smallest structure of all the nucleic acids. There are some 20 known tRNAs, each corresponding to one of the 20 amino acids found in proteins. Transfer RNAs have molecular weights of about 25,000 and contain 75 to 80 mononucleotide residues per molecule. The function of tRNA during protein synthesis is to carry a specific amino acid to a site on the ribosome. For this transportation, an amino acid is joined to a specific tRNA in an enzyme-catalyzed ester formation between the α-carbonyl group of the amino acid and the 3′-hydroxyl of the ribose from the terminal tRNA nucleotide. Amino acids so esterified are said to be "activated." Several tRNAs have been isolated and their base sequences decoded.

Messenger RNA (mRNA) is present in the cell in relatively small amounts and has a high turnover rate. This RNA fraction has an average molecular weight of several hundred thousand and a base composition like that of the DNA of the organism from which it is isolated. For this reason, its discoverers called it DNA-like RNA. Messenger RNA is synthesized in a cell using DNA as a template and the reaction is catalyzed by an enzyme named RNA polymerase. For the synthesis of mRNA the cell requires (1) the four major ribonucleoside triphosphates, (2) DNA as a template, and (3) RNA polymerase.

$$\begin{matrix} n_1\text{ATP} \\ n_2\text{GTP} \\ n_3\text{CTP} \\ n_4\text{UTP} \end{matrix} \xrightarrow[\text{RNA polymerase}]{\text{DNA (template)}} \text{RNA} \begin{cases} \text{AMP} \\ \text{GMP} \\ \text{CMP} \\ \text{UMP} \end{cases} + (n_1 + n_2 + n_3 + n_4)\text{pyrophosphate}$$

The name "messenger" RNA was coined by the French scientists F. Jacob and J. Monod because this RNA, which is made in the nucleus of the cell on a DNA template, migrates into the cytoplasm and to the ribosome where it is involved in protein synthesis.

14.9   The Genetic Code

Given our understanding of the structure of DNA and its function as an archive of genetic information for the entire organism, the next questions to be asked are: In what manner is this genetic information coded on the DNA molecule? What does it code for? How is this code read and expressed? In answer to the first two questions, it is now clear that the sequence of bases in the DNA molecule constitutes the store of genetic information and that this sequence of bases serves to direct protein synthesis. However, the statement that the sequence of bases is the genetic information and that this sequence directs the synthesis of proteins presents a paradox. How can a molecule consisting of only four variable units (adenine, cytosine, guanine, thymine) direct the synthesis of a protein in which there are as many as 20 different units? How can a 4-letter alphabet specify the sequence of the 20-letter alphabet that occurs in proteins?

One obvious answer is that it is not one base but some combination of bases that codes for each amino acid. If the code consists of nucleoside pairs, there are 16 (4^2) combinations, a more extensive code, but still not extensive enough to code for 20 amino acids. If the nucleosides are considered in groups of three, there are 64 (4^3) possible combinations, more than enough to specify the primary sequence of a protein. This appears to be a very simple solution to a system that must have taken eons of evolutionary trial and error to develop. Yet there is convincing evidence that nature does indeed use this simple 3-letter code to store genetic information. One of the triumphs of molecular biology is that the code has been deciphered. But before we analyze the code itself, let us examine how the code directs protein synthesis.

14.10   Protein Synthesis

The process of information transfer in protein synthesis is formulated as follows. Each gene (a section of a DNA molecule) contains a sequence of bases that dictates the sequence of amino acids in a protein. For example, to code for the sequence of 146 amino acids in the β chain of normal adult hemoglobin, there must be a run of 438 nucleotides (146 triplets) on a section of DNA. This section serves as a template for the synthesis of a complementary strand of mRNA. Thus the sequence —CGAATTA— in DNA directs the synthesis of the complementary sequence —GCU-UAAU— in mRNA, a process called transcription.

The newly synthesized mRNA migrates to the cytoplasm and attaches to a ribosome, thus initiating the process in which the information encoded in mRNA is expressed as a precise sequence of amino acids in a polypeptide molecule. This process is called translation. Each of the 20 amino acids is brought to the site of protein synthesis as an "activated" ester of a specific tRNA. Each tRNA has in its structure a triplet complementary to three bases in mRNA. The tRNA triplet that is the complement of a mRNA codon is called an anticodon. These complementary triplets interact by hydrogen bonding, and in so doing position the activated amino acid in its proper sequence along the polypeptide chain. Finally, enzyme-catalyzed peptide bond formation between the carboxyl group of one amino acid and the amino group of the next initiates growth of the polypeptide chain. The strand of mRNA is "read off," triplet by triplet, until the polypeptide chain is complete.

14.11   The Genetic Code Deciphered

The next question of course is: Which particular triplets code for each amino acid? At one time it seemed that the only hope of answering this question was to isolate a section of a gene coding for a particular protein and then compare the sequence of amino acids on the protein with the sequence of bases on the corresponding section of DNA. However, this was not experimentally possible even as late as 1960 because the base sequences of genes were unknown.

Fortunately, the young biochemist Marshall Nirenberg provided a simple and very direct experimental approach to the problem. It was based on the observation that synthetic polynucleotides will direct polypeptide synthesis in much the same manner as mRNA. Therefore Nirenberg incubated ribosomes, amino acids, tRNA, and the appropriate protein-synthesizing enzymes. With only these components, there was essentially no polypeptide synthesis. However, when Nirenberg added synthetic polyuridylic acid (poly U), a polypeptide of high molecular weight was formed. The exciting result of this experiment was that the polypeptide contained only phenylalanine. Poly U had served as a synthetic messenger RNA. With this discovery, the first element of the genetic code had been deciphered. The triplet UUU codes for the amino acid phenylalanine.

This same type of experiment was carried out with different poly-ribonucleotides. It was found that polyadenylic acid (poly A) led to the synthesis of polylysine and that polycytidylic acid (poly C) led to the synthesis of polyproline.

codon	amino acid
UUU	phenylalanine
AAA	lysine
CCC	proline

This strategy was extended and, by 1966, all 64 codons had been

UUU Phe	UCU Ser	UAU Tyr	UGU Cys
UUC Phe	UCC Ser	UAC Tyr	UGC Cys
UUA Leu	UCA Ser	UAA Stop	UGA Stop
UUG Leu	UCG Ser	UAG Stop	UGG Trp
CUU Leu	CCU Pro	CAU His	CGU Arg
CUC Leu	CCC Pro	CAC His	CGC Arg
CUA Leu	CCA Pro	CAA Gln	CGA Arg
CUG Leu	CCG Pro	CAG Gln	CGG Arg
AUU Ile	ACU Thr	AAU Asn	AGU Ser
AUC Ile	ACC Thr	AAC Asn	AGC Ser
AUA Ile	ACA Thr	AAA Lys	AGA Arg
AUG Met	ACG Thr	AAG Lys	AGG Arg
GUU Val	GCU Ala	GAU Asp	GGU Gly
GUC Val	GCC Ala	GAC Asp	GGC Gly
GUA Val	GCA Ala	GAA Glu	GGA Gly
GUG Val	GCG Ala	GAG Glu	GGG Gly

TABLE 14.3 The genetic code: the mRNA codons and the amino acid whose incorporation each codon directs.

deciphered. Table 14.3 lists these codons and the amino acid that each one codes.

A number of features of the genetic code are evident from Table 14.3.

1. Only 61 triplets code for amino acids. The remaining three (UAA, UAG, and UGA) are signals for chain termination, that is, they are signals to the protein-synthesizing machinery of the cell that the primary structure of the protein is complete.

2. The code is degenerate. Since there are 20 amino acids and 61 triplets to code for them, obviously many amino acids must be coded for by more than one triplet. If you count the number of triplets coding for each amino acid, you will find that only methionine and tryptophan are coded by just one triplet. The other 18 are coded by two or more triplets.

3. In all cases where an amino acid is coded by two, three or four triplets, the degeneracy is only in the last base of the triplet. In other words, in the codons for these 15 amino acids, it is only the third letter of the code that varies. For example, glycine is coded by the triplets GGA, GGG, GGC, and GGU.

4. Finally, there is no ambiguity in the code. Each triplet codes for one and only one amino acid.

We must ask one last question about the genetic code, namely, is the code universal—is it the same for all organisms? Every bit of experimental evidence available today from the study of viruses, bacteria, and higher animals including man indicates that the code is the same for all organisms and that it is universal. Furthermore, the fact that it is the same in all these organisms means that it has been the same over billions of years of evolution.

14.1 Examine the structure of purine. Would you predict this molecule to be planar (or nearly so) or puckered? to exist as a number of interconvertible conformations (as in the case of cyclohexane) or to be rigid and inflexible? Explain the basis for your answers.

14.2 Draw structural formulas for 5-methylcytosine, N^6-methyladenine, and N^2-methylguanine.

14.3 An important drug in the chemotherapy of leukemia is 6-mercaptopurine. Draw the structural formula of this compound.

14.4 Compare and contrast the structural formulas of:
(a) ribose and deoxyribose
(b) nucleoside and nucleotide
(c) nucleotide and nucleic acid

14.5 Draw structural formulas for:
(a) uridine monophosphate
(b) guanosine triphosphate
(c) adenosine diphosphate
(d) deoxycytidine monophosphate
(e) deoxythymidylic acid

14.6 Show by structural formulas the hydrogen bonding between thymine and adenine; between uracil and adenine.

14.7 Cyclic-AMP (adenosine-3′,5′-cyclic monophosphate), first isolated in 1959, is involved in many diverse biological processes as a regulator of metabolic and physiological activity. In it, a single phosphate group is esterified with both the 3′- and 5′-hydroxyls of adenosine. Draw a structural formula for this substance.

14.8 Compare and contrast the α-helix found in proteins with the double helix of DNA in the following ways.
(a) The units that repeat in the backbone of the chain.
(b) The projection in space of the backbone substituents (R— groups in the case of amino acids, purine and pyrimidine bases in the case of DNA) relative to the axis of the helix.
(c) The importance of the backbone substituents in stabilizing the helix.
(d) The number of backbone units per complete turn of the helix.

14.9 List the postulates of the Watson–Crick model of DNA structure. This model is based on certain experimental observations of base composition and molecular dimensions. Describe these observations and show how the model accounts for each.

14.10 Describe the process of DNA replication.

14.11 Compare and contrast DNA and RNA in the following ways.
(a) primary structure **(b)** the major purine and pyrimidine bases present
(c) location in the cell **(d)** function in the cell

14.12 Compare and contrast ribosomal RNA, transfer RNA, and messenger RNA as follows:
(a) molecular weight
(b) function in protein synthesis

14.13 List all amino acids that are specified by:
(a) only a single codon **(b)** two codons
(c) three codons **(d)** four codons
(e) five codons **(f)** six codons

14.14 List all amino acids that have either U or C (both pyrimidines) as the second base of a codon. Also list all amino acids that have either A or G (both purines) as the second base of a codon. What generalization can you draw from these lists?

14.15 What peptide sequences are coded for by the following mRNA sequences?

(a) —G-C-U-G-A-A-U-G-G—   (b) —U-C-A-G-C-A-A-U-C—
(c) —G-U-C-G-A-G-G-U-G—   (d) —G-C-U-U-C-U-U-A-A—

14.16 There are two principal types of mutations: (1) substitution of one mononucleotide for another, as for example A for C in mRNA, and (2) insertion or deletion of a mononucleotide. Below is shown an mRNA sequence which is read from left to right beginning with the codon UCC. Below it are three substitution mutations and one insertion mutation. For what polypeptide sequence does the normal mRNA sequence code and what is the effect of each mutation on the resulting polypeptide?

(a) normal   —U-C-C-C-A-G-G-C-U-U-A-C-A-A-A-G-U-A—
(b) substitution   —U-C-C-A-A-G-G-C-U-U-A-C-A-A-A-G-U-A—
      of A for C
(c) substitution   —U-C-C-C-A-G-G-C-U-U-A-A-A-A-A-G-U-A—
      of A for C
(d) substitution   —U-C-A-C-A-G-G-C-U-U-A-C-A-A-A-G-U-A—
      of A for C
(e) insertion   —U-C-C-C-A-G-G-C-U-A-U-A-C-A-A-A-G-U-A—
      of A

14.17 In HbS, the abnormal human hemoglobin found in individuals with sickle-cell anemia, (page 392), glutamic acid at position 6 in the β-chain is replaced by valine.

(a) List the two codons for glutamic acid and the four codons for valine.
(b) Show that a glutamic acid codon can be converted into a valine codon by a single mononucleotide substitution.

14.18 Examine Table 1 of the mini-essay "Abnormal Human Hemoglobins" and show that each amino acid substitution is consistent with the change of a single nucleotide in a codon.

Spectroscopy

15.1 Introduction

The first several chapters of this text describe many of the most important functional groups in organic chemistry, the typical reactions of each, how to convert one functional group to another, and how to determine by chemical tests which functional groups a substance contains. For example, if a substance of unknown structure decolors a solution of bromine in CCl_4, you should suspect immediately that it contains some kind of unsaturation, possibly an alkene or alkyne. If the same substance fails to react with 2,4-dinitrophenylhydrazine, we know that it probably does not contain an aldehyde or ketone. An understanding of the typical reactions of functional groups can help us answer other types of problems as well. For example, suppose you know that a compound is one of the following three substances.

CH₂OH

a primary
alcohol

an
aldehyde

a carboxylic
acid

How can you tell which of the three substances you are dealing with? One is a primary alcohol, the second an aldehyde, the third a carboxylic acid, and each has its own characteristic reactions. Only the aldehyde will react with 2,4-dinitrophenylhydrazine and only the carboxylic acid will react with an aqueous solution of sodium bicarbonate. If the substance fails to react with either of these reagents, then it must be the alcohol. Suppose instead that you are asked to distinguish between *n*-propyl alcohol and isopropyl alcohol.

$$CH_3-CH_2-CH_2-OH \qquad CH_3-\overset{\overset{\displaystyle OH}{|}}{C}H-CH_3$$

n-propyl alcohol   isopropyl alcohol

Each has the same molecular formula, C_3H_8O, and each contains the same functional group. These two substances can be distinguished by chemical tests, but the process is much more time consuming and involved than for the first examples we examined.

The purpose of this chapter is to show how information about the presence or absence of particular functional groups can be obtained from the study of molecular spectra. As you will see, such methods have three major advantages over most chemical tests.

1. Spectral analyses are easier and faster to do than most chemical tests.
2. Spectral analyses generally provide far more detailed information about molecular structure.
3. Spectral analyses are nondestructive and, if necessary, the entire sample can be recovered.

Let us begin our study of molecular spectroscopy by first reviewing some facts about radiant energy and the interaction of energy and matter.

15.2   Electromagnetic Radiation

Recall from general chemistry that we can describe light (visible light, ultraviolet light, radio waves, X-rays, and so on) in a very simple way if we assume that light is a wave which travels through space. We can describe this light wave in the same terms we would use to describe an ocean wave as it moves toward the shore, namely, in terms of its wavelength and its frequency.

Wavelength is the distance between consecutive crests (or troughs).

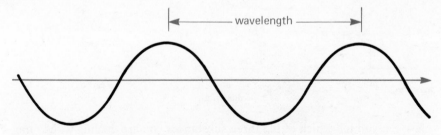

Wavelength is given the symbol λ (lambda) and is generally measured in meters (m) or fractions of meters.

$$1 \text{ centimeter (cm)} = 10^{-2} \text{ meter}$$
$$1 \text{ millimeter (mm)} = 10^{-3} \text{ meter}$$
$$1 \text{ micrometer } (\mu\text{m}) = 10^{-6} \text{ meter}$$
$$1 \text{ nanometer (nm)} = 10^{-9} \text{ meter}$$

Frequency is the number of full cycles of the wave that pass a given point in a fixed period of time. Frequency is given the symbol ν (nu) and is reported in cycles per second (cps) or Hertz (Hz; one Hz = one cps). Wavelength and frequency are inversely proportional to each other, and one can be calculated from the other using the following relationship:

$$\nu = \frac{c}{\lambda}$$

ν = frequency in Hz or cps
λ = wavelength in meters
c = velocity of light, 3×10^8 m/sec

For example, infrared radiation, or heat radiation as it is also called, has a wavelength of about 15×10^{-6} m ($15 \ \mu$m). The frequency of this radiation is 2×10^{13} Hz.

$$\nu = \frac{3 \times 10^8 \text{ m/sec}}{15 \times 10^{-6} \text{ m}} = 2 \times 10^{13} \text{ Hz}$$

A third way to describe light is in terms of its energy. To scientists at the turn of the century, there were a number of puzzling experimental observations about light that could not be explained in terms of wave properties. Einstein discovered that he could explain these results if he first assumed that light has some of the properties of particles. We now call these particles of light photons. The amount of energy in a mole of photons is related to the frequency of the light by the following equation:

$$E = h\nu$$

E = energy, in cal/mole
h = Planck's constant, 6.625×10^{-27} erg-sec
ν = frequency, in Hz

TABLE 15.1 The electromagnetic spectrum.

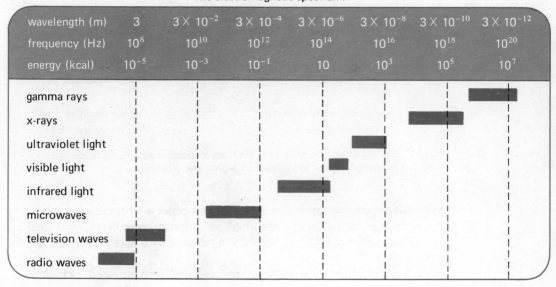

wavelength (m)	3	3×10^{-2}	3×10^{-4}	3×10^{-6}	3×10^{-8}	3×10^{-10}	3×10^{-12}
frequency (Hz)	10^8	10^{10}	10^{12}	10^{14}	10^{16}	10^{18}	10^{20}
energy (kcal)	10^{-5}	10^{-3}	10^{-1}	10	10^3	10^5	10^7

gamma rays
x-rays
ultraviolet light
visible light
infrared light
microwaves
television waves
radio waves

Notice that there is a direct relationship between the frequency of a light wave and its energy—the greater the frequency, the greater the energy. Thus, ultraviolet light (frequency 10^{15} Hz) has a greater energy than infrared radiation (frequency 10^{13} Hz). The wavelengths, frequencies, and energies of the various parts of the electromagnetic spectrum are summarized in Table 15.1.

In this chapter, we will be concerned primarily with three regions of the electromagnetic spectrum: infrared light, ultraviolet-visible light, and radio waves.

15.3   Molecular Spectroscopy

Organic molecules are surprisingly flexible structures. As we have already discussed in Chapter 2, atoms or groups of atoms can rotate about covalent bonds, and a given molecule can have an almost limitless number of conformations. In addition, covalent bonds themselves can bend and stretch just as if the atoms themselves were joined by flexible springs. Furthermore, electrons within molecules can move from one electronic energy level or orbital to another. We know from experimental observations and from theories of molecular structure that these energy changes within molecules are quantized; bonds within a molecule can undergo transitions only between allowed vibrational energy levels, electrons can undergo transitions only between allowed electronic energy levels, and so on.

Organic molecules can be made to undergo a transition from energy state E_1 to a higher state, E_2, by irradiating them with electromagnetic radiation corresponding to the energy difference between states E_2 and E_1.

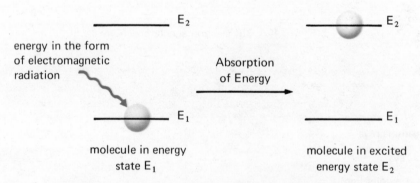

FIGURE 15.1   Absorption of energy in the form of electromagnetic radiation causes a molecule in energy state E_1 to change to a higher energy state E_2.

Molecular spectroscopy is the experimental process of measuring which frequencies of radiation are absorbed by a particular molecule and then attempting to correlate these absorption patterns with details of molecular structure.

We will study how organic molecules absorb infrared radiation, ultraviolet-visible radiation, and microwaves, and we will study what types of information these absorption patterns can give us about details of molecular structure. Specifically, we will see that:

1. Absorption of infrared radiation causes covalent bonds within molecules to be promoted from one vibrational energy level to a higher vibrational energy level.
2. Absorption of ultraviolet-visible radiation causes electrons within molecules to be promoted from one electronic energy level to a higher electronic energy level.
3. Absorption of radiowaves in the presence of a magnetic field causes nuclei within molecules to be promoted from one spin energy level to a higher spin energy level.

15.4   Infrared Spectroscopy

Wavelengths in the region of the infrared spectrum of most interest to us range from about 2.5×10^{-6} to 15×10^{-6} m. In order to simplify the reporting and tabulation of infrared information, chemists generally report absorption frequencies in either micrometers (μm) or in wavenumbers. Expressed in micrometers, this region of the infrared spectrum runs from 2.5 μm to 15 μm. Alternatively, infrared absorption frequencies may be reported in wavenumbers, that is, the number of complete wave cycles per centimeter. Wavenumbers ($\bar{\nu}$) are expressed in cm^{-1} (reciprocal centimeters) and are easily calculated from the wavelength by the following relationship:

$$\bar{\nu} = \frac{1}{\lambda}$$

For example, radiation of wavelength 2.5×10^{-6} m is equivalent to 4000 cm^{-1}. Expressed in wavenumbers, the infrared spectrum runs from 4000 cm^{-1} to 600 cm^{-1}. Chart paper for most infrared spectrophotometers is generally calibrated in both micrometers (μm) and reciprocal centimeters (cm^{-1}).

Virtually all organic molecules are infrared active, because radiation in this region of the spectrum corresponds to the energy required to excite the natural vibrational frequencies of covalent bonds. There are two basic types of bond vibrations: those that correspond to bond stretching and those that correspond to bond bending. In stretching, the distance between bonded atoms increases and decreases in a rhythmic manner, much as the distance between two objects connected by a spring first increases and then decreases as the spring is stretched and then relaxed. In bending, the position of the bonded atoms changes in relation to the original bond axis. You can imagine bond bending as something like the wagging motion of a dog's tail or the flapping motion of the wings of birds in flight. While infrared radiation between 4000 cm^{-1} and 600 cm^{-1} will excite both stretching and bending vibrations, we will look only at bond stretching.

An infrared spectrum is a plot of the percent of radiation transmitted through the sample versus the wavelength of the radiation. (Note that 100% transmission corresponds to 0% absorption.) Figure 15.2 is an infrared spectrum of n-octane, a saturated hydrocarbon.

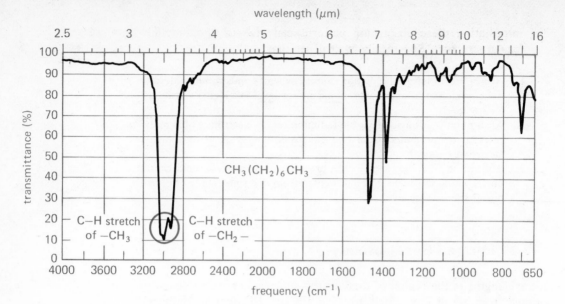

FIGURE 15.2 An infrared spectrum of *n*-octane.

The scale at the left of the chart paper indicates % transmittance. Notice also that the chart paper is calibrated in reciprocal centimeters (cm^{-1}) along the bottom and in micrometers (μm) along the top. The most prominent feature of this spectrum is three closely spaced peaks around $2900 \ cm^{-1}$. It has been determined that this cluster of peaks corresponds to stretchings of the C—H bonds in the —CH_3 and —CH_2— groups. We will not be concerned with how it has been determined which peaks correspond to which bond vibration. Rather, let us concentrate on what to look for in an infrared spectrum and on what types of information such a spectrum can give us about molecular structure.

First, an infrared spectrum can tell us about the presence or absence of particular functional groups. The stretching vibrations of each type of covalent bond (C—H, O—H, N—H, C=O, C=C, and so on) absorb infrared radiation only in certain small regions of the spectrum. Furthermore, the absorption of energy by a particular type of covalent bond is not greatly influenced by the rest of the molecule. For example, the stretching frequencies for the vibration of the carbonyl groups of aldehydes, ketones, carboxylic acids, and esters are all found within the narrow range from $1680 \ cm^{-1}$ to $1750 \ cm^{-1}$. Characteristic infrared stretching frequencies for several types of covalent bonds and organic functional groups are given in Table 15.2.

Second, a comparison of infrared spectra can often tell us about structural similarities between two substances. Although each type of covalent bond has its own characteristic absorption frequencies, no two molecules will have precisely the same spectrum. Although many absorption frequencies may be the same for closely related substances, almost always there will be some differences. In general, these differences appear in the range from $1600 \ cm^{-1}$ to $600 \ cm^{-1}$, and accordingly, this region of the infrared spectrum is often called the "fingerprint" region. By comparing spectra, particularly in the fingerprint region, it is often possible to tell

TABLE 15.2 Characteristic infrared stretching frequencies.

Type of Bond	Found in	Frequency (cm^{-1})
C—H	alkanes	2850–2950
=C—H	alkenes and aromatic hydrocarbons	3000–3200
O—H	alcohols, phenols	3600–3650
O—H	carboxylic acids (hydrogen bonded)	2500–3000
N—H	amines	3200–3500
C—O	esters, alcohols, ethers	1000–1300
C=O	aldehydes, ketones, carboxylic acids, esters, amides	1680–1750

whether or not two compounds are identical. If the spectra are identical, peak for peak, then it is almost certain that the two substances are identical. If the spectra are not identical, then the two substances do not have the same molecular structure.

What should you look for as you attempt to interpret an infrared spectrum? In this course, we will not attempt any detailed analysis of an infrared spectrum, something which requires a great deal of practice. Rather, we will be concerned only with the recognition of the presence or absence of particular functional groups, as indicated by the presence or absence of particular absorption peaks in the infrared spectrum. The stretching frequencies due to O—H, C—H, N—H, C=O, and C—O are the most important for you to recognize. The relative positions of these stretching frequencies are listed in Table 15.2; they are also shown in the form of a chart (Figure 15.3).

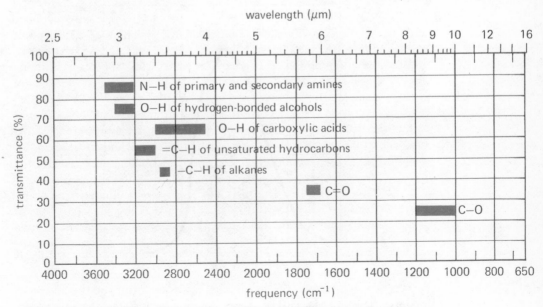

FIGURE 15.3 A correlation chart of the infrared stretching frequencies of several functional groups.

15.5 Ultraviolet-Visible Spectroscopy

The near ultraviolet region of the electromagnetic spectrum extends from 200×10^{-9} m to 400×10^{-9} m, and the visible region extends from 400×10^{-9} m (violet light) to 700×10^{-9} m (red light). To simplify reporting of spectral information, it is common practice to report wavelengths in nanometers (nm). Reported in these units, the ranges of near ultraviolet and visible spectra are:

Type of Spectrum	Nanometers
near ultraviolet	200–400
visible	400–700

Prior to the adoption of the nanometer, it was common to report ultraviolet-visible spectral information in Angstroms (Å; 10Å = 1 nm) or in millimicrons (mμ; 1 mμ = 1 nm). In our discussions, we will use only the nanometer for reporting wavelengths of absorption maximum.

An ultraviolet-visible spectrum is a plot of absorbance versus wavelength. Figure 15.3 is an ultraviolet spectrum of 2,5-dimethyl-2,4-hexadiene, a conjugated diene. The spectrum consists of a single, broad absorption band between 210 and 260 nm with an absorption maximum at 241 nm. In reporting UV spectral information, it is customary to report only the position of the absorption maximum (or maxima, if there are several peaks). Thus, we would report that the absorption maximum for 2,5-dimethyl-2,4-hexadiene is 241 nm.

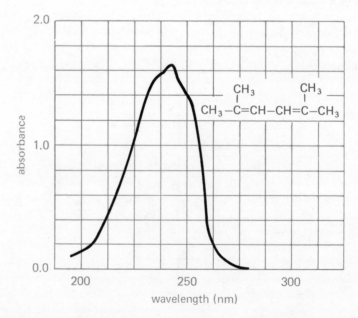

FIGURE 15.4 An ultraviolet absorption spectrum of 2,5-dimethyl-2,4-hexadiene.

Absorption of ultraviolet-visible radiation is accompanied by the promotion of electrons from one energy level or orbital to a higher one. Electrons of sigma bonds are held too tightly to be affected by near ultraviolet-visible radiation. Hence, neither alkanes and other saturated hydrocarbons nor simple alkenes show absorption between 200 nm and 700 nm. However, pi electrons of carbonyl groups ($C=O$), conjugated unsaturated systems, and aromatic rings readily absorb this type of radiation. Butadiene itself shows an absorption maximum at 215 nm. The conjugated carbon-carbon and carbon-oxygen double bond system of 3-pentene-2-one shows an absorption maximum at 227 nm. Benzene, the simplest of the aromatic hydrocarbons, shows an absorption maximum at 257 nm.

$$CH_2=CH-CH=CH_2 \qquad CH_3-CH=CH-\overset{\overset{\textstyle O}{\|}}{C}-CH_3$$

butadiene	3-pentene-2-one	benzene
215 nm	227 nm	257 nm

What should you look for in interpreting an ultraviolet-visible spectrum? These spectra are not as complicated as infrared spectra (compare Figures 15.2 and 15.4) or for that matter nuclear magnetic resonance spectra. For this reason, they do not provide as much information about details of molecular structure as do the other spectral methods. Our use of UV-visible spectra will be quite simple. If an organic molecule does not absorb radiation in the region 200 nm to 700 nm, we conclude that probably it does not contain any type of conjugated unsaturation. If it does absorb in this region, we immediately suspect that it contains some type of unsaturation, for example, a conjugated diene, a carbonyl group, an α,β-unsaturated carbonyl group, or an aromatic ring.

In addition to providing information about molecular structure, ultraviolet-visible spectroscopy can also be used for quantitative analysis, for there is a direct proportionality between absorbance (A), concentration of the sample (c), and the length of the light path through the sample (l). The proportionality constant which relates these three variables is called the molar absorptivity and is given the symbol ε (epsilon). The equation for this relationship is

$$A = \varepsilon \times l \times c \qquad \begin{aligned} A &= \text{absorbance} \\ \varepsilon &= \text{molar absorptivity} \\ l &= \text{length of light path in cm} \\ c &= \text{concentration in mole/liter} \end{aligned}$$

This equation, known as Beer's Law, forms the basis for the use of absorption spectroscopy, including ultraviolet-visible spectroscopy, for quantitative analysis. The molar absorptivity is a constant for any given compound and once the value of this constant has been determined, it can then be used to determine the concentration of the substance in solution. For several examples of quantitative absorption spectroscopy, read the mini-essay, "Clinical Chemistry—The Search for Specificity."

15.6 Nuclear Magnetic Resonance Spectroscopy

Nuclear magnetic resonance spectroscopy involves absorption of electromagnetic radiation in the radiofrequency region of the spectrum, and, as you have already learned from the information in Table 15.1, the energy of this radiation is very small. Radiowaves of frequency 60×10^6 Hz (60 Megahertz or 60 MHz) have a wavelength of 5 meters and an energy of slightly less than 0.01 cal/mole. Absorption of radiowaves is accompanied by a special type of nuclear transition, and, for this reason we call this type of spectroscopy nuclear magnetic resonance (NMR) spectroscopy.

The nuclei of certain elements behave as if they were spinning charges. Any spinning charge creates a magnetic field (Figure 15.5a) and in effect behaves as if it were a tiny bar magnet (Figure 15.5b).

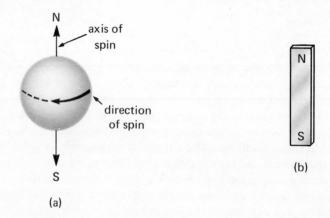

FIGURE 15.5 A spinning charge (a) creates a magnetic field along the axis of spin and (b) behaves as if it were a tiny bar magnet.

Of the three isotopes most common to organic compounds (^{1}H, ^{12}C, and ^{16}O), only the proton behaves in this manner. Therefore, proton magnetic resonance (PMR) spectroscopy is the only type of nuclear magnetic resonance we will study. We should note, however, that isotopes of other elements including ^{13}C, ^{14}N, ^{19}F, and ^{31}P also behave as if they were spinning charges, and the magnetic resonance spectroscopy of these nuclei is a field of active and expanding research at the present time.

When a hydrogen nucleus in an organic molecule is placed in a strong magnetic field, there are only two allowed orientations for its magnetic field: with the applied field or against it (Figure 15.6). In the lower energy state, E_1, the nuclear magnet is aligned with the applied magnetic field. If energy is supplied in the form of radiowaves of exactly the right frequency, radiation will be absorbed and the spinning nuclear magnet will flip and become aligned against the applied magnetic field in the higher energy state, E_2. The most common commercially available nuclear magnetic resonance spectrometers generate magnetic fields of approximately 14,000 Gauss. By way of comparison, note that the magnetic field of the earth is only about 0.5 Gauss. Given this applied field strength, radiofrequency

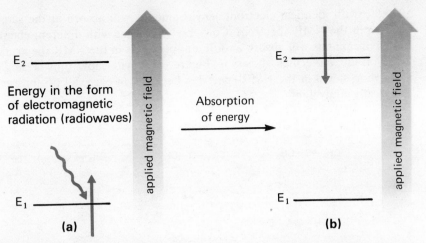

FIGURE 15.6 Orientation of a hydrogen nucleus in an applied magnetic field. (a) Nucleus in the lower energy state, (b) nucleus in the higher energy state. Absorption of electromagnetic radiation causes a transition from the lower to the higher energy state.

radiation in the range of 60 MHz is required to cause transitions of hydrogen nuclei from the lower to the higher energy state. One type of NMR spectrometer supplies radiowaves of frequency 60 MHz and then gradually increases the strength of the applied magnetic field. As the field strength increases, protons within a molecule absorb radiation and produce an NMR signal. An NMR spectrum, then, is a plot of the strength of the applied magnetic field versus the intensity of the absorption signal for the protons within a molecule.

What should you look for as you attempt to analyze an NMR spectrum? We will be concerned with three things:

1. the position of each signal in the spectrum;
2. the relative areas of the various signals;
3. the splitting pattern of each signal.

Let us look at each of these features separately, and learn what types of information each can give us about molecular structure.

15.7 The Chemical Shift

At this point, it might seem that protons are protons, and all protons within a molecule should absorb at exactly the same point in the NMR spectrum. Fortunately, this is not the case, for the position at which any given proton absorbs depends on (1) the strength of the applied magnetic field and (2) the proton's immediate electronic environment. Electrons in the immediate environment of a proton (bonding as well as nonbonding electrons, those in sigma bonds as well as those in pi bonds) serve to shield a proton very slightly from the external magnetic field, altering the position at which it absorbs in the NMR spectrum. All chemically equivalent protons, that is,

421

all those with identical electronic environments, will absorb at the same position in the NMR spectrum. Conversely, protons with different chemical environments will absorb at different positions in the NMR spectrum. Each of the molecules shown in Figure 15.7 will show only a single absorption signal in the NMR spectrum because, in each, all protons are chemically equivalent.

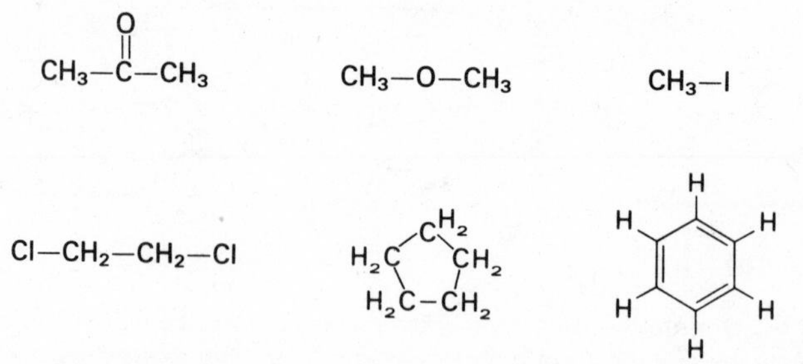

FIGURE 15.7 Molecules showing only a single absorption signal in the NMR spectrum.

Each of the molecules shown in Figure 15.8 will show more than one signal in the NMR spectrum because, in each, there are two or more different sets of chemically equivalent protons.

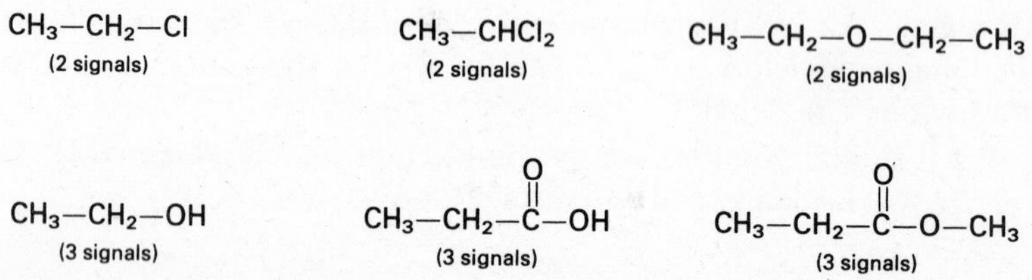

FIGURE 15.8 Molecules showing two or more absorption signals in the NMR spectrum.

In reporting where the various chemically equivalent protons within a molecule absorb, it is common practice to select a reference signal and then to report other signals in terms of how far each is shifted from the reference signal. This shift is called the chemical shift and is given the symbol δ (delta). The most generally used reference standard is the signal due to the twelve chemically equivalent protons in tetramethylsilane, $(CH_3)_4Si$, abbreviated TMS. On the delta scale, the peak for TMS is set at zero. Other signals are given positive delta values depending on how far they are shifted to the left on the chart paper from TMS. Figure 15.9 is the NMR spectrum of the 1,2-dichloroethane. Note that there are two signals in this spectrum. The signal of low intensity on the far right at $\delta = 0$ is that of the TMS reference standard. The signal at $\delta = 3.7$ is a single sharp peak due to the four chemically equivalent protons of 1,2-dichloroethane. Delta values for the types of protons we will encounter fall within the range 0 to 13 and are summarized in Table 15.3.

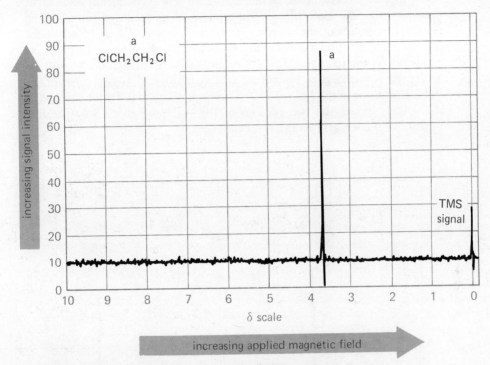

FIGURE 15.9   An NMR spectrum of 1,2-dichloroethane, ClCH₂CH₂Cl.

TABLE 15.3   Chemical shifts of several types of protons, relative to TMS.

Type of Proton	Chemical Shift Relative to TMS	Type of Proton	Chemical Shift Relative to TMS
C—CH₃	0.9–1.0	O=C—H	9.0–10.0
C—CH₂—C	1.20–1.40		
C—CH—C C	1.40–1.60	⟨benzene⟩—H	6.5–8.3
C=C (vinyl)	4.5–6.5	—O—CH₃	3.2–3.3
		—O—CH₂—C	3.2–3.8
O=C—CH₃	2.0–2.8	—O—H	1.0–5.0
O=C—CH₂—	2.2–2.5	—C(=O)—O—H	10.0–13.0

423

15.8   Relative Signal Areas

As we have shown in the previous section, the number of signals in an NMR spectrum corresponds to the number of sets of chemically equivalent protons within the molecule. In addition, an NMR spectrum can also give us information about the number of chemically equivalent protons there are in each set for the relative intensities of the various signals (as measured by the relative areas) are proportional to the number of protons giving rise to each signal. For benzyl acetate (Figure 15.10) the areas of the three signals are in the ratio of $3:2:5$ corresponding to the three protons of the $-CH_3$ group, the two protons of the $-CH_2-$ group, and the five protons of the $-C_6H_5$ group, respectively.

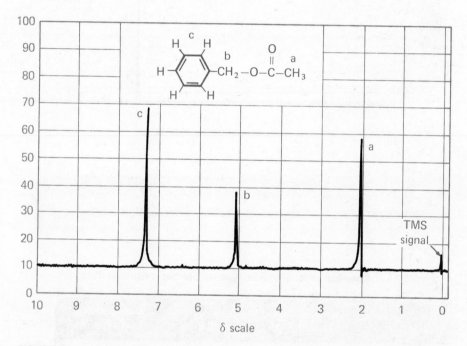

FIGURE 15.10   An NMR spectrum of benzyl acetate. The relative areas of the signals at $\delta = 2.1$, 5.1, and 7.3 are in the ratio of $3:2:5$.

15.9   The Splitting Pattern

A third kind of information can be derived from the splitting pattern of each NMR signal. Consider, for example, the NMR spectrum of 1,1-dichloroethane (Figure 15.11). Note that this molecule is an isomer of 1,2-dichloroethane, a substance whose NMR spectrum we saw in Figure 15.9.

According to what we have said so far, you would predict two signals with relative areas in the ratio $3:1$, corresponding to the three protons of the $-CH_3$ group and the single proton of the $-CHCl_2$ group. Notice from the spectrum that there are in fact two signals, but neither is a single sharp peak. The signal at $\delta = 2.1$ is split into two closely spaced peaks (a

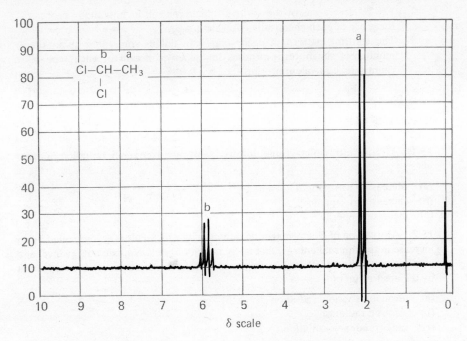

FIGURE 15.11 An NMR spectrum of 1,1-dichloroethane, CH_3CHCl_2. The spectrum consists of two signals, a doublet at $\delta = 2.1$ and a quartet at $\delta = 5.8$.

doublet) and the signal at $\delta = 5.8$ is split into four closely spaced peaks (a quartet). The areas of the doublet and quartet are in the ratio $3:1$, as we would have predicted they should be. But, why is the first signal split into a doublet and the second into a quartet? We can account for this splitting if we remember that protons themselves behave as tiny bar magnets. Therefore, any given proton within a molecule will feel the effect not only of the external magnetic field but also that of the magnetic field generated by neighboring protons. We can predict the signal splitting due to non-chemically equivalent proton neighbors by the so-called $n + 1$ rule. According to this rule, if a proton has n nonchemically equivalent proton neighbors, then its NMR signal will be split into $n + 1$ peaks. If we apply this rule to the spectrum of 1,1-dichloroethane, we would predict that the —CH_3 signal should be split to a doublet by the one proton neighbor. The —$CHCl_2$ signal should be split into a quartet by the three hydrogen neighbors. This is exactly what is observed (Figure 15.8).

It is important to keep in mind that chemically equivalent protons do not split each other. For example, the NMR spectrum of 1,2-dichloro-ethane (Figure 15.9) consists of only a single, sharp peak. All four protons within this molecule are chemically equivalent; therefore, no splitting of the signal occurs.

In summary, an NMR spectrum can provide us with three types of information about molecular structure:

1. The number of signals in the NMR spectrum and the chemical shift of each can tell us the number of different types of chemically equivalent protons within the molecule.

2. The relative areas of the various signals can tell us how many protons there are of each chemically equivalent type.
3. The splitting pattern of each signal can tell us something about the number of neighboring protons in the environment of each type of chemically equivalent proton.

PROBLEMS

15.1 Arrange ultraviolet-visible, infrared, and radiofrequency radiation in order of

(a) increasing wavelength
(b) increasing frequency
(c) increasing energy

15.2 Absorption of electromagnetic radiation by a molecule is accompanied by an increase in the internal energy, that is, by transitions between one energy level and another within the molecule. What types of molecular transitions are brought about by the absorption of

(a) ultraviolet-visible radiation
(b) infrared radiation
(c) radiofrequency radiation

15.3 Define the term *molecular spectroscopy*.

15.4(a) Name the unit in which ultraviolet-visible spectral information is most commonly reported. What is the numerical relationship between this unit and the Angstrom?
(b) Name two units in which infrared spectral information is commonly reported. What is the numerical relationship between these two units?
(c) Name one unit in which nuclear magnetic resonance spectral information is commonly reported.

15.5 State whether you would predict the following molecules to absorb radiation in the ultraviolet-visible region of the spectrum (between 200 nm and 700 nm).

(a) H_2O

(b) CH_3CH_2OH

(c) $CH_2=CH-CH_2-CH_2-CH=CH_2$ **(d)** $CH_3-CH=CH-CH=CH-CH_3$

(e) $CH_3-CH_2-\overset{\overset{O}{\|}}{C}-CH_2-CH_3$ **(f)** $CH_3-CH_2-\overset{\overset{O}{\|}}{C}-CH=CH_2$

(g) $CH_3-CH_2-\overset{\overset{OH}{|}}{CH}-CH=CH_2$ **(h)**

(i) 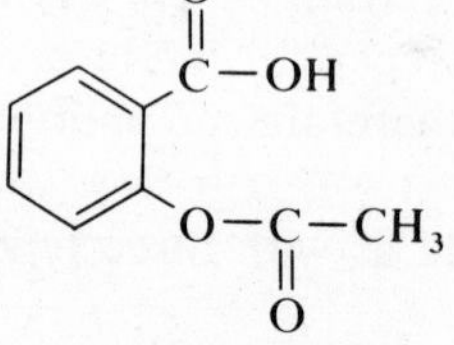 aspirin

(j) 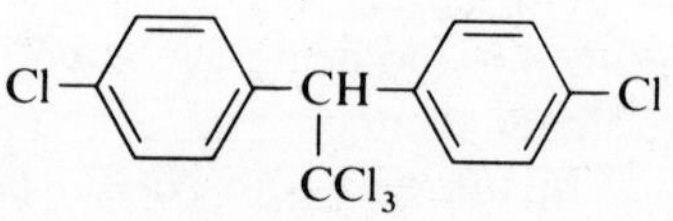 DDT

15.6 Using the following pairs of compounds, for each pair (i) name the functional groups present, and (ii) list one major feature that will appear in the infrared spectrum of the first but not the second molecule. Your answer to part (ii) should

state what type of bond vibration is responsible for the spectral feature you have
listed and its approximate position in the spectrum.

(a) $CH_3\overset{\overset{\textstyle O}{\|}}{C}H$ and CH_3CH_2OH

(b) $CH_3\overset{\overset{\textstyle OH}{|}}{C}HCH_3$ and $CH_3\overset{\overset{\textstyle O}{\|}}{C}CH_3$

(c) $CH_3\overset{\overset{\textstyle O}{\|}}{C}OH$ and $CH_3\overset{\overset{\textstyle O}{\|}}{C}OCH_3$

(d) CH_3CH_2OH and $CH_3CH_2OCH_2CH_3$

(e) $CH_3CH_2CH_2NH_2$ and $CH_3\overset{\overset{\textstyle CH_3}{|}}{N}CH_3$

(f) $CH_3CH_2\overset{\overset{\textstyle O}{\|}}{C}OH$ and $CH_3CH_2\overset{\overset{\textstyle O}{\|}}{C}H$

(g) $CH_3CH_2\overset{\overset{\textstyle O}{\|}}{C}OH$ and $CH_3CH_2CH_2OH$

(h) $CH_3CH_2\overset{\overset{\textstyle O}{\|}}{C}N\overset{\diagup CH_3}{\diagdown CH_3}$ and $CH_3CH_2N\overset{\diagup CH}{\diagdown CH}$

(i) cyclopentane-CH₂OH   and   cyclopentane-CHO

(j) cyclopentane-CH₂OH   and   cyclopentane-COOH

(k) cyclopentane-CH₂OH   and   cyclopentane-COCH₃

(l) cyclopentane-CHO   and   cyclopentane-COOH

(m) cyclohexane-CHO   and   cyclohexane-COCH₃

(n) cyclohexane-COCH₃   and   cyclohexane-CH(OH)CH₃

(o)
$$\text{(CH}_3)_2\text{CH}\overset{\displaystyle O}{\overset{\|}{C}}\text{CH}_3 \quad \text{and} \quad (\text{CH}_3)_2\text{CH}\overset{\displaystyle O}{\overset{\|}{C}}\text{OCH}_3$$

(p)
$$\text{(CH}_3)_2\text{CH}\overset{\displaystyle O}{\overset{\|}{C}}\text{CH}_3 \quad \text{and} \quad (\text{CH}_3)_2\text{CHO}\overset{\displaystyle O}{\overset{\|}{C}}\text{CH}_3$$

(q)
$$\text{(CH}_3)_2\text{CH}\overset{\displaystyle O}{\overset{\|}{C}}\text{CH}_2\text{CH}_3 \quad \text{and} \quad (\text{CH}_3)_2\text{CHOCH}_2\text{CH}_3$$

(r)
$$\text{(CH}_3)_2\text{CH}\overset{\displaystyle O}{\overset{\|}{C}}\text{OH} \quad \text{and} \quad (\text{CH}_3)_2\text{CH}\overset{\displaystyle O}{\overset{\|}{C}}\text{CH}_3$$

(s)   cyclobutane–NH_2 and   cyclobutane–OCH_2CH_3

(t)   cyclobutane–$\overset{O}{\overset{\|}{C}}OH$ and   cyclobutane–$\overset{O}{\overset{\|}{C}}OCH_3$

15.7 State the number of sets of chemically equivalent protons in the following molecules, and the number of protons in each set.

(a) $CH_3\overset{\displaystyle O}{\overset{\|}{C}}CH_3$

(b) CH_3OCH_3

(c) $CH_3CH_2OCH_2CH_3$

(d) $CH_3\overset{\displaystyle O}{\overset{\|}{C}}OCH_3$

(e) $CH_3CH_2\overset{\displaystyle O}{\overset{\|}{C}}OH$

(f) $CH_3\overset{\displaystyle O}{\overset{\|}{C}}-\underset{\underset{\displaystyle CH_3}{|}}{\overset{\overset{\displaystyle CH_3}{|}}{C}}CH_3$

(g) $(CH_3)_2C{=}CH_2$

(h) $ClCH_2CH_2Cl$

(i) CH_3CHCl_2

(j) $CH_3-\underset{\underset{\displaystyle CH_3}{|}}{\overset{\overset{\displaystyle CH_3}{|}}{C}}-CH_3$

(k) ⬠ (cyclopentane)

(l) $CH_3C{\equiv}CH$

(m) $CH_3C{\equiv}CCH_3$

(n) $CH_3O\overset{\displaystyle O}{\overset{\|}{C}}CH_2\overset{\displaystyle O}{\overset{\|}{C}}OCH_3$

(o) $CH_3CHCH_2CH_3$ with CH_3 substituent

15.8 Which of the compounds in Problem 15.7 will show only a single peak in its NMR spectrum?

15.9 Use the $n + 1$ rule to predict the splitting patterns of each set of chemically equivalent protons in the following molecules.

(a) $CH_3CH_2OCH_2CH_3$

(b) CH_3CHCl with CH_3 substituent

(c) CH_3CHCl_2

428

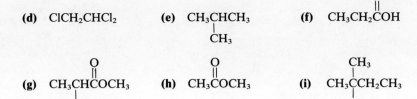

(d) ClCH$_2$CHCl$_2$

(e) CH$_3$CHCH$_3$ | CH$_3$

(f) CH$_3$CH$_2$COH (O)

(g) CH$_3$CHCOCH$_3$ | CH$_3$ (O)

(h) CH$_3$COCH$_3$ (O)

(i) CH$_3$CCH$_2$CH$_3$ | CH$_3$ | Cl

15.10 Following are pairs of constitutional isomers. For each pair, state one major feature in the NMR spectrum which will allow you to distinguish the first isomer from the second.

(a) CH$_3$—O—CH (CH$_3$, CH$_3$)   and   CH$_3$—CH$_2$—O—CH$_2$—CH$_3$

(b) Cl—C—CH$_3$ (Cl, Cl)   and   Cl—CH—CH$_2$—Cl (Cl)

(c) CH$_3$—CH$_2$—C—OH (O)   and   CH$_3$—C—O—CH$_3$ (O)

(d) (CH$_3$)(CH$_3$)C=C(H)(H)   and   (CH$_3$)(CH$_3$)C=CH$_2$

(e) Cl—CH$_2$—CH$_2$—CH$_2$—Cl   and   CH$_3$—C—CH$_3$ (Cl, Cl)

15.11 Draw the structural formula of a substance of the given molecular formula that will show only a single absorption signal in its NMR spectrum.

(a) C$_5$H$_{12}$
(b) C$_5$H$_{10}$
(c) C$_3$H$_6$O
(d) C$_3$H$_6$Cl$_2$
(e) C$_2$H$_6$O
(f) C$_4$H$_8$
(g) C$_2$H$_2$Cl$_4$
(h) C$_4$H$_9$Cl
(i) C$_6$H$_{12}$

15.12 Following are pairs of constitutional isomers. The members of each pair will have the same number of signals in the NMR spectrum and the same splitting pattern of each signal. However, these isomers can be distinguished by chemical shifts. Use the information in Table 15.2 to show how the chemical shifts of the underlined protons can enable you to distinguish between these constitutional isomers.

(a) CH$_3$—O—C—CH$_2$—CH$_3$ (O)   and   CH$_3$—C—O—CH$_2$—CH$_3$ (O)

(b) ⟨cyclohexyl⟩—O—C—CH$_2$—CH$_3$ (O)   and   ⟨cyclohexyl⟩—C—O—CH$_2$—CH$_3$ (O)

Index

433